The Static and Dynamic Continuum Theory of Liquid Crystals

THE LIQUID CRYSTALS BOOK SERIES

Edited by

G.W. GRAY, J.W. GOODBY & A. FUKUDA

The Liquid Crystals book series publishes authoritative accounts of all aspects of the field, ranging from the basic fundamentals to the forefront of research; from the physics of liquid crystals to their chemical and biological properties; and, from their self-assembling structures to their applications in devices. The series will provide readers new to liquid crystals with a firm grounding in the subject, while experienced scientists and liquid crystallographers will find that the series is an indispensable resource.

Volume 1
Introduction to Liquid Crystals
Chemistry and Physics
P. Collings and M. Hird

The Static and Dynamic Continuum Theory of Liquid Crystals

A Mathematical Introduction

Iain W. Stewart

Taylor & Francis Group

LONDON AND NEW YORK

First published 2004 by Taylor & Francis

Published 2021 by Routledge
2 Park Square, Milton Park, Abingdon, Oxon OX14 4RN
605 Third Avenue, New York, NY 10017

*Routledge is an imprint of the Taylor & Francis Group, an informa
business*

Publisher's Note
This book has been prepared from camera-ready copy supplied by the author

British Library Cataloguing in Publication Data
A catalogue record for this book is available from the
British Library

Library of Congress Cataloging in Publication Data
A catalog record for this book has been requested

ISBN 13: 978-0-7484-0896-2 (pbk)
ISBN 13: 978-0-7484-0895-5 (hbk)

To

Kaoru

Hannah and James

my father and the memory of my mother

Elizabeth Josephine Fry Stewart (1920–1991)

Contents

Preface

This book grew out of a perceived need for a text specifically aimed at applied mathematics, theoretical physics and engineering graduates who wish to obtain some basic grounding in the static and dynamic continuum theory of liquid crystals. It is hoped that beginners and more seasoned readers will benefit from the topics to be raised and discussed throughout the book.

Chapter 1 gives a brief introduction to some of the elementary aspects and descriptions of liquid crystals and helps to set the scene for later Chapters. The static theory of nematic liquid crystals is developed in Chapter 2 while Chapter 3 goes on to discuss some applications of this theory which have particular physical relevance. The dynamics of nematics, leading to the celebrated Ericksen–Leslie dynamic equations, are fully derived in Chapter 4, with Chapter 5 providing some detailed accounts of applications of this dynamic theory.

Chapter 6 contains the most advanced mathematics in the book, and introduces a particular nonlinear static and dynamic continuum model for smectic C liquid crystals. This continuum theory is a natural extension of nematic theory, originating from ideas that are familiar from the continuum description of nematics.

The results and applications in Chapters 2 to 5 for nematic liquid crystals are given in fairly full mathematical detail. It has been my experience that the stumbling block for many people comes at the first attempts at the actual calculations: here I will reveal many details and more explanation than is usually given in articles and common texts, in the hope that readers will gain confidence in how to apply the main results from continuum theory to practical problems. These Chapters contain extensive derivations of the static and dynamic nematic theory and applications. Chapter 6, on the other hand, does not give as many detailed computations as those presented in the earlier Chapters: it is my intention that it introduces the reader to a continuum theory of smectic C liquid crystals and it is probably written more in the style of an introductory review. This is partly because some of the calculations are similar for both nematic and smectic C materials, but with different physical parameters and some different physical interpretations. However, despite some of these similarities, smectic liquid crystals have some uniquely different mathematical problems, and these can only be touched upon within the remit of a book such as this.

Some Appendices and Tables appear at the end of the book. Table D.3 presents experimental data which are of use when attempting to model some common nematic liquid crystals.

Many references are given so that those wishing to pursue particular aspects in further depth can do so at their leisure, knowing that they will be equipped with the basic methods illustrated throughout the book.

Iain W. Stewart

January 2004

Glasgow

Acknowledgements

I am indebted to the late Professor Frank Leslie, FRS, from the University of Strathclyde, for introducing me to liquid crystals during the period 1988–1990 when I was his research assistant. It was during this period at Strathclyde University that my interests in liquid crystals were kindled and developed.

I am especially thankful to Professor Ray Atkin, who enlightened me on many aspects of the fundamental continuum theory of liquid crystals and took time to look through a draft of the initial manuscript. Many of my colleagues at the University of Strathclyde have assisted and encouraged me in the theory of liquid crystals, and to them I am very grateful. I particularly wish to thank Dr Brian Duffy, Dr Geoff McKay, Dr Nigel Mottram, Professor Mikhail Osipov, Dr André Sonnet and Dr Patrick Woods, all from the liquid crystal group at Strathclyde. Many worthwhile comments, suggestions and corrections have been gratefully received from my former and present research students, especially Dr David Anderson, Dr Julie Kidd, Mr Andrew Smith and Miss Elizabeth Wigham.

I also wish to thank the editors of the series, Professor George Gray, FRS, Professor John Goodby and Professor Atsuo Fukuda, for their invitation to write this book and for their general suggestions during its preparation.

Chapter 1

Introduction

In this Introduction we record some brief details about the discovery and development of liquid crystals and their continuum descriptions. Kelker [143] has written an intricate and authoritative account of the early history and development of liquid crystals, with some more detailed aspects on the early history being discussed by Kelker and Knoll [144]. Collings [52] has also written an introductory and readable account on the discovery, physics and applications of liquid crystals aimed at those new to the field. The books by de Jeu [137] and Collings and Hird [53] provide suitable technical introductions to the physics and structures of liquid crystals while more advanced accounts of the theory and applications of liquid crystals can be found in the books by de Gennes and Prost [110], Chandrasekhar [38] and Blinov [18], as well as the reviews by Stephen and Straley [258] and Leslie [168]. Treatments on the static theory of liquid crystals have been given in the books by Virga [273] and Barbero and Evangelista [11]. A recent introductory text on the physical properties of liquid crystals, focusing on optical phenomena via elementary static and dynamic theory, has been written by Khoo [146].

The aim of this book is to present a mathematical introduction to the static and dynamic continuum theory of liquid crystals. Before doing so, we outline some points on the discovery and basic description of liquid crystals in Sections 1.1 and 1.2. This is followed by a short summary of the development of the continuum theory of liquid crystals in Section 1.3. The Chapter closes in Section 1.4 with some basic comments on the notation and conventions employed in later Chapters and refers to some sources for those who may require further background on some of these conventions used throughout this book.

1.1 The Discovery of Liquid Crystals

Everyone is familiar with three of the common states of matter: solid, liquid and gas. Some substances exhibit all these states as the temperature is varied. For example, water is solid below 0°C, liquid between 0°C and 100°C, and a gas above 100°C. However, this simplified classification is known to be not generally accurate for many materials. Liquid crystals are states of matter which are sometimes observed to occur between the solid crystal state and the isotropic liquid state. A substance is

1

called *isotropic* when its physical properties are uniform in all directions. Water at room temperature is an example of an isotropic liquid. Liquid crystals are actually examples of *mesophases*, also called *mesomorphic phases*: they are intermediate states of matter. While liquid crystals may flow like fluids or viscous fluids, they also possess features that are characteristic of solid crystals, such as certain optical properties: they therefore display both liquid and crystal properties, hence the name *liquid crystals*. Liquid crystals are also known to be *anisotropic* because they exhibit different physical properties in different directions.

The discovery of liquid crystals is generally attributed to the Austrian botanist Reinitzer [229], who reported his observations in 1888; an English translation of his article is also available [230]. Reinitzer observed what he called two melting points when heating up cholesteryl benzoate, which is solid at room temperature. At the first melting point of 145.5°C this material became a cloudy liquid and, upon further heating, he noted that there appeared to be a second melting point at 178.5°C where the cloudy liquid turned into a clear liquid. The cloudy liquid reported by Reinitzer was what we now call a cholesteric liquid crystal, also called a chiral nematic liquid crystal, with the higher melting point now being termed the clearing point. Reinitzer sent two samples and an accompanying letter to the German physicist Lehmann requesting him to investigate these samples further, recognising some connections between his own observations and the previous work of Lehmann. The term 'flowing crystals' was originally used by Lehmann [160] to describe these materials in 1889 and in 1900 he appears to have established the general term *liquid crystal*, which is now in everyday use.

Vorländer, in 1907 [274], discovered that an essential prerequisite for the occurrence of two melting points was a rod-like molecule. This discovery was to be of crucial importance for the theoretical development of liquid crystals since it allowed theoreticians to mathematically describe the molecular structure as rod-like. This rod-like description, together with more advanced descriptions of molecular shapes, remains in use today for modelling many liquid crystal phases.

In 1922, Friedel [94] described different liquid crystal phases and proposed a classification scheme consisting of three broad categories called *nematic, cholesteric* and *smectic*. This classification has since been widely adopted and is now in common usage.

1.2   Basic Descriptions of Liquid Crystals

Most liquid crystals are organic substances which can be induced to exhibit liquid crystal phases by either of two ways: changing the temperature or changing the concentration in a solvent. Those obtained by changing the temperature are called *thermotropic* liquid crystals and those derived by changes of concentration in a solvent are called *lyotropic* liquid crystals. Liquid crystals are examples of *mesogenic* materials because they give rise to mesophases under appropriate conditions. This book deals only with the isothermal (fixed temperature) description of thermotropic liquid crystals. Only brief details will be given here with further mathematical details being given at appropriate stages in subsequent Chapters.

In the simplest description, liquid crystals can be thought of as consisting of elongated rod-like molecules which have a preferred local average direction. Figure 1.1 is a schematic illustration of the solid crystal, liquid crystal and isotropic fluid (liquid) phases of a possible material as the temperature T increases through a melting point and a clearing point, the short bold lines representing the molecules.

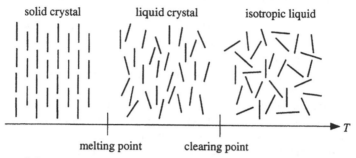

Figure 1.1: A schematic representation of the possible solid crystal, liquid crystal and isotropic liquid phases of a substance as the temperature increases through a melting point and a clearing point. The short bold lines represent molecules.

Nematic Liquid Crystals

In the nematic liquid crystal phase the long axes of the constituent molecules tend to align parallel to each other along some common preferred direction, as depicted schematically in Fig. 1.2. In many nematics this direction is sometimes referred to as the *anisotropic axis*. There is no long-range correlation between the centres of mass of the molecules; that is, they can translate freely while being aligned, on average, parallel to one another. There is a rotational symmetry around the anisotropic axis

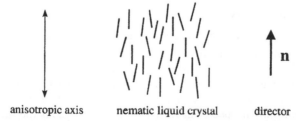

Figure 1.2: A schematic representation of a nematic liquid crystal phase where the short bold lines represent the molecules. A unit vector n, called the director, describes the average direction of the molecular alignment along what is commonly called, in uniaxial nematics, the anisotropic axis.

which means that the nematic phase is uniaxial. This axis of uniaxial symmetry has no polarity: although the constituent molecules may be polar, there is no imposed effect on a larger scale, which is the situation considered in this book based on classical continuum mechanics. In a typical nematic such as *p*-azoxyanisole (PAA), the molecules closely resemble rigid rods of length 20 Å and width 5 Å [110, p.3] (recall that 1 angstrom (Å) equals 10^{-10} m).

It proves convenient to introduce a unit vector **n**, called the *director*, to describe the local direction of the average molecular alignment in liquid crystals, as shown in Fig. 1.2. The absence of polarity means that **n** and −**n** are indistinguishable, in the sense that the sign of **n** has no physical significance.

The word nematic comes from the Greek word νημα, meaning thread, arising from the thread-like textures often seen in nematic samples. These 'threads' correspond to lines of singularity in the director alignment called *disclinations*. Such defects will be discussed below in Section 3.8 after a more detailed mathematical description of nematic liquid crystals is given in Chapter 2.

Before going on to discuss briefly other basic liquid crystal phases, it is worth recording some points about the development of nematic liquid crystals, since a substantial part of this book concerns these substances. The first nematic liquid crystal, *p*-azoxyanisole (PAA), was synthesised by Gattermann and Ritschke [102], who reported their results in 1890: it was the first liquid crystal not based on a naturally occurring substance and went on to be extensively investigated by Lehmann and others. The first relatively stable room temperature nematic liquid crystal, 4-methoxybenzylidene-4′-butylaniline (MBBA), was synthesised by Kelker and Scheurle [142] in 1969, but, for various reasons, this substance was not considered suitable for some applications. The nematic liquid crystal 4-pentyl-4′-cyanobiphenyl (5CB) is an example of a stable room temperature material designed for use in twisted nematic displays (discussed in Section 3.7) and was synthesised by Gray and co-workers in the early 1970s: some details can be found in articles [117, 118].

Some of the physical parameters for the nematic liquid crystals PAA, MBBA and 5CB are given in Table D.3 on page 330, which should guide those interested in the theoretical modelling of nematics. In many practical situations, it should be borne in mind that mixtures of nematic materials are often used in displays.

Cholesteric Liquid Crystals

The cholesteric liquid crystal phase is similar to the nematic phase except that the molecular orientation shows a preferred configuration that is helical, in the sense pictured in Fig. 1.3. This helical structure arises from the chiral properties of the constituent molecules: chiral molecules differ from their mirror image and have a left-hand or right-hand sense and are called *enantiomorphic*. The director is not fixed uniformly in space, but rather rotates throughout the sample as shown in the Figure. The helical axis is in the horizontal direction and the helix itself may be left-handed or right-handed. The director 'twists' perpendicular to the page as an observer moves along the helical axis. The sense of twisting can be left-handed or right-handed and depends on physical conditions. At a given fixed temperature, a sample of cholesteric liquid crystal always produces helices of the same sense, but there are cholesterics that can change their handedness of helix by changing the temperature [110, p.15]. The director alignment tends to vary

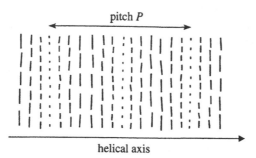

Figure 1.3: A schematic representation of a cholesteric liquid crystal. There is a helical axis in the horizontal direction and the molecules 'rotate' within a plane perpendicular to the page as an observer moves along the axis of the helix. The pitch P is the distance over which the director rotates through 2π radians.

through a given sample with a periodicity $P/2$, where P is the pitch of the helix, defined as the distance measured along the helical axis over which the director rotates through a full 2π radians, as shown in Fig. 1.3. The periodicity is half of this distance because $\mathbf{n}$ and $-\mathbf{n}$ are indistinguishable. The pitch P may vary from around 200 nm upwards, and so the pitch is generally much larger than the molecular dimensions [137].

Many examples of the molecules that form the cholesteric phase are derivatives of cholesterol, and it is for this reason that the word cholesteric is frequently used. We note that cholesterol itself is known not to be a cholesteric liquid crystal. However, as pointed out by Collings [52], there are many cholesteric liquid crystals that are not connected to cholesterol, which is why many prefer the term *chiral nematic*.

A given material may possess either the nematic liquid crystal phase or the cholesteric liquid crystal phase, but none are known to possess both. Cholesterics will not feature greatly in this book; nevertheless, a brief discussion involving them is incorporated into the mathematical description given in Section 2.2.2.

Smectic Liquid Crystals

The smectic liquid crystal phase is present in many substances. The word smectic comes from the Greek word σμηγμα, meaning soap, since smectic liquid crystals display mechanical properties reminiscent of soaps. All smectics are layered structures having a well defined interlayer distance. Smectics are therefore more ordered than nematics and, for many materials, the smectic phase occurs at a temperature below that for which the same material will exhibit a nematic phase. In this book we shall be considering only the smectic A and smectic C liquid crystal phases, although it should be recorded here that other smectic phases have also been classified; the interested reader is referred to de Gennes and Prost [110] for details. A short survey on the development of the understanding and classification of various smectic phases has been written by Sackmann [237].

When the smectic A phase occurs the molecules are arranged in layers as shown
in Fig. 1.4(a) where the director is, on average, aligned perpendicular to the layers and is parallel to the layer normal. In thermotropic liquid crystals the smec-

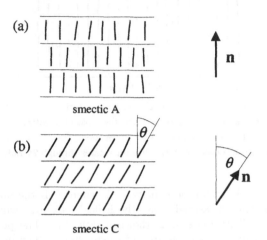

Figure 1.4: Schematic diagrams of (a) smectic A and (b) smectic C liquid crystals.
The molecules are equidistantly spaced in layers as shown. In smectic A the director
n is perpendicular to the layers while in smectic C the director makes an angle θ to
the local layer normal.

tic layer thickness, also called the smectic interlayer distance, may be anything
from something close to the full length of the constituent molecules up to around
twice their length [110, p.19] (typical values are perhaps in the range $20 \sim 80$ Å).
The lyotropic smectic A phase can have layer thicknesses up to several thousand
angstroms [110, p.19]. The director **n**, as before, represents the average molecular
orientation of the molecules and, for our purposes, satisfies the symmetry condition
that **n** and $-$**n** are physically indistinguishable, as is the case in the nematic phase.

In smectic C liquid crystals the director is tilted at an angle θ relative to the layer
normal as shown in Fig. 1.4(b). The angle θ is usually temperature dependent and
is called the *smectic C tilt angle* or *smectic cone angle*. The director **n** continues
to be defined as the average direction of the molecular alignment and, from the
physical point of view, **n** and $-$**n** remain indistinguishable, as before. In non-chiral
smectic C liquid crystals the director, in the absence of external influences, is often
uniformly aligned as shown in Fig. 1.4(b). Further details on the mathematical
description of smectic C can be found in Chapter 6.

Chiral smectic C phases can occur when the constituent molecules are enantiomorphic. Chiral smectic C materials exhibit a helical structure in an analogous
way to cholesterics with the 'twisting' taking place across the smectic layers, the
helical axis being in the direction of the layer normals. Chiral smectic C liquid
crystals are described in more mathematical detail in Section 6.4.

Polymorphism

Some liquid crystal materials can exhibit a considerable range of mesophases, a phenomenon known as *polymorphism*. Some substances, however, exhibit only one type of liquid crystal mesophase. For example, PAA is a solid below 118°C, a nematic liquid crystal between 118°C and 135.5°C, and an isotropic liquid above 135.5°C [137]. This is shown in the phase diagram in Fig. 1.5(a). On the other hand, cholesteryl

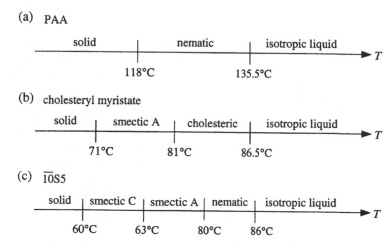

(a) PAA

(b) cholesteryl myristate

(c) $\overline{1}$0S5

Figure 1.5: Phase diagrams of the materials (a) PAA, (b) cholesteryl myristate and (c) $\overline{1}$0S5. As the temperature T increases, different liquid crystal mesophases can occur.

myristate exhibits two liquid crystal mesophases and has the phase diagram shown in Fig. 1.5(b) [137]. An example of a material with three liquid crystal mesophases is the compound 4-*n*-pentylbenzenethio-4'-*n*-decyloxybenzoate, commonly abbreviated to $\overline{1}$0S5, having the rough phase diagram shown in Fig. 1.5(c) [52]. Many other materials exhibit even richer phase diagrams with, for example, nematic and many different smectic or crystal smectic mesophases appearing.

1.3 The Development of the Continuum Theory of Liquid Crystals

The presently accepted continuum theory for liquid crystals has its origins going back to at least the work of Oseen [214, 215], from 1925 onwards, and Zocher [286] in 1927. Oseen derived a static version of the continuum theory for nematics which was to be of instrumental importance, especially when the static theory was further developed and formulated more directly by Frank [91] in 1958. This static theory, introduced in Chapter 2, is based upon the director **n** and its possible distortions.

One important aspect of liquid crystals is the *Freedericksz transition*, which can occur when a finite sample of liquid crystal is subjected to an applied magnetic or electric field. It is well known that the director alignment throughout a sample may be influenced by a magnetic or electric field of sufficiently large magnitude because of the competition between the alignment of the director **n** at boundary surfaces and its orientation within the bulk of the sample. In other words, there may exist, for example, a critical magnetic field of magnitude H_c, such that for $0 \leq H < H_c$ the director is not influenced by the field, while for $H > H_c$ it *is* influenced by the field and begins to adjust its orientation accordingly. This effect is called the Freedericksz transition and H_c is an example of what is called a *critical field strength* or *Freedericksz threshold*. A basic understanding of Freedericksz transitions can be gained from static continuum theory. Freedericksz was the first to observe this phenomenon (see, for example, Freedericksz and Zolina [93]) and Zocher [287], in 1933, successfully applied his earlier static continuum theory to explain a Freedericksz transition induced by a magnetic field. The Freedericksz transition and related phenomena are of immense importance in the 'switching' of liquid crystal displays (LCDs). The twisted nematic device, described in Section 3.7, is based upon a more refined version of the Freedericksz transition and led to applications for displays which were developed in the early 1970s. The classical Freedericksz transitions in nematics can be fairly well identified using static continuum theory and some aspects of these are explored in Sections 3.4 and 3.5.

In 1961, Ericksen [73] proceeded to generalise the static theory of nematics in order to propose balance laws for their dynamical behaviour. Making use of these ideas, Leslie [162, 163], in two articles in 1966 and 1968, successfully managed to formulate constitutive equations and therefore complete a dynamic theory for nematic liquid crystals. This led to the celebrated Ericksen–Leslie dynamic theory for nematic liquid crystals. The isothermal version of this theory is derived in full in Chapter 4, with a convenient summary of the dynamic equations being given in Section 4.2.5. One of the key verifications of the dynamic theory was provided by the experimental observations of Fisher and Frederickson [89] of an unusual scaling law for Poiseuille flow which confirmed a theoretical prediction made by Atkin [5]; this result, discussed in Section 5.8.3, helped to establish the Ericksen–Leslie theory as the generally accepted dynamic theory for nematic liquid crystals. Although not discussed in this introductory book, it should be mentioned that Leslie [164, 165] proposed a dynamic theory for cholesterics in the late 1960s and that he and his co-authors Carlsson and Laverty [31, 33, 34, 176] also worked on biaxial nematic theory in the early 1990s. A detailed account of the development of the mathematical theory of nematics has been written by Carlsson and Leslie [37], who also draw attention to the work of other scientists who played significant rôles in the pursuit of a successful dynamic theory.

Certain defects in nematic liquid crystals were discussed in a mathematical way by Oseen [215] in 1933 and later by Frank [91] in 1958. These defects, and others, are described in some detail in Section 3.8. However, not all defects can be adequately described by the classical continuum theory mentioned above, and this led Ericksen [82] to return to the equilibrium theory of nematic liquid crystals in

1991, when he formulated an extended and modified continuum theory for nematics that is capable of describing defects in a more general setting. This more recent development is beyond the scope of this book, although the interested reader may find the literature based on this work by Ericksen more accessible after becoming familiar with the continuum theory presented here. An account of this theory and some applications have been given in the book by Virga [273].

In 1991, Leslie, Stewart and Nakagawa [173] introduced a nonlinear static and dynamic theory for non-chiral smectic C liquid crystals building upon, in part, earlier work by the Orsay Group [213], Rapini [228] and Martin, Parodi and Pershan [192]. This isothermal continuum theory is based upon the same types of balance laws used to model nematic liquid crystals and appears to be a natural attempt at a fully nonlinear dynamic theory for smectic C. This theory is introduced in Chapter 6 and can be extended to encompass both non-chiral and chiral smectic C liquid crystals. Details of the static and dynamic theory for non-chiral smectic C are given in Sections 6.2 and 6.3, respectively, while some elementary descriptions for chiral smectic C, also called ferroelectric smectic C, are given in Section 6.4. One important test for the static theory of smectic C liquid crystals was to show that there are static equilibrium configurations corresponding to the physically observed Dupin and parabolic cyclides: this proved to be the case in certain situations and is reviewed in Section 6.2.4. Freedericksz transitions also occur in smectic C liquid crystals and an example is given in Section 6.2.5.

The above comments represent only some particularly chosen topics and events in the *mathematical* development of the continuum theory for liquid crystals, focusing especially on nematic and smectic C materials. In a book of this scope it is inevitable that some major topics have not been included, such as soliton-like behaviour in nematics or the flexoelectric effect, and many important contributions to the field have been omitted for the sake of brevity. Nevertheless, readers should have no difficulty in accessing these topics in the current literature if they have been armed with the material presented in subsequent Chapters. Interested readers can find more extensive details on the development of liquid crystals in the historical review by Kelker [143] or the forthcoming volume in this book series by Sluckin, Dunmur and Stegemeyer [253].

1.4   Notation and Conventions

It is assumed throughout that the reader has some familiarity with index notation (also called suffix notation in what is presented here) and the Einstein summation convention. Many suitable introductions are available such as those provided by, for example, the books of Aris [4], Leigh [161], Goodbody [115] or Spencer [256], or the introductory notes by Leslie [174].

In the usual system of basis vectors $\{\mathbf{e}_1, \mathbf{e}_2, \mathbf{e}_3\}$ in $\mathbb{R}^3$, a vector $\mathbf{a} = (a_1, a_2, a_3)$ can be written as

$$\mathbf{a} = a_1\mathbf{e}_1 + a_2\mathbf{e}_2 + a_3\mathbf{e}_3 = \sum_{i=1}^{3} a_i\mathbf{e}_i. \qquad (1.1)$$

The terms a_i, $i = 1, 2, 3$, are called the components of $\mathbf{a}$. The Einstein summation convention allows the sum appearing in (1.1) to be expressed in a more succinct way so that we may write

$$\mathbf{a} = a_i \mathbf{e}_i, \tag{1.2}$$

where it is understood that the repeated index i is summed from, in this example, 1 to 3. In general, this summation convention obeys the following rule: whenever an index appears twice, and *only* twice, in the same term, a summation is implied over all the contributions obtained by letting that particular index assume all its possible values, *unless* an explicit statement is made to the contrary. The summation convention can also be applied to matrices and tensors. For example, if $A = [a_{ij}]$ and $B = [b_{ij}]$ are $n \times n$ matrices, then their product $AB \equiv C = [c_{ij}]$ is the matrix with components

$$c_{ij} = a_{ik} b_{kj}, \tag{1.3}$$

with a summation over the index k from 1 to n being implied. An index on which a summation is carried out is independent of its actual nomenclature. For example, and $a_{ik} b_{kj}$ is equivalent to $a_{is} b_{sj}$. The *trace* of the matrix $A = [a_{ij}]$ is defined by

$$\mathrm{tr}(A) = a_{ii}. \tag{1.4}$$

Two useful quantities that are employed throughout this book are the *Kronecker delta* δ_{ij} and the *alternator* ϵ_{ijk}, defined when i, j and k can each take any of the values 1, 2 or 3. The Kronecker delta and the alternator are defined by, respectively,

$$\delta_{ij} = \begin{cases} 1 & \text{if } i = j, \\ 0 & \text{if } i \neq j, \end{cases} \tag{1.5}$$

and

$$\epsilon_{ijk} = \begin{cases} 1 & \text{if } i, j \text{ and } k \text{ are unequal and in cyclic order,} \\ -1 & \text{if } i, j \text{ and } k \text{ are unequal and in non-cyclic order,} \\ 0 & \text{if any two of } i, j \text{ or } k \text{ are equal.} \end{cases} \tag{1.6}$$

For example, $\delta_{12} = 0$, $\delta_{33} = 1$, $\epsilon_{123} = \epsilon_{231} = \epsilon_{312} = 1$, $\epsilon_{132} = \epsilon_{213} = \epsilon_{321} = -1$, $\epsilon_{112} = 0$, and, following the summation convention, $\delta_{ii} = 3$.

The following two results are particularly useful:

$$\epsilon_{ijk} \epsilon_{ipq} = \delta_{jp} \delta_{kq} - \delta_{jq} \delta_{kp}, \tag{1.7}$$

$$\epsilon_{ijk} \epsilon_{pqr} = \begin{vmatrix} \delta_{ip} & \delta_{iq} & \delta_{ir} \\ \delta_{jp} & \delta_{jq} & \delta_{jr} \\ \delta_{kp} & \delta_{kq} & \delta_{kr} \end{vmatrix}, \tag{1.8}$$

where the right-hand side of (1.8) is a 3×3 determinant. The left-hand sides of these identities employ the summation convention; proofs can be found in any standard textbook, and the results can be verified directly by the reader if required. The result in equation (1.7) occurs frequently and is commonly referred to as the *contraction rule for alternators*.

It is a common convention to denote the partial derivative with respect to the ith variable by a comma followed by the variable. For instance, $p_{,i}$ means the partial derivative of the quantity p with respect to the ith variable. Similarly, $a_{i,j}$ indicates the partial derivative of the ith component of the vector $\mathbf{a}$ with respect to the jth variable.

Scalar and vector products, together with the usual ideas of gradient, divergence and curl, will be required in Cartesian notation. Briefly, the *scalar product* of the vectors $\mathbf{a} = (a_1, a_2, a_3)$ and $\mathbf{b} = (b_1, b_2, b_3)$ is defined by

$$\mathbf{a} \cdot \mathbf{b} = a_i b_i. \tag{1.9}$$

The *magnitude* of $\mathbf{a}$ is defined by $|\mathbf{a}| = \sqrt{\mathbf{a} \cdot \mathbf{a}}$. The scalar product has the property that $\mathbf{a} \cdot \mathbf{a} = 0$ if and only if $\mathbf{a} = \mathbf{0}$. Also, two vectors $\mathbf{a}$ and $\mathbf{b}$ are said to be *orthogonal* if and only if $\mathbf{a} \cdot \mathbf{b} = 0$.

The *vector product* is defined by

$$\mathbf{a} \times \mathbf{b} = \mathbf{e}_i \, \epsilon_{ijk} a_j b_k \,. \tag{1.10}$$

Note that $\mathbf{a} \times \mathbf{b} = -\mathbf{b} \times \mathbf{a}$ and that the vector product of any vector with itself is the zero vector. The vector product is an example of what is called an *axial vector*: it changes sign if we transform to a left-handed system of coordinates [4, p.36].

The *scalar triple product* of the vectors $\mathbf{a}$, $\mathbf{b}$ and $\mathbf{c}$ is defined by

$$\mathbf{a} \cdot (\mathbf{b} \times \mathbf{c}) = a_i \epsilon_{ijk} b_j c_k \,. \tag{1.11}$$

An even permutation may be applied to $\mathbf{a}$, $\mathbf{b}$ and $\mathbf{c}$ without changing the value of the scalar triple product, but an odd permutation will change its sign. Consequently, the scalar triple product satisfies the relations

$$\mathbf{a} \cdot (\mathbf{b} \times \mathbf{c}) = \mathbf{b} \cdot (\mathbf{c} \times \mathbf{a}) = \mathbf{c} \cdot (\mathbf{a} \times \mathbf{b}) = -\mathbf{a} \cdot (\mathbf{c} \times \mathbf{b}) = -\mathbf{b} \cdot (\mathbf{a} \times \mathbf{c}) = -\mathbf{c} \cdot (\mathbf{b} \times \mathbf{a}). \tag{1.12}$$

The *dyadic product*, sometimes called the *tensor product* [256, 273], is denoted by the symbol $\otimes$ and can be defined as having the property that for any given vectors $\mathbf{a}$ and $\mathbf{b}$

$$(\mathbf{a} \otimes \mathbf{b})\mathbf{c} := (\mathbf{b} \cdot \mathbf{c})\mathbf{a} \qquad \text{for all vectors } \mathbf{c}. \tag{1.13}$$

Roughly speaking, in Cartesian components, it means that we can consider $\mathbf{a} \otimes \mathbf{b}$ as a matrix with components given by $[\mathbf{a} \otimes \mathbf{b}]_{ij} = a_i b_j$.

The *gradient* of a scalar field p is defined by

$$\nabla p = \mathbf{e}_i \, p_{,i} \,. \tag{1.14}$$

The *divergence* of the vector $\mathbf{a}$ is given by

$$\nabla \cdot \mathbf{a} = a_{i,i} \,, \tag{1.15}$$

while its *curl* is defined by

$$\nabla \times \mathbf{a} = \mathbf{e}_i \, \epsilon_{ijk} a_{k,j} \,. \tag{1.16}$$

In this text we shall define the *gradient* of a vector field **v** by [161, 273]

$$\nabla\mathbf{v} = v_{i,j}\,\mathbf{e}_i \otimes \mathbf{e}_j\,. \tag{1.17}$$

We can consider $\nabla\mathbf{v}$ as a matrix with components given by $[\nabla\mathbf{v}]_{ij} = v_{i,j}$. There is also a definition in the literature for $\nabla\mathbf{v}$ which happens to be the transpose of that just defined here, for example that used by Goodbody [115].

It is assumed that the reader is familiar with some basic properties of Cartesian tensors, such as those that may be found in the book by Spencer [256]. In this book we shall define the divergence of a second order tensor T_{ij} to be the tensor

$$T_{ij,j}\,, \tag{1.18}$$

this being in accordance with the convention adopted by Truesdell and Noll [270] and Leigh [161]. If the tensor T_{ij} is symmetric, that is, $T_{ij} = T_{ji}$, then this convention coincides with that given by Goodbody [115], among others. However, care needs to be exercised in liquid crystal theory because the stress tensor, a second order tensor, is in general not symmetric: its divergence therefore has to be treated carefully and consistently. Such occasions will be highlighted where appropriate.

The divergence theorem of Gauss will be used frequently. This theorem states that if **a** has continuous first order partial derivatives at all points of a volume V having boundary surface S, then

$$\int_V a_{i,i}\,dV = \int_S a_i \nu_i\,dS\,, \tag{1.19}$$

where $\boldsymbol{\nu}$ is the outward unit normal to S. It is common practice to write dS_i as an abbreviation for $\nu_i\,dS$. The divergence theorem can also be applied to tensors. For example, if t_{ij} is a second order tensor then

$$\int_V t_{ij,j}\,dV = \int_S t_{ij}\nu_j\,dS\,. \tag{1.20}$$

Analogous results hold for tensors of higher order.

Chapter 2

Static Theory of Nematics

2.1 Introduction

The static theory of nematic liquid crystals is introduced in this Chapter. Static continuum theory involves two major steps. The first step is to construct an 'energy' based upon possible distortions of the director n, introduced in Chapter 1. The second step is to minimise this energy in some sense, and this leads to equilibrium equations involving n and its derivatives. The solutions of these differential equations yield possible equilibrium orientations for n: it is these alignments of n that are the ultimate goal of static continuum theory since they indicate the director alignment within a sample of liquid crystal. The solutions with the least energy are interpreted as the physically relevant ones.

We begin by constructing the Frank–Oseen elastic energy, also called the free energy, for nematic and cholesteric liquid crystals in Section 2.2. The key elastic energy results are given below by equation (2.49) for nematics and equation (2.78) for cholesterics. Electric and magnetic fields are introduced in Section 2.3 and the electric energy that arises when a sample of liquid crystal is subjected to an applied electric field is discussed in Section 2.3.1, with the corresponding results for magnetic fields presented in Section 2.3.2. Many applications of continuum theory require some knowledge of the associated physical units and therefore some comments on this topic are included in Section 2.3.3 and in Tables D.1 and D.2 on page 329.

The general equilibrium equations arise from minimising the total energy of any given system (which may incorporate contributions from the elastic, magnetic or electric energies). These equations are formulated in Section 2.4 and are summarised on pages 38 to 38, followed by comments on null Lagrangians and a simplification that is often available. Section 2.5 contains some verifications of certain basic general static solutions to the equilibrium equations in the absence of any external fields.

Liquid crystals in confined geometries are known to be influenced by the orientation of the director at the boundaries. Section 2.6 introduces various ideas on boundary conditions and the concepts of 'strong' and 'weak' anchoring of the director at the boundaries. Finally, Section 2.7 gives a simplifying reformulation of the equilibrium equations and boundary conditions in terms of orientation angles of the director and highlights conditions under which these simplified equations are feasible.

2.2   The Frank–Oseen Elastic Energy

The static theory of nematic and cholesteric liquid crystals employs the unit vector field $\mathbf{n}$, frequently referred to as the director, to describe the mean molecular alignment at a point $\mathbf{x}$ in a given sample volume V. Hence

$$\mathbf{n} = \mathbf{n}(\mathbf{x}), \qquad \mathbf{n} \cdot \mathbf{n} = 1. \tag{2.1}$$

This alignment displays a certain elasticity and it is known that an initial uniform alignment of a nematic liquid crystal commonly returns after the removal of any disturbing influences. It is therefore assumed that there is a free energy density, also called the free energy integrand, associated with distortions of the anisotropic axis of the form

$$w = w(\mathbf{n}, \nabla\mathbf{n}), \tag{2.2}$$

with the total elastic Helmholtz free energy being

$$W = \int_V w(\mathbf{n}, \nabla\mathbf{n}) \, dV. \tag{2.3}$$

For this form of energy it is assumed that the liquid crystal sample is incompressible, that is, the mass density remains constant. A liquid crystal sample exhibiting a completely relaxed configuration in the absence of forces, fields and boundary conditions, is said to be in a natural orientation. The free energy is usually defined to within the addition of an arbitrary constant and it is advantageous to choose this constant such that $w = 0$ for any natural orientation and to suppose that any other state or configuration induced upon the sample produces an energy that is greater than or equal to that for a completely relaxed natural orientation. It is therefore supposed that

$$w(\mathbf{n}, \nabla\mathbf{n}) \geq 0. \tag{2.4}$$

Since liquid crystals generally lack polarity, the vectors $\mathbf{n}$ and $-\mathbf{n}$ are physically indistinguishable; this conclusion is also true if the constituent molecules are polar, that is, they have different 'ends', because they can be thought of as being arranged locally at any given point in V in such a way that they are equally divided into two groups having opposite orientations. It is therefore natural to require

$$w(\mathbf{n}, \nabla\mathbf{n}) = w(-\mathbf{n}, -\nabla\mathbf{n}). \tag{2.5}$$

The free energy per unit volume must also be the same when described in any two frames of reference, that is, it must be frame-indifferent. This means that the energy density must be invariant to arbitrary superposed rigid body rotations and consequently we require

$$w(\mathbf{n}, \nabla\mathbf{n}) = w(Q\mathbf{n}, Q\nabla\mathbf{n}Q^T), \tag{2.6}$$

where Q is any proper orthogonal matrix ($\det Q = 1$), Q^T being its transpose. Frank [91], Oseen [215] and Zocher [287] considered the construction of such an energy for liquid crystals and we shall proceed by deriving the well known Frank–Oseen elastic energy (2.44) below. Once this energy has been constructed we shall consider two special cases: the nematic energy (2.49), or its equivalent (2.50), and the cholesteric energy (2.78). We shall always use the term nematic to refer to non-chiral nematic and reserve the term cholesteric for chiral nematic.

We develop the approach adopted by Frank [91]. Let $\mathbf{n}$ be a unit vector representing the preferred orientation of a uniaxial liquid crystal at any point $\mathbf{x}$ and assume that it varies slowly with position. The sign of $\mathbf{n}$ has no physical significance in most cases. However, for molecules with permanent dipole moments this may not be the case and then the sign of $\mathbf{n}$ becomes important, but this will not be considered here. We introduce a local system of Cartesian coordinates x, y, z with z parallel to

$$\mathbf{n_0} = (0, 0, 1), \tag{2.7}$$

at the origin (see Fig. 2.1). Referred to these axes, small changes Δx, Δy, and Δz from the origin in the x, y and z directions, respectively, induce three possible types of changes in the orientation of $\mathbf{n_0}$, leading to six components of curvature, also called the curvature strains. In the notation of Frank, these 'splay', 'twist' and 'bend' components are

$$\text{splay}: \quad s_1 = \frac{\partial n_x}{\partial x}, \quad s_2 = \frac{\partial n_y}{\partial y},$$

$$\text{twist}: \quad t_1 = -\frac{\partial n_y}{\partial x}, \quad t_2 = \frac{\partial n_x}{\partial y}, \tag{2.8}$$

$$\text{bend}: \quad b_1 = \frac{\partial n_x}{\partial z}, \quad b_2 = \frac{\partial n_y}{\partial z},$$

where the components of the director are given by $\mathbf{n} = (n_x, n_y, n_z)$. The above terms are evaluated at $\mathbf{x} = \mathbf{0}$ and can be identified from the corresponding depictions of splay, twist and bend shown in Fig. 2.1: the left column in the Figure depicts the possible reorientations of $\mathbf{n}$ while the right column shows the corresponding two-dimensional cross-sections of such orientations (for example, in the yz-plane) where the short bold lines represent the director.

The curvature strains can also be obtained by considering a Taylor series expansion for the components of $\mathbf{n}$ about the origin. Firstly, because $\mathbf{n} \cdot \mathbf{n} = 1$, it follows that

$$\mathbf{0} = \nabla(\mathbf{n} \cdot \mathbf{n}) = 2\mathbf{e}_j \, n_i n_{i,j}, \tag{2.9}$$

where $\mathbf{e}_j$, $j = 1, 2, 3$ are the basis vectors introduced in Section 1.4; equivalently, we have

$$n_i n_{i,j} = 0, \qquad j = 1, 2, 3, \tag{2.10}$$

and therefore, by (2.7),

$$n_{z,j}(\mathbf{0}) = 0, \qquad j = 1, 2, 3. \tag{2.11}$$

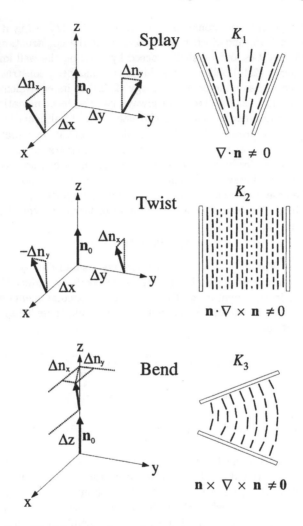

Figure 2.1: The left-hand column above shows the various reorientations of the director $\mathbf{n}$, shown by the bold arrows, when it is originally aligned at $\mathbf{n} = \mathbf{n}_0$ along the z-axis. These reorientations are labelled 'Splay', 'Twist' and 'Bend' and are depictions of the six basic curvature strains in equation (2.8). The right-hand column shows the corresponding two-dimensional cross-sections of such reorientations, for example in the yz-plane, the short lines representing the director. Each reorientation is related to a term in the elastic energy and has its corresponding Frank elastic constant K_1, K_2 or K_3 as indicated: see also the energy formulations for nematics (2.49) and cholesterics (2.78). Each K_i, $i = 1, 2$ or 3, is non-negative, by (2.58).

Here we have adopted the notation contained in Section 1.4 for Cartesian tensors and assumed the summation convention where repeated indices are summed from

1 to 3. Expanding in a Taylor series round the origin gives

$$n_i(\mathbf{x}) = n_i(0) + x_j n_{i,j}(0) + O(|\mathbf{x}|^2), \qquad i = 1, 2, 3, \tag{2.12}$$

which, by (2.7) and (2.11), leads to

$$
\begin{aligned}
n_x &= a_1 x + a_2 y + a_3 z + O(|\mathbf{x}|^2), \\
n_y &= a_4 x + a_5 y + a_6 z + O(|\mathbf{x}|^2), \\
n_z &= 1 + O(|\mathbf{x}|^2),
\end{aligned}
\tag{2.13}
$$

where the constants a_i, $i = 1, 2, ...6$, are related to the notation introduced above in (2.8) by

$$a_1 = s_1, \quad a_2 = t_2, \quad a_3 = b_1, \quad a_4 = -t_1, \quad a_5 = s_2, \quad a_6 = b_2. \tag{2.14}$$

It is anticipated that the elastic free energy W of a liquid crystal, relative to its energy in a state of uniform orientation, is expressible as the volume integral of the free energy density w given in equation (2.2), expanded as a quadratic function of the six curvature strains:

$$w = k_i a_i + \tfrac{1}{2} k_{ij} a_i a_j, \qquad i, j = 1, 2, ...6, \tag{2.15}$$

where k_i and k_{ij} are the curvature elastic constants and repeated indices are summed from 1 to 6; the quadratic terms in (2.15) can always be written as a unique symmetric quadratic form where $k_{ij} = k_{ji}$. Since we are dealing with uniaxial nematics, a rotation about the z-axis shown in Fig. 2.1 makes no change to the physical description, that is, if (x, y, z) is replaced by another equally permissible local coordinate system (x', y', z') then the energy density w should be identical when described by the new curvature strains a_i'. It is therefore a requirement that

$$w = k_i a_i' + \tfrac{1}{2} k_{ij} a_i' a_j', \qquad i, j = 1, 2, ...6, \tag{2.16}$$

for any allowed transformation of coordinates. This condition will impose restrictions on the possible elastic constants k_i and k_{ij}. These restrictions can be obtained by considering suitable changes of coordinates via rotations around the z-axis.

Firstly, to demonstrate the method, consider a rotation of $\frac{\pi}{2}$ around the z-axis in the original coordinate system introduced above. The transformed coordinates and director components are then, respectively,

$$x' = y, \qquad y' = -x, \qquad z' = z, \tag{2.17}$$
$$n_{x'} = n_y, \qquad n_{y'} = -n_x, \qquad n_{z'} = n_z. \tag{2.18}$$

Figure 2.1 shows this to be possible and leads, via (2.8), (2.14), (2.17), (2.18) and simple calculations, to $(a_i) \rightarrow (a_i')$ where

$$(a_i') = (a_5, -a_4, a_6, -a_2, a_1, -a_3). \tag{2.19}$$

This transformation for the a_i can also be obtained by considering the entries in the product $A' = QAQ^T$ where

$$Q = \begin{bmatrix} 0 & 1 & 0 \\ -1 & 0 & 0 \\ 0 & 0 & 1 \end{bmatrix}, \qquad A = \nabla\mathbf{n}(0) = \begin{bmatrix} a_1 & a_2 & a_3 \\ a_4 & a_5 & a_6 \\ 0 & 0 & 0 \end{bmatrix}, \qquad (2.20)$$

obtained via (2.13) and applying the invariance requirement in equation (2.6) for proper orthogonal matrices such as Q. Since we require $k_i a_i = k_i a'_i$ we have, on collecting terms in a_i,

$$a_1(k_1 - k_5) + a_2(k_2 + k_4) + a_3(k_3 + k_6) + a_4(k_2 + k_4)$$
$$+ a_5(k_5 - k_1) + a_6(k_6 - k_3) = 0, \qquad (2.21)$$

and therefore by the linear independence of the a_i,

$$(k_i) = (k_1, k_2, 0, -k_2, k_1, 0). \qquad (2.22)$$

Similarly, considering $k_{ij}a_i a_j = k_{ij}a'_i a'_j$ leads to

$$a_1^2(k_{11} - k_{55}) + a_2^2(k_{22} - k_{44}) + a_3^2(k_{33} - k_{66}) + a_4^2(k_{44} - k_{22})$$
$$+ a_5^2(k_{55} - k_{11}) + a_6^2(k_{66} - k_{33}) + 2a_1 a_2(k_{12} + k_{45})$$
$$+ 2a_1 a_3(k_{13} + k_{56}) + 2a_1 a_4(k_{14} + k_{25}) + 2a_1 a_6(k_{16} - k_{35})$$
$$+ 2a_2 a_3(k_{23} - k_{46}) + 2a_2 a_5(k_{25} + k_{14}) + 2a_2 a_6(k_{26} + k_{34})$$
$$+ 2a_3 a_4(k_{34} - k_{26}) + 2a_3 a_5(k_{35} + k_{16}) + 2a_4 a_5(k_{45} + k_{12})$$
$$+ 4a_3 a_6 k_{36} + 2a_4 a_6(k_{46} + k_{23}) + 2a_5 a_6(k_{56} - k_{13}) = 0, \qquad (2.23)$$

and so by the independence of the products $a_i a_j$

$$k_{11} = k_{55}, \qquad k_{22} = k_{44}, \qquad k_{33} = k_{66}, \qquad (2.24)$$
$$k_{12} = -k_{45}, \qquad k_{14} = -k_{25}, \qquad (2.25)$$
$$k_{13} = k_{16} = k_{23} = k_{26} = k_{34} = k_{35} = k_{36} = k_{46} = k_{56} = 0. \qquad (2.26)$$

One further admissible rotation around the z-axis through an arbitrary angle different from $\frac{\pi}{2}$ will turn out to be sufficient for our purposes in analysing the k_{ij} terms. For simplicity, we choose a rotation through $\frac{\pi}{4}$, accomplished by the transformation

$$x' = \tfrac{1}{\sqrt{2}}(x + y), \qquad y' = \tfrac{1}{\sqrt{2}}(y - x), \qquad z' = z, \qquad (2.27)$$
$$n_{x'} = \tfrac{1}{\sqrt{2}}(n_x + n_y), \qquad n_{y'} = \tfrac{1}{\sqrt{2}}(n_y - n_x), \qquad n_{z'} = n_z, \qquad (2.28)$$

leading to

$$(a'_i) = \Big(\tfrac{1}{2}(a_1 + a_2 + a_4 + a_5), \ \tfrac{1}{2}(a_2 + a_5 - a_1 - a_4), \tfrac{1}{\sqrt{2}}(a_3 + a_6),$$
$$\tfrac{1}{2}(a_4 + a_5 - a_1 - a_2), \tfrac{1}{2}(a_1 + a_5 - a_2 - a_4), \tfrac{1}{\sqrt{2}}(a_6 - a_3) \Big), \qquad (2.29)$$

which, similar to the previous case, can be obtained by employing the matrix A in equation (2.20) but with Q replaced by

$$Q = \begin{bmatrix} \frac{1}{\sqrt{2}} & \frac{1}{\sqrt{2}} & 0 \\ -\frac{1}{\sqrt{2}} & \frac{1}{\sqrt{2}} & 0 \\ 0 & 0 & 1 \end{bmatrix}. \tag{2.30}$$

We are now in a position to discuss some of the consequences of the requirement $k_{ij}a_i a_j = k_{ij}a'_i a'_j$. Collecting together the terms involving a_1^2 and $a_1 a_4$ and using equations (2.24), (2.25), (2.29) and the method employed in obtaining (2.23) after the first rotation introduced above, shows that

$$\frac{1}{2}a_1^2(k_{11} - k_{15} - k_{22} - k_{24}) = 0, \tag{2.31}$$
$$2a_1 a_4(k_{14} + k_{12}) = 0, \tag{2.32}$$

and therefore

$$k_{15} = k_{11} - k_{22} - k_{24}, \tag{2.33}$$
$$k_{14} = -k_{12}. \tag{2.34}$$

Other products $a_i a_j$ lead to relationships for the k_{ij} that have already been derived. Equations (2.24), (2.25), (2.26), (2.33) and (2.34) now demonstrate that of the thirty-six components in k_{ij} eighteen are zero and only five are independent. The resulting matrix k_{ij} is given by

$$[k_{ij}] = \begin{bmatrix} k_{11} & k_{12} & 0 & -k_{12} & (k_{11} - k_{22} - k_{24}) & 0 \\ k_{12} & k_{22} & 0 & k_{24} & k_{12} & 0 \\ 0 & 0 & k_{33} & 0 & 0 & 0 \\ -k_{12} & k_{24} & 0 & k_{22} & -k_{12} & 0 \\ (k_{11} - k_{22} - k_{24}) & k_{12} & 0 & -k_{12} & k_{11} & 0 \\ 0 & 0 & 0 & 0 & 0 & k_{33} \end{bmatrix}. \tag{2.35}$$

The general form of the energy density can now be written via (2.8), (2.14), (2.15), (2.22) and (2.35) as

$$\begin{aligned} w = \; & k_1(s_1 + s_2) + k_2(t_1 + t_2) + \tfrac{1}{2}k_{11}(s_1 + s_2)^2 + \tfrac{1}{2}k_{22}(t_1 + t_2)^2 \\ & + \tfrac{1}{2}k_{33}(b_1^2 + b_2^2) + k_{12}(s_1 + s_2)(t_1 + t_2) - (k_{22} + k_{24})(s_1 s_2 + t_1 t_2). \end{aligned} \tag{2.36}$$

If the constants

$$s_0 = -\frac{k_1}{k_{11}}, \qquad t_0 = -\frac{k_2}{k_{22}}, \tag{2.37}$$

are introduced and the energy is redefined as

$$w_F = w + \tfrac{1}{2}k_{11}s_0^2 + \tfrac{1}{2}k_{22}t_0^2, \tag{2.38}$$

then the energy in equation (2.36) can be written in the more convenient form adopted by Frank [91], namely,

$$\begin{aligned} w_F = \; & \tfrac{1}{2}k_{11}(s_1 + s_2 - s_0)^2 + \tfrac{1}{2}k_{22}(t_1 + t_2 - t_0)^2 + \tfrac{1}{2}k_{33}(b_1^2 + b_2^2) \\ & + k_{12}(s_1 + s_2)(t_1 + t_2) - (k_{22} + k_{24})(s_1 s_2 + t_1 t_2). \end{aligned} \tag{2.39}$$

This version of the energy has its lowest free energy corresponding not to a state of uniform orientation but to an orientation that has the optimum degree of splay and twist, as can be seen by the contributions to the terms in k_{11} and k_{22} and the interpretation offered by (2.8) and Fig. 2.1. This latter form of the energy density can also be written in vector notation. When $\mathbf{n} = \mathbf{n}_0 = (0, 0, 1)$ and $\nabla \mathbf{n}$ is given by equation (2.20) it is simple to use equations (2.11) to (2.14) to verify that

$$\nabla \cdot \mathbf{n} = n_{x,x} + n_{y,y} = s_1 + s_2, \tag{2.40}$$

$$\mathbf{n} \cdot \nabla \times \mathbf{n} = n_{y,x} - n_{x,y} = -(t_1 + t_2), \tag{2.41}$$

$$(\mathbf{n} \times \nabla \times \mathbf{n})^2 = n_{x,z}^2 + n_{y,z}^2 = b_1^2 + b_2^2, \tag{2.42}$$

$$\tfrac{1}{2}\nabla \cdot [(\mathbf{n} \cdot \nabla)\mathbf{n} - (\nabla \cdot \mathbf{n})\mathbf{n}] = n_{y,x}n_{x,y} - n_{x,x}n_{y,y} = -(s_1 s_2 + t_1 t_2). \tag{2.43}$$

These results allow the energy density w_F in (2.39) to be written in vector notation as

$$w_F = \ \tfrac{1}{2}k_{11}(\nabla \cdot \mathbf{n} - s_0)^2 + \tfrac{1}{2}k_{22}(\mathbf{n} \cdot \nabla \times \mathbf{n} + t_0)^2 + \tfrac{1}{2}k_{33}(\mathbf{n} \times \nabla \times \mathbf{n})^2 \\ - k_{12}(\nabla \cdot \mathbf{n})(\mathbf{n} \cdot \nabla \times \mathbf{n}) + \tfrac{1}{2}(k_{22} + k_{24})\nabla \cdot [(\mathbf{n} \cdot \nabla)\mathbf{n} - (\nabla \cdot \mathbf{n})\mathbf{n}]. \tag{2.44}$$

It should be noted that when partial derivatives commute then the last term in (2.44) has an equivalent representation provided by the identity

$$\nabla \cdot [(\mathbf{n} \cdot \nabla)\mathbf{n} - (\nabla \cdot \mathbf{n})\mathbf{n}] = \left[\mathrm{tr}((\nabla\mathbf{n})^2) - (\nabla \cdot \mathbf{n})^2 \right], \tag{2.45}$$

where tr is the trace defined earlier by equation (1.4), which confirms that (2.44) is indeed a quadratic function of the first derivatives of $\mathbf{n}$.

2.2.1   The Nematic Energy

For both nematics and cholesterics the invariance requirement in equation (2.5) must hold. This means, from w_F in (2.44) and the definition of s_0 in (2.37), that $k_1 = k_{12} = 0$. Further, nematic molecules generally resemble rods and remain alike upon reflection within planes containing the director. The orthogonal matrix

$$Q = \begin{bmatrix} 1 & 0 & 0 \\ 0 & -1 & 0 \\ 0 & 0 & 1 \end{bmatrix}, \tag{2.46}$$

is clearly an example of an admissible representation for a reflection in the x-axis for a nematic if the director lies within the xy-plane (this is equivalent to the example considered by Frank [91, Eq.(16)]) and leads to the invariance in equation (2.6) being satisfied for *any* orthogonal tensor Q. The approach used in the previous Section from equation (2.19) to (2.22) can be repeated here but with Q in equation (2.20) replaced by the matrix in equation (2.46). The resulting analogue for (k_i) in (2.22) is, by (2.21) and noting that $k_1 = k_5$ and $k_3 = 0$,

$$(k_i) = (k_1, 0, 0, 0, k_1, 0). \tag{2.47}$$

The expressions in (2.22) and (2.47) must coincide and therefore, in addition to k_1 being zero, we must have $k_2 = 0$, resulting in $t_0 = 0$. Hence, for nematics,

$s_0 = t_0 = k_{12} = 0$ in equation (2.44). It is currently common practice to introduce the notation

$$K_1 = k_{11}, \quad K_2 = k_{22}, \quad K_3 = k_{33}, \quad K_4 = k_{24}, \tag{2.48}$$

where the K_i are often referred to as the *Frank elastic constants* (or moduli). The free energy density for nematics may finally be expressed as

$$\begin{aligned}
w_F = {} & \tfrac{1}{2}K_1(\nabla \cdot \mathbf{n})^2 + \tfrac{1}{2}K_2(\mathbf{n} \cdot \nabla \times \mathbf{n})^2 + \tfrac{1}{2}K_3(\mathbf{n} \times \nabla \times \mathbf{n})^2 \\
& + \tfrac{1}{2}(K_2 + K_4)\nabla \cdot [(\mathbf{n} \cdot \nabla)\mathbf{n} - (\nabla \cdot \mathbf{n})\mathbf{n}],
\end{aligned} \tag{2.49}$$

the last expression in (2.49) sometimes taking the equivalent form introduced by the identity (2.45). This energy can also be expressed in the equivalent Cartesian component form

$$\begin{aligned}
w_F = {} & \tfrac{1}{2}(K_1 - K_2 - K_4)(n_{i,i})^2 + \tfrac{1}{2}K_2 n_{i,j} n_{i,j} + \tfrac{1}{2}K_4 n_{i,j} n_{j,i} \\
& + \tfrac{1}{2}(K_3 - K_2)n_j n_{i,j} n_k n_{i,k},
\end{aligned} \tag{2.50}$$

obtained by employing the relations (1.7) to (1.16). K_1, K_2 and K_3 are called the *splay*, *twist* and *bend* constants respectively, while the combination $(K_2 + K_4)$ is called the *saddle-splay* constant. In (2.50) the usual summation convention has been applied where terms containing repeated indices are summed from 1 to 3. A more detailed mathematical interpretation of these constants can be found in Virga [273, Ch.3.3]. The saddle-splay term in w_F is often omitted since it does not contribute to the bulk equilibrium equations for problems involving strong anchoring (see the comments on null Lagrangians on page 38 and also Section 2.6.2 on page 47 below). Being a divergence, its contribution to $\int_V w_F\, dV$ can be transformed via the divergence theorem (see equation (1.19)) to a surface integral over the boundary, so that it does not contribute to the bulk equilibrium equations. It should also be mentioned that an alternative algebraic derivation of the nematic energy (2.49) has been constructed by Clark [44].

It is a simple exercise to use (2.7), (2.8), (2.11), (2.14) and (2.20) to find

$$\nabla \cdot \mathbf{n} = a_1 + a_5, \tag{2.51}$$

$$\mathbf{n} \cdot \nabla \times \mathbf{n} = a_4 - a_2, \tag{2.52}$$

$$\mathbf{n} \times \nabla \times \mathbf{n} = (-a_3, -a_6, 0), \tag{2.53}$$

$$\operatorname{tr}((\nabla \mathbf{n})^2) - (\nabla \cdot \mathbf{n})^2 = 2(a_2 a_4 - a_1 a_5). \tag{2.54}$$

Inserting these quantities into (2.49) and using the identity (2.45) reveals that

$$\begin{aligned}
2w_F = {} & K_1(a_1^2 + a_5^2) + 2(K_1 - K_2 - K_4)a_1 a_5 \\
& + K_2(a_2^2 + a_4^2) + 2K_4 a_2 a_4 \\
& + K_3(a_3^2 + a_6^2).
\end{aligned} \tag{2.55}$$

The right-hand side of (2.55) is the sum of three independent quadratic forms in the variables a_i each of which may be considered in the form

$$f(x, y) = \alpha(x^2 + y^2) + 2\beta xy = \begin{bmatrix} x & y \end{bmatrix} \begin{bmatrix} \alpha & \beta \\ \beta & \alpha \end{bmatrix} \begin{bmatrix} x \\ y \end{bmatrix}. \tag{2.56}$$

Thus the non-negativity requirement $w_F \geq 0$ arising from (2.4) will hold if and only if all the eigenvalues of the symmetric matrix in (2.56) are non-negative for each quadratic form appearing in (2.55) (see Anton [2, pp.478–480]); this condition is equivalent to requiring that the determinants of all the principal submatrices of the above symmetric matrix are non-negative. This produces the requirement

$$\alpha \geq |\beta| \geq 0. \tag{2.57}$$

Applying this result to (2.55) yields the inequalities

$$K_1 \geq 0, \qquad K_2 \geq 0, \qquad K_3 \geq 0,$$
$$K_2 \geq |K_4|, \qquad K_1 \geq |K_1 - K_2 - K_4|. \tag{2.58}$$

The last inequality in (2.58) may be replaced by

$$2K_1 \geq K_2 + K_4 \geq 0. \tag{2.59}$$

The inequalities in (2.58) and (2.59) were first derived by Ericksen [76] and are known as *Ericksen's inequalities*.

The elastic constants are dependent upon the temperature T and are commonly of the order 10^{-7} dynes to 10^{-6} dynes in cgs units, equivalent to around the order of 10^{-12} N to 10^{-11} N (newtons) in SI units, with K_3 often being two or three times larger than K_1 or K_2 (recall that 1 dyne = 10^{-5} N). For the nematic liquid crystal PAA at $T = 125°C$, the experimental work of Zwetkoff (Tsvetkov), in 1937, established the values in SI units (originally quoted in dynes)

$$K_1 = 4.5 \times 10^{-12} \text{ N}, \quad K_2 = 2.9 \times 10^{-12} \text{ N}, \quad K_3 = 9.5 \times 10^{-12} \text{ N}, \tag{2.60}$$

as indicated in [110]. For the nematic MBBA at around 25°C the constants were measured by Haller [121] as

$$K_1 = 6 \times 10^{-12} \text{ N}, \quad K_2 = 3.8 \times 10^{-12} \text{ N}, \quad K_3 = 7.5 \times 10^{-12} \text{ N}, \tag{2.61}$$

in SI units. The values for these constants for PAA and MBBA obtained by other workers at different temperatures may be found in the tabulated lists provided by de Gennes and Prost [110, pp.104–105]. These constants generally decrease as the temperature increases to the transition temperature T_{NI} where the nematic becomes isotropic. Other values of elastic constants for nematics can be found in Table D.3 on page 330.

The constant K_4 is rarely mentioned in the literature. In many instances, the saddle-splay term does not usually influence the equilibrium equations because of an application of the divergence theorem, as mentioned previously. However, recent investigations on possible experiments for measuring K_4 have been briefly commented upon by Virga [273, p.123].

One-constant Approximation

Sometimes, for example, when the relative values of the K_i are unknown or when the resulting equilibrium equations are complicated, the one-constant approximation

$$K \equiv K_1 = K_2 = K_3, \quad K_4 = 0, \tag{2.62}$$

is made. (Some authors, in the one-constant approximation, equate the splay, twist and bend constants and assume that $K_2 + K_4 = 0$ rather than $K_4 = 0$. In the case of strong anchoring this alternative approximation leads to equilibrium equations that are identical to those obtained using (2.62).) In this case the energy w_F in (2.49) can be simplified, when the partial derivatives commute and use is made of the result in (2.9), via the elementary identity

$$(\nabla \times \mathbf{n})^2 = (\mathbf{n} \cdot \nabla \times \mathbf{n})^2 + (\mathbf{n} \times \nabla \times \mathbf{n})^2, \qquad (2.63)$$

to

$$w_F = \tfrac{1}{2} K \left\{ (\nabla \cdot \mathbf{n})^2 + (\nabla \times \mathbf{n})^2 \right\} + \tfrac{1}{2} K \nabla \cdot \left[(\mathbf{n} \cdot \nabla)\mathbf{n} - (\nabla \cdot \mathbf{n})\mathbf{n} \right]. \qquad (2.64)$$

The last divergence term appearing in (2.64) is frequently omitted when considering strong anchoring, for the reasons indicated earlier in this Section, since it can be transformed to a surface term via the divergence theorem (cf. de Gennes and Prost [110, p.104]). The identity

$$(\nabla \cdot \mathbf{n})^2 + (\nabla \times \mathbf{n})^2 + \nabla \cdot \left[(\mathbf{n} \cdot \nabla)\mathbf{n} - (\nabla \cdot \mathbf{n})\mathbf{n} \right] = \|\nabla \mathbf{n}\|^2, \qquad (2.65)$$

is easily proved in Cartesian component form: the left-hand side of the above is, by the contraction rule (1.7) for alternators,

$$
\begin{aligned}
n_{i,i} n_{j,j} + \epsilon_{ijk} \epsilon_{ipq} n_{k,j} n_{q,p} + (n_j n_{i,j} - n_{j,j} n_i)_{,i} &= (\delta_{jp}\delta_{kq} - \delta_{jq}\delta_{kp}) n_{k,j} n_{q,p} + n_{j,i} n_{i,j} \\
&= n_{k,j} n_{k,j} \\
&= \|\nabla \mathbf{n}\|^2. \qquad (2.66)
\end{aligned}
$$

This allows the one-constant approximation in equation (2.64) to be expressed in an equivalent Cartesian components form as

$$w_F = \tfrac{1}{2} K \|\nabla \mathbf{n}\|^2 = \tfrac{1}{2} K n_{i,j} n_{i,j}. \qquad (2.67)$$

This means that, in Cartesian coordinates, the energy is calculated simply by summing the squares of the components $n_{i,j}$.

2.2.2   The Cholesteric Energy

As indicated in the previous Section, nematic molecules remain alike upon reflection within planes containing the director. However, the molecular alignment of cholesteric liquid crystals, also referred to as chiral nematics, possesses a helical structure which undergoes a change in chirality upon reflection. This means that right-handed helices are transformed into left-handed helices under reflections and vice-versa; this property of cholesterics, where cholesteric molecules are distinguishable from their mirror images, is known as *enantiomorphy*, as mentioned in Section 1.2. Figure 2.2 shows a typical helical structure for a cholesteric, where the director rotates within the xz-plane as the observer moves along the cholesteric helical axis aligned along the y-axis. The *pitch* of the helix is defined to be the distance along the helical axis over which the director rotates through 2π radians.

Since $-\mathbf{n}$ and $\mathbf{n}$ are indistinguishable the period of repetition L is half that of the pitch: see Fig. 2.2. Typical values for L are in the range of $3,000$ Å [110, p.15]. The pitch satisfies

$$P = 2L = \frac{2\pi}{|\tau|}, \tag{2.68}$$

where τ corresponds to the wave vector. The sign of τ changes according to the type of helix present: it is positive for a right-handed helix (for example, cholesteryl chloride) and is negative for a left-handed helix (for example, many of the aliphatic esters of cholesterol) [110, p.264]. In the geometry of Fig. 2.2 with the x-axis pointing towards the reader, the director rotates in a right-handed helical structure in the xz-plane as the observer travels along the y-axis when $\mathbf{n}$ takes the form

$$\mathbf{n}_c = (\sin(\tau y), 0, \cos(\tau y)), \tag{2.69}$$

the director 'rotating' to the right in a clockwise motion when the observer is looking along and travelling in the positive y-direction. This represents a natural equilibrium state for a cholesteric in the absence of external forces, fields and boundary conditions.

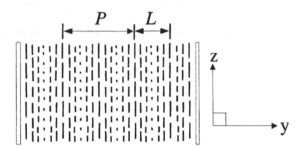

Figure 2.2: The pitch P over which the director rotates through 2π radians in cholesterics; $L = P/2$.

It is known from the previous Section that $k_1 = k_{12} = 0$ for both nematics and cholesterics, and so, by $(2.37)_1$, $s_0 = 0$ also. The invariance condition (2.6) must hold for cholesterics also, showing that the constants k_i must again be of the form given in equation (2.22); but these values for k_i must necessarily be *inequivalent* to those given in (2.47) because of the presence of enantiomorphy in cholesterics. Noting that k_3 must be zero, we are therefore forced to conclude that $k_2 \neq 0$ for cholesterics. Hence, by $(2.37)_2$, $t_0 \neq 0$ and, employing the notation of the previous Section, the energy for cholesterics based upon (2.44) and (2.48) can now be written as

$$\begin{aligned} w_{chol} = {} & \tfrac{1}{2}K_1(\nabla \cdot \mathbf{n})^2 + \tfrac{1}{2}K_2(\mathbf{n} \cdot \nabla \times \mathbf{n} + t_0)^2 + \tfrac{1}{2}K_3(\mathbf{n} \times \nabla \times \mathbf{n})^2 \\ & + \tfrac{1}{2}(K_2 + K_4)\nabla \cdot [(\mathbf{n} \cdot \nabla)\mathbf{n} - (\nabla \cdot \mathbf{n})\mathbf{n}]. \end{aligned} \tag{2.70}$$

It is instructive at this point to insert the natural cholesteric state $\mathbf{n}_c$ from equation (2.69) into w_{chol} to find that

$$w_{chol}(\mathbf{n}_c) = \tfrac{1}{2}K_2(t_0 - \tau)^2. \tag{2.71}$$

Clearly, $w_{chol}(\mathbf{n}_c)$ is minimised, with $w_{chol}(\mathbf{n}_c) = 0$, when

$$\tau = t_0 = -\frac{k_2}{K_2}, \tag{2.72}$$

where we have adopted the notation in $(2.37)_2$ and $(2.48)_2$, supposing for the moment that $K_2 > 0$ for cholesterics. This observation shows that $\tau = t_0$ corresponds to the wave vector of a natural equilibrium configuration given by (2.69) and that the wave vector of a cholesteric and, by (2.68), the pitch P, are determined by the ratio of the two elastic constants k_2 and K_2. (Note that here we have adopted Frank's notation for the definition of k_2, which is opposite in sign to that used by Chandrasekhar [38, p.97] and Collings and Hird [53, p.34].)

The quantities derived above in equations (2.51) to (2.54) can also be used again to gain information about the elastic constants for cholesterics. Inserting these quantities into equation (2.70) gives

$$\begin{aligned}
2w_{chol} = {} & K_1(a_1^2 + a_5^2) + 2(K_1 - K_2 - K_4)a_1 a_5 \\
& + K_2(a_2^2 + a_4^2) + 2K_4 a_2 a_4 + 2K_2 t_0(a_4 - a_2) + K_2 t_0^2 \\
& + K_3(a_3^2 + a_6^2).
\end{aligned} \tag{2.73}$$

Each of the three lines on the right-hand side of (2.73) is a quadratic in the variables a_i and so applying the results from (2.56) and (2.57) to the first and third lines shows that if each line is to represent a non-negative quantity then, as in the nematic case,

$$K_1 \geq |K_1 - K_2 - K_4|, \qquad K_3 \geq 0. \tag{2.74}$$

For $\mathbf{n} = \mathbf{n}_c$ in (2.69) we have from (2.71) and (2.72) that $w_{chol}(\mathbf{n}_c) = 0$ when $\tau = t_0$. Given the inequalities in (2.74) we can conclude that $w_{chol} \geq 0$ provided the general expression on the second line in (2.73) is at its minimum value of zero when $\mathbf{n} = \mathbf{n}_c$. Determining when this is the case is accomplished by considering this expression as a function of the two variables a_2 and a_4, say $f(a_2, a_4)$: it is easily calculated from $\nabla\mathbf{n}_c(0)$ that in this particular example

$$a_2 = \tau, \quad \text{and} \quad a_i = 0 \quad \text{for} \quad i \neq 2. \tag{2.75}$$

It is then a simple exercise to compute the first derivatives of f and use the values in (2.75) to show that a necessary condition for $\mathbf{n}_c$ to provide a stationary value of f is that $2\tau(K_2 + K_4) = 0$; but $\tau \neq 0$ and hence

$$K_2 + K_4 = 0. \tag{2.76}$$

When this is the case, the second line in (2.73) is, for the general situation, $K_2(a_4 - a_2 + t_0)^2$ and so for w_{chol} to be non-negative we require $K_2 \geq 0$. This inequality also means that w_{chol} will achieve its minimum value at one of its natural orientations $\mathbf{n}_c$ with $w_{chol}(\mathbf{n}_c) = 0$. Further, inserting the relation (2.76) into (2.74) shows that $K_1 \geq 0$. Hence, in conjunction with (2.74) we have

$$K_1 \geq 0, \quad K_2 \geq 0, \quad K_3 \geq 0. \tag{2.77}$$

Thus for general cholesterics the elastic constants obey the relation (2.76) and the inequalities (2.77), finally giving the energy

$$w_{chol} = \tfrac{1}{2}K_1(\nabla \cdot \mathbf{n})^2 + \tfrac{1}{2}K_2(\mathbf{n} \cdot \nabla \times \mathbf{n} + t_0)^2 + \tfrac{1}{2}K_3(\mathbf{n} \times \nabla \times \mathbf{n})^2. \qquad (2.78)$$

The values for the cholesteric elastic constants appearing in equation (2.77) are often comparable to those for nematics given by, for example, the values in equations (2.60) and (2.61), and consequently in theoretical investigations they are often assumed to be of the order 10^{-12} N to 10^{-11} N. For a cholesteric produced by making a dilute solution of chiral molecules in a conventional nematic it is known that K_2 is anticipated to be of the order 10^{-11} N [110, p.291].

The form of energy for w_{chol} indicated above in (2.78) and the relations (2.76) and (2.77) for the elastic constants are generally accepted as correct for cholesterics provided $\nabla \mathbf{n}$ and t_0 are small on the molecular scale, which is the case in many practical situations [110, p.287]. It should be noted in particular that Jenkins [134] believed the condition (2.76) to be too restrictive to model cholesterics realistically in general and proposed an alternative but more complex energy constructed by writing $\mathbf{n} = \mathbf{n}_c + \mathbf{u}$ subject to the constraint (2.1) and forming an energy density of the form $w(\mathbf{n}_c, \nabla \mathbf{n}_c; \mathbf{u}, \nabla \mathbf{u})$. Given the comments on the viability of w_{chol} for cholesterics discussed above, we do not pursue this alternative description in this text and refer the reader to the brief review in the book by Virga [273].

2.3   Electric and Magnetic Fields

It has been common practice from early research in liquid crystals to apply a magnetic field to align samples of nematic or cholesteric liquid crystals. The alignment so produced encourages the director to be parallel to the field for the majority of nematics and perpendicular to the field in many cholesterics. Similar effects can be produced by an electric field. In this Section we briefly describe electric and magnetic fields and construct their corresponding energies in the context of liquid crystals. Section 2.3.3 below contains a short discussion on the units used for the theory of electromagnetic fields, particularly the conversion from cgs units to SI units for magnetic properties.

2.3.1   Electric Fields and the Electric Energy

The application of an electric field $\mathbf{E}$ to a liquid crystal sample produces a dipole moment per unit volume called the polarisation, denoted by $\mathbf{P}$. We shall employ a description that uses SI units. For clarity of exposition we begin by assuming that $\mathbf{n}$ is aligned parallel to the z-axis in a Cartesian system of coordinates, that is, $\mathbf{n} = \mathbf{n}_0 = (0, 0, 1)$ as introduced in equation (2.7); such a local description is always possible. The anisotropy of the liquid crystal generally forces $\mathbf{E}$ and $\mathbf{P}$ to have different directions. They are related by the electric susceptibility tensor χ_e via the equation

$$\mathbf{P} = \epsilon_0 \chi_e \mathbf{E}, \qquad \chi_e = \begin{bmatrix} \chi_{e_\perp} & 0 & 0 \\ 0 & \chi_{e_\perp} & 0 \\ 0 & 0 & \chi_{e_\parallel} \end{bmatrix}, \qquad (2.79)$$

where ϵ_0 is the permittivity of free space. In SI units, $\epsilon_0 \doteq 8.854 \times 10^{-12}$ F m^{-1} (see Table D.1 in Appendix D). The quantities $\chi_{e\parallel}$ and $\chi_{e\perp}$ denote the electric susceptibilities parallel and perpendicular to the director, respectively.

The electric displacement $\mathbf{D}$ induced by $\mathbf{E}$ and $\mathbf{P}$ is defined in SI units by

$$\mathbf{D} = \epsilon_0 \mathbf{E} + \mathbf{P}, \qquad (2.80)$$

and so by equation (2.79)

$$\mathbf{D} = \epsilon_0 \epsilon \mathbf{E}, \qquad \epsilon = \mathbf{I} + \chi_e, \qquad (2.81)$$

where $\mathbf{I}$ is the identity tensor and ϵ is called the dielectric tensor, which can be written as

$$\epsilon = \begin{bmatrix} \epsilon_\perp & 0 & 0 \\ 0 & \epsilon_\perp & 0 \\ 0 & 0 & \epsilon_\parallel \end{bmatrix}, \qquad \epsilon_\parallel = 1 + \chi_{e\parallel}, \quad \epsilon_\perp = 1 + \chi_{e\perp}. \qquad (2.82)$$

The coefficients $\epsilon_\parallel$ and $\epsilon_\perp$ denote the (relative) *dielectric permittivities*, also called the *dielectric constants*, of the liquid crystal when the field and director are parallel and perpendicular, respectively. In terms of $\mathbf{n}_0$ it is simple to verify that (2.81) is equivalent to

$$\mathbf{D} = \epsilon_0 \epsilon_\perp \mathbf{E} + \epsilon_0 \epsilon_a (\mathbf{n}_0 \cdot \mathbf{E}) \mathbf{n}_0, \qquad \epsilon_a = \epsilon_\parallel - \epsilon_\perp. \qquad (2.83)$$

Since it is always possible to describe $\mathbf{n}$ locally by $\mathbf{n}_0$ by changing to a suitable Cartesian frame, we have that in general the electric displacement is given by

$$\mathbf{D} = \epsilon_0 \epsilon_\perp \mathbf{E} + \epsilon_0 \epsilon_a (\mathbf{n} \cdot \mathbf{E}) \mathbf{n}. \qquad (2.84)$$

The quantity ϵ_a is called the *dielectric anisotropy* of the liquid crystal, with $\Delta\epsilon$ also being a frequent notation for the same quantity. In the above formulation the dielectric constants $\epsilon_\parallel$ and $\epsilon_\perp$ and the dielectric anisotropy ϵ_a are *unitless* since they are measured relative to ϵ_0. Notice also from the above definitions that $\epsilon_a = \chi_{e\parallel} - \chi_{e\perp}$, sometimes written in the literature as $\Delta\chi_e$. Values for ϵ_a can be negative or positive, depending upon the particular features of each individual nematic or cholesteric. Typical values for these constants for the nematic MBBA at 25°C are $\epsilon_\parallel = 4.7$, $\epsilon_\perp = 5.4$ leading to $\epsilon_a = -0.7$ [258]. Values for the dielectric anisotropy of the nematic 5CB between around 24°C to 34°C lie in the range $\epsilon_a \sim 8$ to 11 [53, p.197]: for example, $\epsilon_a = 9.8$ at 29.7°C [252]. When $\epsilon_a > 0$ the director is attracted to be parallel to the field and when $\epsilon_a < 0$ its preferred orientation is perpendicular to the field. Further information on the physics of the dielectric constants in nematics and cholesterics can be found in references [38, 110]; other examples of ϵ_a for nematics can be found in Table D.3 on page 330. We remark here that values for ϵ_a in the SI description are numerically identical when described relative to Gaussian cgs units: cf. equation (2.87) below and equation (2.111) in Section 2.3.3.

The total electric energy density w_{elec} arising when a fixed voltage is maintained and applied on external conductors is given by (see the formulæ 10.05.4 and 10.08.1 in Guggenheim [120] and de Gennes and Prost [110, p.134])

$$w_{elec} = -\int_0^{\mathbf{E}} \mathbf{D} \cdot d\mathbf{E}, \qquad (2.85)$$

and hence it is seen that

$$w_{elec} = -\tfrac{1}{2}\mathbf{D} \cdot \mathbf{E} = -\tfrac{1}{2}\epsilon_0\epsilon_\perp E^2 - \tfrac{1}{2}\epsilon_0\epsilon_a(\mathbf{n} \cdot \mathbf{E})^2, \qquad (2.86)$$

where $E = |\mathbf{E}|$. The term $-\tfrac{1}{2}\epsilon_0\epsilon_\perp E^2$ is independent of the orientation of $\mathbf{n}$ and is therefore usually omitted. When $\epsilon_a > 0$ the last term in the above electric energy is clearly minimised when $\mathbf{n}$ and $\mathbf{E}$ are parallel, and when $\epsilon_a < 0$ it is minimised when $\mathbf{n}$ is perpendicular to $\mathbf{E}$, as expected from the brief comments above on the physical interpretation of ϵ_a. The most commonly adopted form of the electric energy density used in calculations is therefore

$$w_{elec} = -\tfrac{1}{2}\epsilon_0\epsilon_a(\mathbf{n} \cdot \mathbf{E})^2. \qquad (2.87)$$

An alternative and illuminating derivation of the energy density in (2.87) is given by Collings and Hird [53, pp.198–200] for the case of an electric field applied across two parallel conducting plates connected to a battery. Their starting point is the usual form for the electric energy density related to the volume of liquid crystal, namely $+\tfrac{1}{2}\mathbf{D} \cdot \mathbf{E}$, which is then suitably amended to incorporate a decrease in the total electric energy due to a decrease in the electric energy of the battery source which occurs when the voltage is kept constant.

The electric field $\mathbf{E}$ and the electric displacement $\mathbf{D}$ must satisfy the relevant Maxwell equations. Unlike magnetic fields to be discussed below, electric fields are known to interact strongly with liquid crystals: the electric field itself can be influenced in a liquid crystal sample due to the presence of polarisation. An allowance is therefore often made to incorporate these changes in the electric field due to its interaction with the liquid crystal. The electric field $\mathbf{E}$ is generally not constant even in simple alignments and is subject to the relevant reduced Maxwell equations

$$\nabla \cdot \mathbf{D} = 0, \qquad \nabla \times \mathbf{E} = 0, \qquad (2.88)$$

when it is assumed that there is no free charge present. Very often, however, as a basic preliminary approximation it is assumed that the electric field is not influenced by the liquid crystal, in which case $\mathbf{E}$ must satisfy

$$\nabla \cdot \mathbf{E} = 0, \qquad \nabla \times \mathbf{E} = 0, \qquad (2.89)$$

that is, it is assumed ϵ_a is small. Given such an approximation, these equations are satisfied for constant fields in most simple planar aligned samples. To incorporate the changes in the electric field one must of course return to equations (2.88). Most of the examples in this book employ equations (2.89) for electric fields, although we shall also consider in Section 3.5 an example of a Freedericksz transition that utilises Maxwell's equations (2.88) to take into account the changes in $\mathbf{E}$ as it is influenced by the liquid crystal.

2.3.2   Magnetic Fields and the Magnetic Energy

The application of a magnetic field $\mathbf{H}$ across a liquid crystal sample induces a magnetisation $\mathbf{M}$ in the liquid crystal due to the weak magnetic dipole moments imposed upon the molecular alignment by the magnetic field. The magnetisation induced by $\mathbf{H}$ satisfies

$$\mathbf{M} = \chi_{m_\parallel}\mathbf{H}, \qquad \text{if } \mathbf{H} \text{ is parallel to } \mathbf{n}, \tag{2.90}$$

$$\mathbf{M} = \chi_{m_\perp}\mathbf{H}, \qquad \text{if } \mathbf{H} \text{ is perpendicular to } \mathbf{n}, \tag{2.91}$$

where the coefficients $\chi_{m_\parallel}$ and $\chi_{m_\perp}$ denote the diamagnetic (negative) susceptibilities when the field and the director are parallel and perpendicular, respectively. Relative to the orientation of the director $\mathbf{n}$, we can write, from simple geometric considerations, $\mathbf{H} = \mathbf{H}_\parallel + \mathbf{H}_\perp$ with

$$\mathbf{H}_\parallel = (\mathbf{n}\cdot\mathbf{H})\mathbf{n}, \tag{2.92}$$

$$\mathbf{H}_\perp = \mathbf{H} - \mathbf{H}_\parallel = \mathbf{H} - (\mathbf{n}\cdot\mathbf{H})\mathbf{n}. \tag{2.93}$$

These relations are easily verified by taking their scalar products with $\mathbf{n}$. Assuming a linear dependence upon the field and taking into account the nematic symmetries introduced in previous discussions (in particular the invariance with respect to the sign of $\mathbf{n}$), when $\mathbf{H}$ makes an arbitrary angle with $\mathbf{n}$ the magnetisation defined by $\mathbf{M} = \chi_{m_\perp}\mathbf{H}_\perp + \chi_{m_\parallel}\mathbf{H}_\parallel$ becomes

$$\mathbf{M} = \chi_{m_\perp}\mathbf{H} + (\chi_{m_\parallel} - \chi_{m_\perp})(\mathbf{n}\cdot\mathbf{H})\mathbf{n}. \tag{2.94}$$

This is the generally accepted form for the magnetisation and has been discussed by, among others, Ericksen [74].

The magnetic induction $\mathbf{B}$, which plays a similar rôle to that of the electric displacement $\mathbf{D}$ in the construction of the electric energy density discussed above, is defined in SI units by

$$\mathbf{B} = \mu_0(\mathbf{H} + \mathbf{M}), \tag{2.95}$$

where $\mu_0 = 4\pi \times 10^{-7}$ H m^{-1} is the *permeability of free space*. Inserting the magnetisation (2.94) into (2.95) shows that we can write

$$\mathbf{B} = \mu_0\mu_\perp\mathbf{H} + \mu_0\Delta\chi(\mathbf{n}\cdot\mathbf{H})\mathbf{n}, \tag{2.96}$$

where we have made use of notation which is analogous to that used for the electric field description given above, namely,

$$\mu_\perp = 1 + \chi_{m_\perp}, \qquad \mu_\parallel = 1 + \chi_{m_\parallel}, \qquad \Delta\chi = \mu_\parallel - \mu_\perp = \chi_{m_\parallel} - \chi_{m_\perp}. \tag{2.97}$$

The unitless quantity $\Delta\chi = \chi_{m_\parallel} - \chi_{m_\perp}$ is called the *magnetic anisotropy* and is generally positive and of the order 10^{-6} when described in SI units for most nematics, despite both the diamagnetic susceptibilities $\chi_{m_\perp}$ and $\chi_{m_\parallel}$ in many nematics being negative and of the order 10^{-6} to 10^{-5}; a typical value for the nematic liquid crystal MBBA at 19°C is $\Delta\chi = 1.55 \times 10^{-6}$ [110, p.119] (equivalent to 1.23×10^{-7} when described relative to Gaussian units). Section 2.3.3 contains details on units and their conversion factors: it suffices for the moment to say that care needs to be exercised when converting magnetic anisotropies from Gaussian cgs units to SI units or vice-versa (see especially the relation (2.105) below). There are also nematics which have negative magnetic anisotropy, such as 7CCH (cf. [29]) where $\Delta\chi$ may take the value -4.05×10^{-7} when described via SI units [240] (equivalent to -3.22×10^{-8} in Gaussian units). In cholesterics the magnetic anisotropy is often negative and of a smaller magnitude than in conventional nematics: usually $|\Delta\chi|$ is of the order 10^{-8} for cholesterics [110, p.288]. Analogous to the electric field case, when $\Delta\chi > 0$ the director prefers to be parallel to the magnetic field and when $\Delta\chi < 0$ its preferred orientation is perpendicular to the field. A more detailed discussion on the physics of liquid crystals in magnetic fields can be found in References [38, 110]. Further examples of $\Delta\chi$ for nematic materials are given in Table D.3 on page 330.

The relevant energy density follows by an argument that is completely analogous to that for obtaining the electric energy density w_{elec} in equation (2.86). We have that the magnetic energy density is given by

$$w_{mag} = -\tfrac{1}{2}\mathbf{B}\cdot\mathbf{H} = -\tfrac{1}{2}\mu_0\mu_\perp H^2 - \tfrac{1}{2}\mu_0\Delta\chi(\mathbf{n}\cdot\mathbf{H})^2, \qquad (2.98)$$

where $H = |\mathbf{H}|$. The contribution $-\tfrac{1}{2}\mu_0\mu_\perp H^2$ is independent of the orientation of $\mathbf{n}$ and is therefore often omitted. When $\Delta\chi > 0$ the last term in the magnetic energy above is clearly minimised when $\mathbf{n}$ and $\mathbf{H}$ are parallel, reflecting the aforementioned interpretation of $\Delta\chi$. It is similarly observed that this term is minimised for $\Delta\chi < 0$ when $\mathbf{n}$ and $\mathbf{H}$ are perpendicular. The most generally adopted magnetic energy density employed in calculations involving finite samples of nematic is therefore

$$w_{mag} = -\tfrac{1}{2}\mu_0\Delta\chi(\mathbf{n}\cdot\mathbf{H})^2. \qquad (2.99)$$

However, in some problems, particularly those involving semi-infinite samples, it is sometimes more useful to employ the relation $(2.97)_3$ in equation (2.98) and add the contribution $\tfrac{1}{2}\mu_0\mu_\parallel H^2$ (which is independent of the orientation of the director) rather than neglect $-\tfrac{1}{2}\mu_0\mu_\perp H^2$. This results in the energy

$$w_{mag} = \tfrac{1}{2}\mu_0\Delta\chi(H^2 - (\mathbf{n}\cdot\mathbf{H})^2). \qquad (2.100)$$

Of course, the equilibrium equations will be unaffected, but the tractability of solutions and their energies for appropriate boundary conditions on the director may often be enhanced.

We do not go into details here, but refer the reader to the book by de Jeu [137] where it can be found, by similar arguments to those given above for the magnetic field $\mathbf{H}$, that an alternative magnetic energy analogous to (2.98) can be derived in terms of $\mathbf{B}$ and is given by

$$w_{mag} = -\tfrac{1}{2}\mu_0^{-1}\mu_\perp B^2 - \tfrac{1}{2}\mu_0^{-1}\Delta\chi(\mathbf{n}\cdot\mathbf{B})^2, \qquad (2.101)$$

where $B = |\mathbf{B}|$, which in turn, when the term independent of the director is neglected, leads to the commonly used energy

$$w_{mag} = -\tfrac{1}{2}\mu_0^{-1}\Delta\chi(\mathbf{n}\cdot\mathbf{B})^2. \qquad (2.102)$$

The results quoted here for w_{mag} in (2.101) and (2.102) are valid provided the magnetic susceptibilities $\chi_{m_\parallel}$ and $\chi_{m_\perp}$ are small, an assumption that has been found generally acceptable in many liquid crystal applications.

In general, the magnetic induction **B** and magnetic field **H** must satisfy the relevant Maxwell field equations

$$\nabla \cdot \mathbf{B} = 0, \qquad \nabla \times \mathbf{H} = \mathbf{0}. \tag{2.103}$$

However, in liquid crystals it is known from experiments that in equilibrium situations magnetic fields applied across samples are virtually unaffected by the presence of the liquid crystal. This situation is generally in contrast to the corresponding situation for electric fields, as mentioned above. This leads to the assumption that, when $\Delta\chi$ is small, **B** can be considered as being parallel to the magnetic field **H** when considering Maxwell's equations: see equations (2.94) and (2.95). Therefore the conditions in (2.103) are often replaced by the two equations

$$\nabla \cdot \mathbf{H} = 0, \qquad \nabla \times \mathbf{H} = \mathbf{0}, \tag{2.104}$$

under this approximation. In most planar sample alignments these equations are easily satisfied by constant fields. However, care must be exercised when non-planar samples are investigated and it must be checked that **H** satisfies equations (2.104). Such circumstances can arise, for example, in a cylindrical sample of nematic confined between two coaxial circular cylinders where a radial or azimuthal field **H** can be applied, as has been considered by Atkin and Barratt [7].

2.3.3   Comments on Fields and Units

The electric and magnetic fields appearing in Sections 2.3.1 and 2.3.2 have been described in terms of the SI system of units. The Gaussian system of cgs units has frequently been employed in the theory of liquid crystals when magnetic fields are discussed. Gaussian units are considered by many to be natural units for the calculation and measurement of magnetic field effects in liquid crystals. Since much of the literature contains results in both Gaussian units and SI units, it seems appropriate at this point to make some comments on the conversion from one system of units to the other. A comprehensive account of the points touched upon here may be found in Jackson [132] or Moskowitz [206]. Readers should also be familiar with derived SI units.

It has been common for those who work in electromagnetic theory to use the Gaussian system of units consisting of selected cgs electromagnetic units (emu) and cgs electrostatic units (esu); to allow a comparison with results stated in the literature, we record here for convenience that the general division of use is [54]

emu :   current, magnetic field, induction, resistance and related quantities,

esu :   charge, electric field, displacement, capacity and related quantities.

These units may be classified further as rationalised or unrationalised, depending on the definition of ϵ_0. In Gaussian units, both ϵ_0 and μ_0 are set equal to 1.

The unitless magnetic anisotropy $\Delta\chi$ obeys the transformation [206]

$$\Delta\chi^{SI} = 4\pi\Delta\chi^{cgs}, \tag{2.105}$$

where $\Delta\chi^{SI}$ and $\Delta\chi^{cgs} = \chi_a$ represent the magnetic anisotropy described relative to SI units and cgs units, respectively. The relation (2.105) is important when employing data in theoretical investigations. Effectively, many of the statements involving an SI units description for the magnetic energy can be written relative to the Gaussian system of units by writing χ_a in place of $\mu_0\Delta\chi$ and vice-versa; when using data for calculations, the transformation (2.105) must of course be used in conjunction with any of the necessary conversions mentioned in Tables D.1 and D.2.

The results to be derived in this text will use the notation χ_a when working in cgs units. For example, one possibility for the general magnetic energy density may be given by the analogue of the result in equation (2.98) when the contribution $-\frac{1}{2}\mu_0 H^2$ is disregarded, namely,

$$w_{mag} = -\tfrac{1}{2}\chi_{m_\perp} H^2 - \tfrac{1}{2}\chi_a(\mathbf{n}\cdot\mathbf{H})^2, \tag{2.106}$$

where the susceptibility $\chi_{m_\perp}$ is now measured relative to the Gaussian system of units. If the total contribution independent of the director is neglected, this energy may be taken as

$$w_{mag} = -\tfrac{1}{2}\chi_a(\mathbf{n}\cdot\mathbf{H})^2, \tag{2.107}$$

similar to equation (2.99). Additionally, as mentioned earlier, it may be useful to note that for some semi-infinite sample problems we can formulate the magnetic energy density in the form (cf. equations (3.82) and (3.83) in Section 3.3 and equation (3.382) in Section 3.8.3 below)

$$w_{mag} = \tfrac{1}{2}\chi_a(H^2 - (\mathbf{n}\cdot\mathbf{H})^2), \tag{2.108}$$

analogous to that given by (2.100). The magnetisation in Gaussian units has the form

$$\mathbf{M} = \chi_{m_\perp}\mathbf{H} + \chi_a(\mathbf{n}\cdot\mathbf{H})\mathbf{n}, \tag{2.109}$$

and the magnetic energy in (2.106) may be written as [110, p.119]

$$w_{mag} = -\tfrac{1}{2}\mathbf{M}\cdot\mathbf{H}. \tag{2.110}$$

The reader is free to choose whichever set of units is most suitable for a particular problem when using magnetic fields, bearing in mind the following comments. When working in cgs units the proper unit for the magnitude of $\mathbf{H}$ is oersted (Oe) while the magnitude of $\mathbf{B}$ has units of gauss (G). However, since in free space $\mathbf{H} = \mathbf{B}$ in cgs units, gauss is also frequently used to express the magnitude of $\mathbf{H}$. In SI units the magnitudes of $\mathbf{H}$ and $\mathbf{B}$ are given in units of ampere per metre (A m^{-1}) and tesla (T), respectively, and an identification similar to that for cgs units is not appropriate. The magnetic energy is often given in terms of $\mathbf{B}$ when working in SI units because it is easier to convert gauss arising from the cgs results for $\mathbf{H}$ to tesla in SI units than to convert oersted to A m^{-1}. It is not that $\mathbf{B}$ is more fundamental than $\mathbf{H}$, it is only that it is more convenient to use the units of $\mathbf{B}$. Further details on these points can be found in the article by Moskowitz [206]. The conversion factors in Table D.2 on page 329 may be used when converting results from cgs units to SI units and vice-versa.

For electric fields no conversion problems occur provided the straightforward conversion factors are used in Table D.2. When Gaussian units are used the magnitude of the electric field is in statvolt cm^{-1} which converts to the appropriate units of volts per metre (V m^{-1}) using the standard conversion factor. The only point to take note of is, in the notation of Section 2.3.1, that ϵ_0 equals 1 in Gaussian units and is, as noted above on page 27 and in Table D.1, given by $\epsilon_0 \doteq 8.854 \times 10^{-12}$ F m^{-1} in SI units. The expressions in (2.86) and (2.87) for the electric energy density are in SI units: they remain valid in the Gaussian system of units provided all electrical terms are appropriately, and carefully, converted. For example, the most common electric energy density stated in (2.87) is given in Gaussian units by [110, p.134]

$$w_{elec} = -\frac{1}{8\pi}\epsilon_a(\mathbf{n} \cdot \mathbf{E})^2. \tag{2.111}$$

This particular form of the expression for w_{elec} arises because $\epsilon_0 = 1$ in Gaussian units and the analogue of the electric displacement (2.80) is given by [132, p.781]

$$\mathbf{D} = \mathbf{E} + 4\pi\mathbf{P}. \tag{2.112}$$

It is recommended that SI units be used if both electric and magnetic fields are applied. The elastic, electric and magnetic energies are given in terms of joules (J) in SI units and ergs in Gaussian units, with the corresponding energy densities given as J m^{-3} in SI and erg cm^{-3} in Gaussian units; recall that J = N m. The corresponding conversions from Gaussian to SI are given in Table D.2.

In some applications it is necessary to introduce the idea of power. This is defined to be the rate at which energy is expended or work is done and is measured in watts (W). In SI units, a watt is defined to be one joule per second (J s^{-1}), equivalent to 10^7 erg s^{-1} in cgs units: cf. Table D.2. In electrical circuits one watt is equivalent to the product of one ampere by one volt.

2.4   Equilibrium Equations

We follow the approach adopted by Ericksen [73, 74] for nematic and cholesteric liquid crystals which appeals to a principle of virtual work and employs the calculus of variations. The advantage of this approach over that of Oseen [214] is in its mechanical interpretation. The theory of liquid crystals is one of the few mechanical theories involving what is generally an asymmetric stress tensor. To simplify the presentation it is assumed that the mass density of the liquid crystal sample remains constant, that is, we assume that the material is incompressible. This assumption is generally acceptable since the energies associated with compression are much greater than those associated with the orientational changes of n introduced above in Section 2.2. (When the compressional energies are required it is commonly found that the curvature elastic energy is negligibly small in comparison and may therefore be ignored.) On a first reading, some readers may prefer to postpone reading through the preliminaries and derivation of the equilibrium equations and choose to go directly to the subsection 'The Equilibrium Equations' on page 38.

2.4.1   Preliminaries

Consider a general functional of the form

$$J\left(\mathbf{u}(\mathbf{x})\right) = \int_V f\left(\mathbf{x}, \mathbf{u}(\mathbf{x}), \nabla \mathbf{u}(\mathbf{x})\right) dV, \qquad (2.113)$$

where $\mathbf{x} \in \mathbb{R}^3$ and $\mathbf{u}$ is a three-dimensional vector function, f is a smooth real-valued function, ∇ represents the gradient operator with respect to $\mathbf{x}$ and V is a volume with boundary surface S. Suppose that both $\mathbf{x}$ and $\mathbf{u}(\mathbf{x})$ can vary so that there is a transformation from $\mathbf{x}$ to, say, $\mathbf{x}^*$ and from $\mathbf{u}$ to $\mathbf{u}^*(\mathbf{x}^*)$ with the volume and surface transforming to V^* and S^*, respectively. This will allow us to consider the deformation of liquid crystal samples where the shapes of the surface and volume may alter, even when the material is considered incompressible. Such a transformation will lead to

$$J\left(\mathbf{u}^*(\mathbf{x}^*)\right) = \int_{V^*} f\left(\mathbf{x}^*, \mathbf{u}^*(\mathbf{x}^*), \nabla^* \mathbf{u}^*(\mathbf{x}^*)\right) dV^*, \qquad (2.114)$$

where ∇^* is the gradient operator with respect to $\mathbf{x}^*$. Now suppose that we carry out such a transformation by varying $\mathbf{x}$ and $\mathbf{u}$ so that small changes lead to the expressions

$$\begin{array}{rcl}
x_i^* & = & x_i + \epsilon\phi_i(\mathbf{x}), \qquad\qquad (2.115) \\
u_i^*(\mathbf{x}^*) & = & u_i(\mathbf{x}) + \epsilon\psi_i(\mathbf{x}), \qquad\quad (2.116) \\
u_i^*(\mathbf{x}) & = & u_i(\mathbf{x}) + \epsilon\overline{\psi}_i(\mathbf{x}), \qquad\quad (2.117)
\end{array}$$

where ϵ is a small quantity and ϕ_i, ψ_i and $\overline{\psi}_i$ are differentiable functions, $i = 1, 2$ or 3. It is customary to define

$$\delta x_i = \epsilon\phi_i(\mathbf{x}) \qquad \text{and} \qquad \delta u_i = \epsilon\overline{\psi}_i(\mathbf{x}), \qquad (2.118)$$

and call $\delta\mathbf{x}$ and $\delta\mathbf{u}$ the *variations* of $\mathbf{x}$ and $\mathbf{u}$, respectively. The quantity defined by

$$\Delta u_i = \epsilon\psi_i(\mathbf{x}), \qquad (2.119)$$

will also be of interest. (The notation introduced here is that commonly used in liquid crystal theory: however, in some general texts, such as Gelfand and Fomin [104, p.170], δu_i and Δu_i are denoted by $\overline{\delta u_i}$ and δu_i, respectively.) It is clear that $\delta\mathbf{x}$ expresses the incremental change to the position $\mathbf{x}$, while $\delta\mathbf{u}$ expresses the change in $\mathbf{u}$ as we transform from $(\mathbf{x}, \mathbf{u}(\mathbf{x}))$ to $(\mathbf{x}, \mathbf{u}^*(\mathbf{x}))$: $\delta\mathbf{u}$ is therefore the usual Lagrange form of variation. Variations introduced as in equations (2.115) to (2.117) are sometimes called *non-contemporaneous variations* [218, p.49] because they consider the general form $(\mathbf{x}, \mathbf{u}(\mathbf{x})) \mapsto (\mathbf{x}^*, \mathbf{u}^*(\mathbf{x}^*))$ rather than the usual Lagrange form $(\mathbf{x}, \mathbf{u}(\mathbf{x})) \mapsto (\mathbf{x}, \mathbf{u}^*(\mathbf{x}))$. The quantity $\Delta\mathbf{u}$ provides an expression for the change in $\mathbf{u}$ when we transform from $(\mathbf{x}, \mathbf{u}(\mathbf{x}))$ to $(\mathbf{x}^*, \mathbf{u}^*(\mathbf{x}^*))$. It can be shown that, to first order in ϵ [104, p.170],

$$\Delta u_i = \delta u_i + u_{i,k}\delta x_k, \qquad (2.120)$$

so that Δu can be interpreted as the variational equivalent of the material derivative of **u** (cf. equation (4.5) below). A result which also proves useful is [104, pp.172–173]

$$\Delta(u_{i,j}) = (\delta u_i)_{,j} + u_{i,jk}\delta x_k. \tag{2.121}$$

Making a coordinate transformation such as that introduced here allows us to incorporate the variation of an integral, such as J above, over a *variable* region or volume, as well as the usual variation over a fixed volume (when $\delta \mathbf{x} \equiv \mathbf{0}$). When $\delta \mathbf{x} \equiv \mathbf{0}$, as can happen in many instances where the volume is fixed (which may be the case in many liquid crystal cells), we of course have that $\Delta \mathbf{u} = \delta \mathbf{u}$. The laws for the variations of sums, products, quotients, etc., are largely analogous to those for the corresponding laws for derivatives. For example,

$$\delta(u_i u_i) = 2u_i \delta u_i \quad \text{and} \quad \delta c = 0, \tag{2.122}$$

when c is a constant.

It is convenient to introduce the general notation

$$f_{,i} \equiv \frac{\partial f}{\partial x_i} + \frac{\partial f}{\partial u_j} u_{j,i} + \frac{\partial f}{\partial u_{j,k}} u_{j,ki}, \tag{2.123}$$

for a function f such as that given in equation (2.113). This conforms to the notational convention followed by Ericksen [74, p.374]. The derivative $f_{,i}$ is evaluated when $\mathbf{u}(\mathbf{x})$ is regarded as a function of $\mathbf{x}$, that is, the value of $\mathbf{u}(\mathbf{x})$ is *not* held constant, as might be inferred by what appears to be an ambiguous notation for partial derivatives (cf. [104, p.172]) (we have interchanged the comma and partial derivative notation used in [104] in order to match the work of Ericksen). Of course, terms such as $\partial n_i / \partial x_j$ are equivalent to $n_{i,j}$, and so there is generally no confusion when the context is clear. (This distinction between derivatives also proves important when obtaining the results in equations (2.152) and (6.84) given later.)

We now quote a quite general variational result for variable volumes from Gelfand and Fomin [104, pp.173–176]. The difference

$$\Delta J \equiv J\left(\mathbf{u}^*(\mathbf{x}^*)\right) - J\left(\mathbf{u}(\mathbf{x})\right), \tag{2.124}$$

is of interest, particularly the principal linear part in ϵ, denoted by δJ. It can be shown for the definition in (2.113) that [104, p.175, Eqn.(101)]

$$\delta J = \int_V \left\{ \left[\frac{\partial f}{\partial u_i} - \left(\frac{\partial f}{\partial u_{i,j}} \right)_{,j} \right] \delta u_i + \left(\frac{\partial f}{\partial u_{i,j}} \delta u_i + f \delta x_j \right)_{,j} \right\} dV. \tag{2.125}$$

2.4.2 Derivation of the Equilibrium Equations

It is postulated that the variation of the total energy satisfies the principle of virtual work

$$\delta \int_V w(\mathbf{n}, \nabla \mathbf{n}) \, dV = \int_V (\mathbf{F} \cdot \delta \mathbf{x} + \mathbf{G} \cdot \Delta \mathbf{n}) dV + \int_S (\mathbf{t} \cdot \delta \mathbf{x} + \mathbf{s} \cdot \Delta \mathbf{n}) dS, \tag{2.126}$$

where V is the volume of the liquid crystal sample, S is its boundary surface, w is the energy density under consideration (nematic or cholesteric) and the virtual displacements are denoted by $\delta \mathbf{x}$, using the variational notation introduced above. The quantity $\Delta \mathbf{n}$ is defined in Cartesian component form via (2.120) by

$$\Delta n_i = \delta n_i + n_{i,j}\,\delta x_j. \tag{2.127}$$

The body force per unit volume is denoted by $\mathbf{F}$, $\mathbf{t}$ is the surface force per unit area, and $\mathbf{G}$ and $\mathbf{s}$ are generalised body and surface forces, respectively. The virtual displacement $\delta \mathbf{x}$ and the variations in the director $\mathbf{n}$ are not arbitrary, but are subject to the constraints

$$(\delta x_i)_{,i} = 0, \quad n_i \delta n_i = 0, \quad n_i \Delta n_i = 0. \tag{2.128}$$

The first constraint above is due to the assumption of incompressibility when the mass density $\rho(\mathbf{x})$ is constant. This follows from the general requirement that the variations must satisfy

$$\delta \rho + (\rho \delta x_i)_{,i} = 0, \tag{2.129}$$

reflecting the conservation of mass property (see Appendix A, equations (A.1) to (A.3)). The latter two constraints in (2.128) follow from applying the rules for variations in equation (2.122) to the original constraint (2.1) which forces $n_i n_{i,j} = 0$, as indicated in equation (2.10), and using the definition of $\Delta \mathbf{n}$ in equation (2.127). We now replace $f(\mathbf{x}, \mathbf{u}, \nabla \mathbf{u})$ by $w(\mathbf{n}, \nabla \mathbf{n})$ in formula (2.125) to find, taking into account the constraint $(\delta x_i)_{,i} = 0$ from (2.128)$_1$,

$$\delta \int_V w\, dV = \int_V \left(\frac{\partial w}{\partial n_i} \delta n_i + \frac{\partial w}{\partial n_{i,j}}(\delta n_i)_{,j} + w_{,i}\delta x_i \right) dV. \tag{2.130}$$

Applying the differentiation rule (2.123) to $w_{,i}$ and using the definition of Δ in (2.127) then allows (2.130) to be rewritten as

$$\delta \int_V w\, dV = \int_V \left(\frac{\partial w}{\partial n_i} \Delta n_i + \frac{\partial w}{\partial n_{i,j}}(\delta n_i)_{,j} + \frac{\partial w}{\partial n_{k,j}} n_{k,ji}\delta x_i \right) dV. \tag{2.131}$$

We can integrate by parts and use the relation (2.127) as required to find that the second and third integrals on the right-hand side of (2.131) are equivalent to, respectively,

$$\int_V \frac{\partial w}{\partial n_{i,j}}(\delta n_i)_{,j}\, dV = -\int_V \left(\frac{\partial w}{\partial n_{i,j}} \right)_{,j} \delta n_i\, dV + \int_S \frac{\partial w}{\partial n_{i,j}}(\Delta n_i - n_{i,k}\delta x_k)\, dS_j, \tag{2.132}$$

$$\int_V \frac{\partial w}{\partial n_{k,j}} n_{k,ji}\delta x_i\, dV = \int_V \left[\left(\frac{\partial w}{\partial n_{k,j}} n_{k,i} \right)_{,j} \delta x_i - \left(\frac{\partial w}{\partial n_{i,j}} \right)_{,j} n_{i,k}\delta x_k \right] dV, \tag{2.133}$$

this latter result assuming that partial derivatives commute. In the above we have used the usual convention of writing dS_j for the surface element $\nu_j\, dS$, where ν

is the outward unit normal to the boundary S. Inserting (2.132) and (2.133) into (2.131) gives, upon relabelling indices where appropriate,

$$\delta \int_V w \, dV = \int_V \left\{ \left[\frac{\partial w}{\partial n_i} - \left(\frac{\partial w}{\partial n_{i,j}} \right)_{,j} \right] \Delta n_i + \left(\frac{\partial w}{\partial n_{k,j}} n_{k,i} \right)_{,j} \delta x_i \right\} dV$$

$$+ \int_S \left(\frac{\partial w}{\partial n_{i,j}} \Delta n_i - \frac{\partial w}{\partial n_{k,j}} n_{k,i} \delta x_i \right) dS_j. \tag{2.134}$$

Before deriving the final equilibrium equations it only remains to take into account the constraint in equation (2.1) and the assumption that the mass density $\rho(\mathbf{x}) = \rho_0$, a constant. These constraints may be written as

$$\psi_1 \;\; = \;\; \rho - \rho_0 = 0, \tag{2.135}$$

$$\psi_2 \;\; = \;\; \tfrac{1}{2}(n_i n_i - 1) = 0. \tag{2.136}$$

Since ψ_1 and ψ_2 do not depend on the derivatives of $\mathbf{n}$, we can introduce scalar function Lagrange multipliers λ_1 and λ_2 satisfying, for non-contemporaneous variations obeying the incompressibility condition (2.128)$_1$,

$$0 \equiv \lambda_m \delta \psi_m = \lambda_m \left(\frac{\partial \psi_m}{\partial n_i} \delta n_i + \frac{\partial \psi_m}{\partial x_i} \delta x_i \right), \quad m = 1, 2. \tag{2.137}$$

From the calculations using the method of Ericksen [73, 74] contained in Appendix A, equations (A.5) to (A.8), it can be shown that

$$0 = \int_V [\lambda_2 n_i \Delta n_i + (\lambda_1 \rho)_{,i} \delta x_i] \, dV - \int_S \lambda_1 \rho \delta_{ij} \delta x_i \, dS_j. \tag{2.138}$$

This means that the sum of the integrands appearing on the right-hand side of equation (2.138) can be added appropriately to those on the right-hand side of equation (2.134) since the addition of the integrals of such terms is identical to adding zero to the right-hand side of (2.134). Doing so allows us to reformulate equation (2.134) in the full general incompressible case as [74, Eqn.(2.18)]

$$\delta \int_V w \, dV = - \int_V [t_{ij,j} \delta x_i + (\Pi_i - \lambda_2 n_i) \Delta n_i] \, dV$$

$$+ \int_S (t_{ij} \delta x_i + s_{ij} \Delta n_i) \, \nu_j \, dS, \tag{2.139}$$

where

$$\Pi_i \;\; = \;\; \left(\frac{\partial w}{\partial n_{i,j}} \right)_{,j} - \frac{\partial w}{\partial n_i}, \tag{2.140}$$

$$t_{ij} \;\; = \;\; - \frac{\partial w}{\partial n_{k,j}} n_{k,i} - p \delta_{ij}, \tag{2.141}$$

$$s_{ij} \;\; = \;\; \frac{\partial w}{\partial n_{i,j}}, \tag{2.142}$$

with $p = \lambda_1 \rho$ being an arbitrary pressure arising from the assumed incompressibility. The components t_{ij} form the stress tensor while the terms s_{ij} belong to what Ericksen [74] calls the torque stress. The term Π is seen to be of the usual Euler–Lagrange form.

The Equilibrium Equations

We are now in a position to equate the coefficients of Δn_i and δx_i in the principle of virtual work (2.126) and equation (2.139) to obtain the differential equations for equilibrium. In the volume V it is then seen that

$$\left(\frac{\partial w}{\partial n_{i,j}}\right)_{,j} - \frac{\partial w}{\partial n_i} + G_i + \lambda n_i = 0, \tag{2.143}$$

$$t_{ij,j} + F_i = 0. \tag{2.144}$$

Here, in the commonly accepted notation, we have set the scalar function Lagrange multiplier to be $\lambda = -\lambda_2$. The stress vector $\mathbf{t}$ and generalised stress vector $\mathbf{s}$ take the forms (cf. Truesdell and Noll [270, p.40], Goodbody [115, p.207])

$$t_i = t_{ij}\nu_j, \tag{2.145}$$

$$s_i = s_{ij}\nu_j + \beta n_i, \tag{2.146}$$

as can be seen by equating the coefficients of Δn_i and δx_i in (2.126) and (2.139) over the surface S, which has unit outward normal $\boldsymbol{\nu}$. The arbitrary scalar β arises from the constraint $(2.128)_3$ when taken into consideration in equation (2.139). The two equations (2.145) and (2.146) are related to the boundary conditions briefly mentioned by Ericksen [74]. The equilibrium equations (2.143) and (2.144) are valid for nematic or cholesteric liquid crystal elastic energy densities; they are also valid for the total energy density that arises when electric or magnetic energy densities are appended to w.

It is worth remarking for future reference that when w is the nematic energy w_F in (2.50) we have

$$\left(\frac{\partial w_F}{\partial n_{i,j}}\right)_{,j} - \frac{\partial w_F}{\partial n_i} = (K_1 - K_2)n_{j,ji} + K_2 n_{i,jj} + (K_3 - K_2)(n_j n_k n_{i,k})_{,j}$$

$$-(K_3 - K_2)n_j n_{k,j}n_{k,i}, \tag{2.147}$$

when partial derivatives commute.

Null Lagrangians

Recall that the saddle-splay term in the nematic energy w_F in equation (2.49) is given by

$$w_{ss} = \tfrac{1}{2}(K_2 + K_4)\nabla \cdot [(\mathbf{n} \cdot \nabla)\mathbf{n} - (\nabla \cdot \mathbf{n})\mathbf{n}]. \tag{2.148}$$

A straightforward calculation reveals that this term in the energy density w_F does not contribute to equation (2.147) when partial derivatives commute and so the saddle-splay term does not contribute to the general equilibrium equations in the bulk given by (2.143). Any term in an energy density that does not contribute to the bulk equilibrium equations is called a *null Lagrangian*. The term w_{ss} is an example of a null Lagrangian. More details on this topic can be found in Virga [273, §3.8]. For example, although w_{ss} does not contribute to the bulk equilibrium equations, it does generally enter the boundary conditions and will affect the actual value of the total energy. In particular, when the director is subject to strong anchoring on the boundary (see Section 2.6.2 on page 47) then the null Lagrangian term w_{ss} contributes the same fixed value to the total energy for all solutions to the bulk equilibrium equations. It should also be mentioned that Ericksen [75] considered null Lagrangians in a more general theory of liquid crystals: a convenient summary of his results can be found in [273].

A Simplification

Some simplification of the above equilibrium equations appearing in equations (2.143) and (2.144) is possible if we consider the body forces $\mathbf{F}$ and $\mathbf{G}$ to be specified functions of $\mathbf{x}$ and $\mathbf{n}$ given by equations of the form

$$F_i = \frac{\partial \Psi}{\partial x_i}, \qquad G_i = \frac{\partial \Psi}{\partial n_i}, \tag{2.149}$$

where $\Psi(\mathbf{n}, \mathbf{x})$ is some scalar energy density function, this being the case considered by Ericksen [74]. This is certainly the situation encountered in Section 2.3 when $-\Psi$ may be taken as the magnetic or electric energy: see Remark (ii) at the end of this Section. An obvious special case is when there are no external body forces and $\Psi \equiv 0$. However, in general $\Psi \neq 0$ and substituting $(2.149)_1$ for $\mathbf{F}$ in equation (2.144) and using equation (2.141) gives

$$\frac{\partial \Psi}{\partial x_i} - p_{,i} - \left(\frac{\partial w}{\partial n_{k,j}} \right)_{,j} n_{k,i} - \frac{\partial w}{\partial n_{k,j}} n_{k,ij} = 0. \tag{2.150}$$

Equation (2.143), with $\mathbf{G}$ given by $(2.149)_2$, can then be used to substitute for $(\partial w/\partial n_{k,j})_{,j}$ in the above and, employing the result in equation (2.10), we find that when derivatives commute

$$\frac{\partial \Psi}{\partial n_j} n_{j,i} + \frac{\partial \Psi}{\partial x_i} - p_{,i} - \frac{\partial w}{\partial n_j} n_{j,i} - \frac{\partial w}{\partial n_{k,j}} n_{k,ji} = 0, \tag{2.151}$$

which can be expressed simply as

$$(\Psi - p - w)_{,i} = 0, \tag{2.152}$$

once the general rule for derivatives given by the relation (2.123) has been applied and we recall that $w = w(\mathbf{n}, \nabla \mathbf{n})$. Hence the balance of linear momentum equation (2.144) yields an expression for the pressure p, namely,

$$p + w - \Psi = p_0, \tag{2.153}$$

where p_0 is an arbitrary constant. This has the important consequence that if F_i and G_i satisfy the relations in (2.149), then to determine equilibrium configurations in nematic or cholesteric liquid crystals one has to solve only equation (2.143). In this case we can always neglect equation (2.144), unless there is a desire to compute forces.

Remark (i)

Equation (2.144) is obviously a balance of forces at equilibrium. This can be seen by considering a translation where the variations satisfy

$$\delta \mathbf{x} = \mathbf{a}, \qquad \Delta \mathbf{n} \equiv \mathbf{0}, \tag{2.154}$$

where $\mathbf{a}$ is an arbitrary *constant* vector. These variations, which clearly fulfil the necessary conditions in (2.128), can be inserted into (2.139) to see that

$$\delta \int_V w\,(\mathbf{n}, \nabla \mathbf{n})\; dV = 0, \tag{2.155}$$

after an integration by parts in the volume integral and noting that $a_{i,j} = 0$. Hence the postulate (2.126) becomes, for these variations,

$$\int_V \mathbf{F} \cdot \mathbf{a}\, dV + \int_S \mathbf{t} \cdot \mathbf{a}\, dS = \mathbf{a} \cdot \left(\int_V \mathbf{F}\, dV + \int_S \mathbf{t}\, dS \right) = 0, \tag{2.156}$$

and therefore, since $\mathbf{a}$ is an arbitrary constant vector, we obtain the usual balance of forces given by

$$\int_V \mathbf{F}\, dV + \int_S \mathbf{t}\, dS = \mathbf{0}. \tag{2.157}$$

Equation (2.144) is obtained from (2.157) by using the result in (2.145) and applying the divergence theorem as stated at equation (1.20).

What is less clear is that the equilibrium equation (2.143) actually arises from a balance of moments, as we shall now demonstrate. Consider an infinitesimal, rigid rotation $\boldsymbol{\omega}$ for which

$$\delta \mathbf{x} = \boldsymbol{\omega} \times \mathbf{x}, \qquad \Delta \mathbf{n} = \boldsymbol{\omega} \times \mathbf{n}, \tag{2.158}$$

where $\boldsymbol{\omega}$ is an arbitrary, infinitesimal *constant* vector. These variations clearly satisfy the constraints (2.128). We can insert them into the identity (2.139), integrate by parts where appropriate, make use of the constraints (2.128) and relabel indices to obtain, after some tedious but straightforward algebra,

$$\delta \int_V w\,(\mathbf{n}, \nabla \mathbf{n})\; dV = \omega_i \int_V \epsilon_{ijk} \left(n_j \frac{\partial w}{\partial n_k} + n_{j,p} \frac{\partial w}{\partial n_{k,p}} + n_{p,j} \frac{\partial w}{\partial n_{p,k}} \right) dV. \tag{2.159}$$

However, the following identity, due to Ericksen [73], has been derived in Appendix B:

$$\epsilon_{ijk} \left(n_j \frac{\partial w}{\partial n_k} + n_{j,p} \frac{\partial w}{\partial n_{k,p}} + n_{p,j} \frac{\partial w}{\partial n_{p,k}} \right) = 0. \tag{2.160}$$

Hence (2.159) reduces to

$$\delta \int_V w\,(\mathbf{n}, \nabla \mathbf{n})\; dV = 0, \tag{2.161}$$

for these variations. We can now insert the variations (2.158) into the postulate (2.126) and use the result (2.161) to see that we must have

$$\int_V (\mathbf{F} \cdot (\boldsymbol{\omega} \times \mathbf{x}) + \mathbf{G} \cdot (\boldsymbol{\omega} \times \mathbf{n}))\; dV + \int_S (\mathbf{t} \cdot (\boldsymbol{\omega} \times \mathbf{x}) + \mathbf{s} \cdot (\boldsymbol{\omega} \times \mathbf{n}))\; dS = 0. \tag{2.162}$$

Since $\boldsymbol{\omega}$ is an arbitrary constant vector, the properties of the scalar triple product (see equation (1.12)) then show that we have the balance of moments

$$\int_V (\mathbf{x} \times \mathbf{F} + \mathbf{n} \times \mathbf{G}) \, dV + \int_S (\mathbf{x} \times \mathbf{t} + \mathbf{n} \times \mathbf{s}) \, dS = 0. \tag{2.163}$$

From this relationship it is seen that the generalised body and surface forces are related to the body moment $\mathbf{K}$ and couple stress vector $\mathbf{l}$ by, respectively,

$$\mathbf{K} = \mathbf{n} \times \mathbf{G}, \qquad \mathbf{l} = \mathbf{n} \times \mathbf{s}. \tag{2.164}$$

From equations (2.142) and (2.146) it follows that at equilibrium the couple stress tensor l_{ij} and couple stress vector $\mathbf{l}$ are given by

$$l_{ij} = \epsilon_{ipq} n_p s_{qj}, \tag{2.165}$$
$$l_i = l_{ij} \nu_j = \epsilon_{ipq} n_p s_q. \tag{2.166}$$

With this notation we can use the divergence theorem (1.20), the balance of forces (2.144) and the relations (2.145) and (2.166) to see that the balance of moments (2.163) is, in Cartesians,

$$\begin{aligned}
0 &= \int_V [\epsilon_{ijk} x_j F_k + K_i + \epsilon_{ijk}(x_j t_{kr})_{,r} + l_{ij,j}] \, dV \\
&= \int_V [\epsilon_{ijk} x_j (F_k + t_{kr,r}) + K_i + \epsilon_{ijk} t_{kj} + l_{ij,j}] \, dV \\
&= \int_V (K_i + \epsilon_{ijk} t_{kj} + l_{ij,j}) \, dV. \tag{2.167}
\end{aligned}$$

It is now a simple exercise to use (2.141), (2.142) and (2.165) to substitute for the quantities appearing in the above integrand to find that

$$\epsilon_{ijk} \left[n_j G_k - \frac{\partial w}{\partial n_{p,j}} n_{p,k} + n_{j,p} \frac{\partial w}{\partial n_{k,p}} + n_j \left(\frac{\partial w}{\partial n_{k,p}} \right)_{,p} \right] = 0, \tag{2.168}$$

upon noticing that $\epsilon_{ijk} \delta_{kj} = \epsilon_{ijj} = 0$. The identity due to Ericksen in equation (2.160) can be applied to equation (2.168) to further reduce it to

$$\epsilon_{ijk} n_j \left[\left(\frac{\partial w}{\partial n_{k,p}} \right)_{,p} - \frac{\partial w}{\partial n_k} + G_k \right] = 0, \tag{2.169}$$

suitably relabelling indices as required and applying the general rule that $\epsilon_{ijk} A_{kj} = -\epsilon_{ijk} A_{jk}$. Therefore the balance of moments in equation (2.163) finally reduces via equation (2.169) to

$$\left(\frac{\partial w}{\partial n_{k,j}} \right)_{,j} - \frac{\partial w}{\partial n_k} + G_k + \widehat{\lambda} n_k = 0, \tag{2.170}$$

where $\widehat{\lambda}$ is an arbitrary scalar function, in agreement with the equilibrium equation derived earlier in (2.143).

Remark (ii)

It is known from elementary classical physics that the magnetic torque **K** acting on the magnetisation **M** is, in a cgs units description [110, pp.119–120],

$$\mathbf{K} = \mathbf{M} \times \mathbf{H} = \mathbf{n} \times \mathbf{G}, \tag{2.171}$$

with

$$\mathbf{G} = \chi_a(\mathbf{n} \cdot \mathbf{H})\mathbf{H}. \tag{2.172}$$

Also, if **H** is non-uniform there is an associated body force given by

$$\mathbf{F} = (\mathbf{M} \cdot \nabla)\mathbf{H}, \tag{2.173}$$

but since in equilibrium $\nabla \times \mathbf{H} = 0$ (from Maxwell's equations (2.104) for static fields) this can be re-expressed as

$$F_i = M_j H_{j,i}, \quad i = 1, 2, 3. \tag{2.174}$$

Using the result (2.110), if we set

$$\Psi = -w_{mag} = \tfrac{1}{2}\mathbf{M} \cdot \mathbf{H}, \tag{2.175}$$

with **M** given by equation (2.109), then equations (2.149) are satisfied with the above **F** and **G** (recall that **n** and **x** are treated as independent variables) and consequently the only equation to be solved for equilibrium solutions is the governing equation (2.143). The quantity $\Psi = -w_{mag}$ is sometimes referred to as the *magnetic potential*. From the point of view of the development of the theory of liquid crystals, equations equivalent in essence to (2.172) and (2.175) seem to have been first recorded by Zocher [287]. Any equilibrium solutions found by this approach are equivalent to those obtained by again employing the equilibrium equation (2.143) but with the energy density replaced by $w = w_F + w_{mag}$ and setting $\mathbf{G} \equiv 0$ and $\Psi \equiv 0$, w_F being the elastic energy density for either a nematic or cholesteric liquid crystal given by equations (2.49) or (2.78), respectively. This latter approach will be adopted extensively in subsequent Chapters when dealing with static problems.

2.5   General Equilibrium Solutions

In this introductory textbook we concentrate on specific equilibrium configurations of pertinent interest to those modelling devices. Nevertheless, before proceeding to specific examples in Chapter 3, mention should be made of what are called general solutions. The determination of what orientation alignments are possible in a liquid crystal in the absence of external body forces or fields, irrespective of the actual form of the energy satisfying equation (2.6), has been carried out by Ericksen [77] where solutions for such alignments must satisfy the equilibrium equations (2.143). There are three such general solutions which arise when boundary conditions are not imposed. In the first, the preferred orientation is everywhere perpendicular to a family of concentric spheres, while in the second it is perpendicular to a family of concentric cylinders. In the third type the preferred direction is everywhere tangential to a family of parallel planes, the direction being constant within each plane but having the possibility of changing uniformly with respect to the normal direction of the planes; this type also includes the special case of a family of parallel planes with the director being the same constant unit vector within every plane.

Explicitly, when referred to suitable Cartesian coordinates, we have the general forms:

(i)   Constant solution:

$$\mathbf{n} = \mathbf{n}_0, \qquad \mathbf{n}_0 \cdot \mathbf{n}_0 = 1, \tag{2.176}$$

(ii)   Spherical solution:

$$\mathbf{n} = \frac{\mathbf{x}}{|\mathbf{x}|}, \qquad \mathbf{x} = (x_1, x_2, x_3) \neq \mathbf{0}, \tag{2.177}$$

(iii)   Cylindrical solution:

$$\mathbf{n} = \frac{\mathbf{x}}{|\mathbf{x}|}, \qquad \mathbf{x} = (x_1, x_2, 0) \neq \mathbf{0}, \tag{2.178}$$

(iv)   Twist solution:

$$\mathbf{n} = (\cos(\tau x_3 + x_0), \sin(\tau x_3 + x_0), 0), \qquad \tau, x_0 \text{ constants.} \tag{2.179}$$

Clearly, $\mathbf{n} \cdot \mathbf{n} = 1$ for all these solutions, and the constant solution (i) can be thought of as a special case of the twist solution (iv) with τ set to zero. These general solutions are important because they can be equilibrium solutions for nematic or cholesteric liquid crystals and are often used as starting configurations for the investigation of Freedericksz transitions such as those to be discussed in Chapter 3. For example, the above twisted solution can be exploited in the twisted nematic device to be discussed below in Section 3.7. The derivation and proof that these are the only general solutions require detailed knowledge of differential geometry beyond the scope of this book, and so for our present discussion it will suffice to verify that these general solutions are indeed solutions for nematic liquid crystals by showing explicitly that they satisfy the equilibrium equations (2.143) for the nematic energy w_F in Cartesian component form (2.50); we omit the case for cholesterics since a similar justification is possible involving the energy w_{chol} in equation (2.78). More intricate details and a discussion of these solutions can be found in References [77, 273].

(i)   Constant Solution

The nematic energy becomes $w_F = 0$ when $\mathbf{n}$ is a constant unit vector and, in the absence of body forces, $\mathbf{G} = \mathbf{0}$. The equilibrium equations (2.143) are obviously satisfied when $\mathbf{n} = \mathbf{n}_0$ by setting $\lambda = 0$, obtained by taking the scalar product of the equilibrium equations with $\mathbf{n}_0$.

(ii) Spherical Solution

Routine calculations for w_F in the form provided by equation (2.50) show that for *any* solution $\mathbf{n}$ (cf. equation (2.147))

$$\frac{\partial w_F}{\partial n_i} = (K_3 - K_2)n_j n_{k,j} n_{k,i}, \tag{2.180}$$

$$\left(\frac{\partial w_F}{\partial n_{i,j}}\right)_{,j} = (K_1 - K_2)n_{j,ji} + K_2 n_{i,jj} + (K_3 - K_2)(n_j n_k n_{i,k})_{,j}, \tag{2.181}$$

bearing in mind that partial derivatives are assumed to commute. Also, it is easily verified by the chain rule for partial derivatives that when $\mathbf{n}$ is given by equation (2.177)

$$n_{i,j} = \frac{1}{r}(\delta_{ij} - n_i n_j), \tag{2.182}$$

$$n_{i,jk} = \frac{1}{r^2}(3n_i n_j n_k - \delta_{ij} n_k - \delta_{jk} n_i - \delta_{ki} n_j), \tag{2.183}$$

where, for convenience, r is defined by

$$r = |\mathbf{x}| = \sqrt{x_1^2 + x_2^2 + x_3^2}. \tag{2.184}$$

Using the summation convention in equation (2.183), noting that $\mathbf{n} \cdot \mathbf{n} = 1$ and $\delta_{jj} = 3$, then yields

$$n_{j,ji} = n_{i,jj} = -2\frac{n_i}{r^2}, \tag{2.185}$$

while identity (2.182) delivers

$$n_k n_{i,k} = 0. \tag{2.186}$$

Hence expressions (2.180) and (2.181) reduce to

$$\frac{\partial w_F}{\partial n_i} = 0 \quad \text{and} \quad \left(\frac{\partial w_F}{\partial n_{i,j}}\right)_{,j} = -2K_1\frac{n_i}{r^2}, \tag{2.187}$$

and therefore the equilibrium equations (2.143) become

$$-2K_1\frac{n_i}{r^2} + \lambda n_i = 0, \tag{2.188}$$

recalling that in this discussion $\mathbf{G} = \mathbf{0}$. Taking the scalar product of equation (2.188) with $\mathbf{n}$ then shows that

$$\lambda = 2\frac{K_1}{r^2}. \tag{2.189}$$

The spherical solution defined by (2.177) is therefore a solution to the equilibrium equations when the scalar function λ has been set by equation (2.189).

It may be expected that the spherical solution can be observed throughout a sample in a spherical container if the director $\mathbf{n}$ is aligned perpendicular to the wall of the container. Nevertheless, the singularity at $r = 0$ indicates that a point defect may occur at the centre of the spherical region and that a different orientation alignment may be present near the central part of such a sample.

(iii) Cylindrical Solution

The first observation to make is that for $\mathbf{n}$ given by equation (2.178) the equilibrium equation (2.143) is automatically satisfied for $i = 3$ when $\mathbf{G} = \mathbf{0}$. When $i \neq 3$ the expressions (2.180) to (2.183) above remain valid for $i = 1, 2$ provided r is redefined to be

$$r = |\mathbf{x}| = \sqrt{x_1^2 + x_2^2}. \tag{2.190}$$

Therefore $\mathbf{n}$ in this case is a solution to the equilibrium equations with λ provided via (2.189) with r given by (2.190).

Ths type of cylindrical solution may be expected to occur in a cylindrical container if the orientation of the director $\mathbf{n}$ is perpendicular on the cylinder boundary. The line singularity occurring at $r = 0$ may be observable, given that line singularities (or defects) are commonly seen in liquid crystal samples. However, this scenario and a possible alternative solution for cylinders of finite radius are discussed in more detail in Section 3.8.1 below.

(iv) Twist Solution

Calculations show that equation (2.186) remains valid. The right-hand sides of equations (2.180) and (2.181) show that the equilibrium equations (2.143) to be satisfied are therefore

$$K_2 n_{i,jj} + \lambda n_i = 0, \tag{2.191}$$

assuming $\mathbf{G} = \mathbf{0}$. Taking the scalar product of the above with $\mathbf{n}$ reveals

$$\lambda = -K_2 n_i n_{i,jj} = K_2 \tau^2. \tag{2.192}$$

Thus it is easily verified that the orientation $\mathbf{n}$ in equation (2.179) is a solution to the equilibrium equations with λ provided by equation (2.192).

Experimental observations of nematic liquid crystals frequently report uniform alignment of the director, corresponding to the above twist solution with $\tau = 0$, resulting in a constant solution of the form in (i) above. This type of alignment occurs in samples between two parallel plates when the alignment of the director on both boundary surfaces is identical. This set-up can be taken one stage further: for example, if the director is fixed parallel to both plates and one plate were then to be rotated about its normal, it is anticipated that a twist orientation similar to the above twist solution with $\tau \neq 0$ would be induced. This will be discussed further in the context of the twisted nematic device in Section 3.7.

In cholesteric liquid crystals, alignments of the director corresponding to twist solutions occur naturally (see Fig. 2.2 above on page 24). For a sample of cholesteric between parallel plates, with the director fixed parallel to the plates at the boundaries, the form of the above twist solution is regularly observed where the preferred direction of $\mathbf{n}$ is parallel to the plates throughout the sample and the axis of the twist is coincident with the normal to the bounding plates. In thin samples, rotation of one plate within its own plane may again lead to a new twist solution exhibiting a different value of the twist parameter τ. The natural pitch of a cholesteric does not enter the bulk equilibrium equations explicitly, but it is known that there is an intimate relationship between boundary conditions and the value of τ.

2.6   Anchoring and Boundary Conditions

The bulk energy W in equation (2.3) and the equilibrium equation (2.143) applicable in the bulk of the liquid crystal sample must, in principle, be supplemented by a relevant surface energy and a description of the alignment of the director $\mathbf{n}$ on the boundary surfaces of the sample, these descriptions entering the mathematics via appropriate boundary conditions. The description of how the director is aligned on the boundary surfaces is called *anchoring*. The main types of anchoring which we now discuss, in addition to the case of a 'natural boundary condition' where there is 'no anchoring', are called 'strong', 'conical' and 'weak' anchoring. The equilibrium solutions for the director on the boundary surfaces must satisfy different equations for different anchoring conditions, as we shall demonstrate below. Simpler reformulations of these boundary conditions are available in Section 2.7.2.

In the simplest case, at equilibrium we require the first variation of the energy given by equation (2.139) to equate to zero. From standard results in the calculus of variations this means that both the volume and surface integrands in (2.139) must vanish. In particular, the surface integrand must obey

$$(t_{ij}\delta x_i + s_{ij}\Delta n_i)\nu_j = 0 \qquad \text{on } S, \tag{2.193}$$

where $\boldsymbol{\nu}$ is the unit outward normal to the surface S. The first term in the above expression, when integrated over the surface S, represents the work done on the sample by the limiting boundaries if they are displaced by an amount $\delta\mathbf{x}$, but with no change in the director on the boundaries taken into account. The second term represents the work done by the boundaries when they are displaced in such a way that they impose a change in $\mathbf{n}$ on the surface. Throughout this Section we shall simplify the discussion by assuming that the boundaries of any given sample are immobilised and do not move, as is the case when, for example, the boundary of the sample corresponds to a rigid container or when the sample is bounded between fixed parallel plates. In these circumstances the variations are restricted so that $\delta\mathbf{x} = \mathbf{0}$ on S and $\Delta\mathbf{n} = \delta\mathbf{n}$, which reduce the requirement (2.193) to

$$s_{ij}\nu_j\delta n_i = 0 \qquad \text{on } S, \tag{2.194}$$

s_{ij} being defined in equation (2.142). Boundary conditions for the situation when the boundary of the region occupied by the liquid crystal sample is free to move or adjust to its surrounding environment are not discussed in this book. The reader is referred to Chapter 5 in the book by Virga [273] who discusses such problems in the context of droplets of nematics.

2.6.1   No Anchoring

If there are no anchoring conditions imposed upon $\mathbf{n}$ on the boundary then equation (2.194) must hold for variations $\delta\mathbf{n}$ that are arbitrary on the boundary surface S, subject of course to the variational constraint $(2.128)_2$. It is then clear, using the definition of s_{ij} in equation (2.142), that the requirement (2.194) will hold provided

$$\frac{\partial w}{\partial n_{i,j}}\nu_j = \gamma n_i, \qquad \gamma = n_k\frac{\partial w}{\partial n_{k,j}}\nu_j, \qquad i = 1, 2, 3, \tag{2.195}$$

the scalar multiplier γ being obtained by taking the scalar product of $(2.195)_1$ with $\mathbf{n}$. This is often referred to as the *natural boundary condition* for $\mathbf{n}$. Note that the energy density w may incorporate electric and magnetic energy densities in addition to the usual elastic energy density.

As a simple example, consider the one-constant approximation for $w = w_F$ given by equation (2.67). A routine calculation employing the quite general property (2.10) arising from $(2.1)_2$ then shows that $\gamma = 0$ and hence in these circumstances the boundary condition reduces to

$$\frac{\partial w}{\partial n_{i,j}} \nu_j = 0, \qquad i = 1, 2, 3 \qquad \text{on } S, \tag{2.196}$$

where $\boldsymbol{\nu}$ is the unit outward normal to S. This result is a special case of that considered by Ericksen [73, Eqn.(29)].

2.6.2   Strong Anchoring

It is known from the physics of liquid crystals that in many practical situations the surface forces are sufficiently large to enforce a clearly defined direction for $\mathbf{n}$ on the physical surface at the boundary [110, p.108]. This situation is referred to as *strong anchoring* and in these circumstances $\Delta\mathbf{n} = \delta\mathbf{n} \equiv \mathbf{0}$ for all admissible variations. Therefore by (2.194) no conditions are to be imposed upon $\mathbf{n}$ on the boundary except those dictated by the given anchoring alignment. This then makes it sufficient only to minimise the bulk energy subject to $\mathbf{n}$ taking prescribed fixed values on the boundary, that is, $\mathbf{n} = \mathbf{n}_b$ on the boundary for some fixed $\mathbf{n}_b$. Two examples of

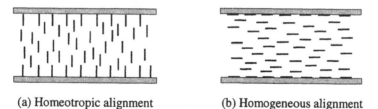

(a) Homeotropic alignment   (b) Homogeneous alignment

Figure 2.3: Strong anchoring of the director $\mathbf{n}$ on the boundaries inducing samples of nematic having (a) a homeotropic alignment and (b) a homogeneous (also called planar) alignment.

strong anchoring between parallel plates are depicted in Fig. 2.3. In Fig. 2.3(a) the director is forced to be perpendicular to the bounding plates thereby inducing the sample also to have the director perpendicular to the plates throughout the entire volume. Sample alignments produced in this way are called *homeotropic* and can be obtained by means of detergents or by chemically treating the surfaces [110, p.110]. When the director is fixed uniformly parallel to the bounding plates as shown in Fig. 2.3(b) then the director can assume an alignment that is parallel to the director on the bounding plates throughout the complete sample. Such samples are called *homogeneous* and are said to have homogeneous or planar alignment at the boundaries. Such samples can be produced by carefully rubbing the surfaces in one direction only, which encourages the director to align along the rubbing direction and therefore induces the sample alignment shown in the Figure. The direction of rubbing is sometimes referred to as the *easy axis* and the director is said to be aligned along an *easy direction*.

2.6.3   Conical Anchoring

Conical anchoring occurs when the director at a boundary makes a fixed angle ψ with the tangent plane of the boundary as shown in Fig. 2.4. When this is the case the director $\mathbf{n}$ can orient itself along the surface of a cone of easy directions, one end of the director placed at the vertex of the cone. The angle ψ is temperature dependent, although in this text we always assume that we treat the isothermal situation where ψ is constant. Examples using conical anchoring are given in Sections 3.8.3 and 3.8.4. If $\boldsymbol{\nu}$ represents the outward unit normal to the boundary surface then $\mathbf{n}$ on the boundary is subject to the constraint

$$(\mathbf{n} \cdot \boldsymbol{\nu})^2 = \sin^2\psi \qquad \text{on } S, \tag{2.197}$$

recalling that $\mathbf{n}$ and $-\mathbf{n}$ are considered to be physically indistinguishable. In the

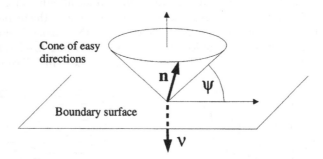

Figure 2.4:   Conical anchoring of the director $\mathbf{n}$ at the boundary surface. The director takes a position on a 'cone' of easy axes making an angle ψ with the tangent plane of the boundary which has outward unit normal $\boldsymbol{\nu}$.

simplest case, we can assume that $\boldsymbol{\nu}$ is prescribed and fixed and that $0 < \psi < \frac{\pi}{2}$. We can then take the first variation in equation (2.197) and apply the rules in (2.122) to find that admissible variations, in addition to satisfying the condition (2.194), must fulfil the requirements

$$\nu_i \delta n_i = 0, \qquad n_i \delta n_i = 0, \tag{2.198}$$

the latter requirement being repeated from (2.128)$_2$. This means that $\delta\mathbf{n}$ is perpendicular to both $\mathbf{n}$ and $\boldsymbol{\nu}$ and hence, since $\psi \neq \frac{\pi}{2}$, we must have $\delta\mathbf{n} = \boldsymbol{\omega} \times \mathbf{n}$ where, by necessity, $\boldsymbol{\omega} = \omega\boldsymbol{\nu}$ for some scalar ω. Therefore

$$\delta\mathbf{n} = \omega \left(\boldsymbol{\nu} \times \mathbf{n} \right), \tag{2.199}$$

and the constraints in (2.198) are clearly satisfied. Thus, by the arbitrariness of ω, the general requirement in equation (2.194) reduces in the case of conical anchoring to

$$\epsilon_{ipq}\nu_p n_q \frac{\partial w}{\partial n_{i,j}}\nu_j = 0 \qquad \text{on } S, \tag{2.200}$$

which is equivalent to saying

$$\frac{\partial w}{\partial n_{i,j}}\nu_j = \lambda\nu_i + \mu n_i \qquad \text{on } S, \tag{2.201}$$

where λ and μ are arbitrary scalar functions (this is easily verified by taking the scalar product of equation (2.201) with $\nu \times \mathbf{n}$). Consequently, for conical anchoring the condition (2.197) has to be supplemented by the additional boundary condition given in equation (2.200) or its equivalent in (2.201). Notice that λ and μ may be identified by taking the scalar product of both sides of equation (2.201) with respect to ν and $\mathbf{n}$, respectively.

As a special case, consider a liquid crystal at an isotropic boundary having, in the obvious Cartesian notation, $\nu = (0,0,1)$. Then, with $\mathbf{n} = (n_x, n_y, n_z)$, the requirement in (2.200) reduces to

$$\epsilon_{i3q}n_q \frac{\partial w}{\partial n_{i,z}} = 0, \tag{2.202}$$

with i and q following the summation convention, which is equivalent to

$$n_x \frac{\partial w}{\partial n_{y,z}} - n_y \frac{\partial w}{\partial n_{x,z}} = 0 \qquad \text{on } S. \tag{2.203}$$

This last result is precisely that obtained by de Gennes and Prost, but in a slightly different nomenclature (see [110, Eqn.(3.90)] and [110, p.160]).

Two cases of conical anchoring are related to the strong anchoring conditions which have been mentioned in the previous Section. The first occurs when $\psi = \pi/2$ in Fig. 2.4, so that $\mathbf{n}$ is parallel to ν, in which case the boundary condition is homeotropic. The second occurs when $\psi = 0$ and the director $\mathbf{n}$ becomes perpendicular to ν, which results in a surface alignment that is degenerately homogeneous.

2.6.4   Weak Anchoring

In the cases of anchoring discussed so far no additional energies were required or introduced in the modelling. However, one can easily visualise a situation arising from Fig. 2.4 where the angle between the director and the layer normal at a boundary or interface may vary under the influence of applied fields. Such anticipated boundary behaviour has to lead to the introduction of an additional surface energy corresponding to what is commonly called a *weak anchoring* energy. The simplest such surface energy density is of a form first proposed by Rapini and Papoular [227] which can be written as

$$w_s = \tfrac{1}{2}\tau_0(1 + \omega(\mathbf{n}\cdot\nu)^2), \tag{2.204}$$

with $\tau_0 > 0$, $\omega > -1$ and ν being, as before, the unit outward normal to the boundary surface or interface. For $-1 < \omega < 0$ the weak anchoring surface energy w_s will favour a director alignment parallel to ν so that the energy is reduced, that is, the favoured orientation of **n** will be homeotropic. When $\omega > 0$, w_s will favour an orientation orthogonal to ν, indicating that a homogeneous alignment tangential to the boundary or interface will be favoured. Rapini and Papoular [227] estimated $\tau_0\omega \sim 1$ erg cm^{-2} (in cgs units), equivalent to 10^{-3} J m^{-2} in SI units, as a general value; for the nematic liquid crystal 5CB, Yokoyama and van Sprang [283] experimentally calculated $\tau_0\omega$ to be around 4×10^{-5} J m^{-2}. Reviews of various techniques for measuring the anchoring strength have been written by Yokoyama [284] and Blinov, Kabayenkov and Sonin [19].

When w_F is any of the bulk elastic energies discussed above the total energy for a sample occupying the region Ω having boundary S is given by

$$W = \int_\Omega w_F \, d\Omega + \int_S w_s \, dS, \qquad (2.205)$$

which will lead in general to two sets of coupled equilibrium equations, one set arising from the bulk energy and the other from the surface energy. More details on this type of problem are given in Section 3.6 below where a specific example is investigated when W also incorporates the magnetic energy integral over Ω. It is perhaps worth remarking that in simple one-dimensional problems the total 'surface' energy per unit area may effectively consist of the surface energy density evaluated at two isolated points, in which case the surface energy per unit area is simply the sum of w_s evaluated at these points via the solution for **n** and appropriately prescribed values for ν.

By repeating the approach used above for establishing the bulk equilibrium equations for the nematic or cholesteric energies, it is possible to derive the additional equilibrium boundary condition

$$\frac{\partial w_F}{\partial n_{i,j}}\nu_j + \frac{\partial w_s}{\partial n_i} = \gamma n_i \qquad \text{on } S, \qquad (2.206)$$

where γ is an arbitrary scalar, first mentioned by Jenkins and Barratt [135]. Details of the derivation of this boundary condition, sometimes called the balance of couple condition, are omitted here, the reader being referred to [135]. Notice that if w_s is zero then this equilibrium condition on the surface S collapses to that given previously by equation (2.195) when there is no anchoring and only the natural boundary condition prevails.

2.7 Reformulation of Equilibrium Equations

It is common in liquid crystal theory to use a representation for the director that automatically ensures **n** is a unit vector, for example,

$$\mathbf{n} = (\cos\theta_1 \cos\theta_2, \cos\theta_1 \sin\theta_2, \sin\theta_1), \qquad -\tfrac{\pi}{2} \le \theta_1 \le \tfrac{\pi}{2}, \quad 0 \le \theta_2 < 2\pi, \quad (2.207)$$

where $\mathbf{n}$ can be interpreted as a point on the unit sphere $\mathbb{S}^2$. Such a representation enables a reformulation which simplifies the equilibrium equations. To illustrate the procedure we follow Ericksen [80] and suppose in general that

$$\mathbf{n} = \mathbf{f}(\theta_1, \theta_2), \qquad \mathbf{f} \cdot \mathbf{f} = 1, \tag{2.208}$$

the constraint that $\mathbf{f}$ is a unit vector leading to

$$n_i \frac{\partial f_i}{\partial \theta_\alpha} = 0, \qquad \alpha = 1, 2. \tag{2.209}$$

We retain the notation $\mathbf{n}$ for the director in the original description and adopt the notation $\mathbf{f}$ to describe the director when using θ_1 and θ_2. It must also be the case that

$$\frac{\partial \mathbf{f}}{\partial \theta_1} \times \frac{\partial \mathbf{f}}{\partial \theta_2} \neq \mathbf{0}, \tag{2.210}$$

to ensure that the tangent vectors in the resulting unit sphere description do not coincide, so that θ_1 and θ_2 can act as local coordinate directions (which necessarily must not be parallel). As Ericksen points out, θ_1 and θ_2 can then be seen as coordinates on the unit sphere $\mathbb{S}^2$ and, since any such characterisation is somewhere singular, some exceptional directions will fail to be represented. Some care must therefore be exercised when using equations that rely upon the conditions (2.208) to (2.210) for their derivation.

2.7.1 Bulk Equilibrium Equations

In an obvious notation, the bulk energy density (2.2) can be considered as

$$w = w(n_i, n_{i,j}) = \widehat{w}(\theta_\alpha, \theta_{\alpha,i}), \qquad \alpha = 1, 2. \tag{2.211}$$

Assuming partial derivatives commute, we first note the identities

$$n_{i,j} = \frac{\partial f_i}{\partial \theta_\beta} \theta_{\beta,j}, \qquad \left(\frac{\partial f_k}{\partial \theta_\alpha} \right)_{,i} = \frac{\partial n_{k,i}}{\partial \theta_\alpha}, \tag{2.212}$$

appropriately summing over repeated indices as required: $(2.212)_1$ follows from an application of the chain rule for partial derivatives, while $(2.212)_2$ can be verified for individual components, if required. Straightforward differentiation using these expressions then leads to

$$\begin{aligned}
\frac{\partial \widehat{w}}{\partial \theta_\alpha} &= \frac{\partial w}{\partial n_i}\frac{\partial f_i}{\partial \theta_\alpha} + \frac{\partial w}{\partial n_{i,j}}\frac{\partial n_{i,j}}{\partial \theta_\alpha} \\
&= \frac{\partial w}{\partial n_i}\frac{\partial f_i}{\partial \theta_\alpha} + \frac{\partial w}{\partial n_{i,j}}\frac{\partial^2 f_i}{\partial \theta_\alpha \partial \theta_\beta}\theta_{\beta,j} \\
&= \frac{\partial w}{\partial n_k}\frac{\partial f_k}{\partial \theta_\alpha} + \frac{\partial w}{\partial n_{k,i}}\frac{\partial^2 f_k}{\partial \theta_\alpha \partial \theta_\beta}\theta_{\beta,i}\,,
\end{aligned} \tag{2.213}$$

$$\begin{aligned}
\frac{\partial \widehat{w}}{\partial \theta_{\alpha,i}} &= \frac{\partial w}{\partial n_{k,j}}\frac{\partial n_{k,j}}{\partial \theta_{\alpha,i}} \\
&= \frac{\partial w}{\partial n_{k,i}}\frac{\partial f_k}{\partial \theta_\alpha}\,,
\end{aligned} \tag{2.214}$$

$$\begin{aligned}
\left(\frac{\partial \widehat{w}}{\partial \theta_{\alpha,i}}\right)_{,i} &= \left(\frac{\partial w}{\partial n_{k,p}}\frac{\partial n_{k,p}}{\partial \theta_{\alpha,i}}\right)_{,i} \\
&= \left(\frac{\partial w}{\partial n_{k,i}}\frac{\partial f_k}{\partial \theta_\alpha}\right)_{,i} \\
&= \left(\frac{\partial w}{\partial n_{k,i}}\right)_{,i}\frac{\partial f_k}{\partial \theta_\alpha} + \frac{\partial w}{\partial n_{k,i}}\frac{\partial n_{k,i}}{\partial \theta_\alpha} \\
&= \left(\frac{\partial w}{\partial n_{k,i}}\right)_{,i}\frac{\partial f_k}{\partial \theta_\alpha} + \frac{\partial w}{\partial n_{k,i}}\frac{\partial^2 f_k}{\partial \theta_\alpha \partial \theta_\beta}\theta_{\beta,i}\,.
\end{aligned} \tag{2.215}$$

Combining the above results shows that

$$\left(\frac{\partial \widehat{w}}{\partial \theta_{\alpha,i}}\right)_{,i} - \frac{\partial \widehat{w}}{\partial \theta_\alpha} = \left[\left(\frac{\partial w}{\partial n_{k,i}}\right)_{,i} - \frac{\partial w}{\partial n_k}\right]\frac{\partial f_k}{\partial \theta_\alpha}, \qquad \alpha = 1,2. \tag{2.216}$$

If we additionally suppose that the body forces $\mathbf{F}$ and $\mathbf{G}$ are specified functions of $\mathbf{x}$ and $\mathbf{n}$ so that the relations in (2.149) hold, then, in particular, we have

$$G_i = \frac{\partial \Psi}{\partial n_i}, \qquad \Psi = \Psi(\mathbf{n},\mathbf{x}). \tag{2.217}$$

Writing

$$\Psi(n_i, x_i) = \widehat{\Psi}(\theta_\alpha, x_i), \qquad \alpha = 1,2, \tag{2.218}$$

we then seen that

$$\frac{\partial \widehat{\Psi}}{\partial \theta_\alpha} = \frac{\partial \Psi}{\partial n_k}\frac{\partial f_k}{\partial \theta_\alpha} = G_k\frac{\partial f_k}{\partial \theta_\alpha}. \tag{2.219}$$

By employing the results in (2.216) and (2.219), the three equilibrium equations appearing in (2.143) may be multiplied by $\frac{\partial f_i}{\partial \theta_\alpha}$ to reduce them to an equivalent simpler set of two equations

$$\left(\frac{\partial \widehat{w}}{\partial \theta_{\alpha,i}}\right)_{,i} - \frac{\partial \widehat{w}}{\partial \theta_\alpha} + \frac{\partial \widehat{\Psi}}{\partial \theta_\alpha} = 0, \qquad \alpha = 1,2, \tag{2.220}$$

with the Lagrange multiplier λ being automatically eliminated because of condition (2.209). Of course, these reformulated equations are readily more tractable but necessarily hold only whenever $\frac{\partial f_i}{\partial \theta_\alpha} \neq 0$ and (2.210) holds, highlighting the previously mentioned concern over the validity of this formulation when special directions of $\mathbf{n}$ may require a more careful treatment. The two equations in (2.220) further simplify to

$$\left(\frac{\partial \hat{w}}{\partial \theta_{\alpha,i}} \right)_{,i} - \frac{\partial \hat{w}}{\partial \theta_\alpha} = 0, \qquad \alpha = 1, 2, \tag{2.221}$$

in the special case when $\mathbf{G} \equiv \mathbf{0}$, a form that will be used frequently in later Chapters.

Curvilinear Coordinates

In some circumstances the total energy density may be dependent upon the spatial variable $\mathbf{x}$ also, as can be the case when considering the magnetic or electric energy as discussed above in Section 2.4, page 39. This situation is a special case of that generally encountered when curvilinear coordinates are introduced so that [80, 170]

$$\mathbf{x} = \mathbf{x}(\boldsymbol{\xi}), \quad \mathbf{n} = \mathbf{f}(\theta_\alpha, \boldsymbol{\xi}), \qquad \alpha = 1, 2, \tag{2.222}$$

and

$$\tilde{w}(\theta_\alpha, \theta_{\alpha,i}, \xi_i) = wJ, \tag{2.223}$$

where J denotes the usual Jacobian

$$J = \left| \frac{\partial(x_1, x_2, x_3)}{\partial(\xi_1, \xi_2, \xi_3)} \right|. \tag{2.224}$$

This allows the total energy W to be evaluated as

$$W = \int_\Omega w \, dx_1 dx_2 dx_3 = \int_{\tilde{\Omega}} \tilde{w} \, d\xi_1 d\xi_2 d\xi_3, \tag{2.225}$$

with $\tilde{\Omega}$ being the transformed spatial domain. Obviously, if $\boldsymbol{\xi} = \mathbf{x}$ then $J \equiv 1$ so that the case when the energy is also dependent upon $\mathbf{x}$ is included in this discussion. Calculations similar to those presented above show that the equilibrium equations become

$$\left(\frac{\partial \tilde{w}}{\partial \theta_{\alpha,i}} \right)_{,i} - \frac{\partial \tilde{w}}{\partial \theta_\alpha} + \frac{\partial \tilde{\Psi}}{\partial \theta_\alpha} = 0, \qquad \alpha = 1, 2, \tag{2.226}$$

where, adopting the above supposition that $\mathbf{G}$ satisfies (2.217), we have set

$$J\Psi(n_i, x_i) = \tilde{\Psi}(\theta_\alpha, \xi_i), \qquad \alpha = 1, 2, \tag{2.227}$$

and a comma now denotes partial differentiation with respect to ξ_i, with the same comments after equation (2.220) also being applicable here.

2.7.2 Reformulation of Boundary Conditions

The reformulated bulk equilibrium equations given in the previous Section should also be supplemented by the appropriate boundary conditions discussed in Section 2.6 which can also be reformulated. We treat separately each of the cases mentioned earlier.

No Anchoring

In the notation introduced above we can employ equations (2.207), (2.208), (2.209) and (2.214) to find that multiplying both sides of the natural boundary condition (2.195) by $\frac{\partial f_i}{\partial \theta_\alpha}$ leads to

$$\frac{\partial \widehat{w}}{\partial \theta_{\alpha,j}} \nu_j = 0 \quad \text{on } S, \quad \alpha = 1, 2, \tag{2.228}$$

where ν is the outward unit normal to the surface or interface S and $\widehat{w}(\theta_\alpha, \theta_{\alpha,i})$ is the bulk free energy as in the previous Section. This result can be presented more concisely in special cases.

As a basic example, consider a sample of nematic liquid crystal in which the orientation of the director is uniform in the x and y directions, so that the sample can be considered as having unit length in these directions, but in which the director may vary in the z-direction. Assume that the sample is arranged such that there is an interface or boundary at $z = 0$ with the liquid crystal occupying the region $z \geq 0$. To simplify matters further, and to facilitate some simpler expositions of special cases to be introduced below, we can choose to set

$$\mathbf{n} = (\cos\theta, 0, \sin\theta), \qquad \theta = \theta(z), \tag{2.229}$$

with

$$\nu = (0, 0, -1) \quad \text{at} \quad z = 0, \tag{2.230}$$

so that the 'surface' mathematically consists of effectively only one point, that is, $S = \{0\}$. Here, $\theta_1 = \theta$ and $\theta_2 \equiv 0$. Equations (2.208), (2.209) and (2.210) become, respectively,

$$\mathbf{n} = \mathbf{f}(\theta), \qquad n_i \frac{\partial f_i}{\partial \theta} = 0, \qquad \frac{\partial \mathbf{f}}{\partial \theta} \neq 0, \tag{2.231}$$

while the bulk energy in (2.211) is replaced by

$$w = w(n_i, n_{i,j}) = \widehat{w}(\theta, \theta'), \tag{2.232}$$

where a prime denotes the derivative with respect to z. We also have

$$\frac{\partial \widehat{w}}{\partial \theta'} = \frac{\partial \widehat{w}}{\partial \theta_{,j}}, \qquad j = 3, \tag{2.233}$$

which, upon using equations (2.230), allows the natural boundary condition in (2.228) to collapse simply to

$$\frac{\partial \widehat{w}}{\partial \theta'} = 0 \quad \text{at} \quad z = 0. \tag{2.234}$$

Strong Anchoring

There are no additional requirements for $\mathbf{n}$ on the boundary except that it must satisfy the imposed prescribed strong anchoring boundary condition which is specified for each problem. There are no relevant reformulations except those for the bulk equilibrium equations.

Conical Anchoring

The conical anchoring requirement in equation (2.197), namely that $(\mathbf{n} \cdot \boldsymbol{\nu})^2 = \sin^2\psi$ on S, ψ being the fixed angle the director makes with the boundary surface as shown in Fig. 2.4, must always hold true and it must also be supplemented by the additional boundary requirement given in equation (2.200). Given the form of this additional boundary condition in (2.200), it is often inconvenient to carry out a reformulation straightaway: it is usually best to use the Cartesian component form of the relevant bulk energy and carry out the necessary differentiations before replacing $\mathbf{n}$ by $\mathbf{f}(\theta)$.

Nevertheless, it is instructive to observe for the example introduced above by equations (2.229) and (2.230) that the condition (2.203) must hold at $z = 0$; but in this example $n_y \equiv 0$ which forces this condition to be satisfied identically. The only remaining boundary condition to be fulfilled is then the aforementioned conical anchoring requirement (2.197) which reduces to

$$|\sin\theta| = \sin\psi \qquad \text{at } z = 0. \tag{2.235}$$

Weak Anchoring

For the form of the weak anchoring surface energy w_s introduced above at equation (2.204) we have

$$w_s = w_s(n_i) = \widehat{w}_s(\theta_\alpha), \qquad \alpha = 1, 2, \tag{2.236}$$

which leads to

$$\frac{\partial \widehat{w}_s}{\partial \theta_\alpha} = \frac{\partial w_s}{\partial n_i}\frac{\partial f_i}{\partial \theta_\alpha}. \tag{2.237}$$

The additional boundary condition stated in equation (2.206) can be multiplied by $\frac{\partial f_i}{\partial \theta_\alpha}$ so that in conjunction with the Frank free energy $w = w_F$ it becomes

$$\left(\frac{\partial w_F}{\partial n_{i,j}}\nu_j + \frac{\partial w_s}{\partial n_i}\right)\frac{\partial f_i}{\partial \theta_\alpha} = \gamma n_i\frac{\partial f_i}{\partial \theta_\alpha}, \tag{2.238}$$

which by equations (2.209), (2.214) and (2.237) finally gives the simplified boundary conditions for weak anchoring as

$$\frac{\partial \widehat{w}_F}{\partial \theta_{\alpha,j}}\nu_j + \frac{\partial \widehat{w}_s}{\partial \theta_\alpha} = 0. \tag{2.239}$$

As a special example, consider again $\mathbf{n}$ and $\boldsymbol{\nu}$ in equations (2.229) and (2.230) and now suppose that the anchoring at the interface or boundary is weak. Straightforward substitutions coupled with the identity given by (2.233) (with $\widehat{w}$ replaced by $\widehat{w}_F$) simplify the condition in (2.239) to

$$\frac{\partial \widehat{w}_F}{\partial \theta'}\nu_3 + \frac{\partial \widehat{w}_s}{\partial \theta} = 0 \quad \text{at} \quad z = 0, \tag{2.240}$$

which, in the situation described by equation (2.230), is equivalent to the equation

$$\frac{\partial \widehat{w}_F}{\partial \theta'} - \frac{\partial \widehat{w}_s}{\partial \theta} = 0 \quad \text{at} \quad z = 0. \tag{2.241}$$

Notice that if, additionally, the liquid crystal sample occupies the region $0 \le z \le d$ where the depth of the sample is d, then the outward unit normal on the surface at $z = d$ is $\boldsymbol{\nu} = (0, 0, 1)$ and the boundary requirement (2.241) is then supplemented via (2.240) by the requirement

$$\frac{\partial \widehat{w}_F}{\partial \theta'} + \frac{\partial \widehat{w}_s}{\partial \theta} = 0 \quad \text{at} \quad z = d. \tag{2.242}$$

Such reformulated boundary conditions will be used in Section 3.6 below in the case of an elementary weak Freedericksz transition. Notice also that $\widehat{w}_F$ appearing in the above formulations may be replaced by any applicable total bulk energy density which may also incorporate any relevant magnetic and/or electric energy densities.

Chapter 3

Applications of Static Theory of Nematics

3.1 Introduction

One of the main interests in the static theory of liquid crystals is in the competition that occurs between the orientation of the director on the boundary and the orientation within the bulk of a sample, particularly when the director is influenced by an electric or magnetic field. We shall concentrate on nematics throughout this Chapter and make passing comments on cholesterics when appropriate. Equilibrium solutions in the absence of electric and magnetic fields will be considered first in Section 3.2 before we go on to discuss the effects of these fields in Sections 3.3 to 3.7. The definition of a magnetic coherence length, an important result for physical observations, will be presented in Section 3.3. The fundamental idea of a Freedericksz transition will be developed in the context of the classical Freedericksz transitions in Section 3.4, together with other related Freedericksz transitions involving director pretilt at the boundaries or tilted fields. Further electric field effects and weak anchoring problems will also be discussed in Sections 3.5 and 3.6. A mathematical description and discussion of the twisted nematic device, widely used in visual display technology, is given in Section 3.7. The Chapter closes in Section 3.8 with an introduction to various defects which can occur in nematic liquid crystals, with particular attention drawn to those discussed by Frank [91] in Section 3.8.1, other defects being described in Section 3.8.2; these defects may occur in the absence of external fields. Line and point surface defects induced by magnetic fields are discussed in mathematical detail in Sections 3.8.3 and 3.8.4.

3.2 Some Equilibrium Solutions

Before going on to consider the effects of electric and magnetic fields, it is of interest first to look at the possible orientations of **n** that can arise throughout a sample solely due to the influence of prescribed surface alignments for **n** on the boundary in the absence of such fields. There are, broadly speaking, two categories of equilibrium solutions. The first type consists of solutions having no singularities or defects; the second type consists of solutions possessing such defects. We defer the discussion on defects until Section 3.8 below and for the moment concentrate on the search for equilibrium solutions with no singularities in the director orientation. Such alignments of the director are relevant to the design of displays; results which involve director 'tilt' and 'twist' (introduced below) have physical significance, as will be pointed out in Section 3.2.2.

3.2.1   Elementary Equilibrium Solutions

As a first example, consider a sample of nematic liquid crystal confined between two parallel glass plates a distance d apart as shown in Fig. 3.1. Suppose that there

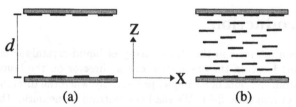

(a)   (b)

Figure 3.1: (a) The initial boundary alignment of the director at the surfaces of a sample cell of depth d. (b) The uniform constant equilibrium solution.

is strong anchoring on these bounding plates so that the director $\mathbf{n}$ satisfies the boundary conditions

$$\mathbf{n}_b = (1, 0, 0) \qquad \text{at} \qquad z = 0, d, \tag{3.1}$$

when referred to Cartesian coordinates as introduced in Fig. 3.1(a). Consider solutions for the director of the form

$$\mathbf{n} = (\cos\phi(z), \sin\phi(z), 0), \tag{3.2}$$

with the boundary conditions to match those for $\mathbf{n}_b$, namely,

$$\phi = 0 \qquad \text{at} \qquad z = 0, d. \tag{3.3}$$

Insertion of $\mathbf{n}$ into the nematic energy integrand w_F in equation (2.49) results in the total energy W per unit area of the plates (in the xy-plane) being given by

$$W = \frac{1}{2}K_2 \int_0^d \left(\frac{d\phi}{dz}\right)^2 dz, \tag{3.4}$$

where $K_2 \geq 0$ is the twist elastic constant. The equilibrium equation is given by equation (2.221) with $\widehat{w}$ being the integrand appearing in W above in the particular case when $\theta_1 \equiv 0$ and $\theta_2 = \phi(z)$ in equation (2.207). Therefore ϕ is required to satisfy the Euler equation

$$\frac{d}{dz}\left(\frac{\partial\widehat{w}}{\partial\phi'}\right) - \frac{\partial\widehat{w}}{\partial\phi} = 0, \tag{3.5}$$

where $' = d/dz$, leading to

$$\frac{d^2\phi}{dz^2} = 0, \tag{3.6}$$

which has the solution

$$\phi(z) = az + b, \qquad a, b \text{ constants.} \tag{3.7}$$

Recall that the fulfilment of the Euler equation (sometimes referred to as the Euler–Lagrange equation when it is discussed in more general settings) is a necessary condition for equilibrium, as is well known from the calculus of variations [218, 238]. Applying the boundary conditions (3.3) reveals that

$$\phi(z) \equiv 0, \quad 0 \le z \le d, \tag{3.8}$$

and the alignment of the director **n** is uniformly constant across the sample as shown in Fig. 3.1(b).

Now consider the same experimental set-up as above, but with the top plate rotated around its normal by $\frac{\pi}{2}$ radians so that the fixed director alignments on the boundaries are at right angles to each other as shown in Fig. 3.2(a). The expressions for the energy and equilibrium equation remain as stated in equations (3.4) and (3.6), respectively. The solution for ϕ is again given by (3.7) but with the boundary conditions changed to

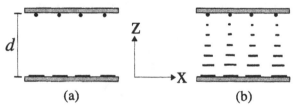

Figure 3.2: (a) The initial director orientation: the director on the upper plate is at right angles to the director on the lower plate, both alignments being within the plane of the plates. (b) The uniformly twisted equilibrium solution (3.13).

$$\begin{aligned}
\mathbf{n}_b &= (0,1,0) \quad \text{at} \quad z = d, \tag{3.9}\\
\mathbf{n}_b &= (1,0,0) \quad \text{at} \quad z = 0, \tag{3.10}
\end{aligned}$$

leading to the requirements

$$\begin{aligned}
\phi &= \frac{\pi}{2} \quad \text{at} \quad z = d, \tag{3.11}\\
\phi &= 0 \quad \text{at} \quad z = 0. \tag{3.12}
\end{aligned}$$

Applying these boundary conditions provides the solution

$$\phi(z) = \frac{\pi}{2d} z. \tag{3.13}$$

Thus the director rotates uniformly about the normal to the plates as shown in Fig. 3.2(b). This is the simplest example of a twisted nematic cell (cf. Section 3.7 below) in the absence of electric and magnetic fields where the director **n** always remains in the plane of the plates, but twists (rotates) around the normal to the plates. The energy per unit area of the plates is given by inserting ϕ from equation (3.13) into (3.4) to find

$$W = K_2 \frac{\pi^2}{8d}. \tag{3.14}$$

Recall that K_2 is the Frank elastic constant related to 'twist' (see Fig. 2.1 on page 16).

3.2.2   Tilt and Twist Equilibrium Solutions

The above simple equilibrium solutions motivate the search for more general solutions which contain these elementary ones as special cases. To this end it is natural to seek equilibrium solutions when the director is of the form

$$\mathbf{n} = (\cos\theta(z)\cos\phi(z), \cos\theta(z)\sin\phi(z), \sin\theta(z)), \tag{3.15}$$

where the angles θ and ϕ depend solely upon the z coordinate and, as before, the sample of nematic liquid crystal is confined between two parallel glass plates a distance d apart, these plate boundaries being parallel to the xy-plane and having their normals parallel to the z-axis. The angle θ represents the tilt of the director

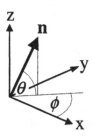

Figure 3.3: Definition of the twist angle ϕ within the plane of the plates (in the xy-plane) and the tilt angle θ, representing the tilt of the director out of the plane of the plates.

out of the plane of the plates while ϕ represents the twist of the director about the normal to the plates (this has the same interpretation as ϕ in the previous Section): see Fig. 3.3. Clearly, setting $\theta \equiv 0$ recovers the form of the director in (3.2). Our discussion now follows those of Leslie [167, 170]. Straightforward calculations show that the nematic energy integrand w_F in equation (2.49) reduces in this case to $\widehat{w}$ defined by

$$
\begin{aligned}
w_F(\mathbf{n}, \nabla\mathbf{n}) &= \widehat{w}(\theta, \phi, \theta', \phi') \\
&= \tfrac{1}{2}f(\theta)(\theta')^2 + \tfrac{1}{2}g(\theta)(\phi')^2, \tag{3.16}
\end{aligned}
$$

where $' = d/dz$ and

$$f(\theta) = K_1 \cos^2\theta + K_3 \sin^2\theta, \tag{3.17}$$
$$g(\theta) = (K_2 \cos^2\theta + K_3 \sin^2\theta)\cos^2\theta. \tag{3.18}$$

Setting $\theta_1 = \theta$ and $\theta_2 = \phi$ in the equilibrium equations (2.221) easily gives the two coupled Euler equations

$$\frac{d}{dz}\left(\frac{\partial\widehat{w}}{\partial\theta'}\right) - \frac{\partial\widehat{w}}{\partial\theta} = 0, \tag{3.19}$$

$$\frac{d}{dz}\left(\frac{\partial\widehat{w}}{\partial\phi'}\right) - \frac{\partial\widehat{w}}{\partial\phi} = 0, \tag{3.20}$$

leading to

$$2f(\theta)\theta'' + \frac{df(\theta)}{d\theta}(\theta')^2 - \frac{dg(\theta)}{d\theta}(\phi')^2 = 0, \tag{3.21}$$

$$[g(\theta)\phi']' = 0. \tag{3.22}$$

These coupled nonlinear equations are the key governing equations.

It is possible to proceed further. First observe that

$$[g(\theta)(\phi')^2]' = \frac{dg(\theta)}{d\theta}\theta'(\phi')^2 + 2g(\theta)\phi'\phi'', \tag{3.23}$$

and, by (3.22),

$$-g(\theta)\phi'' = \frac{dg(\theta)}{d\theta}\theta'\phi'. \tag{3.24}$$

Substituting for ϕ'' from (3.24) into (3.23) eliminates ϕ'' to give

$$[g(\theta)(\phi')^2]' = -\frac{dg(\theta)}{d\theta}\theta'(\phi')^2. \tag{3.25}$$

Hence, multiplying both sides of equation (3.21) by θ' and employing the identity (3.25) allows us to rewrite (3.21) as

$$2f(\theta)\theta''\theta' + \frac{df(\theta)}{d\theta}(\theta')^3 + [g(\theta)(\phi')^2]' = 0, \tag{3.26}$$

which is equivalent to

$$\frac{d}{dz}\left(f(\theta)(\theta')^2 + g(\theta)(\phi')^2\right) = 0. \tag{3.27}$$

Thus the equilibrium equations (3.21) and (3.22) can be integrated to give

$$f(\theta)(\theta')^2 + g(\theta)(\phi')^2 = a, \tag{3.28}$$
$$g(\theta)\phi' = b, \tag{3.29}$$

where a and b are constants.

To simplify the presentation in this Section on tilt and twist equilibrium solutions we shall henceforth assume that the elastic constants K_1, K_2 and K_3 are all strictly positive (recall that they are necessarily non-negative by Ericksen's inequalities (2.58)).

A Tilt Solution

The first simple situation to consider is where the prescribed alignment of the director is the same at both plates with there being no net tilt or twist in the sample. In this event, the strong anchoring boundary conditions are

$$\theta(0) = \theta(d) \;=\; \theta_0, \tag{3.30}$$
$$\phi(0) = \phi(d) \;=\; 0, \tag{3.31}$$

where θ_0 is a given fixed acute angle that the director makes with the plane of the plates at the boundaries. Notice that $f(\theta)$ and $g(\theta)$ are both strictly positive

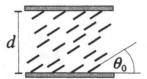

Figure 3.4: The fixed boundary tilt θ_0.

functions for $0 \le \theta < \frac{\pi}{2}$. If the constant $b \neq 0$, then equation (3.29) shows that $\phi'(z)$ is single signed and therefore, invoking the boundary conditions (3.31), we are forced to conclude that $\phi \equiv 0$; this result clearly holds if $b = 0$, and so $\phi \equiv 0$ in all cases. Substituting this result for ϕ into (3.28) shows that $f(\theta)(\theta')^2 = a$. If $a > 0$ then $\theta'(z) \neq 0$ for any $0 \le z \le d$ and therefore the boundary conditions in (3.30) cannot be satisfied. Hence $a = 0$ and, by the positivity of $f(\theta)$, $\phi'(z) = 0$, and so applying the associated boundary conditions (3.30) gives $\theta(z) \equiv \theta_0$. Thus the solutions for θ and ϕ are

$$\theta(z) = \theta_0, \qquad \phi(z) = 0, \qquad 0 \le z \le d, \tag{3.32}$$

which show that the director alignment is uniform across the sample depth as shown in Fig. 3.4. The special case of $\theta_0 = 0$ clearly corresponds to the elementary solution (3.8) displayed in Fig. 3.1(b) which is an example of the homogeneous alignment depicted in Fig. 2.3(b) on page 47.

A Twist Solution

The next simple situation is when the director is aligned parallel to the plates at the boundaries, so that $\theta_0 = 0$ in the above notation, and Fig. 3.2 is considered again where, rather than a rotation of $\frac{\pi}{2}$ of the top plate, a rotation of the top plate through an arbitrary fixed angle $2\phi_0$ is considered. The boundary conditions can then be written as

$$\theta(0) = \theta(d) = 0, \tag{3.33}$$
$$\phi(0) = -\phi_0, \qquad \phi(d) = \phi_0, \tag{3.34}$$

so that there is an initial difference of $2\phi_0$ between the two boundary alignments of the director. Elementary solutions to the equilibrium equations in the forms given by (3.21) and (3.22) which satisfy the boundary conditions (3.33) and (3.34) are easily found to be

$$\theta \equiv 0, \qquad \phi = \phi_0\left(2\frac{z}{d} - 1\right), \qquad 0 \le z \le d, \tag{3.35}$$

which lead to a representation of a simple twist through $2\phi_0$ of the director **n** across the sample, **n** always remaining in the plane of the plates. This is the type of solution first discussed by Ericksen [78] and is of the twist solution form which arose naturally in Section 2.5 at equation (2.179). The energy per unit area of the plates is given, via equations (3.35) and (3.4), by

$$W_p = 2K_2\frac{\phi_0^2}{d}, \tag{3.36}$$

the suffix p indicating that there is only a twisted 'planar' alignment of the director parallel to the plates. The elementary twist solution (3.13) is easily obtained from the general solution (3.35) with $\phi_0 = \frac{\pi}{4}$ by carrying out an obvious rotation of the xy-axes around the z-axis. The energy for this simple twist solution given by equation (3.14) is recovered from (3.36) by simply setting $\phi_0 = \frac{\pi}{4}$.

A Tilt and Twist Solution

The energy W_p increases as ϕ_0 increases. This observation led Leslie [167] to look at the possibility of what are called non-planar twist solutions. These solutions not only twist, as above, but also tilt out of the xy-plane, despite the boundary tilt angle θ_0 being set to zero on both boundaries. It will be of interest to compare the resulting energy W_{np} for the non-planar twist solution with W_p for the planar twist solution. The boundary conditions (3.33) and (3.34) will be retained for this problem and solutions will be sought which satisfy the symmetries

$$\theta(z) = \theta(d - z), \qquad \phi(z) = -\phi(d - z), \qquad 0 \le z \le d, \tag{3.37}$$

motivated by the boundary conditions. The given symmetries dictate that

$$\theta'(\tfrac{d}{2}) = 0, \qquad \theta(\tfrac{d}{2}) = \theta_m > 0, \tag{3.38}$$
$$\phi(\tfrac{d}{2}) = 0, \tag{3.39}$$

regarding, without loss of generality, the maximum tilt angle θ_m to be positive. As above, we shall assume that the elastic constants are positive so that $f(\theta)$ and $g(\theta)$ are also positive for $0 \le \theta < \frac{\pi}{2}$.

Using the conditions (3.38) in equation (3.28) shows

$$a = g(\theta_m)(\phi'(\tfrac{d}{2}))^2, \tag{3.40}$$

and so (3.28) becomes

$$f(\theta)(\theta')^2 = g(\theta_m)(\phi'(\tfrac{d}{2}))^2 - g(\theta)(\phi'(z))^2. \tag{3.41}$$

However, from (3.29),

$$[g(\theta_m)\phi'(\tfrac{d}{2})]^2 = b^2 = [g(\theta)\phi'(z)]^2, \tag{3.42}$$

and insertion of these results into (3.41) shows

$$f(\theta)(\theta')^2 = b^2 \left[\frac{1}{g(\theta_m)} - \frac{1}{g(\theta)} \right] = b^2 \left[\frac{g(\theta) - g(\theta_m)}{g(\theta_m)g(\theta)} \right]. \tag{3.43}$$

The right-hand side of (3.43) must always be non-negative for solutions θ because $f(\theta)$ is positive; this results in restrictions upon possible solutions depending on the relative magnitudes of the Frank elastic constants. Notice also that (3.43) is undefined at $\theta_m = \frac{\pi}{2}$ and cannot be positive at $\theta_m = 0$.

It is seen, by converting cosine terms into terms involving sines, that

$$g(\theta) - g(\theta_m) = (\sin^2\theta_m - \sin^2\theta) \left[2K_2 - K_3 + (K_3 - K_2)(\sin^2\theta_m + \sin^2\theta) \right], \tag{3.44}$$

and therefore it is necessary to consider two separate cases, noticing that $\sin\theta \le \sin\theta_m$. When $K_3 \le 2K_2$, the identity (3.44) leads to

$$\begin{aligned}
g(\theta) - g(\theta_m) &\ge (\sin^2\theta_m - \sin^2\theta) \left[(2K_2 - K_3)\sin^2\theta_m + (K_3 - K_2)(\sin^2\theta_m + \sin^2\theta) \right] \\
&\ge K_2(\sin^2\theta_m - \sin^2\theta)^2 \\
&\ge 0, \tag{3.45}
\end{aligned}$$

and so θ_m is restricted according to

$$0 < \theta_m < \frac{\pi}{2} \qquad \text{if} \quad K_3 \le 2K_2. \tag{3.46}$$

On the other hand, when $K_3 > 2K_2$ the identity (3.44) shows that

$$g(\theta) - g(\theta_m) \ge (\sin^2\theta_m - \sin^2\theta)[2K_2 - K_3 + (K_3 - K_2)\sin^2\theta_m], \tag{3.47}$$

and therefore $g(\theta) - g(\theta_m) \ge 0$ according to the restrictions

$$0 < \theta_c \le \theta_m < \frac{\pi}{2} \qquad \text{if} \quad K_3 > 2K_2, \tag{3.48}$$

where $\theta_c > 0$ is defined by

$$\sin^2\theta_c = \frac{K_3 - 2K_2}{K_3 - K_2}, \tag{3.49}$$

and it is noticed that the right-hand side of the above expression is strictly positive and less than 1. The restriction $K_3 > 2K_2$ in (3.48) is known to be satisfied by, for example, the nematic liquid crystal MBBA at various temperatures [110, p.105].

We are now in a position to find the solutions θ and ϕ implicitly. Accepting the above restrictions upon the parameter θ_m, the solution for $\theta(z)$ is obtained from (3.43) and the boundary conditions (3.33) by separation of variables to find

$$bz = \sqrt{g(\theta_m)} \int_0^\theta \left\{ \frac{g(u)f(u)}{g(u) - g(\theta_m)} \right\}^{\frac{1}{2}} du, \qquad 0 \le z \le \tfrac{d}{2}. \tag{3.50}$$

The solution for $\frac{d}{2} \leq z \leq d$ is obtained by using the symmetry condition $(3.37)_1$ (simply replace z by $d - z$ in (3.50) for z between $\frac{d}{2}$ and d). Similarly, the solution for ϕ is obtained from (3.29) and the boundary conditions (3.34) by substituting for the constant b via (3.43) to give

$$\phi = -\phi_0 + \sqrt{g(\theta_m)} \int_0^\theta \left\{ \frac{f(u)}{g(u)[(g(u) - g(\theta_m)]} \right\}^{\frac{1}{2}} du, \qquad -\phi_0 \leq \phi \leq 0, \qquad (3.51)$$

with the solution for $0 \leq \phi \leq \phi_0$ being provided via the symmetry condition $(3.37)_2$ (simply replace ϕ by $-\phi$ when θ is defined on $\frac{d}{2} \leq z \leq d$). Of course, the solution for ϕ is calculated from (3.51) after substituting the solution for θ obtained from (3.50) into the integral limit in (3.51). These solutions for θ and ϕ are valid provided they satisfy the conditions $(3.38)_2$ and (3.39), which means that the relations

$$b\frac{d}{2} = \sqrt{g(\theta_m)} \int_0^{\theta_m} \left\{ \frac{g(u)f(u)}{g(u) - g(\theta_m)} \right\}^{\frac{1}{2}} du, \qquad (3.52)$$

$$\phi_0 = \sqrt{g(\theta_m)} \int_0^{\theta_m} \left\{ \frac{f(u)}{g(u)[(g(u) - g(\theta_m)]} \right\}^{\frac{1}{2}} du, \qquad (3.53)$$

must be fulfilled. The latter condition dictates the value of θ_m for a given angle ϕ_0: inserting the resulting value for θ_m into equation (3.52) evaluates the constant b. Equations (3.50) to (3.53) then complete the description of the solutions for the *combined* tilt θ and twist ϕ.

A comparison will now be made between the energy W_p in equation (3.36) for the planar twist solution (3.35) and the energy for the non-planar twist solution provided by equations (3.50) and (3.51), the solution with the lesser energy being interpreted as the physically preferred solution. The total energy W_{np} per unit area of the plates is obtained by integrating the energy $\hat{w}$ in (3.16) from $z = 0$ to $z = d$:

$$W_{np} = \frac{1}{2} \int_0^d [f(\theta)(\theta')^2 + g(\theta)(\phi')^2] dz. \qquad (3.54)$$

Inserting the results (3.29), $(3.42)_2$ and (3.43) into W_{np} gives

$$W_{np} = \frac{b^2 d}{2g(\theta_m)}. \qquad (3.55)$$

It follows from the results (3.36) and (3.55) that

$$W_{np} - W_p = \frac{b^2 d}{2g(\theta_m)} - 2K_2 \frac{\phi_0^2}{d}, \qquad (3.56)$$

and therefore $W_{np} < W_p$ if

$$b^2 d^2 < 4K_2 \phi_0^2 g(\theta_m), \qquad (3.57)$$

that is, the non-planar twist solution is preferred to the planar twist solution whenever the inequality (3.57) is satisfied.

It is clear at this point that the fulfilment of the relations (3.52) and (3.53) are important in the evaluation of the criterion (3.57) which determines whether or not the non-planar twist has a lower energy than the planar twist. To gain more insight into the application of this result we follow Leslie [167] and investigate the special case when $K_2 = K_3$. In this scenario, $g(\theta) = K_2 \cos^2 \theta$ and the relations (3.52) and (3.53) simplify to, respectively,

$$b\frac{d}{2} = \sqrt{K_2} \cos \theta_m \int_0^{\theta_m} \frac{\cos u \sqrt{f(u)}}{\sqrt{\sin^2 \theta_m - \sin^2 u}} \, du, \qquad (3.58)$$

$$\phi_0 = \frac{\cos \theta_m}{\sqrt{K_2}} \int_0^{\theta_m} \frac{1}{\cos u} \frac{\sqrt{f(u)}}{\sqrt{\sin^2 \theta_m - \sin^2 u}} \, du, \qquad (3.59)$$

where, as above, it is assumed without loss of generality that $\theta_m > 0$. Using the substitution

$$\sin u = \sin \theta_m \sin \lambda, \qquad (3.60)$$

these expressions can be simplified further to

$$b\frac{d}{2} = \sqrt{K_2} \cos \theta_m \int_0^{\frac{\pi}{2}} \sqrt{f(u)} \, d\lambda, \qquad (3.61)$$

$$\phi_0 = \frac{\cos \theta_m}{\sqrt{K_2}} \int_0^{\frac{\pi}{2}} \frac{\sqrt{f(u)}}{\cos^2 u} \, d\lambda. \qquad (3.62)$$

Firstly, consider (3.61). The change of variable

$$t = \tan \lambda \qquad (3.63)$$

transforms (3.61) into the more tractable form

$$b\frac{d}{2} = \sqrt{K_1 K_2} \cos \theta_m I_1(\theta_m), \qquad (3.64)$$

where I_1 is defined by

$$I_1(\theta_m) = \int_0^\infty \frac{1}{1+t^2} \sqrt{1 + \frac{(K_2 - K_1)}{K_1} \left(\frac{t^2 \sin^2 \theta_m}{1+t^2} \right)} \, dt, \qquad (3.65)$$

recalling that we have set $K_2 = K_3$. The similar change of variable

$$t = \cos \theta_m \tan \lambda \qquad (3.66)$$

transforms (3.62) into, after some tedious but obvious manipulations,

$$\phi_0 = \sqrt{\frac{K_1}{K_2}} I_2(\theta_m), \qquad (3.67)$$

with I_2 defined by

$$I_2(\theta_m) = \int_0^\infty \frac{1}{1+t^2} \sqrt{1 + \frac{(K_2 - K_1)}{K_1} \left(\frac{t^2 \sin^2 \theta_m}{\cos^2 \theta_m + t^2} \right)} \, dt. \qquad (3.68)$$

The results (3.64) and (3.67) now lead to conclusions which depend upon the relative sizes of K_1 and K_2 and it is worthwhile considering the three distinct cases that arise.

Case (i) $K_2 > K_1$

Observe that when $K_2 > K_1$

$$I_2(\theta_m) > I_1(\theta_m) \geq 0, \quad 0 < \theta_m < \frac{\pi}{2}. \tag{3.69}$$

Employing equations (3.18), (3.64), (3.67) and (3.69) then shows that (assuming $K_2 = K_3$ as above)

$$
\begin{aligned}
b^2 d^2 - 4K_2\phi_0^2 g(\theta_m) &= 4K_1 K_2 \cos^2\theta_m [I_1^2(\theta_m) - \frac{g(\theta_m)}{K_2 \cos^2\theta_m} I_2^2(\theta_m)] \\
&= 4K_1 K_2 \cos^2\theta_m \left[I_1^2(\theta_m) - I_2^2(\theta_m) \right] \\
&< 0.
\end{aligned}
\tag{3.70}
$$

Hence the energy difference criterion (3.57) holds when $K_1 > 0$ and therefore we have proved that when $K_2 = K_3$ and $K_2 > K_1 > 0$ then the non-planar twist solution is preferred to the planar twist solution.

Case (ii) $K_1 > K_2$

In this case it is seen that

$$I_1(\theta_m) > I_2(\theta_m) \geq 0, \quad 0 < \theta_m < \frac{\pi}{2}, \tag{3.71}$$

when the appropriate solutions for θ and ϕ exist. By a similar argument to Case (i) above, it easily follows that the energy criterion (3.57) is *not* satisfied. Therefore when $K_2 = K_3$ and $K_1 > K_2 > 0$ then the planar twist solution is preferred to the non-planar twist solution. It should be remarked here that for some materials, for example MBBA, PAA and 5CB at certain temperatures, it is known that $K_1 > K_2$: see Table D.3 and [110, pp.104–105]. Therefore the planar twist solution (3.35) is expected to be relevant when modelling displays such as the twisted nematic device discussed in Section 3.7.

Case (iii) $K_1 = K_2$

In this case, $K_1 = K_2 = K_3$ and the expressions (3.64) and (3.67) reduce to, respectively,

$$b\frac{d}{2} = K_2\frac{\pi}{2}\cos\theta_m, \tag{3.72}$$

$$\phi_0 = \frac{\pi}{2}. \tag{3.73}$$

In this special case it follows that if the overall twist $2\phi_0$ equals π then the non-planar twist solutions for θ and ϕ in equations (3.50) and (3.51) always exist for all $0 < \theta_m < \frac{\pi}{2}$. Surprisingly, as pointed out by Leslie [167], in this case

$$b^2 d^2 - 4K_2\phi_0^2 g(\theta_m) = b^2 d^2 - 4K_2^2\phi_0^2 \cos^2(\theta_m) = 0, \tag{3.74}$$

and it is then seen from (3.56) that $W_{np} = W_p$. This indicates that both the non-planar twist and planar twist solutions always have the same energy.

For more details on the treatment of the general case the reader is referred to Leslie [167]. It is worth mentioning here that the case when the boundary condition (3.33) is replaced by

$$\theta(0) = \theta(d) = \theta_0, \tag{3.75}$$

θ_0 being a given acute angle, has been considered by Porte and Jadot [225], whose analysis is not dissimilar to that presented above. We shall extend our exposition of twisted nematics when we discuss the twisted nematic device in Section 3.7 below where a magnetic field is incorporated into the model.

3.3   Magnetic Coherence Length

This Section will present our first example of how a magnetic field can influence the orientation of the director **n** in a nematic liquid crystal. This example also introduces the idea of a magnetic coherence length. The equations presented here will be refashioned in a natural way, in the next Section, to obtain the equations required for the study of Freedericksz transitions.

A parameter called the magnetic coherence length, denoted by $\xi(H)$, will be introduced as a measure of the thickness of a transition layer near the boundary of a liquid crystal sample where there is a competition between the alignment of the director on the boundary and the direction of the magnetic field **H**. To clarify this concept, consider an idealised semi-infinite sample of nematic liquid crystal occupying the region $z \geq 0$ having a boundary plate placed in the xy-plane at $z = 0$ as shown in Fig. 3.5. Using the Cartesian coordinates introduced in Fig. 3.5, we follow de Gennes and Prost [110, p.120] and suppose that the director is strongly anchored parallel to the plate in the x-direction so that on the boundary **n** takes the form

$$\mathbf{n}_b = (1, 0, 0), \tag{3.76}$$

and we consider the director in the bulk to be represented by

$$\mathbf{n} = (\cos\phi, \sin\phi, 0), \qquad \phi = \phi(z). \tag{3.77}$$

The boundary condition $\mathbf{n}_b$ is then recast as

$$\phi(0) = 0. \tag{3.78}$$

Let

$$\mathbf{H} = H(0, 1, 0), \tag{3.79}$$

be a fixed magnetic field of magnitude H aligned parallel to the y-axis as shown in Fig. 3.5, and suppose that χ_a, the magnetic anisotropy of the nematic, is positive, indicating that the director desires to be parallel to **H**. For the geometric set-up just described, there is then a clear competition between the alignment of **n** on the boundary and the anticipated desired alignment of **n** within the bulk of the sample; it is this competition that will now be quantified mathematically.

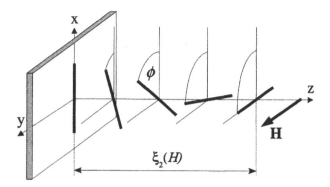

Figure 3.5: The magnetic coherence length $\xi_2(H)$ (defined by equation (3.87)) related to the twist elastic constant K_2. The director is strongly anchored at a boundary surface where the orientation angle ϕ of the director parallel to the xy-plane is zero. For a nematic having magnetic anisotropy $\chi_a > 0$, this angle increases into the bulk of the nematic as the director tries to align parallel to the applied magnetic field H. Most of the reorientation of $\mathbf{n}$ occurs over the length $\xi_2(H)$.

It is easily checked that

$$\nabla \cdot \mathbf{n} = 0, \quad \mathbf{n} \cdot \nabla \times \mathbf{n} = -\frac{d\phi}{dz}, \quad \mathbf{n} \times \nabla \times \mathbf{n} = 0, \quad (\mathbf{n} \cdot \nabla)\mathbf{n} = 0, \qquad (3.80)$$

and so the bulk free energy density for the nematic is, by equation (2.49),

$$w_F = \frac{1}{2} K_2 \left(\frac{d\phi}{dz} \right)^2. \qquad (3.81)$$

To this must be added the appropriate magnetic energy density. Since we are considering a semi-infinite sample, it is convenient in this situation to choose the form of magnetic energy provided by equation (2.108), giving

$$w_{mag} = \tfrac{1}{2}\chi_a(H^2 - (\mathbf{n} \cdot \mathbf{H})^2) = \tfrac{1}{2}\chi_a H^2 \cos^2\phi. \qquad (3.82)$$

The total energy per unit area in the xy-plane is therefore

$$\begin{aligned} W &= \int_V (w_F + w_{mag})dV \\ &= \frac{1}{2} \int_0^\infty \left\{ K_2 \left(\frac{d\phi}{dz} \right)^2 + \chi_a H^2 \cos^2\phi \right\} dz. \end{aligned} \qquad (3.83)$$

The resulting Euler equation providing the equilibrium equation is given by (2.221), in this case reducing to

$$\frac{d}{dz}\left(\frac{\partial \widehat{w}}{\partial \phi'} \right) - \frac{\partial \widehat{w}}{\partial \phi} = 0, \qquad (3.84)$$

where $' = d/dz$ and $\widehat{w}$ is the integrand appearing in W above. Thus the equilibrium equation is

$$K_2\phi'' + \chi_a H^2 \cos\phi\sin\phi = 0, \tag{3.85}$$

which can be written as

$$\xi_2^2(H)\phi'' + \cos\phi\sin\phi = 0, \tag{3.86}$$

where ξ_2 as a function of H is defined by

$$\xi_2(H) = \frac{1}{H}\sqrt{\frac{K_2}{\chi_a}}, \tag{3.87}$$

a length scale involving the twist elastic constant K_2. The boundary conditions are

$$\phi(0) = 0, \qquad \phi(z) \to \frac{\pi}{2} \quad \text{as} \quad z \to \infty, \tag{3.88}$$

the former coming from (3.78) and the latter from the expected alignment of $\mathbf{n}$ as $z \to \infty$ (see Fig. 3.5). For the energy W to be finite, it is also required that

$$\frac{d\phi}{dz} \to 0 \quad \text{as} \quad z \to \infty. \tag{3.89}$$

Multiplying equation (3.86) throughout by ϕ' shows that

$$\frac{d}{dz}\left[\xi_2^2(\phi')^2 - \cos^2\phi\right] = 0, \tag{3.90}$$

and therefore integrating gives

$$\xi_2^2(\phi')^2 = \cos^2\phi + c, \quad c \text{ a constant.} \tag{3.91}$$

Applying the boundary conditions $(3.88)_2$ and (3.89) shows that $c = 0$ and so

$$\xi_2\frac{d\phi}{dz} = \pm\cos\phi, \tag{3.92}$$

where the plus sign refers to ϕ increasing in a 'right-handed' way, as shown in Fig. 3.5, while the minus sign refers to ϕ increasing in a 'left-handed' way (the mirror reflection in the xz-plane of the orientation shown in the Figure). Both signs are equally feasible and therefore without loss of generality we choose the positive sign. Separating the variables and integrating (3.92) gives

$$\int_0^\phi \frac{d\overline{\phi}}{\cos\overline{\phi}} = \int_0^z \frac{d\overline{z}}{\xi_2}, \tag{3.93}$$

and therefore, from standard results for integration [116, 2.526.9]

$$\ln\left|\cot\left(\frac{\pi}{4} - \frac{\phi}{2}\right)\right| = \frac{z}{\xi_2}, \tag{3.94}$$

that is,

$$\phi(z) = \frac{\pi}{2} - 2\tan^{-1}\left\{\exp\left(-\frac{z}{\xi_2}\right)\right\}. \tag{3.95}$$

Clearly, the boundary conditions (3.88) and (3.89) are all satisfied and the energy (3.83) becomes, after substituting for ϕ' and using (3.92),

$$
\begin{aligned}
W &= \chi_a H^2 \int_0^\infty \cos^2\phi\, dz \\
&= \chi_a H^2 \int_0^{\frac{\pi}{2}} \cos^2\phi\, \frac{dz}{d\phi}\, d\phi \\
&= \chi_a H^2 \xi_2 \int_0^{\frac{\pi}{2}} \cos\phi\, d\phi \\
&= H\sqrt{K_2 \chi_a}.
\end{aligned} \tag{3.96}
$$

The solution $\phi(z)$ provided by (3.95) shows that the main effect of the 'transition' is in a layer of thickness $\xi_2(H)$ as defined in equation (3.87), in the z-direction, as indicated in Fig. 3.5. Other solutions for different geometrical set-ups and alignments of $\mathbf{H}$ lead to the introduction of the *characteristic lengths* [110, p.122]

$$\xi_1(H) = \frac{1}{H}\sqrt{\frac{K_1}{\chi_a}}, \qquad \xi_3(H) = \frac{1}{H}\sqrt{\frac{K_3}{\chi_a}}, \tag{3.97}$$

these lengths being of comparable magnitude to $\xi_2(H)$.

In the one-constant approximation $K_1 = K_2 = K_3 = K$, all these lengths coincide to equal

$$\xi(H) = \frac{1}{H}\sqrt{\frac{K}{\chi_a}}. \tag{3.98}$$

The quantity $\xi(H)$ is called the *magnetic coherence length* of the nematic. Taking as typical values (in Gaussian cgs units)

$$K = 10^{-6}\,\text{dynes}, \quad \chi_a = 10^{-7}, \quad H = 10^4\,\text{oersteds}, \tag{3.99}$$

we have

$$\xi \approx 3 \times 10^{-4}\text{cm}, \tag{3.100}$$

and so $\xi(H)$ is expected to be of the order of a micron (1 μm equals 10^{-6} m). The physical significance of this result is that if the sample depth d is much greater than $\xi(H)$, then most of the nematic will have the director aligned in the direction of the applied magnetic field $\mathbf{H}$ for $z > \xi(H)$. Additionally, there will be a transition layer of depth $\xi(H)$ extending from the boundary at $z = 0$ into the bulk where most of the reorientation of $\mathbf{n}$ induced by $\mathbf{H}$ will occur as $\mathbf{n}$ reorients from its alignment at the boundary to its alignment in the bulk. This transition layer will, by (3.100), be comparable to an optical wavelength, showing that it may be optically observable (recall that the wavelength of light is approximately in the range 0.39×10^{-4} cm $\sim 0.74 \times 10^{-4}$ cm).

3.4   Freedericksz Transitions

It is well known that in a finite sample of liquid crystal there is a competition between the alignment of the director **n** at a surface or boundary and the orientation of **n** induced within the sample by an externally applied magnetic or electric field. It is this most fascinating feature of liquid crystals to which we now apply the static continuum theory.

In a thin sample of nematic liquid crystal which has, for example, magnetic anisotropy $\chi_a > 0$, a magnetic field **H** may be in a position to attract the director **n** and cause it to begin to align in the bulk to be parallel to **H**, and different from its initial sample alignment, when $H = |\mathbf{H}|$ is greater than some critical value H_c, which is often, but not always, greater than zero. In other words, the director alignment throughout the sample will not be influenced by the magnetic field whenever $0 \leq H < H_c$, but *will* be influenced and begin to adjust its orientation to become more parallel to **H** when $H \geq H_c$. This change in the director orientation after the magnitude of the field increases through the value H_c is called the *Freedericksz transition* and H_c is called the *critical field strength* or *Freedericksz threshold*. Similar critical phenomena occur under the influence of electric fields, the main difference being that electric field effects generally require more care in the analysis since it is known that the electric field itself can be influenced by the presence of the liquid crystal (see Section 2.3.1). Therefore, the onset of Freedericksz transitions and critical thresholds can be derived theoretically for magnetic and electric fields in precisely analogous ways, but post-critical threshold behaviour of the director requires more care in applications involving electric fields. In this Section we concentrate on magnetic fields and go on to investigate electric field effects in Section 3.5 below.

We first discuss the classical Freedericksz transitions and critical thresholds for a nematic. The understanding of these phenomena is crucial to the basic traditional idea of 'switching' liquid crystal cells by fields having magnitudes above the critical threshold. The commercial exploitation of these results in liquid crystal display devices, especially the twisted nematic display to be discussed in Section 3.7 below, has greatly increased the general interest in theoretical and experimental aspects of Freedericksz transitions, and vice-versa.

3.4.1   The Classical Freedericksz Transitions in Nematics

The three classical Freedericksz transitions will be considered in what are commonly called the splay, twist and bend geometries. Full details will be given for the splay geometry, with the corresponding results for the twist and bend geometries being stated since their analysis is analogously similar. The Section ends with a qualitative discussion for the one-constant approximation.

Case (i): Splay Geometry

Consider a nematic liquid crystal sample confined between two parallel glass plates a distance d apart and suppose that, in the geometry shown in Fig. 3.6, the director is strongly anchored parallel to the bounding plates so that **n** satisfies

$$\mathbf{n}_b = (1, 0, 0), \qquad \text{at} \quad z = 0, d. \tag{3.101}$$

It was shown in Section 3.2.1 via equations (3.2) and (3.8) that in the absence of

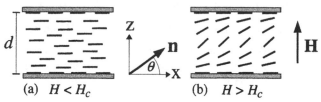

(a) $H < H_c$ (b) $H > H_c$

Figure 3.6: The geometrical set-up of the classical Freedericksz transition which measures the splay elastic constant K_1 via the critical field strength $H_c = \frac{\pi}{d}\sqrt{\frac{K_1}{\chi_a}}$, assuming that $\chi_a > 0$. For $H < H_c$ the original uniform sample alignment parallel to the plates remains intact. As H increases through the critical field strength H_c the sample undergoes a Freedericksz transition where, for $H > H_c$, the director orientation changes and adjusts in response to the influence of the applied magnetic field **H**, as discussed in the text.

fields

$$\mathbf{n} = (1, 0, 0), \qquad 0 \leq z \leq d, \tag{3.102}$$

gives the uniform equilibrium solution displayed in Fig. 3.6(a). Now suppose the nematic has $\chi_a > 0$ and let a uniform magnetic field **H** be applied perpendicular to the plates so that

$$\mathbf{H} = H(0, 0, 1), \qquad H = |\mathbf{H}|. \tag{3.103}$$

It is natural to seek solutions for the director of the form

$$\mathbf{n} = (\cos\theta(z), 0, \sin\theta(z)), \tag{3.104}$$

given that **n** is expected to reorient itself in an attempt to be parallel to **H** along the direction of the z-axis because $\chi_a > 0$ (this is a special case of the form proposed in equation (3.15) with $\phi \equiv 0$). It is assumed that any reorientation of **n** when it occurs will be uniform in the x- and y-directions. The boundary conditions (3.101) become

$$\theta(0) = \theta(d) = 0. \tag{3.105}$$

Inserting **n** and **H** into the Frank elastic energy (2.49) and magnetic energy (2.107) gives, with $' = d/dz$ as before,

$$w_F = \tfrac{1}{2}\left[K_1 \cos^2\theta + K_3 \sin^2\theta\right](\theta')^2, \tag{3.106}$$

$$w_{mag} = -\tfrac{1}{2}\chi_a H^2 \sin^2\theta, \tag{3.107}$$

and hence the total energy per unit area of the bounding plates over a sample volume V is

$$W = \int_V (w_F + w_{mag})\,dV \tag{3.108}$$

$$= \tfrac{1}{2}\int_0^d \left\{\left[K_1\cos^2\theta + K_3\sin^2\theta\right](\theta')^2 - \chi_a H^2 \sin^2\theta\right\}dz. \tag{3.109}$$

The relevant equilibrium equation is obtained from equation (2.221) and is

$$\frac{d}{dz}\left(\frac{\partial \widehat{w}}{\partial \theta'}\right) - \frac{\partial \widehat{w}}{\partial \theta} = 0, \tag{3.110}$$

where $\widehat{w} = w_F + w_{mag}$ is the integrand appearing in (3.108) and (3.109). Hence, for the problem considered here, the equilibrium equation for the director $\mathbf{n}$ is

$$[K_1 \cos^2 \theta + K_3 \sin^2 \theta]\theta'' + (K_3 - K_1)(\theta')^2 \sin\theta \cos\theta + \chi_a H^2 \sin\theta \cos\theta = 0. \tag{3.111}$$

(If K_1 is set equal to K_3 then this equation essentially reduces to the form considered by Freedericksz and Zolina [93] and Zocher [286, 287].) One obvious solution to the above equilibrium equation satisfying the boundary conditions (3.105) is

$$\theta \equiv 0, \tag{3.112}$$

giving the solution mentioned above at equations (3.101) and (3.102); this solution is clearly valid for *all* values of H and we shall call it the undistorted solution to the present problem. However, from the anticipated symmetry of any non-zero solution, we can also expect symmetric solutions of the form

$$\theta(z) = \theta(d - z), \qquad 0 \le z \le d, \tag{3.113}$$

satisfying

$$\theta'(\tfrac{d}{2}) = 0, \quad \theta(\tfrac{d}{2}) = \theta_m > 0. \tag{3.114}$$

The constant θ_m represents the maximum angle through which the director 'distorts' and, without loss of generality, it can be regarded as being positive ($\theta_m < 0$ gives the energetically equivalent scenario consisting of the mirror image in the yz-plane of the solution for $\theta_m > 0$ shown in Fig. 3.6). Multiplying the equilibrium equation (3.111) throughout by θ' and integrating gives the first integral

$$\left[K_1 \cos^2 \theta + K_3 \sin^2 \theta\right](\theta')^2 + \chi_a H^2 \sin^2 \theta = c_1, \qquad c_1 \text{ a constant.} \tag{3.115}$$

Applying the conditions (3.114) to (3.115) evaluates c_1 to give

$$\left[K_1 \cos^2 \theta + K_3 \sin^2 \theta\right](\theta')^2 = \chi_a H^2 \left[\sin^2 \theta_m - \sin^2 \theta\right], \qquad 0 \le \theta \le \theta_m, \tag{3.116}$$

leading to

$$\left[K_1 \cos^2 \theta + K_3 \sin^2 \theta\right]^{\frac{1}{2}} \frac{d\theta}{dz} = \pm\sqrt{\chi_a} H \left[\sin^2 \theta_m - \sin^2 \theta\right]^{\frac{1}{2}}. \tag{3.117}$$

On the interval $0 \le z \le \frac{d}{2}$, θ is expected to increase monotonically from zero to θ_m and so the plus sign is adopted in (3.117) which then integrates to reveal the solution

$$\sqrt{\chi_a} H z = \int_0^\theta \frac{[K_1 \cos^2 \widehat{\theta} + K_3 \sin^2 \widehat{\theta}]^{\frac{1}{2}}}{[\sin^2 \theta_m - \sin^2 \widehat{\theta}]^{\frac{1}{2}}} \, d\widehat{\theta}, \tag{3.118}$$

which defines θ implicitly as a function of z with, of course, $\theta(0) = 0$ to match the boundary condition stated in equation (3.105). We shall refer to this solution as the distorted solution. The condition (3.114)$_2$ leads to the requirement

$$\sqrt{\chi_a} H \frac{d}{2} = \int_0^{\theta_m} \frac{[K_1 \cos^2 \widehat{\theta} + K_3 \sin^2 \widehat{\theta}]^{\frac{1}{2}}}{[\sin^2 \theta_m - \sin^2 \widehat{\theta}]^{\frac{1}{2}}} \, d\widehat{\theta}, \tag{3.119}$$

which provides a relationship between θ_m, d and H that must be fulfilled; for a fixed depth d, it provides θ_m as a function of H. The solution for $\frac{d}{2} \leq z \leq d$ is, by (3.113), $\theta(d-z)$, obtained from (3.118) by simply replacing z by $d-z$. Thus a complete solution for θ is furnished by (3.118) provided the condition (3.119) is satisfied. The critical threshold will now be obtained in order to facilitate a further investigation of this distorted solution.

Use of the substitution (cf. Zocher [287])

$$\sin \widehat{\theta} = \sin \theta_m \sin \lambda, \tag{3.120}$$

which obeys the rule

$$\frac{d\widehat{\theta}}{d\lambda} = \frac{\sin \theta_m \cos \lambda}{\sqrt{1 - \sin^2 \theta_m \sin^2 \lambda}}, \tag{3.121}$$

allows the solution (3.118) and the requirement (3.119) to be reformulated in the convenient forms, respectively,

$$\sqrt{\chi_a} H z = \int_0^\phi G(\theta_m, \lambda) d\lambda, \qquad \phi = \sin^{-1}\left(\frac{\sin \theta}{\sin \theta_m}\right), \tag{3.122}$$

$$\sqrt{\chi_a} H \frac{d}{2} = \int_0^{\frac{\pi}{2}} G(\theta_m, \lambda) \, d\lambda, \tag{3.123}$$

where $G(\theta_m, \lambda)$ is the integrand defined by

$$G(\theta_m, \lambda) = \left[\frac{K_1 + (K_3 - K_1)\sin^2 \theta_m \sin^2 \lambda}{1 - \sin^2 \theta_m \sin^2 \lambda}\right]^{\frac{1}{2}}. \tag{3.124}$$

It is seen that $G(\theta_m, \lambda)$ is a monotonic increasing function of θ_m for $0 < \theta_m < \frac{\pi}{2}$ when $K_3 > 0$, because

$$\frac{\partial}{\partial \theta_m} G(\theta_m, \lambda) = \frac{1}{2}\left[\frac{K_1 + (K_3 - K_1)\sin^2 \theta_m \sin^2 \lambda}{1 - \sin^2 \theta_m \sin^2 \lambda}\right]^{-\frac{1}{2}} \frac{K_3 \sin^2 \lambda \sin(2\theta_m)}{[1 - \sin^2 \theta_m \sin^2 \lambda]^2} > 0, \tag{3.125}$$

for $0 < \theta_m < \frac{\pi}{2}$ and $0 < \lambda < \frac{\pi}{2}$. A straightforward application of the well known dominated convergence theorem allows the limit as $\theta_m \to 0$ to be taken inside the integral in (3.123) to find that the reformulated requirement (3.123) and solution (3.122) first become available at a critical field strength H_c given by

$$H_c = \frac{\pi}{d}\sqrt{\frac{K_1}{\chi_a}}. \tag{3.126}$$

Further, from (3.123) and (3.124),

$$\theta_m \to \frac{\pi}{2} \quad \text{as} \quad H \to \infty. \tag{3.127}$$

This means, for example, that for a fixed depth d and a given field magnitude $H > H_c$, the corresponding value of $\theta_m > 0$ can be determined from the relation

(3.123). This value of θ_m is then inserted into the solution (3.122) to obtain the distorted solution θ as a function of z: this reformulated solution is often more suitable for numerical calculations than the original solution (3.118). Recall that at $H = H_c$, $\theta_m = 0$ and the solution coincides with the undistorted solution.

There are therefore two solutions available for $H > H_c$ satisfying the boundary conditions (3.105), namely, the undistorted solution $\theta \equiv 0$ from (3.112) and the distorted solution $\theta(z)$ provided by (3.118) and (3.119), or its equivalent form given via (3.122) and (3.123). To verify that H_c is indeed the critical threshold and that a Freedericksz transition occurs, we need to check that the distorted solution is energetically favoured over the undistorted solution for $H > H_c$; the only solution available for $0 \le H \le H_c$ is the undistorted solution. For $H > H_c$, consider the difference in energies per unit area of the plates

$$\Delta W = W(\theta) - W(0), \tag{3.128}$$

where θ is the distorted solution and 0 is the undistorted solution. From the energy (3.109), the result for $\frac{d\theta}{dz}$ (with the plus sign) in (3.117), and the symmetry in θ from (3.113), we have

$$\begin{aligned}
\Delta W &= \tfrac{1}{2}\int_0^d \left\{ \left[K_1 \cos^2 \theta + K_3 \sin^2 \theta\right](\theta')^2 - \chi_a H^2 \sin^2 \theta \right\} dz \\
&= \chi_a H^2 \int_0^{\frac{d}{2}} (\sin^2 \theta_m - 2\sin^2 \theta) dz \\
&= \chi_a H^2 \int_0^{\theta_m} (\sin^2 \theta_m - 2\sin^2 \theta)\frac{dz}{d\theta} d\theta \\
&= \sqrt{\chi_a} H \int_0^{\theta_m} (\sin^2 \theta_m - 2\sin^2 \theta)\frac{[K_1 \cos^2 \theta + K_3 \sin^2 \theta]^{\frac{1}{2}}}{[\sin^2 \theta_m - \sin^2 \theta]^{\frac{1}{2}}} d\theta.
\end{aligned} \tag{3.129}$$

Utilising the substitution (3.120) (with $\widehat{\theta}$ replaced by θ) and the rule (3.121) in the above integral then gives

$$\Delta W = \sqrt{\chi_a} H \sin^2 \theta_m \int_0^{\frac{\pi}{2}} \cos(2\lambda) G(\theta_m, \lambda)\, d\lambda, \tag{3.130}$$

with $G(\theta_m, \lambda)$ as defined above at equation (3.124). From the symmetry of G in θ_m and λ, a similar computation to that given for equation (3.125) shows, when $K_3 > 0$,

$$\frac{\partial}{\partial \lambda} G(\theta_m, \lambda) > 0 \qquad \text{for}\ \ 0 < \lambda < \tfrac{\pi}{2}\ \ \text{and}\ \ 0 < \theta_m < \tfrac{\pi}{2}. \tag{3.131}$$

Thus, integrating (3.130) by parts demonstrates that

$$\Delta W = -\tfrac{1}{2}\sqrt{\chi_a} H \sin^2 \theta_m \int_0^{\frac{\pi}{2}} \sin(2\lambda)\frac{\partial}{\partial \lambda} G(\theta_m, \lambda)\, d\lambda < 0, \tag{3.132}$$

this last inequality being evident from the behaviour of the integrand on the interval $0 < \lambda < \tfrac{\pi}{2}$. Hence

$$\Delta W < 0 \qquad \text{for}\ \ \ H > H_c, \tag{3.133}$$

indicating that the distorted solution is energetically preferred for $H > H_c$. There is therefore a Freedericksz transition and the critical threshold is given by H_c defined by the value stated at equation (3.126). For $0 \le H \le H_c$ only the undistorted zero solution is available while for $H > H_c$ the distorted solution $\theta(z)$ provided via equations (3.122) and (3.123) is energetically preferred to the undistorted solution, as shown schematically in Fig. 3.6. Experimentally, the detection of H_c in (3.126) allows the measurement of the splay elastic constant K_1 if χ_a is known.

Case (ii): Twist Geometry

The second case to consider is that when the initial director alignment and the magnetic field are parallel to the glass plates but mutually orthogonal to each other as shown in Fig. 3.7(a). The initial uniform alignment is given by $\phi \equiv 0$ when solutions in this case take the form

$$\mathbf{n} = (\cos\phi(z), \sin\phi(z), 0), \tag{3.134}$$

with ϕ satisfying the boundary conditions and the symmetry and maximum distor-

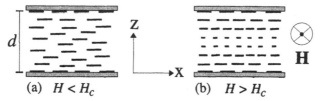

Figure 3.7: The classical Freedericksz transition which measures the twist elastic constant K_2 via the critical field strength $H_c = \frac{\pi}{d}\sqrt{\frac{K_2}{\chi_a}}$, assuming that $\chi_a > 0$. For $H < H_c$ the original uniform sample alignment parallel to the plates remains intact. As H increases through the critical field strength H_c the sample undergoes a Freedericksz transition where, for $H > H_c$, the director orientation begins to twist out of the page in response to the influence of the applied magnetic field $\mathbf{H}$, the director always lying parallel to the boundary plates.

tion assumptions assumed previously for θ in equations (3.105), (3.113) and (3.114). The magnetic field in this case is

$$\mathbf{H} = H(0, 1, 0). \tag{3.135}$$

The equilibrium equation for this 'twist geometry' can be derived in a similar fashion to that above for the 'splay geometry' and is found to be

$$K_2 \frac{d^2\phi}{dz^2} + \chi_a H^2 \sin\phi \cos\phi = 0, \tag{3.136}$$

with the boundary conditions

$$\phi(0) = \phi(d) = 0. \tag{3.137}$$

Symmetric solutions of the form

$$\phi(z) = \phi(d - z), \qquad 0 \le z \le d, \tag{3.138}$$

satisfying

$$\phi'(\tfrac{d}{2}) = 0, \quad \phi(\tfrac{d}{2}) = \phi_m > 0, \tag{3.139}$$

can be found as before and an analysis identical to that for the previous case in the splay geometry can be carried out again for this case: the results can be obtained by simply setting $K_1 = K_3 \equiv K_2$ in the working set out above with, of course, ϕ taking the rôle of θ. The resulting critical threshold is

$$H_c = \frac{\pi}{d}\sqrt{\frac{K_2}{\chi_a}}, \tag{3.140}$$

which allows the experimental measurement of the twist elastic constant K_2 if χ_a is known.

Case (iii): Bend Geometry

The third case occurs when the magnetic field is parallel to the glass plates while the initial uniform alignment of the sample is perpendicular to the plates, corresponding to one of the simple general solutions giving the homeotropic alignment shown in Fig. 3.8. Solutions take the form

$$\mathbf{n} = (\sin\theta(z), 0, \cos\theta(z)), \tag{3.141}$$

with the initial alignment corresponding to $\theta \equiv 0$, as before, and

$$\mathbf{H} = H(1, 0, 0). \tag{3.142}$$

The analysis of this case is identical to the first case except that the rôles of K_1 and K_3 are interchanged, the critical field being given by

$$H_c = \frac{\pi}{d}\sqrt{\frac{K_3}{\chi_a}}, \tag{3.143}$$

which allows the measurement of the bend elastic constant K_3 if χ_a is known.

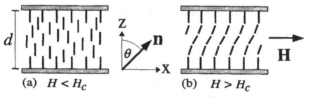

(a) $H < H_c$ (b) $H > H_c$

Figure 3.8: The classical Freedericksz transition which measures the bend elastic constant K_3 via the critical field strength $H_c = \frac{\pi}{d}\sqrt{\frac{K_3}{\chi_a}}$, assuming that $\chi_a > 0$. The original uniform sample alignment perpendicular to the plates remains intact for $H < H_c$. As H increases through the critical field strength H_c, the sample undergoes a Freedericksz transition where, for $H > H_c$, the director begins to change its orientation within the xz-plane in response to the applied magnetic field $\mathbf{H}$.

The above three critical fields H_c in equations (3.126), (3.140) and (3.143) give the classical results for measurements of the three Frank elastic constants K_1, K_2 and K_3. Notice that K_1 and K_3 both contribute to the post-threshold behaviour in the splay and bend geometries. The constant K_4 is not amenable to these set-ups.

One-constant Approximation

In order to expand upon the qualitative features of the favoured distorted solution above the critical threshold H_c it is convenient to simplify the presentation by making the one-constant approximation $K_1 = K_2 = K_3 = K$ and normalise the magnetic field magnitude to $\overline{H}$, defined by

$$\overline{H} = \frac{H}{H_c}, \qquad H_c = \frac{\pi}{d}\sqrt{\frac{K}{\chi_a}}. \tag{3.144}$$

In this context the requirement (3.123) linking d, H and θ_m reduces to, in all three cases above,

$$\sqrt{\chi_a}H\frac{d}{2} = \sqrt{K}\int_0^{\frac{\pi}{2}} \frac{d\lambda}{\sqrt{1 - \sin^2\theta_m\sin^2\lambda}}, \tag{3.145}$$

(cf. Zocher [287]) from which it easily follows that

$$\overline{H} = \frac{2}{\pi}\int_0^{\frac{\pi}{2}} \frac{d\lambda}{\sqrt{1 - \sin^2\theta_m\sin^2\lambda}} = \frac{2}{\pi}K(\sin\theta_m), \tag{3.146}$$

where $K(k)$, $k = \sin\theta_m$, is the complete elliptic integral of the first kind of modulus k [116]. This gives θ_m as a function of the normalised field $\overline{H}$. Figure 3.9 shows how θ_m varies as $\overline{H}$ increases; it is clear that θ_m attains values very close to $\frac{\pi}{2}$ once H is

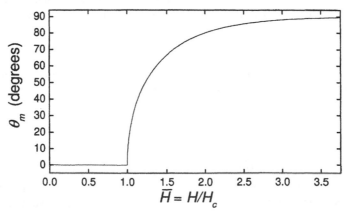

Figure 3.9: Graph of the dependence of the maximum distortion angle θ_m upon the dimensionless magnetic field $\overline{H}$ for the one-constant approximation K discussed in the text. $H_c = \frac{\pi}{d}\sqrt{\frac{K}{\chi_a}}$ with $\chi_a > 0$.

three or four times the value of H_c. This observation can be verified mathematically since for values of θ_m close to $\frac{\pi}{2}$ the asymptotic formula from [116, 8.113.3] can be used to find that

$$\overline{H} \sim \frac{2}{\pi} \ln \left[\frac{4}{\sqrt{1 - \sin^2 \theta_m}} \right], \tag{3.147}$$

which shows, using some other basic approximations, that

$$\sin \theta_m \sim 1 - 8 \exp[-\pi H / H_c], \tag{3.148}$$

which, upon using trigonometric approximations, gives

$$\theta_m \sim \frac{\pi}{2} - 4 \exp\left[-\frac{\pi}{2} \frac{H}{H_c} \right], \tag{3.149}$$

indicating that θ_m is close to $\frac{\pi}{2}$ when H is a small integral multiple of H_c. (The relation (3.149) can be verified by taking sines of both sides and expanding by a series.)

For a given $\overline{H}$ we can determine θ_m from equation (3.146) or from Fig. 3.9. These values for $\overline{H}$ and θ_m are then inserted into the solution given by (3.122) to determine $\theta(z)$. In the one-constant approximation being considered here, the solution (3.122) is (cf. Zocher [287])

$$\sqrt{\chi_a} H z = \sqrt{K} \int_0^\phi \frac{d\lambda}{\sqrt{1 - \sin^2 \theta_m \sin^2 \lambda}}, \tag{3.150}$$

where, as before, we have set

$$\phi = \sin^{-1} \left(\frac{\sin \theta}{\sin \theta_m} \right), \tag{3.151}$$

for convenience. This solution to the one-constant approximation problem can now be put into a non-dimensional form by introducing the rescaled variable

$$\overline{z} = \frac{z}{d}. \tag{3.152}$$

Employing the dimensionless variables $\overline{H}$ and $\overline{z}$, the solution θ can be formulated in the one-constant approximation by the solution

$$\begin{aligned}
\overline{z} &= \frac{1}{\overline{H}\pi} \int_0^\phi \frac{d\lambda}{\sqrt{1 - \sin^2 \theta_m \sin^2 \lambda}} \\
&= \frac{1}{\overline{H}\pi} F(\phi, \sin \theta_m), \qquad 0 \leq \overline{z} \leq \tfrac{1}{2},
\end{aligned} \tag{3.153}$$

where $F(\phi, k)$, $k = \sin \theta_m$, is the elliptic integral of the first kind [116] having modulus k. The solution for $\frac{1}{2} \leq \overline{z} \leq 1$ is obtained from (3.153) via the symmetry requirement (3.113): simply replace $\overline{z}$ by $1 - \overline{z}$. Clearly, for $\theta = \theta_m$ and $\overline{z} = \frac{1}{2}$, $\phi = \frac{\pi}{2}$ and we recover the condition (3.146). Figure 3.10 displays how the solution $\theta(\overline{z})$ changes with increasing field magnitude $H > H_c$, that is, $\overline{H} > 1$.

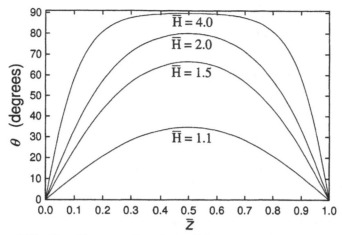

Figure 3.10: Plots illustrating the solution $\theta(\bar{z})$, $0 \le \bar{z} \le 1$, for different values of the dimensionless magnetic field strength $\overline{H} = H/H_c$ using the one-constant approximation K discussed in the text. $H_c = \frac{\pi}{d}\sqrt{\frac{K}{\chi_a}}$ with $\chi_a > 0$.

Freedericksz Threshold via Linearisation

It is quite common in the literature to obtain the critical Freedericksz threshold more swiftly by linearising the governing equilibrium equation and inserting an appropriate sinusoidal ansatz for the solution. Since the orientation angle of the director, as discussed above, will be relatively small when H is just above H_c, this approach is acceptable for identifying H_c. However, the disadvantage is that only H_c may be determined: in general, the post-Freedericksz threshold behaviour for $H > H_c$ cannot be deduced from the solution found by linearisation, in contrast to the solution obtained to the nonlinear equation, such as that displayed in Fig. 3.10 for the one-constant approximation.

As an example, take the Freedericksz transition involving the splay constant K_1, as pictured in Fig. 3.6. The relevant governing equilibrium equation (3.111) can be linearised around $\theta = 0$ to give

$$K_1\theta'' + \chi_a H^2 \theta = 0, \tag{3.154}$$

with the accompanying boundary conditions (3.105) and requirements (3.114) given by

$$\theta(0) = \theta(d) = 0, \qquad \theta'(\tfrac{d}{2}) = 0, \qquad \theta(\tfrac{d}{2}) = \theta_m > 0. \tag{3.155}$$

By inserting the ansatz

$$\theta(z) = \theta_m \sin\left(\frac{\pi}{d}z\right), \tag{3.156}$$

which satisfies all the conditions in (3.155), into equation (3.154), it is seen that $\theta(z)$ is a solution provided $H = H_c = \frac{\pi}{d}\sqrt{\frac{K_1}{\chi_a}}$. We therefore recover the result for the critical field strength stated at equation (3.126).

This method can be easily adapted for other geometrical set-ups by introducing suitably modified forms for the ansatz. More generally, equation (3.154) subject to the conditions (3.155) can be solved by elementary methods to find that there are solutions of the form

$$\theta(z) = \theta_m \sin\left((4n+1)\frac{\pi}{d}z\right), \qquad n = 0, 1, 2, \dots, \tag{3.157}$$

provided

$$H = (4n+1)\frac{\pi}{d}\sqrt{\frac{K_1}{\chi_a}}, \qquad n = 0, 1, 2, \dots, \tag{3.158}$$

and hence the least positive value of H for which a non-zero solution is available occurs at $n = 0$, giving $H = H_c$ as before.

3.4.2   Pretilt at the Boundaries

In the above transitions it is possible to have different domains forming after the Freedericksz transitions because two equally possible energetically favourable distortions can occur, depending on whether $\theta_m < 0$ or $\theta_m > 0$. This can be avoided by imposing a small tilt of the director on the boundaries. Consider the case first discussed in the previous Section on page 73, but with the director strongly anchored at the surfaces with a pretilt angle θ_0 as shown in Fig. 3.11. The magnetic anisotropy χ_a is again assumed positive. The solution for the director profile follows equations (3.103) to (3.111) precisely, except that the boundary conditions (3.105) are replaced by

$$\theta(0) = \theta(d) = \theta_0 > 0. \tag{3.159}$$

For $H = 0$, the equilibrium equation (3.111) has the obvious constant solution

$$\theta(z) = \theta_0, \qquad 0 \le z \le d, \tag{3.160}$$

giving the initial profile depicted in Fig. 3.11(a). For $H > 0$, the search for symmetric solutions follows equations (3.113) to (3.119) except that now θ must be

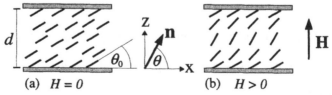

(a) $H = 0$ (b) $H > 0$

Figure 3.11: The Freedericksz transition when the director is strongly anchored on the boundaries with a fixed pretilt of θ_0 as shown in (a). There is a Freedericksz transition at $H_c = 0$ but, for small θ_0, the difference between θ_0 and the distortion angle θ_m remains relatively small until $H_c \approx H_{c_2} = \frac{\pi}{d}\sqrt{\frac{K_1}{\chi_a}}$, as demonstrated in equations (3.171) to (3.175). See also Fig. 3.12.

restricted to values in the range $\theta_0 \leq \theta \leq \theta_m$, rather than $0 \leq \theta \leq \theta_m$. The analogues of the solution (3.118) and the requirement (3.119) become, respectively,

$$\sqrt{\chi_a} H z = \int_{\theta_0}^{\theta} \frac{[K_1 \cos^2 \widehat{\theta} + K_3 \sin^2 \widehat{\theta}]^{\frac{1}{2}}}{[\sin^2 \theta_m - \sin^2 \widehat{\theta}]^{\frac{1}{2}}} \, d\widehat{\theta}, \qquad 0 \leq z \leq \tfrac{d}{2}, \qquad (3.161)$$

$$\sqrt{\chi_a} H \frac{d}{2} = \int_{\theta_0}^{\theta_m} \frac{[K_1 \cos^2 \widehat{\theta} + K_3 \sin^2 \widehat{\theta}]^{\frac{1}{2}}}{[\sin^2 \theta_m - \sin^2 \widehat{\theta}]^{\frac{1}{2}}} \, d\widehat{\theta}. \qquad (3.162)$$

Using the transformation (3.120) again shows that the requirement (3.162) can be rewritten as

$$\sqrt{\chi_a} H \frac{d}{2} = \int_{\lambda_0}^{\frac{\pi}{2}} G(\theta_m, \lambda) \, d\lambda, \qquad (3.163)$$

$$\lambda_0 = \sin^{-1}\left(\frac{\sin \theta_0}{\sin \theta_m}\right), \qquad (3.164)$$

with $G(\theta_m, \lambda)$ defined as before by equation (3.124). The function G is a monotonic increasing function of θ_m because it satisfies

$$\frac{\partial}{\partial \theta_m} G(\theta_m, \lambda) > 0, \quad \text{for} \quad \theta_0 < \theta_m < \tfrac{\pi}{2}, \quad \lambda_0 < \lambda < \tfrac{\pi}{2}, \qquad (3.165)$$

by the inequality derived in (3.125) whenever $K_3 > 0$. By symmetry, it is also an increasing function of λ and therefore

$$G(\theta_m, \lambda) \leq G(\theta_m, \tfrac{\pi}{2}) \leq \sqrt{K_1 + K_3} \sec \theta_m. \qquad (3.166)$$

For $\theta \leq \theta_m < \tfrac{\pi}{2}$,

$$\lambda_0 \to \tfrac{\pi}{2} \qquad \text{as} \qquad \theta_m \to \theta_0. \qquad (3.167)$$

It is then clear from (3.163) that H must tend to zero as $\theta_m \to \theta_0$ because

$$\int_{\lambda_0}^{\frac{\pi}{2}} G(\theta_m, \lambda) \, d\lambda \leq \sqrt{K_1 + K_3} \sec \theta_m \int_{\lambda_0}^{\frac{\pi}{2}} d\lambda \to 0 \qquad \text{as} \qquad \theta_m \to \theta_0. \qquad (3.168)$$

Therefore the distorted solution (3.161) first becomes available at the critical field strength

$$H_c = 0. \qquad (3.169)$$

For $H > 0$ there is in general no constant solution to the equilibrium equation (3.111) satisfying the pretilt boundary conditions (3.159) for small θ_0. Nevertheless, the energy of the distorted solution satisfies

$$W(\theta) < 0 \qquad \text{for} \quad H > 0, \qquad (3.170)$$

by an argument analogous to that used to obtain (3.133), with the lower limit of integration in (3.130) replaced by λ_0, as defined above. There is therefore a Freedericksz transition and the critical field strength is given by $H_c = 0$.

However, this distorted solution remains relatively small for $0 < \theta_0 \ll 1$ until the field magnitude reaches another critical magnitude H_{c_2}, as we now demonstrate. Consider

$$0 < \theta_0 \ll 1, \qquad \theta_m = \theta_0 + \epsilon, \quad |\epsilon| \ll 1. \tag{3.171}$$

Then, from (3.163) and (3.164),

$$\sqrt{\chi_a} H \frac{d}{2} \doteq \int_{\lambda_0}^{\frac{\pi}{2}} \sqrt{K_1} \, d\lambda = \sqrt{K_1} \left(\frac{\pi}{2} - \lambda_0 \right), \tag{3.172}$$

$$\sin \lambda_0 \doteq \frac{\theta_0}{\theta_0 + \epsilon} \tag{3.173}$$

Substituting (3.173) into (3.172) and rearranging shows that

$$\epsilon \doteq \theta_0 \left[\sec \left(\sqrt{\frac{\chi_a}{K_1}} H \frac{d}{2} \right) - 1 \right]. \tag{3.174}$$

The distorted solution θ for values of $\theta_0 \ll 1$ therefore remains relatively small until H increases close to

$$H_{c_2} = \frac{\pi}{d} \sqrt{\frac{K_1}{\chi_a}}, \tag{3.175}$$

which coincides with the critical field strength obtained earlier in (3.126). In other words, when the pretilt θ_0 is small, the effect of the Freedericksz transition will essentially not be noticed until H is close to H_{c_2}, which is the classical threshold obtained earlier in (3.126); the advantage for pretilt boundary conditions is that only one domain of reoriented liquid crystal ought to occur after the Freedericksz transition.

Figure 3.12 shows the dependency of θ_m upon θ_0 in the special case of $K = K_1 = K_3$ where, from equation (3.163), we have for the classical value of $H_c = H_{c_2}$

$$\overline{H} = \frac{H}{H_c} = \frac{2}{\pi} \int_{\lambda_0}^{\frac{\pi}{2}} \frac{1}{\sqrt{1 - \sin^2 \theta_m \sin^2 \lambda}} \, d\lambda. \tag{3.176}$$

Rapini and Papoular [227] used a graph similar to Fig. 3.12 to estimate the difference in measurements of the elastic constant at the critical field magnitude when it is first noticeably observed as $\overline{K}$ via $\overline{H}_c$ at $\theta_0 \neq 0$, and as K via H_c at $\theta_0 = 0$. They estimated that the difference between these observed critical magnetic fields near the classical value of H_c for $\theta_0 = 0$ and $\theta_0 \sim 2^\circ$ is roughly -10% by estimating

$$\frac{\Delta H_c}{H_c} = \frac{\overline{H}_c - H_c}{H_c} \doteq -0.1. \tag{3.177}$$

Therefore

$$\frac{\overline{K}}{K} \doteq 0.81, \tag{3.178}$$

and so, despite the observed threshold value $\overline{H}_c$ being close to H_c, the error which is introduced to the measurement of K may be around 20%. This can be interpreted as indicating that a small misalignment of the director at the boundary surface may noticeably affect the measurement of K.

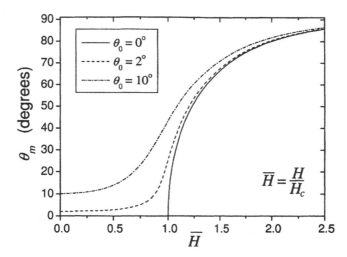

Figure 3.12: Graphs obtained via equation (3.176) showing the dependency of the maximum distortion angle θ_m upon the dimensionless magnetic field $\overline{H}$ when the prescribed pretilt θ_0 of the director at the boundaries takes the values $0°$, $2°$ and $10°$ and $K_1 = K_3 = K$. $H_c = \frac{\pi}{d}\sqrt{\frac{K}{\chi_a}}$ is the classical value of the critical threshold when $\theta_0 = 0$, given by equation (3.126), when $K = K_1$.

3.4.3 Tilted Fields

The Freedericksz transition for a tilted magnetic field is similar to the classical Freedericksz transitions. We restrict our attention to the case when $\chi_a > 0$ and $\mathbf{H}$ is applied across the sample at a fixed angle α, as shown in Fig. 3.13. The director and magnetic field now take the forms, respectively,

$$\mathbf{n} = (\cos\theta(z), 0, \sin\theta(z)), \tag{3.179}$$

$$\mathbf{H} = H(\cos\alpha, 0, \sin\alpha), \quad H = |\mathbf{H}|, \quad 0 < \alpha < \tfrac{\pi}{2}, \tag{3.180}$$

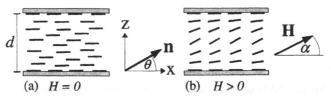

(a) $H = 0$ (b) $H > 0$

Figure 3.13: The Freedericksz transition when the director is strongly anchored parallel to the boundaries and the magnetic field is tilted to the boundaries by a fixed angle α as shown. The Freedericksz transition occurs at $H_c = 0$ but, for $\alpha \approx \frac{\pi}{2}$, the maximum distortion angle θ_m remains relatively small until $H_c \approx H_{c_2} = \frac{\pi}{d}\sqrt{\frac{K_1}{\chi_a}}$, as derived from equation (3.206).

where α is the given fixed acute angle made between the magnetic field and the x-axis. The boundary conditions are

$$\theta(0) = \theta(d) = 0. \tag{3.181}$$

The energy integrand consists of the sum of the Frank elastic energy and the magnetic energy given by equations (2.49) and (2.107). The total energy per unit area of the bounding plates over a sample of volume V is then given by

$$\begin{aligned} W &= \int_V (w_F + w_{mag})dV \\ &= \tfrac{1}{2}\int_0^d \left\{ [K_1 \cos^2\theta + K_3 \sin^2\theta]\,(\theta')^2 - \chi_a H^2 \cos^2(\alpha - \theta) \right\}dz. \end{aligned} \tag{3.182}$$

A similar calculation to that involved in deriving the equilibrium equation (3.111) for the first classical Freedericksz transition shows that the equilibrium equation for this example is

$$[K_1 \cos^2\theta + K_3 \sin^2\theta]\theta'' + (K_3 - K_1)(\theta')^2 \sin\theta\cos\theta$$
$$+ \chi_a H^2 \sin(\alpha - \theta)\cos(\alpha - \theta) = 0. \tag{3.183}$$

As before, this equation can be multiplied throughout by θ' and integrated to find

$$[K_1 \cos^2\theta + K_3 \sin^2\theta]\,(\theta')^2 + \chi_a H^2 \cos^2(\alpha - \theta) = c_1, \qquad c_1 \text{ a constant.} \tag{3.184}$$

Assuming a symmetric solution satisfying (3.113), that is,

$$\theta(z) = \theta(d - z), \quad 0 \le z \le d, \tag{3.185}$$

and supposing

$$\theta'(\tfrac{d}{2}) = 0, \quad \theta(\tfrac{d}{2}) = \theta_m > 0, \tag{3.186}$$

the constant c_1 can be evaluated in equation (3.184) to see that

$$[K_1 \cos^2\theta + K_3 \sin^2\theta]\,(\theta')^2 = \chi_a H^2 \left[\cos^2(\alpha - \theta_m) - \cos^2(\alpha - \theta)\right], \tag{3.187}$$

with θ obeying the restrictions

$$0 \le \theta \le \theta_m \le \alpha. \tag{3.188}$$

Notice that if $\theta = \theta_m = \alpha$, then the alignment of the director is parallel to the applied field, in which case the alignment represents the optimum preferred orientation, given that $\chi_a > 0$. Taking positive square roots and integrating (3.187) gives the solution

$$\sqrt{\chi_a}Hz = \int_0^\theta \frac{[K_1 \cos^2\widehat{\theta} + K_3 \sin^2\widehat{\theta}]^{\frac{1}{2}}}{[\cos^2(\alpha - \theta_m) - \cos^2(\alpha - \widehat{\theta})]^{\frac{1}{2}}}\,d\widehat{\theta}, \quad 0 \le z \le \tfrac{d}{2}, \tag{3.189}$$

with the solution for $\frac{d}{2} \leq z \leq d$ obtained via (3.185). The condition (3.186)$_2$ leads to the requirement

$$\sqrt{\chi_a}H\frac{d}{2} = \int_0^{\theta_m} \frac{[K_1 \cos^2 \widehat{\theta} + K_3 \sin^2 \widehat{\theta}]^{\frac{1}{2}}}{[\cos^2(\alpha - \theta_m) - \cos^2(\alpha - \widehat{\theta})]^{\frac{1}{2}}} \, d\widehat{\theta}, \tag{3.190}$$

which gives, similar to the classical Freedericksz transitions, a relationship between θ_m and H. Employing the substitution

$$\cos(\alpha - \widehat{\theta}) = \cos(\alpha - \theta_m) \sin \lambda, \tag{3.191}$$

which satisfies

$$\frac{d\widehat{\theta}}{d\lambda} = \frac{\cos(\alpha - \theta_m)\cos\lambda}{\sqrt{1 - \cos^2(\alpha - \theta_m)\sin^2\lambda}}, \tag{3.192}$$

the requirement (3.190) becomes

$$\sqrt{\chi_a}H\frac{d}{2} = \int_{\lambda_0}^{\frac{\pi}{2}} \frac{[K_1 \cos^2 \widehat{\theta} + K_3 \sin^2 \widehat{\theta}]^{\frac{1}{2}}}{[1 - \cos^2(\alpha - \theta_m)\sin^2\lambda]^{\frac{1}{2}}} \, d\lambda, \tag{3.193}$$

$$\lambda_0 = \sin^{-1}\left(\frac{\cos\alpha}{\cos(\alpha - \theta_m)}\right), \tag{3.194}$$

where $\widehat{\theta}$ appearing in the integrand is, of course, now a function of λ, which can be obtained via (3.191) if required. The integrand in (3.193) is clearly well behaved as a function of λ and for $0 \leq \theta_m < \alpha$ it is always continuous and bounded for $0 \leq \lambda \leq \frac{\pi}{2}$. It is further observed that

$$\lambda_0 \to \frac{\pi}{2} \quad \text{as} \quad \theta_m \to 0, \tag{3.195}$$

and therefore, by an argument similar to that involving (3.165) to (3.169) above, taking the limit as $\theta_m \to 0$ on the right-hand side of equation (3.193) shows that H must tend to zero and so

$$H_c = 0. \tag{3.196}$$

As in the pretilt boundary example in the previous Section, for $H > 0$ there is generally no constant solution to the equilibrium equation (3.183) which satisfies the boundary conditions (3.181) for $\alpha \neq 0$. However, it is straightforward to use the result in (3.187) and suitably adapt the method used in equations (3.129) to (3.133) to find that

$$W(\theta) = \sqrt{\chi_a}H\cos^2(\alpha - \theta_m)\int_{\lambda_0}^{\frac{\pi}{2}} \frac{\cos(2\lambda)\sqrt{f(\theta)}}{\sqrt{1 - \cos^2(\alpha - \theta_m)\sin^2\lambda}} \, d\lambda, \tag{3.197}$$

where, for convenience, we have set

$$f(\theta) = K_1 \cos^2 \theta + K_3 \sin^2 \theta. \tag{3.198}$$

Integrating (3.197) by parts, using the relations (3.194) and (3.191) (with $\widehat{\theta}$ replaced by θ), and observing that $\theta \to 0$ as $\lambda \to \lambda_0$, gives

$$W(\theta) = -\frac{1}{2}\sqrt{\chi_a}H\cos^2(\alpha - \theta_m)\int_{\lambda_0}^{\frac{\pi}{2}}\sin(2\lambda)\frac{\partial}{\partial\lambda}\left[\frac{f(\theta)}{\sin^2(\alpha - \theta)}\right]^{\frac{1}{2}}d\lambda$$

$$-\frac{1}{2}\sqrt{\chi_a K_1}H\mathrm{cosec}(\alpha)\cos^2(\alpha - \theta_m)\sin(2\lambda_0). \quad (3.199)$$

A simple calculation reveals that

$$\frac{\partial}{\partial\lambda}\left[\frac{f(\theta)}{\sin^2(\alpha - \theta)}\right]^{\frac{1}{2}} = [f(\theta)]^{-\frac{1}{2}}\frac{d\theta}{d\lambda}\mathrm{cosec}^2(\alpha - \theta)[K_1\cos\theta\cos\alpha + K_3\sin\theta\sin\alpha].$$

$$(3.200)$$

Given the supposition that $0 < \alpha < \frac{\pi}{2}$ and the restrictions in (3.188), it is clear from (3.192) and the form of the right-hand side in (3.200) that

$$\frac{\partial}{\partial\lambda}\left[\frac{f(\theta)}{\sin^2(\alpha - \theta)}\right]^{\frac{1}{2}} > 0 \qquad \text{for} \qquad \lambda_0 < \lambda < \frac{\pi}{2}. \quad (3.201)$$

It now follows from (3.199), (3.200) and (3.201) that

$$W(\theta) < 0 \qquad \text{for} \qquad H > 0, \quad (3.202)$$

and so $H_c = 0$ is indeed the critical threshold for a Freedericksz transition; this means, as in the pretilted boundary example, that the distorted solution is always available for any $H > 0$.

Nevertheless, it is worth noting that this distorted solution is always small whenever α is close to $\frac{\pi}{2}$, that is, whenever the magnetic field is applied nearly perpendicular to the boundaries so that the geometrical set-up in Fig. 3.13 becomes similar to Fig. 3.6. Consider setting

$$\alpha = \frac{\pi}{2} - \epsilon, \qquad 0 < \epsilon \ll 1, \quad (3.203)$$

and supposing $0 < \theta_m \ll 1$ (so that $\theta \ll 1$ also). Then from (3.193) and (3.194) we have

$$\sqrt{\chi_a}H\frac{d}{2} \doteq \sqrt{K_1}\left(\frac{\pi}{2} - \lambda_0\right), \quad (3.204)$$

$$\sin\lambda_0 \doteq \frac{\epsilon}{\epsilon + \theta_m}. \quad (3.205)$$

Substituting (3.204) into (3.205) gives

$$\theta_m \doteq \epsilon\left[\sec\left(\sqrt{\frac{\chi_a}{K_1}}H\frac{d}{2}\right) - 1\right]. \quad (3.206)$$

The interpretation is that the distorted solution, always available for $H > 0$, remains relatively small for $\alpha \approx \frac{\pi}{2}$ until H is close to $H_{c_2} = \frac{\pi}{d}\sqrt{\frac{K_1}{\chi_a}}$, which, as for the pretilt boundary case, coincides with the classical critical field strength H_c in equation (3.126) for the set-up in Fig. 3.6.

3.5 Electric Field Effects

Freedericksz transitions also occur under the influence of electric fields. All of the critical threshold magnetic field magnitudes derived in the previous Section have their analogues for electric fields when the electric energy is approximated by that given by, for example, equation (2.87). The simple identification

$$\chi_a H^2 \quad \longleftrightarrow \quad \epsilon_0 \epsilon_a E^2, \tag{3.207}$$

where $E = |\mathbf{E}|$ is the magnitude of the electric field, yields the critical fields (for $\epsilon_a > 0$)

$$E_c = \frac{\pi}{d}\sqrt{\frac{K_1}{\epsilon_0 \epsilon_a}}, \qquad E_c = \frac{\pi}{d}\sqrt{\frac{K_2}{\epsilon_0 \epsilon_a}}, \qquad E_c = \frac{\pi}{d}\sqrt{\frac{K_3}{\epsilon_0 \epsilon_a}}, \tag{3.208}$$

for the analogous problems described above for magnetic fields in Figs. 3.6, 3.7 and 3.8, respectively; the critical voltages are of the form $V_c = E_c d$ for a sample of depth d. For typical values of the constants (in SI units), we can approximate the values given by equation (2.61) and choose to set the one-constant approximation K, ϵ_0 and ϵ_a as

$$K = 4 \times 10^{-12} \text{ N}, \qquad \epsilon_0 = 8.854 \times 10^{-12} \text{ F m}^{-1}, \qquad \epsilon_a = 1, \tag{3.209}$$

(cf. Section 2.3.1 on page 27). This gives a critical voltage V_c of around 2.1 V. The Figs. 3.9 and 3.10 can be repeated with $\overline{E} = E/E_c$ if desired. Of course, the analogous solutions obtained for the orientation of the director are not expected to be as accurate in modelling electrically induced Freedericksz transitions as those obtained for magnetic fields because it is known that the electric field is influenced by the liquid crystal; magnetic fields are generally considered not to be influenced by the liquid crystal. Nevertheless, critical thresholds such as those in (3.208) serve as good guides to the onset of electrically induced Freedericksz transitions in many circumstances.

In an attempt to model more accurately the electric field effects, especially when $E > E_c$, Deuling [63] took the reduced Maxwell equations (2.88) into account, rather than use the approximation to these equations given by (2.89). We now follow the work of Deuling, noting that there are some misprints in [63].

Consider the geometry of Fig. 3.14 where the director is strongly anchored on the boundaries parallel to the x-axis and the electric field is applied across the boundary plates as indicated, with the electric field now dependent upon z. Choose

$$\mathbf{n} = (\cos\theta(z), 0, \sin\theta(z)), \tag{3.210}$$
$$\mathbf{E} = (0, 0, E(z)), \tag{3.211}$$

with the boundary conditions

$$\theta(0) = \theta(d) = 0. \tag{3.212}$$

The elastic energy integrand w_F is again given by (3.106), while for the electric energy we adopt w_{elec} in equation (2.86). This leads to the total energy per unit

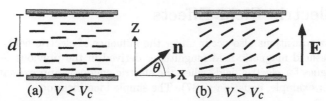

(a) $V < V_c$ (b) $V > V_c$

Figure 3.14: The Freedericksz transition when the director is strongly anchored parallel to the boundaries and the electric field is applied as shown. The Freedericksz transition threshold occurs at $V_c = E_c d = \pi\sqrt{\frac{K_1}{\epsilon_0 \epsilon_a}}$ when $\epsilon_a > 0$. For the post-threshold solution see Fig. 3.15, which differs from the analogue with the magnetic field case due to the interaction of the electric field and the liquid crystal fluid for $V > V_c$.

area of the bounding plates

$$W = \tfrac{1}{2}\int_0^d \left\{ \left[K_1 \cos^2\theta + K_3 \sin^2\theta\right](\theta')^2 - \mathbf{D}\cdot\mathbf{E}\right\} dz, \qquad (3.213)$$

with, using the result for $\mathbf{D}$ given by equation (2.84),

$$\mathbf{D} = \epsilon_0\epsilon_\perp\mathbf{E} + \epsilon_0\epsilon_a(\mathbf{n}\cdot\mathbf{E})\mathbf{n}, \qquad (3.214)$$

and $\mathbf{E}$ satisfying Maxwell's equations

$$\nabla\cdot\mathbf{D} = 0, \qquad \nabla\times\mathbf{E} = \mathbf{0}. \qquad (3.215)$$

Writing $\mathbf{D} = (D_1, D_2, D_3)$, the requirement $(3.215)_1$ shows that

$$D_3 = D, \qquad D \text{ a constant}, \qquad (3.216)$$

because it has been assumed that $\mathbf{n}$ and $\mathbf{E}$ can only have a dependence upon z; the requirement $(3.215)_2$ is automatically satisfied by the form of $\mathbf{E}$. It then follows easily from equations (3.210), (3.211) and (3.214) that

$$D = \epsilon_0(\epsilon_\perp + \epsilon_a \sin^2\theta)E(z). \qquad (3.217)$$

The voltage V is given by

$$V = \int_0^d E(z)dz = D\epsilon_0^{-1}\int_0^d (\epsilon_\perp + \epsilon_a \sin^2\theta)^{-1}dz, \qquad (3.218)$$

which allows D to be written as a functional of $\theta(z)$

$$D = \epsilon_0 V \left\{ \int_0^d (\epsilon_\perp + \epsilon_a \sin^2\theta)^{-1}dz \right\}^{-1}. \qquad (3.219)$$

It is worth remarking here that D is a constant, having its value fixed by a *given* solution θ. On the other hand, V is always a prescribed fixed constant independent of the actual solution θ. It follows that

$$
\begin{aligned}
\frac{1}{2}\int_0^d \mathbf{D}\cdot\mathbf{E}\,dz &= \frac{1}{2}D\int_0^d E(z)\,dz \\
&= \frac{1}{2}DV \\
&= \frac{1}{2}\epsilon_0 V^2 \left\{\int_0^d (\epsilon_\perp + \epsilon_a \sin^2\theta)^{-1}dz\right\}^{-1}.
\end{aligned}
\tag{3.220}
$$

Inserting this result into the total energy (3.213) reveals that the required energy is

$$
W = \frac{1}{2}\int_0^d \left[K_1\cos^2\theta + K_3\sin^2\theta\right](\theta')^2 dz - \frac{1}{2}\epsilon_0 V^2 \left\{\int_0^d (\epsilon_\perp + \epsilon_a \sin^2\theta)^{-1}dz\right\}^{-1}.
\tag{3.221}
$$

The equilibrium equation for W is obtained from setting the first variation to zero. The first variation is defined by [238, p.26]

$$
\delta W = \frac{d}{dt}W(\theta + th)\Big|_{t=0}
\tag{3.222}
$$

where the admissible variations $h \in C^1[0,d]$ satisfy

$$
h(0) = h(d) = 0.
\tag{3.223}
$$

Carrying out the necessary differentiation for W gives

$$
\begin{aligned}
\delta W &= \int_0^d \left[K_1\cos^2\theta + K_3\sin^2\theta\right]\theta' h'dz + \int_0^d (K_3 - K_1)(\theta')^2 h\sin\theta\cos\theta\,dz \\
&\quad - \epsilon_a\epsilon_0 V^2 \left\{\int_0^d (\epsilon_\perp + \epsilon_a\sin^2\theta)^{-1}dz\right\}^{-2}\int_0^d (\epsilon_\perp + \epsilon_a\sin^2\theta)^{-2}h\sin\theta\cos\theta\,dz.
\end{aligned}
\tag{3.224}
$$

Integrating the first integral in (3.224) by parts and applying the boundary conditions (3.223) on $h(z)$, and replacing V in terms of D via (3.219), shows that (3.224) is equivalent to

$$
\begin{aligned}
\delta W &= \int_0^d \left\{ -D^2\epsilon_a\epsilon_0^{-1}\frac{\sin\theta\cos\theta}{(\epsilon_\perp + \epsilon_a\sin^2\theta)^2} - \left[K_1\cos^2\theta + K_3\sin^2\theta\right]\theta'' \right. \\
&\quad \left. -(K_3 - K_1)(\theta')^2\sin\theta\cos\theta \right\} h(z)\,dz.
\end{aligned}
\tag{3.225}
$$

By the arbitrariness of h it follows that the equilibrium equation is

$$
\left[K_1\cos^2\theta + K_3\sin^2\theta\right]\theta'' + (K_3 - K_1)(\theta')^2\sin\theta\cos\theta = -D^2\epsilon_a\epsilon_0^{-1}\frac{\sin\theta\cos\theta}{(\epsilon_\perp + \epsilon_a\sin^2\theta)^2}.
\tag{3.226}
$$

Multiplying throughout by θ' and integrating gives

$$[K_1 \cos^2 \theta + K_3 \sin^2 \theta] \left(\frac{d\theta}{dz}\right)^2 = D^2 \epsilon_0^{-1} (\epsilon_\perp + \epsilon_a \sin^2 \theta)^{-1} + C, \qquad (3.227)$$

C being a constant of integration. As in the classical Freedericksz transitions, it is assumed that

$$\theta'(\tfrac{d}{2}) = 0, \qquad \theta(\tfrac{d}{2}) = \theta_m > 0. \qquad (3.228)$$

This determines C as

$$C = -D^2 \epsilon_0^{-1} (\epsilon_\perp + \epsilon_a \sin^2 \theta_m)^{-1}, \qquad (3.229)$$

and by introducing the constants

$$\gamma = \epsilon_a \epsilon_\perp^{-1}, \qquad \kappa = (K_3 - K_1) K_1^{-1}, \qquad (3.230)$$

equation (3.227) becomes

$$\frac{d\theta}{dz} = \frac{D\sqrt{\gamma}}{\sqrt{\epsilon_0 \epsilon_\perp K_1}} \left\{ \frac{\sin^2 \theta_m - \sin^2 \theta}{(1 + \kappa \sin^2 \theta)(1 + \gamma \sin^2 \theta)(1 + \gamma \sin^2 \theta_m)} \right\}^{\frac{1}{2}}, \qquad (3.231)$$

adopting the positive square root for this choice of $\theta_m > 0$, for the same reasons indicated in equations (3.114) to (3.118). This equation can be integrated to obtain the solution

$$z = D^{-1} \gamma^{-\frac{1}{2}} (1 + \gamma \sin^2 \theta_m)^{\frac{1}{2}} \sqrt{\epsilon_0 \epsilon_\perp K_1} \int_0^\theta \left\{ \frac{(1 + \kappa \sin^2 \widehat{\theta})(1 + \gamma \sin^2 \widehat{\theta})}{\sin^2 \theta_m - \sin^2 \widehat{\theta}} \right\}^{\frac{1}{2}} d\widehat{\theta}. \quad (3.232)$$

For $z = \frac{d}{2}$ we must have $\theta = \theta_m$. Hence, from (3.232), the constant D must satisfy

$$D = \gamma^{-\frac{1}{2}} (1 + \gamma \sin^2 \theta_m)^{\frac{1}{2}} \sqrt{\epsilon_0 \epsilon_\perp K_1} \frac{2}{d} \int_0^{\theta_m} \left\{ \frac{(1 + \kappa \sin^2 \widehat{\theta})(1 + \gamma \sin^2 \widehat{\theta})}{\sin^2 \theta_m - \sin^2 \widehat{\theta}} \right\}^{\frac{1}{2}} d\widehat{\theta}. \quad (3.233)$$

Inserting this value of D into (3.232) now gives the solution as

$$z \frac{2}{d} \int_0^{\theta_m} \left\{ \frac{(1 + \kappa \sin^2 \widehat{\theta})(1 + \gamma \sin^2 \widehat{\theta})}{\sin^2 \theta_m - \sin^2 \widehat{\theta}} \right\}^{\frac{1}{2}} d\widehat{\theta} = \int_0^\theta \left\{ \frac{(1 + \kappa \sin^2 \widehat{\theta})(1 + \gamma \sin^2 \widehat{\theta})}{\sin^2 \theta_m - \sin^2 \widehat{\theta}} \right\}^{\frac{1}{2}} d\widehat{\theta}.$$
$$(3.234)$$

The voltage is given by (3.218), which can be expressed, via (3.231) and the definition of γ, as

$$\begin{aligned}
V &= 2D\epsilon_0^{-1} \int_0^{\frac{d}{2}} (\epsilon_\perp + \epsilon_a \sin^2 \theta)^{-1} dz \\
&= 2D\epsilon_0^{-1} \epsilon_\perp^{-1} \int_0^{\theta_m} (1 + \gamma \sin^2 \theta)^{-1} \frac{dz}{d\theta} d\theta \\
&= 2\sqrt{\frac{K_1}{\epsilon_0 \epsilon_a}} (1 + \gamma \sin^2 \theta_m)^{\frac{1}{2}} \int_0^{\theta_m} \left\{ \frac{1 + \kappa \sin^2 \theta}{(\sin^2 \theta_m - \sin^2 \theta)(1 + \gamma \sin^2 \theta)} \right\}^{\frac{1}{2}} d\theta.
\end{aligned}$$
$$(3.235)$$

A further application of the substitution (3.120) and the rule (3.121) (with $\hat{\theta}$ replaced by θ) finally gives the voltage in the convenient form

$$
V = 2\sqrt{\frac{K_1}{\epsilon_0 \epsilon_a}} (1 + \gamma \sin^2 \theta_m)^{\frac{1}{2}} \int_0^{\frac{\pi}{2}} \left\{ \frac{1 + \kappa \sin^2 \theta_m \sin^2 \lambda}{(1 + \gamma \sin^2 \theta_m \sin^2 \lambda)(1 - \sin^2 \theta_m \sin^2 \lambda)} \right\}^{\frac{1}{2}} d\lambda.
\tag{3.236}
$$

For a given fixed voltage V, the maximum distortion angle θ_m can be calculated numerically from (3.236); this value of θ_m is then inserted into (3.234) to finally yield the solution for θ implicitly as a function of z.

Similar to the classical Freedericksz transitions, it is now possible to take the limit as $\theta_m \to 0$ in the above expression for V to find that $V \to V_c$ where

$$
V_c = \pi \sqrt{\frac{K_1}{\epsilon_0 \epsilon_a}}.
\tag{3.237}
$$

Notice that this critical voltage coincides with that obtained earlier for the electric field case stated in equation $(3.208)_1$ where $V_c = E_c d$: the crucial difference between the analogy for solutions and Freedericksz transitions drawn from the earlier results, which use the substitution (3.207), and the present results, which incorporate the effect of the liquid crystal on the electric field, arises when $V > V_c$, as will be evident below.

For illustrative and computational purposes it is convenient to non-dimensionalise the voltage and the variable z by setting (cf. Deuling [63, Eqn.(2.21)])

$$
\overline{V} = \frac{V}{V_c} = \frac{2}{\pi} (1 + \gamma \sin^2 \theta_m)^{\frac{1}{2}} \int_0^{\frac{\pi}{2}} \left\{ \frac{1 + \kappa \sin^2 \theta_m \sin^2 \lambda}{(1 + \gamma \sin^2 \theta_m \sin^2 \lambda)(1 - \sin^2 \theta_m \sin^2 \lambda)} \right\}^{\frac{1}{2}} d\lambda,
\tag{3.238}
$$

and carrying out another application of the substitution (3.120) and the rule (3.121) to the solution (3.234), with z rescaled to $\overline{z} = \frac{z}{d}$. This enables the solution to be expressed for $\theta_m > 0$ as

$$
\overline{z} = \frac{1}{2} \frac{\int_0^\phi \{ G(\theta_m, \kappa, \gamma, \lambda) \}^{\frac{1}{2}} d\lambda}{\int_0^{\frac{\pi}{2}} \{ G(\theta_m, \kappa, \gamma, \lambda) \}^{\frac{1}{2}} d\lambda}, \qquad \phi = \sin^{-1} \left(\frac{\sin \theta}{\sin \theta_m} \right), \qquad 0 \le \overline{z} \le \tfrac{1}{2}, \tag{3.239}
$$

where

$$
G(\theta_m, \kappa, \gamma, \lambda) = \frac{(1 + \kappa \sin^2 \theta_m \sin^2 \lambda)(1 + \gamma \sin^2 \theta_m \sin^2 \lambda)}{1 - \sin^2 \theta_m \sin^2 \lambda}.
\tag{3.240}
$$

The solution for $\frac{1}{2} \le \overline{z} \le 1$ is obtained by the symmetry condition $\theta(\overline{z}) = \theta(1 - \overline{z})$. The analysis of other aspects of this solution can proceed as indicated in the previous sections on Freedericksz transitions, the consequent modifications resulting in similar, but more complex, calculations.

For a given voltage $\overline{V} > 1$, the maximum distortion θ_m is numerically calculated from (3.238). This value is inserted into the solution (3.240) to obtain $\theta(z)$ implicitly via numerical calculations. To enable a rough comparison with Fig. 3.10 on page 81, we have carried out this numerical scheme for the cases when $\kappa = 0$ (that is,

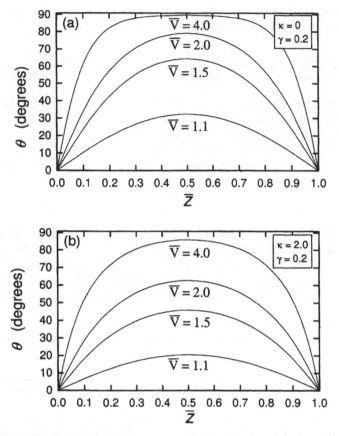

Figure 3.15: The solution $\theta(\bar{z})$, $0 \leq \bar{z} \leq 1$, obtained from equations (3.238) to (3.240) when $\gamma = \epsilon_a \epsilon_\perp^{-1} = 0.2$ for different values of the dimensionless voltage $\overline{V} = V/V_c$ where $V_c = \pi \sqrt{\frac{K_1}{\epsilon_0 \epsilon_a}}$ and $\kappa = (K_3 - K_1)K_1^{-1}$: (a) $\kappa = 0$, that is, $K_1 = K_3$, and (b) $\kappa = 2.0$.

$K_1 = K_3$) and $\kappa = 2$ and chosen $\gamma = 0.2$ as a typical value for the computations; Deuling [63] has examples using these and other values of κ and γ. The resulting plots are shown in Fig. 3.15. The main feature is that the maximum distortion angle, compared to the normalised plots in Fig. 3.10, is lowered by the influence of ϵ_a not being small. This effect is more pronounced when κ increases from zero to 2, as shown in Fig. 3.15(b). Increasing the value of γ also lowers the maximum distortion angle θ_m, as discussed by Deuling [63].

3.6   Weak Anchoring Effects

As an elementary example of a weak Freedericksz transition consider the same situation first encountered in Section 3.4.1 for a nematic but now with the director weakly anchored to both boundary plates, so that $\mathbf{n}$ satisfies

$$\mathbf{n}_b = (\cos\theta_0, 0, \sin\theta_0) \quad \text{at} \quad z = 0, d, \tag{3.241}$$

with the constant angle θ_0 to be determined. The relevant geometry is illustrated in Fig. 3.16 where initially $\theta_0 = 0$ in the absence of a magnetic field. As the field magnitude H increases, the sample is expected to exhibit a Freedericksz transition, with θ_0 (the angle the director makes with the boundary surface) varying as a function of H, as will now be demonstrated.

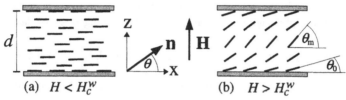

(a) $H < H_c^w$ (b) $H > H_c^w$

Figure 3.16:   The Freedericksz transition when the director is weakly anchored parallel to the boundary plates. The critical field strength is H_c^w given by (3.263). For $H < H_c^w$ the director remains parallel to the plates throughout the sample, but for $H > H_c^w$ the distorted solution $\theta(z)$ for the director tilts at the surface by the angle θ_0 and reaches a maximum distortion angle of θ_m in the middle of the sample as shown. The magnitudes of θ_0 and θ_m depend upon the value of H, with both of these angles tending to zero as $H \to H_c^w$.

As in the classical Freedericksz transitions, solutions of the form

$$\mathbf{n} = (\cos\theta(z), 0, \sin\theta(z)), \qquad 0 \le z \le d, \tag{3.242}$$

will be sought, except that on this occasion the boundary conditions in (3.105) are replaced by

$$\theta(0) = \theta(d) = \theta_0. \tag{3.243}$$

The bulk Frank elastic energy and magnetic energy are given, as before, by equations (3.106) and (3.107), while the weak anchoring surface energy is obtained by inserting $\mathbf{n}$ and $\boldsymbol{\nu} = (0, 0, \pm 1)$ into equation (2.204) in Section 2.6.4 to obtain the integrand

$$w_s = \tfrac{1}{2}\tau_0(1 + \omega\sin^2\theta_0) \quad \text{at} \quad z = 0, d. \tag{3.244}$$

In this particular example the total surface energy per unit area actually reduces to the sum of the energy integrand evaluated at the two points $z = 0$ and $z = d$, that is,

$$\int_S w_s dS = \tfrac{1}{2}\tau_0\omega\left\{\sin^2\theta(0) + \sin^2\theta(d)\right\} + \tau_0. \tag{3.245}$$

The total energy is therefore

$$
W = \frac{1}{2} \int_0^d \left\{ \left[K_1 \cos^2 \theta + K_3 \sin^2 \theta \right] (\theta')^2 - \chi_a H^2 \sin^2 \theta \right\} dz
$$
$$
+ \frac{1}{2} \tau_0 \omega \left\{ \sin^2 \theta(0) + \sin^2 \theta(d) \right\} + \tau_0. \quad (3.246)
$$

The surface energy contribution introduced here only differs by a constant from that introduced by Rapini and Papoular [227], taking into account the slightly different notation.

The bulk equilibrium equation for the director is exactly the same as (3.111), and if symmetric solutions satisfying

$$
\theta'(\tfrac{d}{2}) = 0, \quad \theta(\tfrac{d}{2}) = \theta_m > 0, \quad (3.247)
$$

are again supposed, then the first integral of the equilibrium equation is given by (3.117), with the plus sign chosen on account of θ_m being positive, leading to

$$
\left[K_1 \cos^2 \theta + K_3 \sin^2 \theta \right]^{\frac{1}{2}} \frac{d\theta}{dz} = \sqrt{\chi_a} H \left[\sin^2 \theta_m - \sin^2 \theta \right]^{\frac{1}{2}}, \quad 0 \leq z \leq \tfrac{d}{2}. \quad (3.248)
$$

This equation must be supplemented by the relevant boundary equilibrium condition for weak anchoring mentioned in equation (2.241), giving

$$
\frac{\partial w}{\partial \theta'} - \frac{\partial w_s}{\partial \theta} = 0 \quad \text{at} \quad z = 0, \quad (3.249)
$$

where w_s is given by (3.244) and $w = w_F + w_{mag}$, equivalent to the integrand appearing in (3.246), is given by equations (3.106) and (3.107). This leads to the boundary requirement

$$
\left[K_1 \cos^2 \theta_0 + K_3 \sin^2 \theta_0 \right] \theta'(0) - \tau_0 \omega \sin \theta_0 \cos \theta_0 = 0. \quad (3.250)
$$

It is worth mentioning here that the two coupled equilibrium equations (3.111) and (3.250) can also be obtained by applying basic results from the calculus of variations: see, for example, Courant and Hilbert [55, pp.209–210].

Notice that for a symmetric solution [227]

$$
\theta'(0) = -\theta'(d), \quad (3.251)
$$

which indicates that in this particular instance the other boundary condition at $z = d$ given by (2.242) will also result in precisely the same form of boundary condition as (3.250). Making use of the first integral (3.248) of the bulk equilibrium equation evaluated at $z = 0$, we can write the boundary requirement (3.250) as

$$
\left[K_1 \cos^2 \theta_0 + K_3 \sin^2 \theta_0 \right]^{\frac{1}{2}} \sqrt{\chi_a} H (\sin^2 \theta_m - \sin^2 \theta_0)^{\frac{1}{2}} = \tau_0 \omega \sin \theta_0 \cos \theta_0. \quad (3.252)
$$

This determines a relationship between H, θ_m and θ_0, and it is noted that

$$
0 \leq \theta_0 \leq \theta(z) \leq \theta_m \quad \text{for} \quad 0 \leq z \leq d. \quad (3.253)
$$

The solution $\theta \equiv 0$ is a solution to the bulk equilibrium equation (3.248) and the boundary condition (3.252) when $\theta_0 = 0$. However, as in the classical Freedericksz transitions, there is also the possibility of a distorted solution for $H \neq 0$. Integrating (3.248) and using the conditions (3.243) and (3.247) provides the solution

$$\sqrt{\chi_a} H z = \int_{\theta_0}^{\theta} \frac{[K_1 \cos^2 \widehat{\theta} + K_3 \sin^2 \widehat{\theta}]^{\frac{1}{2}}}{[\sin^2 \theta_m - \sin^2 \widehat{\theta}]^{\frac{1}{2}}} \, d\widehat{\theta}, \qquad 0 \leq z \leq \tfrac{d}{2}, \qquad (3.254)$$

which defines θ implicitly as a function of z. The condition $(3.247)_2$ then leads to

$$\sqrt{\chi_a} H \frac{d}{2} = \int_{\theta_0}^{\theta_m} \frac{[K_1 \cos^2 \widehat{\theta} + K_3 \sin^2 \widehat{\theta}]^{\frac{1}{2}}}{[\sin^2 \theta_m - \sin^2 \widehat{\theta}]^{\frac{1}{2}}} \, d\widehat{\theta}. \qquad (3.255)$$

For a given H and d, the two relations (3.252) and (3.255) can be solved simultaneously to determine θ_m and θ_0. Employing the substitution

$$\sin \widehat{\theta} = \sin \theta_m \sin \lambda, \qquad (3.256)$$

the solution (3.254) and the requirement (3.255) become, respectively,

$$\sqrt{\chi_a} H z = \int_{\lambda_0}^{\phi} G(\theta_m, \lambda) \, d\lambda, \qquad \phi = \sin^{-1}\left(\frac{\sin \theta}{\sin \theta_m}\right), \qquad (3.257)$$

$$\sqrt{\chi_a} H \frac{d}{2} = \int_{\lambda_0}^{\frac{\pi}{2}} G(\theta_m, \lambda) \, d\lambda, \qquad (3.258)$$

where $G(\theta_m, \lambda)$ and λ_0 are defined by

$$G(\theta_m, \lambda) = \left[\frac{K_1 + (K_3 - K_1) \sin^2 \theta_m \sin^2 \lambda}{1 - \sin^2 \theta_m \sin^2 \lambda}\right]^{\frac{1}{2}}, \qquad (3.259)$$

$$\lambda_0 = \sin^{-1}\left(\frac{\sin \theta_0}{\sin \theta_m}\right), \qquad (3.260)$$

in a similar fashion to that for the case of pretilt at the boundaries in Section 3.4.2. These forms for the solution and the condition at $z = \frac{d}{2}$ are more amenable to numerical computations, θ_m and θ_0 being obtained from solving the two simultaneous equations (3.252) and (3.258) in these values: once θ_m and θ_0 have been found, they can then be inserted into (3.257) to calculate the full solution. For example, a substitution for $\sin \theta_m$ can be made from (3.252) and inserted into (3.258) to obtain θ_0: this value for θ_0 then delivers θ_m via insertion into (3.252). Notice that both θ_0 and θ_m depend upon H and d. It will now be shown that this numerical procedure may be deployed provided H is above a certain critical value.

The boundary requirement (3.252) can also be written in the form

$$\tan \lambda_0 = \frac{\sqrt{\chi_a} H}{\tau_0 \omega} \sec \theta_0 \left[K_1 \cos^2 \theta_0 + K_3 \sin^2 \theta_0\right]^{\frac{1}{2}}. \qquad (3.261)$$

As $\theta_m \to 0$ and $\theta_0 \to 0$ we must have $H \to H_c^w$ where H_c^w is the weak critical Freedericksz threshold. From (3.261) we then see that λ_0 at H_c^w must be given by

$$\tan \lambda_0 = \frac{\sqrt{K_1 \chi_a}}{\tau_0 \omega} H_c^w. \qquad (3.262)$$

Further, from (3.258), when $H \to H_c^w$ the critical field H_c^w must also satisfy the relation

$$H_c^w = \frac{2}{d}\sqrt{\frac{K_1}{\chi_a}} \left(\frac{\pi}{2} - \lambda_0\right). \tag{3.263}$$

Substituting (3.263) for H_c^w in (3.262) delivers the relation

$$\tau_0 \omega \tan \lambda_0 = K_1 \frac{2}{d}\left(\frac{\pi}{2} - \lambda_0\right), \tag{3.264}$$

which determines λ_0. Inserting the resulting value for λ_0 into (3.263) then determines the value of H_c^w. It is clear from the value given by (3.263) that for weak anchoring the value of H_c^w is always lower than the value of H_c obtained for the analogous strong anchoring case in equation (3.126) in Section 3.4.1. The effect of the weak anchoring is apparent from the relation (3.264) and H_c^w in (3.263). For example, if the anchoring strength $\tau_0 \omega$ is high and satisfies $dK_1^{-1}\tau_0\omega \gg 1$ then $\lambda_0 \sim 0$ and H_c^w becomes close to the classical value for strong anchoring given by equation (3.126), whereas if the anchoring strength is low and satisfies $dK_1^{-1}\tau_0\omega \ll 1$ then $\lambda_0 \sim \frac{\pi}{2}$ and H_c^w is greatly lowered. A more illuminating comparison can be obtained by introducing

$$h_c = \frac{H_c^w}{H_c}, \quad H_c = \frac{\pi}{d}\sqrt{\frac{K_1}{\chi_a}}, \quad \rho = \frac{K_1\pi}{d\tau_0\omega}, \tag{3.265}$$

where H_c is the corresponding classical critical field magnitude for the strong anchoring case and ρ is a rescaled reciprocal of the anchoring strength. Eliminating λ_0 from the two equations (3.262) and (3.263) shows that h_c must satisfy

$$\rho h_c = \cot\left(\frac{\pi}{2}h_c\right). \tag{3.266}$$

Figure 3.17 illustrates the graph of h_c as a function of ρ and shows that $H_c^w \to H_c$ as ρ decreases, that is, as the anchoring strength $\tau_0\omega$ increases.

In common with the classical Freedericksz transitions, the zero solution $\theta \equiv 0$ is always available for any $H \geq 0$, while the distorted solution is only possible when $H > H_c^w$. To determine which solution is energetically favoured for $H > H_c^w$ we consider the difference in energies per unit area of the plates for the total energy W in equation (3.246)

$$\Delta W = W(\theta) - W(0), \tag{3.267}$$

where θ is the distorted solution. Using the results (3.248), (3.252) and (3.243) and an argument similar to that used to obtain ΔW in equation (3.130), we obtain

$$\Delta W = \sqrt{\chi_a}H \sin^2\theta_m \int_{\lambda_0}^{\frac{\pi}{2}} \cos(2\lambda)G(\theta_m, \lambda)\, d\lambda + \tau_0\omega \sin^2\theta_0, \tag{3.268}$$

for G defined at equation (3.259). From (3.260) and (3.261) we have

$$\tau_0\omega \sin^2\theta_0 = \sin^2\theta_m \sin\lambda_0 \cos\lambda_0\sqrt{\chi_a}H \sec\theta_0 \left[K_1 \cos^2\theta_0 + K_3 \sin^2\theta_0\right]^{\frac{1}{2}}, \tag{3.269}$$

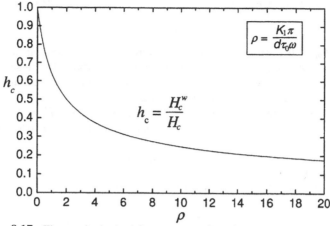

Figure 3.17: The graph obtained from equation (3.266) showing the influence of the anchoring strength $\tau_0\omega$ upon the critical threshold H_c^w for the weak Freedericksz transition by considering the dimensionless quantity h_c. Clearly, $H_c^w \to H_c$ as $\tau_0\omega \to \infty$ where $H_c = \frac{\pi}{d}\sqrt{\frac{K_1}{\chi_a}}$ is the corresponding usual classical critical threshold for a Freedericksz transition under strong anchoring conditions.

which enables ΔW to be reformulated as, using (3.260) as appropriate,

$$\Delta W = \sqrt{\chi_a}H\sin^2\theta_m \left\{ \int_{\lambda_0}^{\frac{\pi}{2}} \cos(2\lambda)G(\theta_m, \lambda)\, d\lambda + \sin\lambda_0\cos\lambda_0 G(\theta_m, \lambda_0) \right\}. \quad (3.270)$$

To progress further, it is easiest to expand G as a series in $\sin^2\theta_m$ to observe what happens when $H \gtrsim H_c^w$. It is simple to check that

$$G(\theta_m, \lambda) = \sqrt{K_1}\left[1 + \frac{1}{2}\frac{K_3}{K_1}\sin^2\theta_m\sin^2\lambda\right] + O(\sin^4\theta_m), \quad (3.271)$$

and

$$G(\theta_m, \lambda_0) = \sqrt{K_1}\left[1 + \frac{1}{2}\frac{K_3}{K_1}\sin^2\theta_m\sin^2\lambda_0\right] + O(\sin^4\theta_m). \quad (3.272)$$

Inserting (3.271) and (3.272) into (3.270) gives

$$\Delta W = -\sqrt{\chi_a}\sin^4\theta_m\frac{H}{8}\frac{K_3}{\sqrt{K_1}}\left[\frac{\pi}{2} - \lambda_0 + \frac{1}{4}\sin(4\lambda_0)\right] + O(\sin^6\theta_m), \quad (3.273)$$

obtained by integrating and carefully rearranging the trigonometric terms. The quantity

$$M(\lambda_0) = \frac{\pi}{2} - \lambda_0 + \frac{1}{4}\sin(4\lambda_0), \quad (3.274)$$

is strictly decreasing for $0 < \lambda_0 < \frac{\pi}{2}$ and satisfies $M(0) = \frac{\pi}{2}$ and $M(\frac{\pi}{2}) = 0$. Therefore

$$M(\lambda_0) > 0 \quad \text{for} \quad 0 < \lambda_0 < \frac{\pi}{2}. \quad (3.275)$$

It now suffices to notice that $\lambda_0 > 0$ for $H > H_c^w$, and hence $M(\lambda_0)$ is positive. By (3.273) this ensures that

$$\Delta W < 0, \tag{3.276}$$

whenever $\theta_m \neq 0$ is small. There is therefore a Freedericksz transition with the critical threshold given by H_c^w defined in equation (3.263) via the solution for λ_0 obtained by solving the relation (3.264). A more detailed discussion of the weak Freedericksz transition can be found in Virga [273].

We close this Section with an observation. If we define $\Delta H_c = H_c^w - H_c$ to be the difference between the two critical magnetic fields being discussed here, then

$$\frac{\Delta H_c}{H_c} = h_c - 1. \tag{3.277}$$

Also, from the relation (3.266), when $h_c \approx 1$ (cf. Fig. 3.17)

$$\rho \approx \frac{\pi}{2}(1 - h_c). \tag{3.278}$$

Coupling this with (3.277) shows that

$$\frac{\Delta H_c}{H_c} \approx -\frac{2}{\pi}\rho. \tag{3.279}$$

Using the approximations suggested by Rapini and Papoular [227] (in cgs units)

$$K_1 = 10^{-6} \text{ dyn}, \quad d = 10^{-3} \text{ cm}, \quad \tau_0\omega = 1 \text{ erg cm}^{-2}, \tag{3.280}$$

it is then found from $(3.265)_3$ and (3.279) that

$$\frac{\Delta H_c}{H_c} \approx -2 \times 10^{-3}. \tag{3.281}$$

This result shows that near the critical field H_c^w the difference between the effects of weak and strong anchoring may be relatively small, of the order of 0.2%. Obviously, as the anchoring strength weakens, this discrepancy becomes appreciably larger. However, the error in measurements of the elastic constant is not expected to be as great as that which can arise when the director is strongly anchored but perhaps misaligned by a small pretilt angle at the boundaries: recall that in Section 3.4.2 it was shown that when there is strong anchoring with a pretilt the resulting error in the measurement of the elastic constant may be around 20% (cf. equations (3.177) and (3.178)). Nevertheless, some caution needs to be exercised and the reader may find the recent review by Sugimura [68, pp.493–502] on the evaluation of the anchoring strength $\tau_0\omega$ useful. Further, work by Jerome, Pieranski and Boix [136] suggests that the surface anchoring can be more complicated than that in equations (3.244) and (3.245) since surface anchoring in nematics can be bistable. Attempts to model this bistability by introducing more complex forms for the surface energy have been made by Sergan and Durand [248].

Attention should also be drawn to the work of Kini [147, 148] where combined magnetic and electric fields have been considered in the context of periodic deformations in nematics. This more advanced treatment examines various possibilities under different conditions, including that of weak anchoring [148].

3.7   The Twisted Nematic Device

We now extend the analysis of tilt and twist equilibrium solutions developed in Section 3.2.2 to incorporate the effect of an applied magnetic field. Our aim is to model the Freedericksz transition in what is commonly called the twisted nematic (TN) display or device. This was studied theoretically by Leslie [166] in 1970 and was applied by Schadt and Helfrich [125, 241] in the early 1970s to twisted nematic liquid crystal devices using an electric rather than a magnetic field. The resulting type of Freedericksz transition and its variants are widely employed in nematic liquid crystal displays. A brief review of liquid crystal displays, including some details on the twisted nematic device and the related ideas of the supertwisted nematic (STN) device, has been written by Schadt [242]. A short survey of developments on liquid crystal flat panel displays has also been given by Wada and Koden [275], including the switching of liquid crystal devices by thin film transistors (TFT). Such flat panel displays are now commonly used as computer screens and are built-in to other devices such as mobile telephones, digital cameras, video cameras, projectors, etc.

The notation used in Section 3.2.2 based on Fig. 3.3 on page 60 to obtain the tilt and twist equilibrium solutions in the absence of any fields will be used again here. Figure 3.18, which follows that given in the short review by Scheuble [243], describes the geometrical set-up and illustrates the principle mechanism by which a twisted nematic display works: further comments on this type of display will be made at the end of this Section.

As a starting point, consider a nematic liquid crystal sample having, in the absence of a magnetic field, a uniformly parallel alignment confined between two parallel bounding plates a distance d apart and assume that the director is strongly anchored parallel to the surfaces of the plates. Suppose that one of the plates is then rotated about its normal by an angle of $2\phi_0$ radians to produce the equilibrium solution (3.35) derived in Section 3.2.2. In twisted nematic devices the angle $2\phi_0$ is usually taken to be $\frac{\pi}{2}$, whereas in the STN device it is often within the range π to $\frac{3}{2}\pi$. In the geometry of Fig. 3.18, recall from Section 3.2.2 that the director is of the general form

$$\mathbf{n} = (\cos\theta(z)\cos\phi(z), \cos\theta(z)\sin\phi(z), \sin\theta(z)), \qquad (3.282)$$

and that the twisted planar solution is given by (3.35), namely,

$$\theta \equiv 0, \qquad \phi = \phi_0\left(2\frac{z}{d} - 1\right), \qquad 0 \le z \le d, \qquad (3.283)$$

the overall twist of the director through the sample being $2\phi_0$, as illustrated in Fig. 3.18(a). Notice that this static solution is a special case of a more general twist equilibrium solution identified in Section 2.5 at equation (2.179). We now follow the work of Leslie [166] and show that a magnetic field applied across the plates can induce a Freedericksz transition where a solution exhibiting both tilt and twist becomes available when the strength of the field is above a certain critical value, this solution being energetically preferred over the simple planar twist solution (3.283).

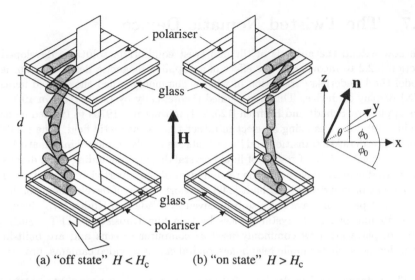

(a) "off state" $H < H_c$ (b) "on state" $H > H_c$

Figure 3.18: Schematic diagram of how a twisted nematic display operates. The grey cylinders represent the director alignment through the cell. The magnetic field is applied in the z-direction across the glass plates. (a) When $H < H_c$ (cf. equation (3.314)), or $V < V_c$ in the electric field case (cf. equation (3.331)), the twisted planar solution is preferred where the director is always parallel to the plates but twists through $2\phi_0$ radians as indicated. (b) When $H > H_c$, or $V > V_c$, then the director tilts out of the plane as indicated. When light enters through the polariser on the top plate the resulting polarisation vector follows the path of the director as shown leading to an 'off state' or 'on state'.

Consider the general form of $\mathbf{n}$ introduced above and now suppose that we apply a magnetic field across the plates of the form

$$\mathbf{H} = H(0,0,1), \tag{3.284}$$

where $H = |\mathbf{H}|$, as shown in Fig. 3.18(b). The boundary conditions are as before when tilt and twist equilibrium solutions were discussed, namely,

$$\theta(0) = \theta(d) = 0, \tag{3.285}$$

$$\phi(0) = -\phi_0, \qquad \phi(d) = \phi_0, \tag{3.286}$$

where, without loss of generality, $\phi_0 > 0$ is assumed. The relevant elastic energy integrand is simply $w_F = \hat{w}$ derived earlier in equation (3.16), which must now be supplemented by the appropriate magnetic energy to take into account the effect of the magnetic field, obtained by inserting $\mathbf{n}$ into w_{mag} in equation (2.107). Thus the full energy integrand to be considered here is given by

$$\hat{w} = \tfrac{1}{2}f(\theta)(\theta')^2 + \tfrac{1}{2}g(\theta)(\phi')^2 - \tfrac{1}{2}\chi_a H^2 \sin^2\theta, \tag{3.287}$$

where $' = d/dz$ and we have employed the definitions of $f(\theta)$ and $g(\theta)$ stated in equations (3.17) and (3.18) above. It will be assumed throughout that $\chi_a > 0$. Calculations which parallel those used to obtain the equilibrium equations (3.21) and (3.22) and their first integrals (3.28) and (3.29) reveal that the equilibrium equations for the present $\hat{w}$ are

$$2f(\theta)\theta'' + \frac{df(\theta)}{d\theta}(\theta')^2 - \frac{dg(\theta)}{d\theta}(\phi')^2 + 2\chi_a H^2 \sin\theta\cos\theta = 0 \qquad (3.288)$$

$$[g(\theta)\phi']' = 0, \qquad (3.289)$$

having first integrals

$$f(\theta)(\theta')^2 + g(\theta)(\phi')^2 + \chi_a H^2 \sin^2\theta = a, \qquad (3.290)$$

$$g(\theta)\phi' = b, \qquad (3.291)$$

where a and b are constants of integration. These equations are the same as the aforementioned first integrals except for the additional term involving the magnetic field.

It is simple to verify that, for any H, one solution of the equilibrium equations (3.288) and (3.289) which satisfies the boundary conditions (3.285) and (3.286) is indeed the twisted planar solution given by equation (3.283). However, another possibility is to adopt the methods of Section 3.2.2 to search for symmetric solutions satisfying

$$\theta(z) = \theta(d - z), \qquad \phi(z) = -\phi(d - z), \qquad 0 \le z \le \tfrac{d}{2}, \qquad (3.292)$$

with, as before,

$$\theta'(\tfrac{d}{2}) = 0, \qquad \theta(\tfrac{d}{2}) = \theta_m > 0, \qquad (3.293)$$

$$\phi(\tfrac{d}{2}) = 0, \qquad (3.294)$$

θ_m being considered positive. Using the conditions (3.293) in equation (3.290) gives, at $z = \tfrac{d}{2}$,

$$a = g(\theta_m)(\phi'(\tfrac{d}{2}))^2 + \chi_a H^2 \sin^2\theta_m, \qquad (3.295)$$

and hence equation (3.290) becomes

$$f(\theta)(\theta')^2 = g(\theta_m)(\phi'(\tfrac{d}{2}))^2 - g(\theta)(\phi'(z))^2 + \chi_a H^2(\sin^2\theta_m - \sin^2\theta). \qquad (3.296)$$

The conditions (3.293) also show that, from (3.291),

$$[g(\theta_m)\phi'(\tfrac{d}{2})]^2 = b^2 = [g(\theta)\phi'(z)]^2, \qquad (3.297)$$

and putting these two results into (3.296) reveals

$$f(\theta)(\theta')^2 = b^2\left[\frac{1}{g(\theta_m)} - \frac{1}{g(\theta)}\right] + \chi_a H^2(\sin^2\theta_m - \sin^2\theta). \qquad (3.298)$$

Integrating (3.298) and using the boundary conditions (3.285) gives the solution

$$z = \int_0^\theta \frac{\sqrt{f(u)}}{\left[\chi_a H^2(\sin^2\theta_m - \sin^2 u) + b^2(1/g(\theta_m) - 1/g(u))\right]^{\frac{1}{2}}}\, du, \quad 0 \le z \le \tfrac{d}{2}, \quad (3.299)$$

which delivers θ implicitly as a function of z. Now notice that by equation (3.297)

$$\frac{d\theta}{dz} = \frac{d\theta}{d\phi}\frac{d\phi}{dz} = \frac{b}{g(\theta)}\frac{d\theta}{d\phi}. \tag{3.300}$$

Inserting this result into (3.298) and integrating using the boundary conditions $(3.286)_1$ gives the solution for the angle ϕ as a function of the solution θ, namely,

$$\phi = -\phi_0 + \int_0^\theta \frac{\sqrt{f(u)}}{\left[\chi_a H^2(\sin^2\theta_m - \sin^2 u) + b^2(1/g(\theta_m) - 1/g(u))\right]^{\frac{1}{2}}}\frac{b}{g(u)}\, du, \quad (3.301)$$

for $0 \le z \le \tfrac{d}{2}$, this solution being calculated for the values of θ obtained from the solution (3.299). For example, z may be calculated from (3.299) for a particular θ in the interval $0 \le \theta \le \theta_m$: this value of θ then gives the value for ϕ via equation (3.301) at the previously calculated value of z.

The constants θ_m and b are obtained from the two relations obtained by the symmetry conditions (3.293) and (3.294) applied to the above solutions. These are

$$\frac{d}{2} = \int_0^{\theta_m} \frac{\sqrt{f(u)}}{\left[\chi_a H^2(\sin^2\theta_m - \sin^2 u) + b^2(1/g(\theta_m) - 1/g(u))\right]^{\frac{1}{2}}}\, du, \tag{3.302}$$

$$\phi_0 = \int_0^{\theta_m} \frac{\sqrt{f(u)}}{\left[\chi_a H^2(\sin^2\theta_m - \sin^2 u) + b^2(1/g(\theta_m) - 1/g(u))\right]^{\frac{1}{2}}}\frac{b}{g(u)}\, du. \tag{3.303}$$

These two equations determine θ_m and b for a given field strength H.

Now make the change of variable

$$\sin u = \sin\theta_m \sin\lambda. \tag{3.304}$$

The requirements (3.302) and (3.303) then become

$$\frac{d}{2} = \int_0^{\frac{\pi}{2}} G(\theta_m, \lambda)\, d\lambda, \tag{3.305}$$

$$\phi_0 = \int_0^{\frac{\pi}{2}} G(\theta_m, \lambda)\frac{b}{g_2(\theta_m, \lambda)}\, d\lambda, \tag{3.306}$$

where

$$G(\theta_m, \lambda) = \frac{\sin\theta_m \cos\lambda}{\sqrt{1 - \sin^2\theta_m \sin^2\lambda}}$$

$$\times \frac{\left[K_1 + (K_3 - K_1)\sin^2\theta_m \sin^2\lambda\right]^{\frac{1}{2}}}{\left[\chi_a H^2 \sin^2\theta_m \cos^2\lambda + b^2\{1/g_2(\theta_m, \tfrac{\pi}{2}) - 1/g_2(\theta_m, \lambda)\}\right]^{\frac{1}{2}}}, \quad (3.307)$$

with $g_2(\theta_m, \lambda)$ defined by

$$g_2(\theta_m, \lambda) = K_2 + (K_3 - 2K_2)\sin^2\theta_m \sin^2\lambda - (K_3 - K_2)\sin^4\theta_m \sin^4\lambda. \quad (3.308)$$

These requirements are considerably more complicated than the analogous requirement derived for the classical Freedericksz transition in equation (3.123). Nevertheless, we can take the limit in $g_2(\theta_m, \lambda)$ and $G(\theta_m, \lambda)$ as $\theta_m \to 0$ to see that

$$g_2(\theta_m, \lambda) \rightarrow K_2, \quad (3.309)$$

$$G(\theta_m, \lambda) \rightarrow \frac{\sqrt{K_1}}{\sqrt{\chi_a H^2 - b^2(K_3 - 2K_2)/K_2^2}}, \quad (3.310)$$

and hence the requirement (3.305) leads to anticipating that $H \to H_c$ where

$$\frac{d}{2} = \frac{\pi}{2}\frac{\sqrt{K_1}}{\sqrt{\chi_a H_c^2 - b^2(K_3 - 2K_2)/K_2^2}}, \quad (3.311)$$

which can be written as

$$\chi_a d^2 H_c^2 = \pi^2 K_1 + \frac{b^2 d^2}{K_2^2}(K_3 - 2K_2). \quad (3.312)$$

Taking the limit as $\theta_m \to 0$ in (3.306), and using (3.311), shows that the constant b (which must actually depend on the constant θ_m) satisfies

$$b = 2K_2\frac{\phi_0}{d}. \quad (3.313)$$

Hence the constant b can be eliminated from (3.312) by (3.313) to reveal that H_c satisfies

$$\chi_a d^2 H_c^2 = \pi^2 K_1 + 4\phi_0^2(K_3 - 2K_2), \quad (3.314)$$

provided

$$K_3 - 2K_2 \geq 0 \quad \text{or} \quad \phi_0 \leq \frac{\pi}{2}\sqrt{\frac{K_1}{2K_2 - K_3}}, \quad (3.315)$$

when $K_3 - 2K_2 < 0$. (This result is identical to Leslie [166, Eqn.(4.19)] when d is replaced by $2l$.)

Leslie [166] has shown that, by writing

$$\beta = \sin^2\theta_m, \quad (3.316)$$

suitable differentiation of the requirements (3.305) and (3.306) with respect to β gives

$$2d^2\chi_a\left(\frac{dH^2}{d\beta}\right)_{\beta=0} = K_3\pi^2 - 4(K_2^2 + K_3^2 - K_2 K_3)\frac{\phi_0^2}{K_2}, \quad (3.317)$$

$$d\left(\frac{db}{d\beta}\right)_{\beta=0} = \phi_0(K_3 - 2K_2). \quad (3.318)$$

These results, combined with (3.312), (3.313) and (3.314), reveal that

$$H^2 = \frac{1}{\chi_a d^2}\left(\pi^2 K_1 + 4\phi_0^2(K_3 - 2K_2)\right)$$

$$+\frac{\beta}{2\chi_a d^2}\left(K_3\pi^2 - 4(K_2^2 + K_3^2 - K_2K_3)\frac{\phi_0^2}{K_2}\right) + O(\beta^2), \quad (3.319)$$

$$b = 2K_2\frac{\phi_0}{d} + \frac{\beta}{d}(K_3 - 2K_2)\phi_0 + O(\beta^2). \quad (3.320)$$

One method of verifying these conditions is to use the following procedure: the technical details are left to the interested reader. Firstly, expand the integrands in (3.305) and (3.306) in terms of β and then integrate to first order in β. This gives two equations in the two variables H^2 and b. Secondly, solve these simultaneous equations to find the solutions for H^2 and b as functions of β to first order. The desired results in equations (3.317) to (3.320) follow directly. This process can be carried out relatively swiftly using a symbolic computer package. Consequently, if either of the inequalities in (3.315) holds and also, from (3.317), if

$$\phi_0 < \overline{\phi} := \frac{\pi}{2}\sqrt{\frac{K_2K_3}{K_2^2 + K_3^2 - K_2K_3}}, \quad (3.321)$$

then it follows that

$$\left(\frac{dH}{d\beta}\right)_{\beta=0} > 0. \quad (3.322)$$

Notice that $K_2^2 + K_3^2 - K_2K_3 > 0$ for all $K_2 > 0$ and $K_3 > 0$. Therefore suitably restricted values of ϕ_0, depending on the values of the elastic constants, ensure that H increases monotonically as θ_m increases, or vice-versa. This result tends to suggest that H_c in equation (3.314) is indeed the critical threshold for the onset of a Freedericksz transition.

  To further justify H_c as a critical threshold, Leslie [166] examined the difference in energies for $H > H_c$ between the non-planar solution provided via equations (3.299) and (3.301) and the planar twist solution (3.283), both of these solutions being available. The total energy difference ΔW between the corresponding energies per unit area of the bounding plates is given, via the energies (3.36) and (3.287), by

$$\Delta W = \frac{1}{2}\int_0^d \left[f(\theta)(\theta')^2 + g(\theta)(\phi')^2 - \chi_a H^2\sin^2\theta - K_2\frac{4}{d^2}\phi_0^2\right]dz. \quad (3.323)$$

With the aid of equations (3.297) and (3.298), and the symmetry conditions on the solutions, this can be written as

$$\Delta W = \int_0^{\frac{d}{2}}\left[\chi_a H^2(\sin^2\theta_m - 2\sin^2\theta) + \frac{b^2}{g(\theta_m)} - K_2\phi_0^2\left(\frac{2}{d}\right)^2\right]dz, \quad (3.324)$$

and changing the variable using (3.304) (with u replaced by θ) and employing calculations similar to those used earlier to obtain (3.307) leads to

$$\Delta W = \int_0^{\frac{\pi}{2}} \frac{\sqrt{f(\theta)} \left[\chi_a H^2 \sin^2\theta_m \cos(2\lambda) + b^2/g(\theta_m) - K_2 \phi_0^2 (2/d)^2\right]}{\sqrt{1 - \sin^2\theta_m \sin^2\lambda} \left[\chi_a H^2 - b^2 P(\theta_m, \lambda)/\left(g_2(\theta_m, \frac{\pi}{2})g_2(\theta_m, \lambda)\right)\right]^{\frac{1}{2}}} \, d\lambda,$$
(3.325)

where

$$P(\theta_m, \lambda) = K_3 - 2K_2 - (K_3 - K_2)\sin^2\theta_m(1 + \sin^2\lambda).$$
(3.326)

Notice that $\Delta W \to 0$ as $\beta \to 0$ by (3.313), where β is defined at equation (3.316). This result can be combined with the calculations by Leslie [166] for the derivatives of ΔW with respect to β, employing the relations (3.317) and (3.318), to find that

$$(\Delta W)_{\beta=0} = 0,$$
(3.327)

$$\left(\frac{d\Delta W}{d\beta}\right)_{\beta=0} = 0,$$
(3.328)

$$\left(\frac{d^2\Delta W}{d\beta^2}\right)_{\beta=0} = -\frac{1}{2d}\left[K_3\left(\frac{\pi}{2}\right)^2 - (K_2^2 + K_3^2 - K_2 K_3)\frac{\phi_0^2}{K_2}\right].$$
(3.329)

These results can be derived in a similar fashion to those found in equations (3.317) and (3.318), bearing in mind that the expansions for H^2 and b have to be taken to second order in β rather than to the first order derived in equations (3.319) and (3.320). Hence for small θ_m and using $\overline{\phi}$ defined by equation (3.321)

$$\Delta W = -\frac{\theta_m^4}{4d}\left[K_3\left(\frac{\pi}{2}\right)^2 - (K_2^2 + K_3^2 - K_2 K_3)\frac{\phi_0^2}{K_2}\right] + O(\theta_m^6)$$

$$= -\frac{\theta_m^4}{4dK_2}(K_2^2 + K_3^2 - K_2 K_3)\left[\overline{\phi}^2 - \phi_0^2\right] + O(\theta_m^6),$$
(3.330)

and therefore the non-planar solution is energetically preferred provided the non-zero value of θ_m is sufficiently small, $\phi_0 < \overline{\phi}$ and ϕ_0 satisfies either of the inequalities in (3.315). This analysis proves that a Freedericksz transition occurs at H_c: for $0 \leq H < H_c$ the planar twist solution occurs while for $H > H_c$ the non-planar solution is preferred; for this non-planar solution the director realigns slightly as it attempts to make itself more parallel to the magnetic field $\mathbf{H}$. Of course, this present analysis is only valid just above H_c. Although a full nonlinear analysis was not carried out by Leslie [166], the above results are sufficient to conclude that if ϕ_0 is small, or obeys suitable restrictions determined by the elastic constants, then a Freedericksz transition occurs at H_c given by (3.314). The situation for cholesteric liquid crystals has also been discussed by Leslie [166].

The identification (3.207) allows the critical threshold value for the Freedericksz transition to be calculated in terms of the analogous electric field problem. For instance, for $\epsilon_a > 0$, the critical voltage is given by $V_c = dE_c$ and so from (3.314) we have

$$V_c^2 = (\epsilon_0 \epsilon_a)^{-1}\left[K_1 \pi^2 + 4\phi_0^2(K_3 - 2K_2)\right].$$
(3.331)

In the special case $2\phi_0 = \frac{\pi}{2}$, where one plate is rotated through $\frac{\pi}{2}$ radians, we have

$$V_c^2 = \frac{\pi^2}{\epsilon_0 \epsilon_a} \left[K_1 + \frac{1}{4}(K_3 - 2K_2) \right]. \tag{3.332}$$

Further, in this special case, the inequalities $(3.315)_2$ and (3.321) become, respectively,

$$\frac{1}{4} \leq \frac{K_1}{2K_2 - K_3}, \qquad \frac{1}{4} < \frac{K_2 K_3}{K_2^2 + K_3^2 - K_2 K_3}. \tag{3.333}$$

These inequalities define constraints on the possible values of the elastic constants which validate the above analysis for showing that V_c is the critical threshold: they are clearly always satisfied in the one-constant approximation $K_1 = K_2 = K_3 \equiv K$. Further, if $K_3 - 2K_2 \geq 0$ then knowing that the condition (3.321) is satisfied is sufficient to validate V_c mathematically as the critical threshold for a Freedericksz transition; on the other hand, if $K_3 - 2K_2 < 0$ then both the inequalities in $(3.315)_2$ and (3.321) need to be fulfilled mathematically to justify V_c as the critical threshold.

With the help of Fig. 3.18, we are now in a position to illustrate briefly how a twisted nematic display operates. For more details on this and other effects related to liquid crystal displays the reader is referred to the review articles by Schadt [242] and Scheuble [243]. This type of device, based on the experiments involving electric fields carried out by Schadt and Helfrich [241], was patented for technological applications by these authors [125] in 1970 and by Fergason [84] in 1971. Fergason also went on to develop other patents on twisted nematic displays, most notably one issued in 1973 [85].

A sample of nematic is placed between two glass plates as shown in Fig. 3.18 to form a cell. The plates are treated so that the strongly anchored orientation of **n** is uniformly parallel to the glass at the internal surfaces. The plates are then rotated through $\frac{\pi}{2}$ radians so that the overall planar twist solution imposed upon the director is given by $2\phi_0$ with $\phi_0 = \frac{\pi}{4}$ in the notation of this Section and Fig. 3.18. Polarisers are placed on the outer surfaces of the two glass plates such that the axes of the polarisers are parallel to the direction of the director **n** on the inner glass surfaces, as shown in the Figure. When light enters the cell through the top polariser vibrating perpendicular to the plates the resulting polarisation vector entering the sample of liquid crystal will undergo a total reorientation of $\frac{\pi}{2}$ radians across the cell as it follows the orientation of **n**. The light will then propagate through the lower plate of the cell because the polarisers are also rotated by $\frac{\pi}{2}$ radians, as indicated in Fig. 3.18(a). This means that the cell will appear transparent when no electric or magnetic field is present: this will also be the situation described by the mathematical analysis discussed earlier when $H < H_c$ (cf. equation (3.314) when $\phi_0 = \frac{\pi}{4}$) or, in the electric field case, when the voltage V is less than V_c given by equation (3.332). This situation is sometimes referred to as the 'off state' in relation to the switching of display devices. However, when $H > H_c$, or $V > V_c$, the non-planar solution for **n** discussed above is energetically favoured over the original planar twist solution and the director prefers to align parallel to the field in the centre of the cell. It is anticipated that if the magnitude of the field is sufficiently high then the director alignment will mostly be parallel to the field, and therefore perpendicular to the glass plates, throughout most of the sample, except at small regions close to the plates. In this case the polarisation vector does not twist round as in Fig. 3.18(a) but rather travels across the cell virtually unhindered as shown in Fig. 3.18(b). The polarisation vector is then orthogonal to the polariser axis on the lower plate after it travels across the cell. This means that the light will not propagate through the lower plate and the cell will then appear opaque for $H > H_c$ or $V > V_c$. This situation is often referred to as the 'on state' when discussing the switching of liquid crystal display devices. The 'switch-on' and 'switch-off' times, which are defined in general terms later in Section 5.9.1, are stated at equation (5.426).

If the axes of the two polarisers were mutually parallel then the cell would appear opaque for $H < H_c$ or $V < V_c$, and transparent for $H > H_c$ or $V > V_c$. Schadt and Helfrich [241] carried out the electric field experiment on a twisted nematic cell having $2\phi_0 = \frac{\pi}{2}$ with parallel polarisers for two liquid crystal substances having positive dielectric anisotropy ϵ_a. The cell depth d was 10 μm and both dc and ac voltages were applied. One substance, a (racemic) mixture, was in the nematic phase for temperatures in the range 20–94°C. For this mixed nematic at around room temperature they found that $V_c \sim 3$V at 1kHz with around a 90% transparency when V was increased to about 6V. For further details the reader is naturally referred to their paper [241], while for a brief physically based review of the details the reader may consult Chandrasekhar [38].

It should be mentioned that Fraser [92] has carried out a theoretical investigation of the twisted nematic device that incorporates surface pretilt of the director and electric field effects. Much of the analysis is naturally extended from the ideas presented on pretilt in Section 3.4.2 and electric field effects in Section 3.5.

3.8 Defects

Liquid crystal samples may contain defects where the director **n** is undefined, that is, where there is a discontinuity in the orientation of **n**. Discontinuities can be located at a point, line, or on a surface and are referred to as point, line or sheet defects, respectively. From physical arguments, sheet defects are considered unstable since they tend to smear out into structures called walls and are therefore omitted from the discussion presented here: information on walls can be found in the book by de Gennes and Prost [110]. The most common defects in liquid crystals are therefore point and line defects, usually called point and line *disclinations*, this terminology being slightly modified from that introduced by Frank [91], who originally called them 'disinclinations'. Attention is first focused on axial line (wedge) disclinations and their associated Frank index and strength before moving on to perpendicular (twist) disclinations, boundary line disclinations and point disclinations (point defects) at a surface; loop disclinations require more extensive modelling and for these the reader is referred to the article by Nehring [211]. We do not consider general point defects or the interaction between disclinations: for a concise review of these topics the reader is referred to Chandrasekhar [38, Ch.3.5]. The discussion will be restricted to nematics although results for cholesterics will be mentioned briefly when appropriate. A concise and accessible account of defects can be found in the review by Stephen and Straley [258] while more details on the physics of defects can be seen in de Gennes and Prost [110]. There is also a review of results for general liquid crystal defects by Bouligand [23].

3.8.1   Axial Line Disclinations

The nematic state is named after the thread-like lines observed in these materials, the word nematic being derived from the Greek word for thread. These lines are actually line singularities such that the direction of the preferred orientation of **n** changes by a multiple of π as an observer travels a complete circuit taken around one of these lines: this characterisation will become clearer when the equilibrium solutions are discussed and pictured below. In axial disclinations we shall suppose that the line disclination lies along the z-axis and that the director **n** always lies parallel to the xy-plane, effectively making the geometry two-dimensional. Such disclination lines have been called axial by Friedel and de Gennes [95] since the rotation of **n** required to form the disclination line occurs about an axis parallel to the disclination. Axial disclinations, sometimes called 'wedge' disclinations, were considered by Oseen [215] and later by Frank [91] and Dzyaloshinskii [70].

In order to simplify the presentation the one-constant approximation will be assumed for the nematic energy density w_F, given by equation (2.67). To analyse axial disclinations we consider configurations in which the director referred to Cartesian axes takes the form

$$\mathbf{n} = (\cos\theta, \sin\theta, 0), \qquad \theta = \theta(x, y), \qquad (3.334)$$

so that it always lies in the xy-plane perpendicular to the z-axis which represents the disclination. Figure 3.19 shows the orientation of **n** around such a disclination placed at the origin, described by Cartesian (x, y) or polar (r, ϕ) coordinates. With these assumptions an easy calculation reveals that the total energy over a unit length in z is

$$W = \tfrac{1}{2}K \int_A \left(\theta_x^2 + \theta_y^2\right) dx dy, \qquad (3.335)$$

where A is an area in the xy-plane and θ_x and θ_y denote the partial derivatives of θ with respect to x and y, respectively. Since the energy integrand only depends upon the single angle θ, the equilibrium equation is given by the one angle case of equation (2.221), resulting in the usual Euler–Lagrange equation which, in this example, reduces to Laplace's equation

$$\theta_{xx} + \theta_{yy} = 0. \qquad (3.336)$$

It will be instructive to solve this equation in the cylindrical coordinates introduced in Fig. 3.19 by setting

$$x = r\cos\phi, \quad y = r\sin\phi, \quad \text{with} \quad r = \sqrt{x^2 + y^2}, \quad \tan\phi = \frac{y}{x}. \qquad (3.337)$$

The equilibrium equation (3.336) then becomes Laplace's equation in cylindrical coordinates, which is, in the obvious notation (cf. equation (C.7) in Appendix C),

$$\theta_{rr} + \frac{1}{r}\theta_r + \frac{1}{r^2}\theta_{\phi\phi} = 0. \tag{3.338}$$

For axial disclinations, θ is expected to be independent of r and so in this case equation (3.338) collapses to

$$\theta_{\phi\phi} = 0, \tag{3.339}$$

which is easily integrated to find that

$$\theta = c\phi + \phi_0, \tag{3.340}$$

where c and ϕ_0 are constants. Since $\mathbf{n}$ and $-\mathbf{n}$ are indistinguishable it follows that when ϕ increases by 2π then θ must change by an integral multiple of π in order to preserve continuity of the director away from the disclination. This clearly shows that c must be of the form $\frac{n}{2}$ for some integer n. Hence, by equations $(3.337)_4$ and (3.340), we arrive at the solutions indicated by Frank [91], namely,

$$\theta = \frac{n}{2}\phi + \phi_0, \qquad \tan\phi = \frac{y}{x}. \tag{3.341}$$

The integer n is called the *Frank index* of the disclination. It is worth noting that in some articles, for example Bouligand [22] or Chandrasekhar [38], the order or strength of a disclination has been defined by $s = \frac{n}{2}$.

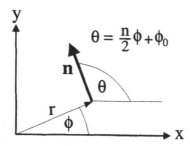

Figure 3.19: The Cartesian coordinates (x, y) and polar coordinates (r, ϕ) used to specify the orientation of the director $\mathbf{n}$ around a disclination which is perpendicular to the page and coincident with the z-axis placed at the origin. The director makes an angle θ with the x-axis and only depends upon the azimuthal angle ϕ. By equation (3.341), θ is of the form indicated where the integer n is the Frank index of the disclination.

An interpretation of the Frank index can be given as follows. An observer carries a unit vector $\mathbf{u}$, say, and begins at $\phi = 0$ to travel one complete circuit in the positive sense (anti-clockwise) around the origin in Fig. 3.19, always ensuring that $\mathbf{u}$ remains parallel to $\mathbf{n}$ throughout the journey. As ϕ increases, the direction of $\mathbf{u}$ changes continuously as it moves around the disclination while following the alignment of the director. Once the observer has travelled through 2π radians it will be found that **u** has either returned to its original orientation or it will have the reverse of its original orientation. The absolute value of the Frank index, $|n|$, indicates the number of times the director **n** (or the vector **u**) 'overlaps' with itself during the complete circuit taken by the observer, bearing in mind that **n** and $-\mathbf{n}$ are indistinguishable. When n is even, **u** overlaps exactly with itself upon its return, while if n is odd the orientation of **u** is reversed. When n is positive, **u** rotates anti-clockwise during the observer's journey; if n is negative then **u** will rotate clockwise.

A disclination line can be formed by a physical process reviewed by Stephen and Straley [258, p.635]. Details on the physics of this Volterra process, in which the sign of the Frank index plays a rôle, can be found in [258].

The pattern of the 'flux lines', that is, the lines which are tangent to **n**, is obtained by first finding the slope of such lines. In Cartesians the slope is given by the second component of the director divided by the first component so that by (3.334) they are solutions of

$$\frac{dy}{dx} = \tan\theta. \tag{3.342}$$

Let $\{\mathbf{e}_1, \mathbf{e}_2, \mathbf{e}_3\}$ and $\{\mathbf{e}_r, \mathbf{e}_\phi, \mathbf{e}_z\}$ denote the usual sets of basis vectors for the Cartesian and cylindrical polar coordinate systems, respectively. It is well known that

$$
\begin{aligned}
\mathbf{e}_1 &= \cos\phi\,\mathbf{e}_r - \sin\phi\,\mathbf{e}_\phi \\
\mathbf{e}_2 &= \sin\phi\,\mathbf{e}_r + \cos\phi\,\mathbf{e}_\phi, \\
\mathbf{e}_3 &= \mathbf{e}_z,
\end{aligned}
\tag{3.343}
$$

and hence **n** may be written in cylindrical coordinates as

$$\mathbf{n} = \cos(\theta - \phi)\,\mathbf{e}_r + \sin(\theta - \phi)\,\mathbf{e}_\phi. \tag{3.344}$$

This allows the cylindrical polar equivalent of equation (3.342) to be given by

$$\frac{1}{r}\frac{dr}{d\phi} = \cot(\theta - \phi), \tag{3.345}$$

recalling the influence of the usual scale factors. From this equation it follows that there are singular radial flux lines emanating from the origin and having direction determined by the relation

$$\theta - \phi = \mu\pi, \qquad \mu \text{ an integer.} \tag{3.346}$$

Hence, employing the solution (3.341) for θ, the singular radial lines originate from the origin whenever

$$\phi = 2\frac{\mu\pi - \phi_0}{n - 2}, \qquad n \neq 2, \qquad \mu \text{ an integer.} \tag{3.347}$$

The number of such lines is clearly $|n - 2|$. In the special case $n = 2$ we can solve equation (3.345) using (3.341) to find that

$$r = c\exp\{\phi\cot\phi_0\}, \qquad c \text{ a constant.} \tag{3.348}$$

We are now in a position to picture the flux lines following the orientation of **n** around an axial disclination for various values of the Frank index and constant ϕ_0. To find the flux lines in particular cases we solve the differential equation (3.342) after substituting for θ in the solution (3.341) for fixed values of the arbitrary constant ϕ_0. In all cases, except the solution for $n = 2$ in equation (3.348), changing the constant ϕ_0 merely rotates the flux lines shown in Fig. 3.20 below, and so we set $\phi_0 \equiv 0$ except for two of the instances when $n = 2$. It suffices to demonstrate the technique for a few cases: the others are in a similar style. For $n = -2$ and $\phi_0 = 0$, equation (3.342) can be solved to find that the flux lines away from the disclination are given by

$$y = \frac{c}{x}, \qquad c \text{ an arbitrary constant.} \tag{3.349}$$

Also, the number of singular radial lines emanating from the origin is $|-2-2| = 4$ and by the result in equation (3.347) they occur in this example at $\phi = -\mu\pi/2$ for integers μ. Clearly, we need only take the lines occurring at $\phi = 0, \pi/2, \pi$ and $3\pi/2$. The resulting orientation is shown in the first diagram in Fig. 3.20, the director always being tangential to the displayed flux lines; the bold lines represent the singular radial lines. In some of the examples it is easier to use the polar form in equation (3.345) for identifying the flux lines; this is particularly true when $n = 2$ and $\phi_0 \neq 0$ where the solutions are given by equation (3.348) (when $n = 2$ and $\phi_0 = 0$ then $\theta = \phi$ and the orientation displayed in Fig. 3.20 is then obvious from the specifications in Fig. 3.19). For example, the flux lines in the last diagram in Fig. 3.20 for $n = 4$ and $\phi_0 = 0$ are given by the equations

$$r = c|\sin(\phi)|, \tag{3.350}$$

with c being an arbitrary positive constant. Further, in this case, the number of singular radial lines emanating from the origin is $|4 - 2| = 2$ and they occur at $\phi = 0$ and $\phi = \pi$, by (3.347).

The energy W in equation (3.335) for an axial disclination can be evaluated most easily by transforming to cylindrical coordinates. For θ given by (3.341) we have, upon using equation (3.337) and the usual chain rule for partial derivatives,

$$\theta_x^2 + \theta_y^2 = \frac{1}{4}\frac{n^2}{r^2}. \tag{3.351}$$

The total energy over a circular annulus having inner radius $r_0 > 0$ and outer radius r_1 placed around the disclination at $r = 0$ is therefore

$$W = \frac{1}{8}K\int_0^{2\pi}\int_{r_0}^{r_1}\frac{n^2}{r}\,drd\phi = \frac{1}{4}K\pi n^2 \ln\left(\frac{r_1}{r_0}\right). \tag{3.352}$$

This energy clearly diverges as $r_0 \to 0$, that is, the energy becomes unbounded as the disclination is approached. Another observation is that all solutions sharing the same absolute value for their Frank index also have the same energy. The region occupying the volume with $0 \leq r \leq r_0$ is often called the *core* and the inner radius r_0 is called the *core radius*. The core radius is usually estimated to be of the order of molecular dimensions [110, p.171].

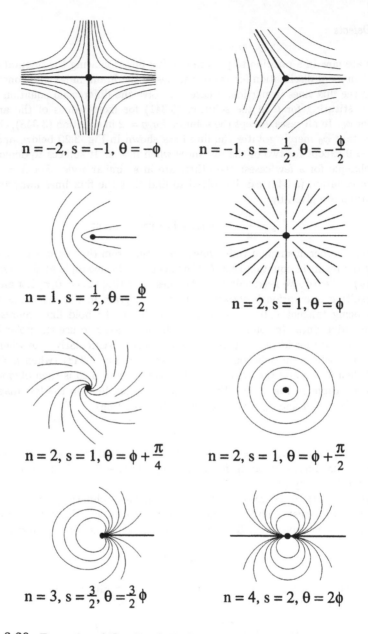

Figure 3.20: Examples of 'flux lines' which are tangential to the orientation of the director **n** around an axial line disclination located perpendicular to the page and passing through the point indicated by a black dot. Various cases of Frank index n are given together with the associated solution θ provided by equation (3.341). The strength of such disclinations is often defined by $s = \frac{n}{2}$. The bold lines represent the singular radial lines obtained from equation (3.347). The constant ϕ_0 has been set to zero except for the examples of Frank index $n = 2$.

An 'Escape' Solution

The occurrence of a divergent energy has led to investigations for alternative solutions having finite energies. It is possible in certain cases to find an appropriate solution possessing a finite energy and to this end we now go on to discuss a celebrated result due to Cladis and Kléman [40] and Meyer [199]. Consider a nematic liquid crystal confined to a long cylindrical capillary of radius R with the z-axis coincident with the cylinder axis and suppose that the director is fixed perpendicular to the surface of the capillary as shown in Fig. 3.21. In the cylindrical coordinates

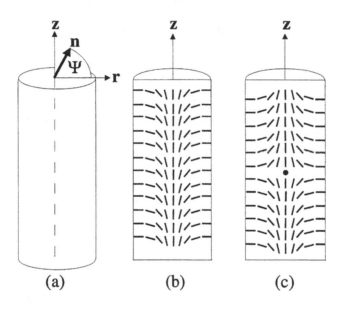

Figure 3.21: (a) The orientation angle Ψ for n given by equation (3.353) when referred to cylindrical coordinates used to describe a circular cylindrical capillary. (b) A cross-section through any plane containing the z-axis, showing a representation of the director for the escape solution $\Psi(r)$ given by equation (3.369). (c) A schematic representation of a cross-section showing a physical interpretation of the solutions $\pm\Psi(r)$ around a point defect, represented by a bold dot.

introduced above at equation (3.337), solutions of the form

$$\mathbf{n} = \cos\Psi\,\mathbf{e}_r + \sin\Psi\,\mathbf{e}_z, \qquad \Psi = \Psi(r), \tag{3.353}$$

can be sought. For simplicity, we adopt the one-constant approximation for the energy given by equation (2.64). Routine calculations using the standard formulae for cylindrical polar coordinates given by equations (C.5) and (C.6) in Appendix C reveal that

$$\nabla \cdot \mathbf{n} = \frac{1}{r} \cos \Psi - \frac{d\Psi}{dr} \sin \Psi, \qquad (3.354)$$

$$(\nabla \times \mathbf{n}) = -\frac{d\Psi}{dr} \cos \Psi \mathbf{e}_\phi, \qquad (3.355)$$

while, for current and future use, an application of the formula (C.4) from Appendix C shows

$$\nabla \cdot [(\mathbf{n} \cdot \nabla)\mathbf{n} - (\nabla \cdot \mathbf{n})\mathbf{n}] = \frac{2}{r} \frac{d\Psi}{dr} \sin \Psi \cos \Psi. \qquad (3.356)$$

These quantities allow the total energy given by equation (2.64) over a sample of unit depth in z to be written as

$$W = \pi K \int_0^R \left[r \left(\frac{d\Psi}{dr} \right)^2 + \frac{\cos^2 \Psi}{r} \right] dr, \qquad (3.357)$$

after an obvious integration in ϕ over the range $0 \leq \phi \leq 2\pi$. (Alternatively, the formulae (C.8) to (C.10) could be used in equation (2.67) to obtain (3.357).) The resulting Euler–Lagrange equilibrium equation for Ψ is provided by equation (2.226) (with Ψ corresponding to only one 'θ_α' dependence) and is

$$r^2 \frac{d^2\Psi}{dr^2} + r \frac{d\Psi}{dr} + \sin \Psi \cos \Psi = 0, \qquad (3.358)$$

with the accompanying boundary conditions

$$\Psi(R) = 0, \qquad \Psi(0) = \Psi_0, \qquad (3.359)$$

where Ψ_0 is a constant that will be determined below. It is worth pointing out here that if $\Psi_0 = 0$ then $\Psi \equiv 0$ solves equations (3.358) and (3.359): this solution corresponds to the solution around a disclination along the z-axis displayed in Fig. 3.20 for the case $n = 2$, $\theta = \phi$, taking into account the orientation defined in Fig. 3.19. It has the energy given by equation (3.352) with $n = 2$, which diverges when $r \to 0^+$ (this of course coincides with the energy given by equation (3.357) if its upper and lower limits of integration are replaced by r_1 and $r_0 > 0$, respectively). In an attempt to overcome this difficulty of having a divergent energy, equation (3.358) is most easily analysed for non-zero solutions via the change of variable

$$s = \ln r, \qquad s_0 = \ln R, \qquad (3.360)$$

which allows it to be formulated as

$$\frac{d^2\Psi}{ds^2} + \sin \Psi \cos \Psi = 0, \qquad (3.361)$$

with the boundary conditions becoming

$$\Psi(s_0) = 0, \qquad \Psi(-\infty) = \Psi_0. \qquad (3.362)$$

Equation (3.361) can be multiplied by $\frac{d\Psi}{ds}$ and integrated once to yield

$$\left(\frac{d\Psi}{ds}\right)^2 + \sin^2 \Psi = c, \tag{3.363}$$

where c is a constant of integration.

The energy per unit length in z of the capillary for this solution is simply given by equation (3.357) using the change of variable in equation (3.360). This gives

$$W = \pi K \int_{-\infty}^{s_0} \left[\left(\frac{d\Psi}{ds}\right)^2 + \cos^2 \Psi\right] ds, \tag{3.364}$$

after noting that $s \to -\infty$ as $r \to 0^+$. For this energy to be finite it is necessary that each term in the integrand tends to zero as $s \to -\infty$ and it is therefore required that

$$\cos \Psi \to 0, \quad \frac{d\Psi}{ds} \to 0, \quad \text{as} \quad s \to -\infty. \tag{3.365}$$

It is therefore possible to set

$$\Psi_0 = \frac{\pi}{2}, \quad c = 1, \tag{3.366}$$

these constants being consistent with what is required in equations (3.362), (3.363) and (3.365). Equation (3.363) is then

$$\frac{d\Psi}{ds} = -\cos \Psi, \tag{3.367}$$

the negative sign being chosen because the anticipated solution will decrease as s increases from $-\infty$, that is, as r increases from zero. This equation integrates to give the solution via equation (3.360) in terms of r as [116, 2.526.9]

$$-\ln r + c_2 = \ln \left|\tan\left(\frac{\pi}{4} + \frac{\Psi}{2}\right)\right|, \tag{3.368}$$

with c_2 being a constant of integration. Applying the zero boundary condition at $r = R$ stated in equation (3.359) shows that $c_2 = \ln R$ and then the solution in (3.368) can finally be rewritten as

$$\Psi(r) = 2 \tan^{-1}\left(\frac{R}{r}\right) - \frac{\pi}{2}. \tag{3.369}$$

This solution has no singularity and clearly solves the equilibrium equation (3.358) with boundary conditions (3.359) when $\Psi_0 = \frac{\pi}{2}$; it is pictured schematically in Fig. 3.21(b), the short bold lines representing the director alignment in a cross-section of the cylindrical capillary. A similar solution occurs if the positive sign is chosen in equation (3.367): the solution then becomes $-\Psi(r)$ for Ψ given above in equation (3.369), with the boundary condition (3.359)$_2$ changed to $\Psi_0 = -\frac{\pi}{2}$.

The energy per unit length in z is easily found by inserting the derivative of Ψ given by equation (3.367) into the energy in equation (3.364) and using the conditions (3.362) with $\Psi_0 = \frac{\pi}{2}$ from equation (3.366). This gives

$$W = 2\pi K \int_{-\infty}^{s_0} \cos^2 \Psi \, ds = 2\pi K \int_0^{\frac{\pi}{2}} \cos \Psi \, d\Psi = 2\pi K. \qquad (3.370)$$

This energy is πK less than that obtained by Cladis and Kléman [40], due to their different one-constant approximation in which $K_2 + K_4$ was set to zero rather than simply K_4 being set to zero, as has been adopted here (see the comments after equation (2.62)). The solution Ψ in (3.369) has finite energy and exhibits what is commonly called 'escape into the third dimension'. The escape solutions for various cases when the elastic constants are not equal have also been discussed in [40]. It is worth remarking that if all the elastic constants are set equal to K except for a non-zero K_4, rather than the one-constant approximation (2.62), then $W = \pi(2K - K_4)$, by calculations similar to the above; this result is easily obtained by observing, via (3.356) and the energy in equation (2.49), that the corresponding additional K_4 contribution to the total energy is

$$
\begin{aligned}
\pi K_4 \int_0^R \nabla \cdot [(\mathbf{n} \cdot \nabla)\mathbf{n} - (\nabla \cdot \mathbf{n})\mathbf{n}] r \, dr &= 2\pi K_4 \int_{-\infty}^{s_0} \frac{d\Psi}{ds} \sin \Psi \cos \Psi \, ds \\
&= -\pi K_4 \int_{-\infty}^{s_0} \frac{d}{ds} \cos^2 \Psi \, ds \\
&= -\pi K_4, \qquad (3.371)
\end{aligned}
$$

using the conditions (3.362) and (3.366)$_1$. If $K_2 + K_4 = 0$ with $K_2 = K$ then $W = 3\pi K$ and the result of Cladis and Kléman [40] and Meyer [199] is recovered. (Recall that as long as the splay, twist and bend constants are set equal then the equilibrium equations remain the same for any value of K_4 when strong anchoring is assumed: only the total finite energy will be modified by an identifiable constant contribution, as indicated by the above.)

The alternative solution for $-\Psi$ mentioned above also has exactly the same energy as that given by equation (3.370). This means that there are two equivalent directions for escape into the third dimension and this can be interpreted as leading to point defects appearing along the axis of the capillary as observed by Williams, Pieranski and Cladis [279]: a schematic diagram of such a director orientation is given in Fig. 3.21(c). This topic is beyond our present discussion and interested readers are referred to the review by Cladis and van Saarloos [43].

The procedure outlined here for an escape solution which overcomes the problem of having a divergent energy for a line defect is not always available and has led to a search for alternative models for some types of disclinations.

3.8.2   Perpendicular Disclinations

Perpendicular disclinations, sometimes called 'twist' disclinations, occur in nematics and have been discussed by de Gennes [105]. In a perpendicular disclination it can be supposed that the line disclination lies along the y-axis and that the director $\mathbf{n}$ always lies parallel to the xy-plane. To achieve such an orientation we can take, in Cartesian coordinates,

$$\mathbf{n} = (\cos\theta, \sin\theta, 0), \qquad \theta = \theta(x, z). \tag{3.372}$$

To simplify matters we again make the one-constant approximation to the nematic energy density w_F in equation (2.64) or its equivalent form in (2.67). The total energy over a unit length in y is then

$$W = \tfrac{1}{2}K \int_A \left(\theta_x^2 + \theta_z^2\right) dx dz, \tag{3.373}$$

where A is an area in the xz-plane perpendicular to the disclination along the y-axis. This leads, via (2.221), to the equilibrium equation

$$\theta_{xx} + \theta_{zz} = 0. \tag{3.374}$$

By arguments completely analogous to those for obtaining equations (3.336) to (3.341) above, we arrive at the solution

$$\theta = \frac{n}{2}\phi + \phi_0, \qquad \tan\phi = \frac{z}{x}, \tag{3.375}$$

with ϕ_0 an arbitrary constant and n an integer, where r in equation (3.337) is replaced by $r = \sqrt{x^2 + z^2}$. Here, ϕ represents the angle between the x-axis and a line drawn within the xz-plane emanating from the origin to another point in the plane, as shown in Fig. 3.22. The orientation of $\mathbf{n}$ within the xy-planes is described by θ.

As an elementary example, consider a perpendicular disclination of Frank index $n = 1$, that is, of strength $s = \tfrac{1}{2}$, in its simplest configuration having $\phi_0 = 0$ in equation (3.375). Then, in the plane $z = 0$, we can clearly take $\phi = 0$ for $x > 0$ and $\phi = \pi$ for $x < 0$ which, by equation (3.375), leads to

$$\theta(x, 0) = \begin{cases} 0 & \text{if } x > 0, \\ \frac{\pi}{2} & \text{if } x < 0, \end{cases} \tag{3.376}$$

recalling that θ is the angle depicted in the xy-plane, shown in Fig. 3.19, the main difference being that $\mathbf{n}$ now varies with x and z rather than x and y. The resulting flux lines representing the orientation of the director for such a perpendicular 'twist' disclination are shown in Fig. 3.22, the bold line along the y-axis representing the disclination. Typical representations of the alignment in the xy-plane are shown for values of $z < 0$, $z = 0$, and $z > 0$ as indicated. The disclination corresponding to $n = -1$, $s = -\tfrac{1}{2}$, is obtained by reflection in the xy-plane at $z = 0$ in this Figure. Changing the constant ϕ_0 from zero simply rotates the depicted alignment inside each xy-planar cross-section by a constant angle ϕ_0.

The total elastic energy W per unit length in y for a circular annulus in the xz-plane centred around the disclination along the y-axis at $r = 0$ is analogous

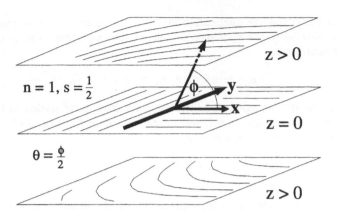

Figure 3.22: Examples of flux lines parallel to the orientation of the director **n** around a perpendicular disclination represented by a bold line placed along the y-axis. This particular case is for Frank index $n = 1$, that is, for strength $s = \frac{1}{2}$. The solution for θ is given by equation (3.375) with $\phi_0 = 0$. The director alignment is shown for cross-sections in the xy-plane for $z < 0$, $z = 0$ (derived from equation (3.376)) and $z > 0$. The interpretation of θ is similar to that depicted in Fig. 3.19.

to that given by equations (3.351) and (3.352), with z now taking the rôle of y; the inner and outer radii of the annulus are again given by r_0 and r_1, respectively. Hence

$$W = \frac{1}{4}K\pi n^2 \ln\left(\frac{r_1}{r_0}\right),\qquad(3.377)$$

as for axial disclinations. The comments immediately following the axial disclination energy at equation (3.352) are also applicable here.

Examples of perpendicular disclinations of Frank index $n = 2$, $s = 1$, can be found in Fig. 11 in the review by Stephen and Straley [258]. Perpendicular disclinations can also occur in cholesterics and are important to the understanding of Grandjean planes. These and other related topics are beyond the scope of this introductory text and the reader is therefore referred to the aforementioned review by Stephen and Straley.

3.8.3   Boundary Line Disclinations

It is instructive to examine boundary disclination lines at a nematic surface or interface by introducing some additional modelling and approximations that incorporate surface tension and the effect due to gravity. Many of these aspects introduced here in this Section are common throughout the literature on liquid crystals and will also form a basis for the discussion on point defects at a free surface of nematic discussed in the next Section. The results presented below are based on those derived by de Gennes [107] and have been further elucidated in physical terms by de Gennes and Prost [110].

Consider a horizontal magnetic field **H** applied at the surface of a nematic liquid crystal when it is assumed that the director **n** exhibits conical anchoring and therefore makes a fixed angle ψ with the tangent plane to the surface. It is assumed that initially the surface of the nematic is located in the plane $z = 0$ and that the sample occupies the semi-infinite region $z \leq 0$ with **n** parallel to **H** as $z \to -\infty$, the magnetic anisotropy χ_a being assumed positive. As the magnitude of **H** increases it is anticipated that the original flat nematic surface will distort, bearing in mind that there may be a critical magnitude of **H** which must be reached before the surface will be displaced or distorted. This will indeed turn out to be the case and will lead to possible disclination lines on the boundary surface.

Let the displacement of the liquid crystal surface in the z-direction be uniform throughout the sample in the y-direction so that it can be represented by the function $\zeta(x)$. Figure 3.23 shows the expected geometry of the two types of possible local distortions of the surface with ψ being the fixed (positive) conical anchoring angle for **n** at the surface, as introduced in Fig. 2.4 on page 48; ϵ denotes the small angle that the distorted surface makes locally with the x-axis and the magnetic field **H** is applied in the x-direction, as shown in Fig. 3.23. It is assumed that

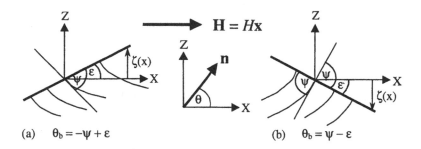

(a) $\theta_b = -\psi + \epsilon$ (b) $\theta_b = \psi - \epsilon$

Figure 3.23: The two possible local distortions of the surface of a nematic with conical anchoring. The curved lines represent the alignment of the director and θ_b is the boundary condition for θ at the surface, ψ being the fixed conical anchoring angle. The displacement of the surface from its original $z = 0$ flat orientation is denoted by $\zeta(x)$. The small angle the distorted surface makes locally with the x-axis is denoted by ϵ, greatly exaggerated here for illustration.

$$\mathbf{n} = (\cos\theta, 0, \sin\theta), \qquad \theta = \theta(z), \tag{3.378}$$

and

$$\mathbf{H} = H(1, 0, 0), \tag{3.379}$$

with boundary conditions on θ given by

$$\begin{aligned}
\theta_b &= -\psi + \epsilon && \text{in Fig. 3.23(a)}, \tag{3.380} \\
\theta_b &= \psi - \epsilon && \text{in Fig. 3.23(b)}. \tag{3.381}
\end{aligned}$$

It should be noted here that, for example, θ_b in Fig. 3.23(a) could be replaced by the condition $\widehat{\theta}_b = \psi + \epsilon$, which is also compatible with the same conical anchoring condition (see de Gennes and Prost [110, p.131]). However, this means that for $0 < \epsilon \ll 1$ we have $|\widehat{\theta}_b| > |\theta_b|$ and so the energy, which must be related to the magnitude of the distortion θ, could be greater for $\widehat{\theta}_b$ as a boundary condition rather than θ_b. Hence $\theta_b = -\psi + \epsilon$ is expected to be an energetically more favourable boundary condition. Similar comments apply to Fig. 3.23(b). These observations motivate the boundary conditions depicted here and will be relevant to Fig. 3.24 below.

We begin by determining the surface energy arising from the one-constant approximation for the nematic energy and the magnetic energy, returning later to examine additional energies involving surface tension and gravity. Substituting $\mathbf{n}$ and $\mathbf{H}$ given by equations (3.378) and (3.379) into equations (2.64) (or (2.67)) and (2.108) gives a contribution to the energy per unit length in y which incorporates the distorted surface, namely,

$$
W = \frac{1}{2} \int\!\!\int_{-\infty}^{\zeta(x)} \left[K\left(\frac{d\theta}{dz}\right)^2 + \chi_a H^2 \sin^2\theta \right] dz\,dx, \tag{3.382}
$$

where $H = |\mathbf{H}|$ and it is assumed that $\chi_a > 0$. The Euler–Lagrange equilibrium equation is therefore

$$
\xi^2 \frac{d^2\theta}{dz^2} = \sin\theta\cos\theta, \qquad \xi = \frac{1}{H}\sqrt{\frac{K}{\chi_a}}, \tag{3.383}
$$

where ξ is now the one-constant approximation to the magnetic coherence length of the nematic (cf. Section 3.3 and equation (3.98)). Motivated by the geometry of Fig. 3.23, the solution for θ must satisfy the lower boundary conditions

$$
\frac{d\theta}{dz} \to 0 \quad \text{and} \quad \theta \to 0 \qquad \text{as} \quad z \to -\infty, \tag{3.384}
$$

conditions which are clearly compatible with the integrand in equation (3.382). The first integral of equation (3.383) is easily obtained by multiplying throughout by $\frac{d\theta}{dz}$ and integrating to find

$$
\xi^2\left(\frac{d\theta}{dz}\right)^2 = \sin^2\theta + c, \tag{3.385}
$$

where c is a constant. Applying the lower boundary conditions (3.384) shows that $c = 0$ and hence the solution for θ must satisfy

$$
\xi\frac{d\theta}{dz} = \sin\theta, \tag{3.386}
$$

with the sign in front of $\sin\theta$ chosen positive to match the expected orientation of θ in Fig. 3.23; for example, $\frac{d\theta}{dz}$ has the same sign as θ, which can be negative or positive.

It is now possible to derive the surface energy contributions arising from the usual nematic and magnetic energies. Attention will first be restricted to the energy relevant to Fig. 3.23(a) when $\theta_b = -\psi + \epsilon$. Inserting equation (3.386) into the energy (3.382) and employing the derivative $\frac{dz}{d\theta}$ which is easily obtained from (3.386), the energy becomes

$$
\begin{aligned}
W &= \frac{1}{2}K \int \int_{-\infty}^{\zeta(x)} \left[\left(\frac{d\theta}{dz} \right)^2 + \frac{\sin^2 \theta}{\xi^2} \right] dzdx \\
&= \frac{K}{\xi^2} \int \int_{-\infty}^{\zeta(x)} \sin^2 \theta \, dzdx \\
&= \frac{K}{\xi} \int \int_0^{\theta_b = -\psi + \epsilon} \sin \theta \, d\theta dx \\
&= \frac{K}{\xi} \int [1 - \cos(\psi - \epsilon)] \, dx.
\end{aligned}
\tag{3.387}
$$

To first order in ϵ this energy, for a finite domain in x, is

$$
W = \frac{K}{\xi} \int (1 - \cos \psi - \epsilon \sin \psi) dx = -\sin \psi \frac{K}{\xi} \int \epsilon \, dx \; + \; \text{constant}, \tag{3.388}
$$

recalling that ψ is a fixed constant angle. From Fig. 3.23(a) it is seen that for $0 < \epsilon \ll 1$, $\zeta(x) \approx \epsilon x$ and so

$$
\epsilon \approx \frac{d\zeta}{dx}. \tag{3.389}
$$

Similar arguments apply for the case pictured in Fig. 3.23(b) for $\theta_b = \psi - \epsilon$ except that in the above approximation $\epsilon \approx -\frac{d\zeta}{dx}$. Hence, ignoring constant contributions to the energy in (3.388), we easily arrive at the relevant term in the surface energy, namely,

$$
F_1 = \pm \sin \psi \frac{K}{\xi} \int \frac{d\zeta}{dx} \, dx, \tag{3.390}
$$

the minus sign being adopted when $\theta_b = -\psi + \epsilon$ in Fig. 3.23(a) is considered and the plus sign when $\theta_b = \psi - \epsilon$ in Fig. 3.23(b) is considered.

To the energy F_1 above must be added the conventional surface terms for a fluid at an interface where another medium is above it. From Landau and Lifshitz [159, p.232] this energy is written as

$$
F_2 = \gamma \int_S dS + \rho g \int_V z \, dxdydz, \tag{3.391}
$$

where γ is the surface tension (measured as erg cm^{-2} = dyn cm^{-1} in cgs units, or J m^{-2} = N m^{-1} in SI units), dS is a surface element at the interface, ρ is the difference in density between the fluid and the medium above it, g is the magnitude of the acceleration due to gravity, and z is the vertical position of the interface or surface above the initial undistorted position of the surface when it is coincident with the plane $z = 0$. The first term in F_2 is the energy due to the surface tension γ and the second term is the energy due to the effect of gravity. Since the particular problem described here is independent of y, the integrals in F_2 can be evaluated over a unit depth in y. Also, by a well known formula, the first integral in F_2 is effectively linked with an arc length and is therefore

$$\gamma \int_S dS = \gamma \int \sqrt{1 + \left(\frac{d\zeta}{dx}\right)^2}\, dx \doteq \frac{\gamma}{2} \int \left(\frac{d\zeta}{dx}\right)^2 dx + c, \qquad (3.392)$$

with c a constant, when the integration is over a finite domain in x and $\zeta(x)$ is assumed small and varying only slowly with respect to x. The second integral in F_2 becomes

$$\rho g \int\int_0^{\zeta(x)} z\,dz\,dx = \tfrac{1}{2}\rho g \int \zeta^2(x)dx. \qquad (3.393)$$

Hence F_2 in this present example is the sum of the results from equations (3.392) and (3.393) and can be written as

$$F_2 = \frac{\gamma}{2} \int \left\{ \left(\frac{d\zeta}{dx}\right)^2 + \frac{\zeta^2(x)}{\lambda^2} \right\} dx, \qquad \lambda = \sqrt{\frac{\gamma}{\rho g}}, \qquad (3.394)$$

when the constant contribution from equation (3.392) is neglected.

The total surface energy per unit length in y is the sum of F_1 and F_2 arising in equations (3.390) and (3.394) and is finally given by

$$
\begin{aligned}
F_{surf} &= F_1 + F_2 \\
&= \frac{1}{2} \int \left[\gamma \left\{ \left(\frac{d\zeta}{dx}\right)^2 + \frac{\zeta^2(x)}{\lambda^2} \right\} \pm 2\sin\psi \frac{K}{\xi}\frac{d\zeta}{dx} \right] dx, \qquad (3.395)
\end{aligned}
$$

the appropriate plus or minus sign being chosen according to the anticipated distortion in Fig. 3.23: minus for Fig. 3.23(a) and plus for Fig. 3.23(b).

A solution will now be sought which will correspond to a periodic array of disclination lines separated by the distance L as indicated in Fig. 3.24. The disclinations are perpendicular to the page and are located at the peaks and troughs appearing in the Figure. The Euler–Lagrange equation for F_{surf} provides the equilibrium equation

$$\frac{d^2\zeta}{dx^2} - \frac{\zeta}{\lambda^2} = 0. \qquad (3.396)$$

Suitable boundary and initial conditions must supplement this equation. Consider the interval of length L shown in Fig. 3.24 described by $-\frac{L}{2} \leq x \leq \frac{L}{2}$ where the initial undistorted surface originated midway between the peaks and troughs at $z = 0$. By simple geometric approximations we can then set

$$\zeta(0) = 0, \qquad (3.397)$$

$$\frac{d\zeta}{dx} = \epsilon, \quad \text{at } x = \pm\frac{L}{2}. \qquad (3.398)$$

The solution to equation (3.396) is then

$$\zeta(x) = \lambda\epsilon\frac{\sinh(x/\lambda)}{\cosh(L/2\lambda)}, \qquad -\frac{L}{2} \leq x \leq \frac{L}{2}. \qquad (3.399)$$

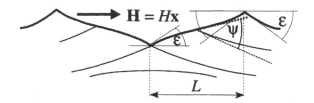

Figure 3.24: A periodic array of boundary line disclinations oriented perpendicular to the page, each separated by the distance L and located at the peaks and troughs of the nematic surface (in bold). The fixed conical anchoring angle is ψ and ϵ is the angle of the slope at a peak (or trough) measured relative to the x-axis. The director is parallel to the other curved lines.

To describe the array of disclinations we need to further impose the condition

$$\zeta(x + L) = -\zeta(x). \tag{3.400}$$

The relevant energy is provided by F_{surf} in (3.395) when the *minus* sign is chosen (compare Fig. 3.24 with Fig. 3.23(a)). Routine calculations for this solution on the interval $-\frac{L}{2} \leq x \leq \frac{L}{2}$ yield

$$F_{surf} = \lambda\gamma\tanh(u)\left[\epsilon^2 - 2\epsilon\frac{K}{\xi\gamma}\sin\psi\right], \tag{3.401}$$

$$u = \frac{L}{2\lambda}. \tag{3.402}$$

This energy is clearly minimised with respect to ϵ when

$$\epsilon = \frac{K}{\xi\gamma}\sin\psi, \tag{3.403}$$

and therefore the energy per unit area is $\widehat{F}_{surf} = F_{surf}/L$ for this value of ϵ, resulting in

$$\widehat{F}_{surf} = -\lambda\epsilon^2\frac{\gamma}{L}\tanh u. \tag{3.404}$$

The energy of a disclination at one of the 'cusp' points must also be added to $\widehat{F}_{surf}$ [107]. This can be estimated by treating the disclination as an axial disclination of strength $s = 1$ (see Fig. 3.20 on p.114), which is limited to a half-space, since we are examining a disclination at a surface. The integration in ϕ in the energy equation (3.352) therefore only takes place over the interval $0 \leq \phi \leq \pi$ with Frank index $n = 2$. The relevant energy per unit area around the disclination is then

$$\widehat{F}_{disc} = \frac{1}{2}\frac{K}{L}\int_0^\pi\int_{r_0}^{r_1}\frac{1}{r}\,dr\,d\phi = \frac{\pi}{2}\frac{K}{L}\ln\left(\frac{r_1}{r_0}\right), \tag{3.405}$$

where r_1 is some cut-off point away from the disclination and r_0 is close to the disclination. Defining the constant c_1 by

$$c_1 = \frac{\pi}{2}\ln\left(\frac{r_1}{r_0}\right), \tag{3.406}$$

allows the complete energy $\mathcal{F}_b$ to be given as

$$\mathcal{F}_b = \widehat{F}_{surf} + \widehat{F}_{disc}$$

$$= -\lambda\epsilon^2\frac{\gamma}{L}\tanh u + c_1\frac{K}{L}, \tag{3.407}$$

with $u = L/2\lambda$. This final form for the energy must now be minimised over L. Differentiation of $\mathcal{F}_b$ with respect to L reveals that it has a minimum whenever L satisfies the requirement

$$\frac{c_1 K}{\lambda\gamma\epsilon^2} = \tanh u - u\,\text{sech}^2(u). \tag{3.408}$$

The left-hand side of this expression can be re-expressed in terms of the field magnitude H via the definitions for ξ, λ and ϵ in equations $(3.383)_2$, $(3.394)_2$ and (3.403) leading to a condition that relates L and H, namely,

$$\left(\frac{H_1}{H}\right)^2 = \tanh u - u\,\text{sech}^2(u), \tag{3.409}$$

with H_1 defined by

$$H_1 = \left(\frac{c_1}{\chi_a}\right)^{\frac{1}{2}}\frac{(\gamma\rho g)^{\frac{1}{4}}}{\sin\psi}. \tag{3.410}$$

The right-hand side of expression (3.409) is always less than unity: it tends to 1 as $L \to \infty$ and to zero as $L \to 0$. Thus H_1 is a critical threshold value such that for $H > H_1$ the above solution (3.399) exists with the distance L between disclinations being determined by the value of H; L can be calculated implicitly via the expression (3.409) once values have been inserted for the various physical constants χ_a, γ, ρ, g, ψ and c_1. It is clear that $L \to 0$ as $H \to \infty$ and $L \to \infty$ as $H \to H_1^+$: hence increasing H decreases the distance between disclinations on the boundary surface. In the special case for a free surface with $\psi = \frac{\pi}{2}$ we can use the approximate values for the physical parameters (in cgs units) considered by de Gennes [107], namely,

$$\chi_a = 10^{-6}, \quad \gamma = 10 \text{ dyn cm}^{-1}, \quad \rho = 1 \text{ g cm}^{-3}, \quad g = 980 \text{ cm s}^{-2}, \quad c_1 = 10. \tag{3.411}$$

Putting these into equation (3.410) gives

$$H_1 \doteqdot 3.1 \times 10^4 \text{ oersteds}, \tag{3.412}$$

and so expected typical values for the threshold are of the order 10^4 oersteds. As remarked by de Gennes and Prost [110, p.176], L is typically of the order λ for $H > H_1$. In the above example, using equation (3.409) at $H = 6 \times 10^4$ oersteds, $\lambda \sim 0.10$ by (3.394) and $L \sim 0.18$ by (3.409). It will be seen from rough calculations in the next Section that point defects at the surface may be energetically preferred to boundary line disclinations when the same surface and boundary conditions are applied. Some more details on the physical interpretations of boundary line disclinations, including references to experiments, can be found in de Gennes and Prost [110, pp.174–176].

3.8.4   Point Defects at a Surface

We return to the formulae for F_1 and F_2 in equations (3.390) and (3.394) with the aim of suitably adapting them to describe a point defect at a free surface of a nematic liquid crystal sample. A radially symmetric solution for the boundary surface is envisaged in cylindrical polar coordinates as pictured in Fig. 3.25, where, as before, it is assumed that there is conical anchoring at the free surface with $\theta_b = \psi - \epsilon$, the angle θ being defined as shown and measured relative to the radial direction $\mathbf{r}$, analogous to that given in Fig. 3.23. Recall that ψ is a fixed positive angle. The point defect is located at the 'peak' of a small conical 'hill'. An

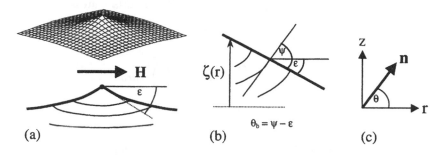

Figure 3.25: (a) A qualitative plot of the solution surface $\zeta(r)$ around a point defect given by equation (3.433) and, below, a cross-section through the conical 'hill' showing the point defect at the 'peak' and the slope of the surface around the defect. The curved lines represent the director alignment and $\mathbf{H}$ is the magnetic field. (b) An exaggerated view of the conical alignment at the boundary near the point defect. (c) The definition of the director orientation angle θ relative to the radial direction $\mathbf{r}$.

equally valid possibility is for the defect to be placed at the lower extremity of a conical 'valley', but this scenario is mathematically similar to the 'hill' case, which is the only one we shall consider. This set-up allows the description and machinery developed earlier for boundary line disclinations to be easily exploited for point defects at a surface, thereby enabling a viable comparison to be drawn between the results for boundary disclinations and those to be derived here for point defects. As for boundary disclinations, the magnetic field is again applied parallel to the surface; similar results hold when the magnetic field is applied perpendicular to the sample and $\theta = \frac{\pi}{2}$ on the lower boundary, as discussed by de Gennes [107], and these results can be readily obtained by amending the relevant equations in Section 3.8.3.

The relevant surface energy is again given by F_{surf} in equation (3.395) by the sum of F_1 and F_2 in cylindrical polar coordinates where $\zeta(r)$ is the displacement of the liquid crystal surface in the z-direction. For instance, the slope of the surface at the peak shown in Fig. 3.25 is clearly $-\epsilon$ and so, analogous to equation (3.389) for $\epsilon \ll 1$,

$$\epsilon \approx -\frac{d\zeta}{dr}, \tag{3.413}$$

and therefore F_1 with the plus sign in (3.390) is modified in this example via (3.388) to

$$F_1 = \sin\psi\frac{K}{\xi}\int_0^{2\pi}\int \frac{d\zeta}{dr}\,r\,dr\,d\phi = 2\pi\sin\psi\frac{K}{\xi}\int \frac{d\zeta}{dr}\,r\,dr, \tag{3.414}$$

after employing the usual change of variable rules when transforming to cylindrical polar coordinates. Recall that ξ is the characteristic length defined by (3.383)$_2$. Also note that the integral over the interval $0 \le \phi \le 2\pi$ is the analogue of taking the energy per unit length in the y-direction in the previous work on boundary line disclinations. Further, employing well known results for surface areas,

$$
\begin{aligned}
\gamma\int_S dS \; &= \; \gamma\int\sqrt{1+(\nabla\zeta)^2}\,dx\,dy \\
&= \; \gamma\int_0^{2\pi}\int\sqrt{1+\left(\frac{d\zeta}{dr}\right)^2}\,r\,dr\,d\phi \\
&\doteq \; \gamma\pi\int\left(\frac{d\zeta}{dr}\right)^2 r\,dr + c,
\end{aligned} \tag{3.415}
$$

with c a constant, when the integration is over a finite domain in r and $\zeta(r)$ is small, and

$$\rho g\int\int\int_0^{\zeta(x,y)} z\,dz\,dx\,dy = \rho g\pi\int \zeta^2(r)\,r\,dr, \tag{3.416}$$

where $\zeta(x,y)$ represents $\zeta(r)$ in Cartesian coordinates. The formula for F_2 given by equation (3.394) is therefore modified via equations (3.391), (3.415) and (3.416) to

$$F_2 = \gamma\pi\int\left\{\left(\frac{d\zeta}{dr}\right)^2 + \frac{\zeta^2(r)}{\lambda^2}\right\} r\,dr, \qquad \lambda = \sqrt{\frac{\gamma}{\rho g}}. \tag{3.417}$$

The total surface energy is the sum of F_1 and F_2 arising in equations (3.414) and (3.417) and is

$$
\begin{aligned}
F_{surf} \; &= \; F_1 + F_2 \\
&= \; \gamma\pi\int\left[\left(\frac{d\zeta}{dr}\right)^2 + \frac{\zeta^2(r)}{\lambda^2} + 2\epsilon\frac{d\zeta}{dr}\right] r\,dr, \qquad \epsilon = \sin\psi\frac{K}{\xi\gamma},
\end{aligned} \tag{3.418}
$$

the constant $-\epsilon$ giving the slope of the surface at the point defect, ξ being the aforementioned characteristic length defined in equation (3.383)$_2$. This value for ϵ may be obtained from geometric considerations, as explained by de Gennes and Prost [110, p.175].

The Euler–Lagrange equation for F_{surf} provides the equilibrium equation

$$r^2\frac{d^2\zeta}{dr^2} + r\frac{d\zeta}{dr} - \frac{r^2}{\lambda^2}\zeta = -\epsilon r, \tag{3.419}$$

with the boundary conditions

$$\frac{d\zeta}{dr} = -\epsilon \quad \text{at} \quad r = 0, \qquad \zeta(r) \to 0 \quad \text{as} \quad r \to \infty. \tag{3.420}$$

Using the transformation

$$\zeta = \lambda \epsilon i \frac{\pi}{2} v(x), \qquad x = i \frac{r}{\lambda}, \tag{3.421}$$

with $i^2 = -1$, equation (3.419) becomes

$$x^2 \frac{d^2 v}{dx^2} + x \frac{dv}{dx} + x^2 v = \frac{2}{\pi} x. \tag{3.422}$$

This is an inhomogeneous Bessel equation and its general solution is [1, p.496]

$$v(x) = a J_0(x) + b Y_0(x) + H_0(x), \quad a, b \text{ constants}, \tag{3.423}$$

where J_0, Y_0 are, respectively, Bessel functions of order zero of the first and second kinds and H_0 is the Struve function. Employing the relations [1, pp.375,498]

$$
\begin{aligned}
J_0(ir/\lambda) &= I_0(r/\lambda), \\
Y_0(ir/\lambda) &= i I_0(r/\lambda) - \frac{2}{\pi} K_0(r/\lambda), \\
H_0(ir/\lambda) &= i L_0(r/\lambda),
\end{aligned}
\tag{3.424}
$$

where I_0, K_0 are the associated modified Bessel functions of order zero of the first and second kinds, respectively, and L_0 is the associated modified Struve function, we can write the solution ζ via equations (3.421) and (3.423) as

$$\zeta(r) = \bar{a} I_0(r/\lambda) + \bar{b} K_0(r/\lambda) - \lambda \epsilon \frac{\pi}{2} L_0(r/\lambda), \tag{3.425}$$

with $\bar{a}$ and $\bar{b}$ arbitrary constants. Since [1]

$$I_0(0) = 1, \quad L_0(0) = 0, \tag{3.426}$$

and

$$K_0(r/\lambda) \to \infty \quad \text{as} \quad r \to 0, \tag{3.427}$$

the solution $\zeta(r)$ in (3.425) can be finite at $r = 0$ only when the constant $\bar{b}$ is set to zero. Also, for large r [1, p.498,12.2.6],

$$L_0(r/\lambda) \sim I_0(r/\lambda) - \frac{2\lambda}{\pi r} \quad \text{as} \quad r \to \infty, \tag{3.428}$$

and therefore

$$\zeta(r) \sim I_0(r) \left[\bar{a} - \lambda \epsilon \frac{\pi}{2} \right] + \epsilon \frac{\lambda^2}{r}. \tag{3.429}$$

Since $I_0(r/\lambda) \to \infty$ as $r \to \infty$, the requirement $\zeta \to 0$ as $r \to \infty$ forces the condition

$$\bar{a} = \lambda \epsilon \frac{\pi}{2}. \tag{3.430}$$

Substituting this value for $\bar{a}$ and setting $\bar{b}$ to zero in equation (3.425) results in the solution

$$\zeta(r) = \lambda\epsilon\frac{\pi}{2}\left[I_0(r/\lambda) - L_0(r/\lambda)\right].\tag{3.431}$$

This solution can be given an alternative formulation via the identity [278, p.338]

$$I_0(\eta) - L_0(\eta) = \frac{1}{\pi}\int_{-\frac{\pi}{2}}^{\frac{\pi}{2}}\exp(-\eta\cos t)dt = \frac{2}{\pi}\int_0^{\frac{\pi}{2}}\exp(-\eta\cos t)dt,\tag{3.432}$$

to give the simplified form

$$\zeta(r) = \lambda\epsilon\int_0^{\frac{\pi}{2}}\exp(-\frac{r}{\lambda}\cos t)dt,\tag{3.433}$$

this being the result stated by Stephen and Straley [258, p.641]. Clearly,

$$\frac{d\zeta}{dr} = -\epsilon \quad\text{at}\quad r = 0,\tag{3.434}$$

verifying the condition $(3.420)_1$ so that the solution for $\zeta(r)$ in (3.431), or equivalently (3.433), is indeed the one being sought.

The height of the point defect above the initial flat surface alignment is $\zeta(0) = \lambda\epsilon\frac{\pi}{2}$. Further, using the expansion available at (3.428), it is seen from (3.431) that $\zeta(r)$ can be approximated by

$$\zeta(r) \sim \lambda^2\frac{\epsilon}{r}, \qquad r > \lambda.\tag{3.435}$$

For a sample of radius $R \gg \lambda$ the energy F_{surf} evaluated over the region $0 \leq r \leq R$ can be approximated by integrating over the range $\lambda \leq r \leq R$ and using equation (3.435). In this case we can insert (3.435) into equation (3.418) to find that

$$\begin{aligned} F_{surf} &\doteqdot \gamma\pi\int_\lambda^R\left[\left(\frac{d\zeta}{dr}\right)^2 + \frac{\zeta^2}{\lambda^2} + 2\epsilon\frac{d\zeta}{dr}\right]rdr \\ &\sim -\gamma\pi\lambda^2\epsilon^2\left[\ln(R/\lambda) + \frac{1}{2}\frac{\lambda^2}{R^2} - \frac{1}{2}\right].\end{aligned}\tag{3.436}$$

Hence for $R \gg \lambda$ we have the result stated by de Gennes [107] and Stephen and Straley [258], namely,

$$F_{surf} = -\gamma\pi\lambda^2\epsilon^2\ln(R/\lambda).\tag{3.437}$$

An expression for the energy around the point defect must also be incorporated into the complete energy. An observer looking down upon the defect will see a director alignment reminiscent of that pictured in Fig. 3.20 for Frank index $n = 2$ (strength $s = 1$) with $\theta = \phi$ (cf. Meyer [198, p.362]). Hence one possibility for an approximate energy for a region consisting of a circular annulus placed around the defect point having the characteristic length ξ as the inner radius and R as the outer radius is ξW where W is given by equation (3.352) with $n = 2$, $r_0 = \xi$ and $r_1 = R$, giving

$$F_{point} = \xi K\pi\ln(R/\xi).\tag{3.438}$$

The factor ξ appearing in the construction of this energy reflects the assumption that the influence of the point defect extends to a depth in the nematic sample having the order of the characteristic length ξ.

The total energy $\mathcal{F}_p$ is the sum of F_{surf} and F_{point} in equations (3.437) and (3.438) leading to

$$\mathcal{F}_p = \xi K \pi \ln(R/\xi) - \gamma \pi \lambda^2 \epsilon^2 \ln(R/\lambda). \tag{3.439}$$

Substituting for ξ, λ and ϵ via equations (3.383)$_2$, (3.394)$_2$ and (3.418)$_2$, respectively, gives

$$\mathcal{F}_p = \frac{\pi K^{\frac{3}{2}}}{H\sqrt{\chi_a}} \left[\ln(R/\xi) - \sin^2\psi \frac{H^3 \chi_a^{\frac{3}{2}}}{\rho g K^{\frac{1}{2}}} \ln(R/\lambda) \right], \tag{3.440}$$

showing that it becomes energetically favourable to form a point defect whenever H exceeds the threshold value H_2 defined by the relation

$$H_2 = K^{\frac{1}{6}} \chi_a^{-\frac{1}{2}} \left(\frac{\rho g \ln(R H_2 \sqrt{\chi_a}/\sqrt{K})}{\sin^2\psi \ln(R/\lambda)} \right)^{\frac{1}{3}}. \tag{3.441}$$

To allow a comparison with the earlier work on boundary line disclinations, we can take the special case for $\psi = \frac{\pi}{2}$ (corresponding to the director being fixed perpendicularly at the surface) and approximate the physical constants by (in cgs units)

$$K = 10^{-6} \text{ dyn}, \quad \chi_a = 10^{-6}, \quad \gamma = 10 \text{ dyn cm}^{-1}, \quad \rho = 1 \text{ g cm}^{-3}, \quad g = 980 \text{ cm s}^{-2} \tag{3.442}$$

These values give $\lambda = 0.10$ cm and so as an example we can take $R = 4$ and substitute these values into equation (3.441) to find

$$H_2 = 1315 \sim 1.3 \times 10^3 \text{ oersteds}. \tag{3.443}$$

This threshold is always of the order 10^3 oersteds for $R \gg \lambda$. As mentioned previously, the height of the point defect above the initial flat surface alignment is $\zeta(0) = \lambda \epsilon \frac{\pi}{2}$, where ϵ is given by (3.418)$_2$; for $H = 5H_2$, $\zeta(0)$ is of the order of 1 μm in this example.

It appears from the thresholds H_1 and H_2 in equations (3.412) and (3.443) that for the same surface conical anchoring condition ($\psi = \frac{\pi}{2}$) it is more favourable for the sample to exhibit point defects at the surface rather than boundary line disclinations because the threshold H_2 for the occurrence of point defects is generally an order of magnitude less than H_1 for boundary line disclinations. It is left to the reader to investigate other values of ψ to reach similar conclusions. An observation of this type led de Gennes [107] to conclude that boundary line disclinations may be less stable than point defects. For $H > H_2$ some sort of square two-dimensional lattice arrangement of point defects corresponding to 'hills' and 'valleys' may be expected in an idealised situation (see [110, p.185]). Photographs and a detailed discussion on point defects at a surface or interface, especially those having Frank index $n = \pm 2$ ($s = \pm 1$), may be found in the article by Meyer [198].

Chapter 4

Dynamic Theory of Nematics

4.1 Introduction

The earliest attempt at a dynamic theory for nematic liquid crystals was made by Anzelius [3] in 1931, although the first widely accepted dynamic theory was formulated much later by Ericksen [73] in 1961 using balance laws from the classical theory of continuum mechanics based upon a generalisation of the equilibrium theory discussed in Chapter 2. This dynamic theory was later completed in 1968 by Leslie [163], who derived suitable constitutive equations and thereby proposed expressions for the various dynamic contributions, after having considered constitutive equations for anisotropic fluids in 1966 [162]. Recall that constitutive equations, sometimes called constitutive relations, describe the mechanical properties that are particular to a given medium; it is clear that these equations will change from one medium to another. For nematic liquid crystals the resulting theory is often referred to as the Ericksen–Leslie dynamic theory and is one of the most successful theories used to model many dynamic phenomena in nematics. A comprehensive review of this theory has been written by Leslie [168] and examples involving applications can be found in the books by Chandrasekhar [38] and de Gennes and Prost [110]. An excellent and informative historical account of how the Ericksen–Leslie dynamic theory of nematics was developed has been written by Carlsson and Leslie [37]. Other continuum theories have been put forward, but the Ericksen–Leslie theory is the most extensively used in the nematic liquid crystal literature. It is also worth noting that Leslie proposed a continuum theory for the dynamics of cholesteric liquid crystals in the late 1960s [164, 165], but we do not pursue this area in this introductory text.

In 1992 Leslie [175] published an alternative derivation of the Ericksen–Leslie theory in the isothermal and incompressible case, which led to a simpler presentation of the results originally derived by Ericksen [73] and Leslie [162, 163]. It is this more recent approach that we partly adopt here; it is a more concise exposition of the original theory and allows the derivation of the constitutive theory to be discussed in the more traditional continuum theory variables such as the rate of strain tensor and the relative angular velocity without recourse to generalised forces. The Ericksen–Leslie theory will be derived in the next Section and a convenient summary will be presented in Section 4.2.5. Section 4.3 contains a reformulation of the dynamic equations, similar in style to that carried out for statics in Section 2.7. This reformulation is often extremely convenient for calculations in practical problems. The Chapter ends with a discussion in Section 4.4 on the Leslie viscosities which result from the dynamic constitutive theory, including a comparison with the earlier Miesowicz viscosities. For the full thermodynamic theory of nematics (which allows a description of thermal effects) the reader is referred to the review by Leslie [168].

4.2   The Ericksen–Leslie Dynamic Equations

4.2.1   Kinematics and Material Frame-Indifference

Kinematics is the study of motion without reference to the forces involved or how the motion is brought about. Dynamics is concerned with how motion arises and how the forces involved change or produce motion. The results of the kinematic studies in this subsection will form a basis for the ensuing construction of dynamic equations. It is useful to introduce here various kinematic quantities and terminology. Many of the terms to be discussed frequently arise in the context of fluids that possess a microstructure which is mechanically significant, such as in the theory of polar fluids, for example. Despite nematic liquid crystals not fitting directly into such general theories, identical concepts of flow and differing angular velocities are important and pertinent to nematics. The terminology we introduce can be found in the review by Cowin [56].

The Eulerian description of the instantaneous motion of a fluid with microstructure employs two independent vector fields. The first is the usual velocity $\mathbf{v}(\mathbf{x}, t)$ and the second is an axial vector $\mathbf{w}(\mathbf{x}, t)$ which, in the case of polar fluids, represents the angular velocity of the polar fluid particle at position $\mathbf{x}$ at time t. In the context of liquid crystals, $\mathbf{w}$ is interpreted as being the *local angular velocity* of the liquid crystal material element, that is, it represents the local angular velocity of the director $\mathbf{n}$. In ordinary continuum theory the only independent field is the velocity $\mathbf{v}$ of the fluid because the angular velocity in such theories equals one half of the curl of the velocity. We denote this particular angular velocity by $\widehat{\mathbf{w}}$ defined by

$$\widehat{\mathbf{w}} = \tfrac{1}{2}\nabla \times \mathbf{v}, \tag{4.1}$$

and refer to it as the *regional angular velocity*, to distinguish it from other angular velocities. It is a measure of the average rotation of the fluid over a neighbourhood of the material element. The angular velocity of the material element relative to the regional angular velocity in which the material element is embedded is denoted by ω and is defined by

$$\omega = \mathbf{w} - \widehat{\mathbf{w}} = \mathbf{w} - \tfrac{1}{2}\nabla \times \mathbf{v}. \tag{4.2}$$

The quantity ω is called the *relative angular velocity* and is introduced to measure the difference between the local angular velocity $\mathbf{w}$ of the liquid crystal director and the regional angular velocity $\widehat{\mathbf{w}}$ of the fluid in the neighbourhood of the director.

It will always be assumed that the director **n** satisfies the constraint

$$\mathbf{n} \cdot \mathbf{n} = 1. \tag{4.3}$$

Since **w** represents the angular velocity of the director, we have that **n**, being a unit vector, also satisfies

$$\dot{\mathbf{n}} = \mathbf{w} \times \mathbf{n}, \tag{4.4}$$

where a superposed dot represents the usual *material time derivative*

$$\frac{D}{Dt} = \frac{\partial}{\partial t} + \mathbf{v} \cdot \frac{\partial}{\partial \mathbf{x}} . \tag{4.5}$$

This derivative can also be expressed using the usual index notation and summation convention as employed in earlier Chapters, giving

$$\frac{D}{Dt} = \frac{\partial}{\partial t} + v_i \frac{\partial}{\partial x_i} . \tag{4.6}$$

Sometimes the notation $(\mathbf{v} \cdot \nabla)$ is used for what is represented in Cartesian components by the operator $v_i \partial / \partial x_i$. Note that the operators defined in (4.5) and (4.6) may be applied to a scalar or vector quantity. Details on the physical interpretation of the material time derivative can be found in Aris [4, pp.77–79] where, in particular, it is shown that the velocity is given by $\mathbf{v} = \dot{\mathbf{x}}$.

The *rate of strain tensor* **A** and *vorticity tensor* **W** are second order tensors introduced and defined in the usual way, in an obvious nomenclature, by

$$A_{ij} = \tfrac{1}{2}(v_{i,j} + v_{j,i}), \qquad W_{ij} = \tfrac{1}{2}(v_{i,j} - v_{j,i}). \tag{4.7}$$

Notice that **A** is symmetric and **W** is skew-symmetric. The vorticity tensor is also sometimes called the *spin tensor* or *angular velocity tensor*. Following Leslie [175], we introduce the vector **N** defined by

$$\mathbf{N} = \omega \times \mathbf{n}. \tag{4.8}$$

By equations (4.2), (4.4) and (4.7) it is observed that, upon contracting the alternators using (1.7) when required,

$$
\begin{aligned}
N_i &= \epsilon_{ijk} \omega_j n_k \\
&= \epsilon_{ijk} w_j n_k - \tfrac{1}{2}(v_{i,j} - v_{j,i}) n_j \\
&= \dot{n}_i - W_{ij} n_j,
\end{aligned}
\tag{4.9}
$$

and hence the definition in equation (4.8) is equivalent to

$$\mathbf{N} = \dot{\mathbf{n}} - \mathbf{W}\mathbf{n}, \tag{4.10}$$

this being the form discussed originally by Ericksen and Leslie. In the terminology used by Truesdell and Noll [270, p.526], **N** is called the *co-rotational time flux* of the director **n**; by (4.8) it is obviously connected with the relative angular velocity

$\boldsymbol{\omega}$ and is a measure of the rotation of $\mathbf{n}$ relative to the fluid (in some texts $\mathbf{N}$ is denoted by $\mathbf{\mathring{n}}$). The constraint that $\mathbf{n}$ is a unit vector leads to the observation that

$$n_i \dot{n}_i = 0, \tag{4.11}$$

which easily leads to showing

$$n_i N_i = 0, \tag{4.12}$$

a result that also follows from (4.8). This result will be required later.

A fundamental principle of classical physics is that material properties must be independent of the frame of reference or observer. This axiom is commonly called the 'principle of material frame-indifference' or 'objectivity' [270, pp.41–44]. This principle states that constitutive equations (discussed in greater detail in Section 4.2.3) must be invariant under changes of frame of reference. Under the motion defined by

$$\mathbf{x}^*(t^*) = \mathbf{c}(t) + \mathbf{Q}(t)\mathbf{x}(t), \qquad t^* = t - a, \tag{4.13}$$

where a is any real number, $\mathbf{c}(t)$ is an arbitrary vector function of time and $\mathbf{Q}(t)$ is an arbitrary rotation represented by a second order proper orthogonal tensor function of time, the vector $\mathbf{a}$ and second order tensor $\mathbf{B}$ are called *frame-indifferent* or *objective* if they transform according to the rules

$$\mathbf{a}^* = \mathbf{Q}\mathbf{a}, \qquad \mathbf{B}^* = \mathbf{Q}\mathbf{B}\mathbf{Q}^T, \tag{4.14}$$

a starred symbol representing the corresponding quantity derived under the motion (4.13). For the purposes of this text we shall denote the components of $\mathbf{Q}$ by Q_{ij} and identify Q_{ij} with the entries of a matrix; also, recall that $\mathbf{Q}$ is called a proper orthogonal matrix provided

$$Q_{ik}Q_{jk} = Q_{kj}Q_{ki} = \delta_{ij}, \qquad \det \mathbf{Q} = +1. \tag{4.15}$$

In Cartesian component form the transformations in equation (4.14) are

$$a_i^* = Q_{ij}a_j, \qquad B_{ij}^* = Q_{ip}B_{pq}Q_{jq}. \tag{4.16}$$

It also follows from equation (4.15) that by multiplying the ith components of both sides of (4.13) by Q_{ik}, and appropriately relabelling indices, we have

$$x_i = Q_{ki}(x_k^* - c_k). \tag{4.17}$$

It will be necessary in Section 4.2.3 to identify quantities that satisfy the transformation rules (4.16). It will always be assumed that the director $\mathbf{n}$ is frame-indifferent so that

$$n_i^* = Q_{ij}n_j. \tag{4.18}$$

We examine the transformations under the motion (4.13) for the quantities v_i, $v_{i,j}$, A_{ij}, W_{ij}, $\dot{n}_i$, N_i, and proceed to show that of these only A_{ij} and N_i are frame-indifferent, that is, they are objective. This result will be of crucial importance in the construction of suitable constitutive equations. Define $\boldsymbol{\Omega}$ by

$$\Omega_{ij} = \dot{Q}_{ik}Q_{jk}, \tag{4.19}$$

and note that this implies

$$\dot{Q}_{ij} = Q_{kj}\Omega_{ik}, \tag{4.20}$$

obtained by relabelling indices after multiplying (4.19) by Q_{jr} and using the property (4.15)$_1$. Further, differentiating (4.15)$_1$ leads, via (4.19), to

$$\Omega_{ij} + \Omega_{ji} = 0. \tag{4.21}$$

Taking the material time derivative of equation (4.13) gives

$$v_i^* \equiv \dot{x}_i^* = \dot{c}_i + \dot{Q}_{ik}x_k + Q_{ik}v_k, \tag{4.22}$$

which yields, via the relations (4.15), (4.17) and the definition of Ω_{ij},

$$\begin{aligned}
\frac{\partial v_i^*}{\partial x_j^*} &= \dot{Q}_{ik}\frac{\partial x_k}{\partial x_j^*} + Q_{ik}\frac{\partial v_k}{\partial x_p}\frac{\partial x_p}{\partial x_j^*} \\
&= \Omega_{ij} + Q_{ik}v_{k,p}Q_{jp}.
\end{aligned} \tag{4.23}$$

Hence by using this equation in conjunction with (4.21) we have

$$\begin{aligned}
A_{ij}^* &= \tfrac{1}{2}\left(\frac{\partial v_i^*}{\partial x_j^*} + \frac{\partial v_j^*}{\partial x_i^*}\right) \\
&= \tfrac{1}{2}(\Omega_{ij} + Q_{ik}v_{k,p}Q_{jp} + \Omega_{ji} + Q_{jk}v_{k,p}Q_{ip}) \\
&= \tfrac{1}{2}(Q_{ik}v_{k,p}Q_{jp} + Q_{jp}v_{p,k}Q_{ik}) \\
&= Q_{ik}A_{kp}Q_{jp}.
\end{aligned} \tag{4.24}$$

Therefore A_{ij} satisfies the principle of material frame-indifference, that is, **A** is objective. Similar reasoning leads to the result

$$W_{ij}^* = Q_{ik}W_{kp}Q_{jp} + \Omega_{ij}, \tag{4.25}$$

which shows that **W** is not objective.

From the assumed director transformation (4.18),

$$\dot{n}_i^* = \dot{Q}_{ij}n_j + Q_{ij}\dot{n}_j, \tag{4.26}$$

and therefore $\dot{n}_i$ is not objective. However, by equations (4.9), (4.15), (4.20), (4.25) and (4.26),

$$\begin{aligned}
N_i^* &= \dot{n}_i^* - W_{ij}^* n_j^* \\
&= \dot{Q}_{ij}n_j + Q_{ij}\dot{n}_j - (\Omega_{ij} + Q_{ik}W_{kp}Q_{jp})Q_{jq}n_q \\
&= \Omega_{ij}Q_{jq}n_q + Q_{ij}\dot{n}_j - (\Omega_{ij} + Q_{ik}W_{kp}Q_{jp})Q_{jq}n_q \\
&= Q_{ij}(\dot{n}_j - W_{jk}n_k) \\
&= Q_{ij}N_j,
\end{aligned} \tag{4.27}$$

and therefore N_i satisfies the principle of material frame-indifference, and so **N** is objective.

To summarise, we have from equations (4.22), (4.23), (4.25) and (4.26) that the quantities $\mathbf{v}$, $\nabla\mathbf{v}$, $\mathbf{W}$ and $\dot{\mathbf{n}}$ do not satisfy the principle of material frame-indifference, while from equations (4.18), (4.24) and (4.27) we see that the director $\mathbf{n}$, the rate of strain tensor $\mathbf{A}$ and the co-rotational time flux $\mathbf{N}$ are frame-indifferent and therefore objective. For later ease of reference, we record that

$$n_i, \quad N_i, \quad A_{ij} \qquad \text{are frame-indifferent.} \tag{4.28}$$

4.2.2   Balance Laws

We shall always assume isothermal conditions and therefore ignore thermal effects. In these circumstances, as in any classically based continuum theory, conservation laws for mass, linear momentum and angular momentum must hold. The balance law for linear momentum, given below, is basically similar to that for an isotropic fluid, except that the resulting stress tensor (to be derived later) need not be symmetric. The balance law for angular momentum is also suitably augmented to include explicit external body and surface moments.

For a volume V of nematic liquid crystal bounded by the surface S the three conservation laws for mass, linear momentum and angular momentum are, respectively,

$$\frac{D}{Dt} \int_V \rho \, dV = 0, \tag{4.29}$$

$$\frac{D}{Dt} \int_V \rho \mathbf{v} \, dV = \int_V \rho \mathbf{F} \, dV + \int_S \mathbf{t} \, dS, \tag{4.30}$$

$$\frac{D}{Dt} \int_V \rho(\mathbf{x} \times \mathbf{v}) \, dV = \int_V \rho(\mathbf{x} \times \mathbf{F} + \mathbf{K}) \, dV + \int_S (\mathbf{x} \times \mathbf{t} + \mathbf{l}) \, dS, \tag{4.31}$$

where ρ denotes the density, $\mathbf{x}$ is the position vector, $\mathbf{v}$ is the velocity, $\mathbf{F}$ is the external body force per unit mass, $\mathbf{t}$ is the surface force per unit area, $\mathbf{K}$ is the external body moment per unit mass, $\mathbf{l}$ is the surface moment per unit area (also called the couple stress vector) and D/Dt represents the material time derivative defined by equation (4.5). No 'director inertial term' has been incorporated to the angular momentum balance since it is generally considered to be negligible in nematic liquid crystal flow problems. (Nevertheless, some comments on the inclusion of such an inertial contribution are contained on pages 147 to 149). There also appears to be a convention in the liquid crystal literature that F_i appears in static theory while ρF_i appears in dynamic theory: this means that in statics F_i represents the external body force per unit volume while it represents the external body force per unit mass in dynamics. A similar statement applies to the external body moment K_i.

We can apply Reynolds' transport theorem, as stated in Appendix B, to the mass conservation law (4.29). If we set $\mathfrak{F} = \rho$ in equation (B.7) then the equation for mass conservation leads to the relation

$$\int_V [\dot{\rho} + \rho(\nabla \cdot \mathbf{v})] \, dV = 0. \tag{4.32}$$

Since V is an arbitrary volume, we can write this result in point form as

$$\dot{\rho} + \rho(\nabla \cdot \mathbf{v}) = 0. \tag{4.33}$$

From the definition of the material derivative (4.5), this result can be recast into the more familiar form known as the 'equation of continuity' [159, p.2]

$$\frac{\partial \rho}{\partial t} + \nabla \cdot (\rho \mathbf{v}) = 0. \tag{4.34}$$

A fluid is said to be incompressible whenever its mass density ρ is constant, resulting in $\dot{\rho} = 0$. Since we shall always suppose that the nematic is incompressible, the result in (4.33) allows us to replace the mass conservation law (4.29) by

$$\nabla \cdot \mathbf{v} = 0, \tag{4.35}$$

and the density ρ can be considered as being homogeneously constant throughout any given volume V. This implies that the trace of the rate of strain tensor $\mathbf{A}$ is zero, that is,

$$A_{ii} = 0. \tag{4.36}$$

Equations such as (4.33) and (4.35) that arise from the consideration of the integrands of relations such as those stated above in equations (4.29), (4.30) and (4.31), are said to be in point form, since they must hold true at all points of the fluid.

If $\boldsymbol{\nu}$ denotes the outward unit normal to the surface S, then the usual tetrahedron argument [161, p.129] shows that the surface force t_i and surface moment l_i are expressible in terms of the stress tensor t_{ij} and couple stress tensor l_{ij}, respectively, through the relations

$$t_i = t_{ij}\nu_j, \qquad l_i = l_{ij}\nu_j. \tag{4.37}$$

We recall here the physical interpretation of the stress tensor in Cartesian coordinates [161, pp.131–132]. Let $\mathbf{t}_j$ be the stress vector (surface force) representing the force per unit area exerted by the material outside the jth coordinate surface upon the material inside (where the unit outward normal to this surface is in the direction $\mathbf{e}_j$). The component t_{ij} then represents the ith component of this stress vector at a point on the jth coordinate surface. For example, if the x coordinate surface has unit outward normal $\boldsymbol{\nu} = (1, 0, 0)$ then the stress vector at a point on this coordinate surface is simply $\mathbf{t}_1 = \mathbf{e}_i t_{ij}\nu_j = \mathbf{e}_i t_{i1} = (t_{11}, t_{21}, t_{31})$. A similar interpretation arises for the couple stress tensor. The components t_{11}, t_{22} and t_{33} are called the *normal stresses* or *direct stresses* and the components t_{12}, t_{21}, t_{13}, t_{31}, t_{23}, t_{32} are called the *shear stresses*.

By equation (4.37)$_1$ and the divergence theorem in the form of equation (1.20),

$$\int_S t_i \, dS = \int_S t_{ij}\nu_j \, dS = \int_V t_{ij,j} \, dV. \tag{4.38}$$

When the equation of continuity (4.34) holds, as it does in this instance, Reynolds' transport theorem leads to the result derived in Appendix B at equation (B.8), which states that

$$\frac{D}{Dt} \int_V \rho F \, dV = \int_V \rho \dot{F} \, dV, \tag{4.39}$$

for any function $\mathcal{F}(\mathbf{x}, t)$. With $\mathcal{F} = v_i$, this result, combined with that in (4.38), allows the balance of linear momentum (4.30) to be formulated as

$$\int_V (\rho \dot{v}_i - \rho F_i - t_{ij,j}) dV = 0. \tag{4.40}$$

Since V is any arbitrary volume of the sample, the above integrand must itself be zero and consequently the balance law for linear momentum can be given in point form as

$$\rho \dot{v}_i = \rho F_i + t_{ij,j}. \tag{4.41}$$

From the version of Reynolds' transport theorem stated in equation (4.39), it is straightforward to deduce that (see [4, p.103])

$$\frac{D}{Dt} \int_V \rho \epsilon_{ijk} x_j v_k \, dV = \int_V \rho \epsilon_{ijk} x_j \dot{v}_k \, dV, \tag{4.42}$$

since $\epsilon_{ijk} \dot{x}_j v_k = \epsilon_{ijk} v_j v_k = 0$, while an application of the divergence theorem (1.20) using the relation $(4.37)_1$ shows

$$\int_S \epsilon_{ijk} x_j t_k \, dS = \int_V (\epsilon_{ijk} x_j t_{kp})_{,p} \, dV = \int_V \epsilon_{ijk} (x_j t_{kp,p} + t_{kj}) dV. \tag{4.43}$$

Inserting the two results (4.42) and (4.43) into the balance law (4.31) for angular momentum gives

$$\int_V \epsilon_{ijk} x_j (\rho \dot{v}_k - \rho F_k - t_{kp,p}) dV = \int_V (\rho K_i + \epsilon_{ijk} t_{kj} + l_{ij,j}) dV, \tag{4.44}$$

with the aid of the relation $(4.37)_2$ and, once again, the divergence theorem. Nevertheless, the left-hand side of this equation is zero by the result obtained for linear momentum in equation (4.41) and so by the arbitrariness of the volume V the balance of angular momentum given in (4.44) reduces to

$$\rho K_i + \epsilon_{ijk} t_{kj} + l_{ij,j} = 0. \tag{4.45}$$

It is perhaps worth remarking here that for *isotropic* fluids $\mathbf{K}$ and $\mathbf{l}$ are generally absent, so that the angular momentum balance law (4.45) further reduces to $\epsilon_{ijk} t_{kj} = 0$, indicating that for isotropic fluids the stress tensor is symmetric. Fluids in which couple stresses and body couples occur are frequently called polar fluids, and those in which they are absent are called non-polar. The forms given by equations (4.35), (4.41) and (4.45) for the balances of mass, linear momentum and angular momentum will eventually lead to the main dynamic equations for nematics.

Following Leslie [175], a rate of work hypothesis will now be invoked. This assumes that the rate at which forces and moments do work on a volume of nematic will be absorbed into changes in the nematic energy w_F (cf. equations (2.49) and (2.50)) or the kinetic energy, or will be lost by means of viscous dissipation. The rate of work postulate is taken to be

$$\int_V \rho(\mathbf{F} \cdot \mathbf{v} + \mathbf{K} \cdot \mathbf{w}) dV + \int_S (\mathbf{t} \cdot \mathbf{v} + \mathbf{l} \cdot \mathbf{w}) dS = \frac{D}{Dt} \int_V (\tfrac{1}{2} \rho \mathbf{v} \cdot \mathbf{v} + w_F) dV + \int_V \mathcal{D} \, dV, \tag{4.46}$$

where $\mathbf{w}$ is the local angular velocity of the director introduced in Section 4.2.1 at equations (4.2) and (4.4), and $\mathcal{D}$ is the rate of viscous dissipation per unit volume, also called the dissipation function. It will be assumed that $\mathcal{D}$ is always positive: we shall derive the most familiar form for $\mathcal{D}$ for nematics, given by the general result in equation (4.85) below and, when the Parodi relation (discussed later on page 146) holds, equation (4.97). For an elementary description of the dissipation function the reader is referred to Landau and Lifshitz [159, pp.55–56] and, for the asymmetric stress tensor case, to Aris [4, pp.117–118]. (It is worth pointing out here that the $\mathbf{w}$ used in Leslie's original paper [163] refers to $\dot{\mathbf{n}}$ and not to the $\mathbf{w}$ used in this present context and introduced by Leslie in 1992 [175].) The relations (4.37) and the divergence theorem (1.20) allow the surface integral in the postulate (4.46) to be formulated as

$$\int_S (t_i v_i + l_i w_i)dS = \int_V (t_{ij}v_{i,j} + l_{ij}w_{i,j} + v_i t_{ij,j} + w_i l_{ij,j})\, dV. \qquad (4.47)$$

Now the point form of the balance laws in (4.41) and (4.45) can be used to substitute for the terms $t_{ij,j}$ and $l_{ij,j}$ to find that (4.47) becomes

$$\int_S (t_i v_i + l_i w_i)dS = \int_V (t_{ij}v_{i,j} + l_{ij}w_{i,j} + \rho\dot{v}_i v_i - \rho F_i v_i - \rho w_i K_i - w_i \epsilon_{ijk}t_{kj})\, dV. \quad (4.48)$$

We can use the form of the Reynolds' transport theorem in (4.39), with $\mathcal{F}$ set to $\frac{1}{2}\mathbf{v}\cdot\mathbf{v}$ to see that

$$\frac{D}{Dt}\int_V \frac{1}{2}\rho\mathbf{v}\cdot\mathbf{v}\, dV = \int_V \rho\dot{\mathbf{v}}\cdot\mathbf{v}\, dV. \qquad (4.49)$$

Also, in the special case for an incompressible material (where, necessarily, div $\mathbf{v}$ = 0), Reynolds' transport theorem reduces to equation (B.9), as derived in Appendix B, and if we set $\mathfrak{F} = w_F$ in (B.9) we obtain

$$\frac{D}{Dt}\int_V w_F\, dV = \int_V \dot{w}_F\, dV. \qquad (4.50)$$

We can insert the results (4.48), (4.49) and (4.50) into the postulate (4.46) for arbitrary volumes V to reduce it to the point form

$$t_{ij}v_{i,j} + l_{ij}w_{i,j} - w_i\epsilon_{ijk}t_{kj} = \dot{w}_F + \mathcal{D}, \qquad (4.51)$$

a result that will be exploited to obtain expressions for the stress and couple stress.

It is seen from the definition (4.5) of material time derivative that

$$\overline{n_{i,j}} = (\dot{n}_i)_{,j} - n_{i,k}v_{k,j}\,, \qquad (4.52)$$

when partial derivatives commute. Therefore, since $w_F = w_F(\mathbf{n}, \nabla\mathbf{n})$, we can use the expressions (4.4), (4.52) and the chain rule for partial derivatives to find

$$\begin{aligned}
\dot{w}_F &= \frac{\partial w_F}{\partial n_p}\dot{n}_p + \frac{\partial w_F}{\partial n_{p,k}}\overline{\dot{n}_{p,k}} \\
&= \epsilon_{iqp}\left[\left(n_q\frac{\partial w_F}{\partial n_p} + n_{q,k}\frac{\partial w_F}{\partial n_{p,k}}\right)w_i + n_q\frac{\partial w_F}{\partial n_{p,k}}w_{i,k}\right] - \frac{\partial w_F}{\partial n_{p,k}}n_{p,q}v_{q,k}, \quad (4.53)
\end{aligned}$$

relabelling indices as required. Employing the Ericksen identity (B.6) from Appendix B to substitute for the coefficient of w_i in the relation (4.53) allows $\dot{w}_F$ to be given as

$$\dot{w}_F = \epsilon_{iqp}\left(n_q\frac{\partial w_F}{\partial n_{p,j}}w_{i,j} - n_{k,q}\frac{\partial w_F}{\partial n_{k,p}}w_i\right) - \frac{\partial w_F}{\partial n_{p,j}}n_{p,i}v_{i,j}. \qquad (4.54)$$

Inserting this result into the point form of the balance law (4.51) gives

$$\left(t_{ij} + \frac{\partial w_F}{\partial n_{p,j}}n_{p,i}\right)v_{i,j} + \left(l_{ij} - \epsilon_{iqp}n_q\frac{\partial w_F}{\partial n_{p,j}}\right)w_{i,j} - w_i\epsilon_{iqp}\left(t_{pq} - \frac{\partial w_F}{\partial n_{k,p}}n_{k,q}\right) = \mathcal{D}. \qquad (4.55)$$

The rate of viscous dissipation $\mathcal{D}$ is necessarily positive. Given that the signs of w_i, $w_{i,j}$ and $v_{i,j}$ may be arbitrarily chosen, it follows that the coefficients of any terms linear in w_i, $w_{i,j}$ and $v_{i,j}$ which occur in (4.55) must equate to zero so that $\mathcal{D}$ remains positive. This leads to the conclusion that the stress t_{ij} and couple stress l_{ij} may take the forms [175]

$$t_{ij} = -p\,\delta_{ij} - \frac{\partial w_F}{\partial n_{p,j}}n_{p,i} + \tilde{t}_{ij}, \qquad (4.56)$$

$$l_{ij} = \epsilon_{ipq}n_p\frac{\partial w_F}{\partial n_{q,j}} + \tilde{l}_{ij}, \qquad (4.57)$$

where p is an arbitrary pressure arising from the assumed incompressibility and $\tilde{t}_{ij}$ and $\tilde{l}_{ij}$ denote possible dynamic contributions. In the terminology of Leslie, $\tilde{t}_{ij}$ is called the *viscous stress* [174, p.38], [182, p.36]. If, for the moment as a rough check, the supposed dynamic contributions are ignored in the above expressions then it is easy to see that their insertion into the left-hand side of (4.55) gives zero, observing that the first bracketed term reduces to $-pv_{i,i}$, which is zero by equation (4.35), while the last bracketed term becomes zero because $\epsilon_{iqp}X_{qp} = 0$ for any second order symmetric tensor X_{pq}. Also, it is worth drawing attention to the observation that when $\tilde{t}_{ij}$ and $\tilde{l}_{ij}$ are absent then t_{ij} and l_{ij} reduce to the static forms given in Section 2.4 via equations (2.141), (2.142) and (2.165). The proposed relations (4.56) and (4.57) reduce the form of the balance law (4.55) to

$$\tilde{t}_{ij}v_{i,j} + \tilde{l}_{ij}w_{i,j} - w_i\epsilon_{ijk}\tilde{t}_{kj} = \mathcal{D} \geq 0, \qquad (4.58)$$

given that $\mathcal{D}$ is positive. This inequality will be of crucial importance when discussing the constitutive theory and will impose restrictions upon the forms of the dynamic terms. (We remark here that t_{ij} and l_{ij} are as defined in Leslie [175], equivalent to σ_{ji} and π_{ji}, respectively, originally used by Leslie [163] in 1968.)

4.2.3   Constitutive Equations

To proceed further we have to make more specific assumptions about the dynamic contributions $\tilde{t}_{ij}$ and $\tilde{l}_{ij}$ to the stress and couple stress tensors. This means that some relations between these stresses and the motion of the material will have to be introduced; such relations result in what are called constitutive equations, sometimes called constitutive relations. The natural continuum variables to use are the director, the velocity gradients and the local angular velocity of the director and therefore it will be assumed that at any material point and at any instant

$$\tilde{t}_{ij} \quad \text{and} \quad \tilde{l}_{ij} \qquad \text{are functions of} \qquad n_i, \, w_i \text{ and } v_{i,j}, \qquad (4.59)$$

evaluated at that point and instant. Since $\tilde{l}_{ij}$ is assumed not to depend upon the gradients $w_{i,j}$ of the local angular velocity $\mathbf{w}$ of the director, it follows from the inequality (4.58) that

$$\tilde{l}_{ij} = 0, \qquad (4.60)$$

because $w_{i,j}$, which may be of arbitrary sign, can only enter the inequality (4.58) linearly with $\tilde{l}_{ij}$. This is analogous to the result of Leslie [163]. The inequality (4.58) then reduces to

$$\tilde{t}_{ij} v_{i,j} - w_i \epsilon_{ijk} \tilde{t}_{kj} = \mathcal{D} \ge 0, \qquad (4.61)$$

which restricts the contributions to the viscous stress $\tilde{t}_{ij}$ (cf. Aris [4, pp.117–118] in his discussion of asymmetric stress tensors and the dissipation function).

In a rigid body motion the dynamic variables w_i and $v_{i,j}$ are not necessarily zero and therefore we require combinations of these quantities that do vanish in such types of motion. In a rigid body motion [24, p.81]

$$v_i = c_i(t) + \epsilon_{ipq} w_p(t) x_q, \qquad (4.62)$$

where $\mathbf{c}$ is a vector function of t, so that

$$v_{i,j} = \epsilon_{ipj} w_p(t). \qquad (4.63)$$

These relations imply that

$$v_{i,j} + v_{j,i} = 0 \quad \text{and} \quad w_i - \tfrac{1}{2} \epsilon_{ijk} v_{k,j} = 0, \qquad (4.64)$$

since

$$\epsilon_{ijk} v_{k,j} = \epsilon_{ijk} \epsilon_{kpj} w_p(t) = 2\delta_{ip} w_p(t) = 2w_i(t). \qquad (4.65)$$

In the light of equations (4.60) and (4.64), the assumption (4.59) can now be amended to

$$\tilde{t}_{ij} \qquad \text{is a function of} \qquad n_i, \, w_i \text{ and } A_{ij}, \qquad (4.66)$$

where the relative angular velocity w_i and the rate of strain tensor A_{ij} are defined by equations (4.2) and $(4.7)_1$ above, respectively. At this point we slightly differ from the approach in [175] and, given the definition (4.8) for $\mathbf{N}$, replace this assumption by the equivalent assumption

$$\tilde{t}_{ij} \qquad \text{is a function of} \qquad n_i, \, N_i \text{ and } A_{ij}. \qquad (4.67)$$

Material frame-indifference of the viscous stress $\tilde{t}_{ij}$ means that we require it to be a hemitropic function of the above named variables, that is,

$$\tilde{t}_{ij}^*(n_i^*, N_i^*, A_{ij}^*) = Q_{ip} \tilde{t}_{pq}(n_i, N_i, A_{ij}) Q_{jq}, \qquad (4.68)$$

where Q is any proper orthogonal second order tensor, since we know that the quantities n_i, N_i and A_{ij} are frame-indifferent, as noted in Section 4.2.1 at equation (4.28). Additionally, as in static theory, all constitutive equations must be independent of a change of sign in $\mathbf{n}$. Further, given the symmetries of nematics, the constitutive equations must also be invariant under reflections within planes containing the director $\mathbf{n}$ (cf. Section 2.2.1 on page 20): this means that $\tilde{t}_{ij}$ must actually obey the transformation (4.68) for *all* orthogonal tensors Q. Hence $\tilde{t}_{ij}$ must be an isotropic function of the variables n_i, N_i and A_{ij}, and must be an even function of the components n_i.

The experiments of Miesowicz [201] and Zwetkoff [288] suggest that $\tilde{t}_{ij}$ has a linear dependence upon $\mathbf{N}$ and $\mathbf{A}$, and accepting such a linear dependence we can write

$$\tilde{t}_{ij} = \mathcal{A}_{ij} + \mathcal{B}_{ijk}N_k + \mathcal{C}_{ijkp}A_{kp}, \tag{4.69}$$

where, since (4.68) must hold for any orthogonal tensor Q, we must have

$$\mathcal{A}_{ij}^* = Q_{ip}Q_{jq}\mathcal{A}_{pq} \quad , \mathcal{B}_{ijk}^* = Q_{iq}Q_{jm}Q_{kp}\mathcal{B}_{qmp} \quad \text{and} \quad \mathcal{C}_{ijkp}^* = Q_{iq}Q_{jm}Q_{kr}Q_{ps}\mathcal{C}_{qmrs}, \tag{4.70}$$

the coefficients $\mathcal{A}_{ij}$, $\mathcal{B}_{ijk}$ and $\mathcal{C}_{ijkp}$ being functions of n_i. Nematic symmetry also makes the further imposition that these coefficients are transversely isotropic with respect to n_i. As pointed out by Ericksen [71], tensors of this type have been studied by Smith and Rivlin [255], who showed that any such tensors are expressible as a linear combination of products of the components n_i and $\delta_{ij} - n_i n_j$, or, equivalently, n_i and δ_{ij}. This indicates that the coefficients can be expanded in the general forms

$$\begin{aligned}
\mathcal{A}_{ij} &= \mu_1\delta_{ij} + \mu_2 n_i n_j, &&\text{(4.71)}\\
\mathcal{B}_{ijk} &= \mu_3\delta_{ij}n_k + \mu_4\delta_{jk}n_i + \mu_5\delta_{ki}n_j, &&\text{(4.72)}\\
\mathcal{C}_{ijkp} &= \mu_6\delta_{ij}\delta_{kp} + \mu_7\delta_{ik}\delta_{jp} + \mu_8\delta_{ip}\delta_{jk} + \mu_9\delta_{ij}n_k n_p &&\\
&\quad + \mu_{10}\delta_{jk}n_i n_p + \mu_{11}\delta_{ik}n_j n_p + \mu_{12}\delta_{ip}n_j n_k &&\\
&\quad + \mu_{13}\delta_{jp}n_i n_k + \mu_{14}\delta_{kp}n_i n_j + \mu_{15}n_i n_j n_k n_p. &&\text{(4.73)}
\end{aligned}$$

Inserting these expansions into equation (4.69) and recalling, from $(4.7)_1$, (4.12) and (4.36), that $A_{ij} = A_{ji}$, $n_i N_i = 0$ and $A_{ii} = 0$, it is found that

$$\begin{aligned}
\tilde{t}_{ij} &= (\mu_1 + \mu_9 n_k A_{kp}n_p)\delta_{ij} + (\mu_2 + \alpha_1 n_k A_{kp}n_p)n_i n_j + \alpha_2 N_i n_j\\
&\quad + \alpha_3 N_j n_i + \alpha_4 A_{ij} + \alpha_5 n_j A_{ik}n_k + \alpha_6 n_i A_{jk}n_k. \tag{4.74}
\end{aligned}$$

where

$$\begin{aligned}
\alpha_1 &= \mu_{15}, \quad \alpha_2 = \mu_5, \quad \alpha_3 = \mu_4,\\
\alpha_4 &= \mu_7 + \mu_8, \quad \alpha_5 = \mu_{11} + \mu_{12} \quad \alpha_6 = \mu_{10} + \mu_{13}. \tag{4.75}
\end{aligned}$$

In order to progress in the derivation of $\tilde{t}_{ij}$ the inequality (4.61) can now be developed further by employing the notation introduced in Section 4.2.1 and introducing some terminology which will be valuable later. We can write, using the expression (4.74) and noting that $\epsilon_{ijk}A_{kj} = 0$ because $\mathbf{A}$ is symmetric,

$$\begin{aligned}
\epsilon_{ijk}\tilde{t}_{kj} &= \epsilon_{ijk}(\alpha_2 N_k n_j + \alpha_3 N_j n_k + \alpha_5 n_j A_{kp}n_p + \alpha_6 n_k A_{jp}n_p)\\
&= \epsilon_{ijk}n_j \tilde{g}_k, \tag{4.76}
\end{aligned}$$

where

$$\tilde{g}_i = -\gamma_1 N_i - \gamma_2 A_{ip} n_p, \tag{4.77}$$

$$\gamma_1 = \alpha_3 - \alpha_2, \tag{4.78}$$

$$\gamma_2 = \alpha_6 - \alpha_5. \tag{4.79}$$

It easily follows from the definitions of $\mathbf{W}$ and $\hat{\mathbf{w}}$ in equations (4.1) and $(4.7)_2$ and the contraction rule for alternators (1.7) that

$$\epsilon_{ipj}\hat{w}_p = W_{ij}. \tag{4.80}$$

Therefore, by the definitions of the relative angular velocity $\boldsymbol{\omega}$ and the co-rotational time flux $\mathbf{N}$ in equations (4.2) and (4.8), respectively, equation (4.76) leads to an equivalent formulation of the left-hand side of inequality (4.61) given by

$$
\begin{aligned}
\tilde{t}_{ij}v_{i,j} - w_i\epsilon_{ijk}\tilde{t}_{kj} &= \tilde{t}_{ij}(A_{ij} + W_{ij}) - w_i\epsilon_{ijk}\tilde{t}_{kj} \\
&= \tilde{t}_{ij}A_{ij} - \epsilon_{ijk}\tilde{t}_{kj}(w_i - \hat{w}_i) \\
&= \tilde{t}_{ij}A_{ij} - \omega_i\epsilon_{ijk}\tilde{t}_{kj} \\
&= \tilde{t}_{ij}A_{ij} - \omega_i\epsilon_{ijk}n_j\tilde{g}_k \\
&= \tilde{t}_{ij}A_{ij} - N_i\tilde{g}_i,
\end{aligned}
\tag{4.81}
$$

and hence the inequality (4.61) can be replaced by the more relevant and equivalent inequality

$$\tilde{t}_{ij}A_{ij} - N_i\tilde{g}_i = \mathcal{D} \geq 0. \tag{4.82}$$

Inserting $\tilde{t}_{ij}$ and $\tilde{g}_i$ into this expression shows that we must have

$$
\begin{aligned}
\mathcal{D} = \mu_2 n_i A_{ij} n_j &+ \alpha_1(n_i A_{ij} n_j)^2 + (\alpha_2 + \alpha_3 + \gamma_2) N_i A_{ij} n_j \\
&+ \alpha_4 A_{ij} A_{ij} + (\alpha_5 + \alpha_6) n_i A_{ij} A_{jk} n_k + \gamma_1 N_i N_i \geq 0,
\end{aligned}
\tag{4.83}
$$

recalling the result (4.36) and making use of the symmetry of $\mathbf{A}$. This inequality must hold for all A_{ij} and N_i The first term on the left-hand side of (4.83) is linear in A_{ij} and therefore we must have

$$\mu_2 = 0, \tag{4.84}$$

which reduces the inequality to its most familiar form

$$
\begin{aligned}
\mathcal{D} = \alpha_1(n_i A_{ij} n_j)^2 &+ (\alpha_2 + \alpha_3 + \gamma_2) N_i A_{ij} n_j \\
&+ \alpha_4 A_{ij} A_{ij} + (\alpha_5 + \alpha_6) n_i A_{ij} A_{jk} n_k + \gamma_1 N_i N_i \geq 0,
\end{aligned}
\tag{4.85}
$$

a form which will be exploited below. The term involving δ_{ij} in the equation for $\tilde{t}_{ij}$ in (4.74) can be incorporated into the arbitrary pressure p appearing in t_{ij} in equation (4.56) and so, in conjunction with the result (4.84), the viscous stress $\tilde{t}_{ij}$ can finally now be given in its most widely adopted and well known form

$$
\begin{aligned}
\tilde{t}_{ij} &= \alpha_1 n_k A_{kp} n_p n_i n_j + \alpha_2 N_i n_j + \alpha_3 n_i N_j + \alpha_4 A_{ij} \\
&\quad + \alpha_5 n_j A_{ik} n_k + \alpha_6 n_i A_{jk} n_k.
\end{aligned}
\tag{4.86}
$$

The coefficients α_1, α_2,..., α_6, are called the Leslie viscosity coefficients, or simply the Leslie viscosities.

Restrictions on the Leslie viscosities follow from inequality (4.85). For example, setting

$$\mathbf{n} = (1,0,0), \qquad \mathbf{N} = (0, N, 0), \tag{4.87}$$

in (4.85) gives the constraint

$$\alpha_1 A_{11}^2 + (\alpha_2 + \alpha_3 + \gamma_2)NA_{12} + \alpha_4 A_{ij}A_{ij} + (\alpha_5 + \alpha_6)A_{1j}A_{1j} + \gamma_1 N^2 \geq 0. \tag{4.88}$$

Since $A_{11} + A_{22} + A_{33} = 0$, terms involving A_{33} can be eliminated and then the constraint reduces to

$$\gamma_1 N^2 + (\alpha_2 + \alpha_3 + \gamma_2)NA_{12} + (\alpha_5 + \alpha_6 + 2\alpha_4)A_{12}^2$$
$$+(\alpha_1 + \alpha_5 + \alpha_6 + 2\alpha_4)A_{11}^2 + 2\alpha_4 A_{11}A_{22} + 2\alpha_4 A_{22}^2$$
$$+(2\alpha_4 + \alpha_5 + \alpha_6)A_{13}^2 + 2\alpha_4 A_{23}^2 \;\geq\; 0. \tag{4.89}$$

The left-hand side of the above expression is a quadratic form in the variables N, A_{12}, A_{11}, A_{22}, A_{13} and A_{23}. It is therefore associated with a unique symmetric matrix, and for this quadratic form to be positive semi-definite it is necessary and sufficient that all the leading minors (determinants of the principal submatrices) of the associated matrix be non-negative [2, pp.479–480]. It is clear from the form of (4.89) that the coefficients of A_{13}^2 and A_{23}^2 must be non-negative, while the first two lines on the left-hand side are associated with the two symmetric matrices

$$\begin{bmatrix} \gamma_1 & \frac{1}{2}(\alpha_2 + \alpha_3 + \gamma_2) \\ \frac{1}{2}(\alpha_2 + \alpha_3 + \gamma_2) & \alpha_5 + \alpha_6 + 2\alpha_4 \end{bmatrix}, \qquad \begin{bmatrix} \alpha_1 + \alpha_5 + \alpha_6 + 2\alpha_4 & \alpha_4 \\ \alpha_4 & 2\alpha_4 \end{bmatrix}, \tag{4.90}$$

which have their first entries and determinants as their leading minors. It easily follows that for positive semi-definiteness we require

$$\gamma_1 = \alpha_3 - \alpha_2 \;\geq\; 0, \tag{4.91}$$
$$\alpha_4 \;\geq\; 0, \tag{4.92}$$
$$2\alpha_4 + \alpha_5 + \alpha_6 \;\geq\; 0, \tag{4.93}$$
$$2\alpha_1 + 3\alpha_4 + 2\alpha_5 + 2\alpha_6 \;\geq\; 0, \tag{4.94}$$
$$4\gamma_1(2\alpha_4 + \alpha_5 + \alpha_6) \;\geq\; (\alpha_2 + \alpha_3 + \gamma_2)^2. \tag{4.95}$$

The Parodi Relation

Parodi [217] proposed, via Onsager relations, that the viscosity coefficients should be further restricted by the relation

$$\gamma_2 = \alpha_6 - \alpha_5 = \alpha_2 + \alpha_3, \tag{4.96}$$

a result that was subsequently obtained by Currie [58] by a stability argument. This result reduces the number of independent viscosities to five rather than six and often leads to some simplification of the theory. We do not pursue the arguments resulting in Parodi's relation (4.96) here, but note that it is a generally accepted relation in the theory. More details about the viscosities and their relation to flow will be discussed in Section 4.4

When the Parodi relation (4.96) holds, the viscous dissipation inequality given by (4.85) reduces to

$$
\begin{aligned}
\mathcal{D} \;=\;& \alpha_1 (n_i A_{ij} n_j)^2 + 2\gamma_2 N_i A_{ij} n_j + \alpha_4 A_{ij} A_{ij} \\
& + (\alpha_5 + \alpha_6) n_i A_{ij} A_{jk} n_k + \gamma_1 N_i N_i \;\geq\; 0.
\end{aligned}
\tag{4.97}
$$

4.2.4   The Dynamic Equations

We are now in a position to complete the dynamic theory and return to the balance laws for linear and angular momentum. From the constitutive relations (4.56), (4.57) and the result (4.60), the balance law for angular momentum (4.45) becomes

$$
\rho K_i + \epsilon_{ipq} \left[\tilde{t}_{qp} + \left(n_p \frac{\partial w_F}{\partial n_{q,j}} \right)_{,j} - n_{k,q} \frac{\partial w_F}{\partial n_{k,p}} \right] = 0.
\tag{4.98}
$$

Using the definition (4.76) and the Ericksen identity (B.6), this equation can be further rearranged (after suitable relabelling of indices) into

$$
\rho K_i + \epsilon_{ipq} n_p \left[\left(\frac{\partial w_F}{\partial n_{q,j}} \right)_{,j} - \frac{\partial w_F}{\partial n_q} + \tilde{g}_q \right] = 0.
\tag{4.99}
$$

If we suppose that the external body moment **K** per unit mass is related to the generalised body force **G** via the relationship $\rho \mathbf{K} = \mathbf{n} \times \mathbf{G}$, that is,

$$
\rho K_i = \epsilon_{ipq} n_p G_q,
\tag{4.100}
$$

then equation (4.99) becomes

$$
\epsilon_{ipq} n_p \left[\left(\frac{\partial w_F}{\partial n_{q,j}} \right)_{,j} - \frac{\partial w_F}{\partial n_q} + \tilde{g}_q + G_q \right] = 0,
\tag{4.101}
$$

from which we can conclude that the final form we adopt for the balance law of angular momentum is

$$
\left(\frac{\partial w_F}{\partial n_{i,j}} \right)_{,j} - \frac{\partial w_F}{\partial n_i} + \tilde{g}_i + G_i = \lambda n_i,
\tag{4.102}
$$

where λ is an arbitrary scalar function. Notice that this is effectively the Euler–Lagrange type equilibrium equation of static theory with the addition of the dynamic term $\tilde{g}$ defined in (4.77) (cf. equation (2.143) on page 38).

Although, as mentioned earlier, we have not considered the director inertial term in the derivation of equation (4.102), it is worth noting that when it is included this equation should be replaced by [168, Eqn.(53)₂]

$$
\left(\frac{\partial w_F}{\partial n_{i,j}} \right)_{,j} - \frac{\partial w_F}{\partial n_i} + \tilde{g}_i + G_i = \lambda n_i + \sigma \ddot{n}_i,
\tag{4.103}
$$

where σ is an inertial constant, measured in terms of kg m^{-1}, that arises from the consideration of a rotational kinetic energy of the material element. The term $\sigma\ddot{n}$ is accepted as being negligible in most circumstances, but, as remarked by Leslie [168], it could conceivably play a rôle when the anisotropic axis is subjected to large accelerations, and for this reason we state equation (4.103) for the sake of completeness.

Taking the scalar product of equation (4.102) with $n_{i,k}$ gives

$$\left(\frac{\partial w_F}{\partial n_{i,j}}\right)_{,j} n_{i,k} - \frac{\partial w_F}{\partial n_i} n_{i,k} + \tilde{g}_i n_{i,k} + G_i n_{i,k} = 0, \tag{4.104}$$

noting that $n_i n_{i,k} = 0$. Inserting the constitutive relation (4.56) for t_{ij} into the linear momentum balance law (4.41) and making use of the result (4.104), with the indices suitably reassigned, gives

$$
\begin{aligned}
\rho\dot{v}_i &= \rho F_i - p_{,i} - \frac{\partial w_F}{\partial n_{k,j}} n_{k,ji} - \left(\frac{\partial w_F}{\partial n_{k,j}}\right)_{,j} n_{k,i} + \tilde{t}_{ij,j} \\
&= \rho F_i - p_{,i} - \frac{\partial w_F}{\partial n_{k,j}} n_{k,ji} - \frac{\partial w_F}{\partial n_k} n_{k,i} + \tilde{g}_k n_{k,i} + G_k n_{k,i} + \tilde{t}_{ij,j}. \quad (4.105)
\end{aligned}
$$

Since $w_F = w_F(\mathbf{n}, \nabla\mathbf{n})$, the above reduces to the general form for the balance of linear momentum using the notation (2.123) for partial derivatives, giving

$$\rho\dot{v}_i = \rho F_i - (p + w_F)_{,i} + \tilde{g}_j n_{j,i} + G_j n_{j,i} + \tilde{t}_{ij,j}. \tag{4.106}$$

The main dynamic equations for nematics are then (4.35), (4.102) and (4.106). These are the equations to be found in the review by Leslie [168] (equations (53)$_2$, (56) and (57)) when the inertial terms involving $\ddot{n}$ are ignored, noting that Leslie chose to absorb ρ into his definition of $\mathbf{F}$ in [168].

As remarked by Ericksen [74], in ordinary circumstances we expect gravitational or electromagnetic fields to produce the forces $\mathbf{F}$ and $\mathbf{G}$. For example, in the case of an external gravitational field in the absence of electromagnetic fields

$$F_i = -\frac{\partial\Psi_g}{\partial x_i}, \quad G_i = 0, \quad \Psi_g = \Psi_g(\mathbf{x}), \tag{4.107}$$

where ψ_g is the gravitational potential.

In many instances of practical interest when the effect of gravity is neglected, an external body force $\mathbf{F}$ and external body moment $\mathbf{K}$ can arise from the presence of a magnetic field. The body moment due to an external magnetic field is assumed to take the form [74]

$$\rho K_i = \epsilon_{ipq} n_p G_q, \quad G_i = \chi_a n_k H_k H_i, \tag{4.108}$$

where $\mathbf{H}$ is the magnetic field and χ_a is the magnetic anisotropy. By analogy with Remark (ii) on page 42, equations (2.171) to (2.175) are valid except that in this present context K_i and F_i are replaced by ρK_i and ρF_i. Therefore

$$\rho F_i = M_k H_{k,i}, \tag{4.109}$$

where F_i is the associated body force and $\mathbf{M}$ is the magentisation (2.109), and a calculation reveals

$$\rho F_i = \frac{\partial \Psi_m}{\partial x_i}, \quad G_i = \frac{\partial \Psi_m}{\partial n_i}, \quad \Psi_m(\mathbf{n}, \mathbf{x}) = \tfrac{1}{2}\mathbf{M} \cdot \mathbf{H}, \tag{4.110}$$

with $\partial/\partial x_i$ denoting the usual notion of partial derivative of Ψ_m with respect to x_i when $\mathbf{n}$ and $\mathbf{x}$ are treated as independent variables. There are analogous expressions for an electric field $\mathbf{E}$ when, in the notation of Section 2.3.1,

$$\rho F_i = D_k E_{k,i}, \quad G_i = \epsilon_0 \epsilon_a n_k E_k E_i, \quad \Psi_e(\mathbf{n}, \mathbf{x}) = \tfrac{1}{2}\mathbf{D} \cdot \mathbf{E}, \tag{4.111}$$

where $\mathbf{D}$ is the electric displacement (2.84), bearing in mind that when dealing with electric fields these assumptions may be subject to more scrutiny, given the comments in Section 3.5. Note that if either of the fields $\mathbf{H}$ or $\mathbf{E}$ is assumed to be uniform and independent of $\mathbf{x}$ then the associated force $\mathbf{F}$ is zero.

It follows from equations (4.107) and (4.110) that in general we can incorporate the effects of gravity and magnetic fields by considering

$$\rho F_i = \frac{\partial \Psi}{\partial x_i}, \quad G_i = \frac{\partial \Psi}{\partial n_i}, \quad \Psi = -\rho \Psi_g + \Psi_m, \tag{4.112}$$

when ρ is constant. Further, since $\Psi = \Psi(\mathbf{n}, \mathbf{x})$, it is seen that, using the definition (2.123),

$$\rho F_i + G_j n_{j,i} = \frac{\partial \Psi}{\partial x_i} + \frac{\partial \Psi}{\partial n_j} n_{j,i} = \Psi_{,i}, \tag{4.113}$$

which then allows the balance of linear momentum (4.106) to be written as

$$\rho \dot{v}_i = \tilde{g}_j n_{j,i} - \tilde{p}_{,i} + \tilde{t}_{ij,j}, \tag{4.114}$$

where

$$\tilde{p} = p + w_F - \Psi, \tag{4.115}$$

with p being the pressure introduced in (4.56).

We state for the sake of completeness, as in the angular momentum case, that if the director inertial contribution is incorporated then the two forms for the linear momentum balance equation given by equations (4.106) and (4.114) are replaced by, respectively [168, Eqns.(57),(58)],

$$\rho \dot{v}_i + \sigma \ddot{n}_j n_{j,i} = \rho F_i - (p + w_F)_{,i} + \tilde{g}_j n_{j,i} + G_j n_{j,i} + \tilde{t}_{ij,j}, \tag{4.116}$$

$$\rho \dot{v}_i + \sigma \ddot{n}_j n_{j,i} = \tilde{g}_j n_{j,i} - \tilde{p}_{,i} + \tilde{t}_{ij,j}, \tag{4.117}$$

bearing in mind the comments mentioned after equation (4.103).

It now follows that for dynamic problems in nematics the key governing equations in many practical situations are the constraint (4.35) and the balance laws given by (4.102) and (4.114), together with the constitutive equations (4.77) and (4.86).

It is important to point out that the number of unknowns in the dynamic theory equals the number of available equations. The unknowns are the velocity $\mathbf{v}(\mathbf{x}, t)$, the director $\mathbf{n}(\mathbf{x}, t)$, the pressure $p(\mathbf{x}, t)$ and the Lagrange multiplier λ. Therefore the total number of unknown quantities is clearly eight: there are three arising from $\mathbf{v}$, three from $\mathbf{n}$, one from p and one from λ. The constraint (4.3) and the incompressibility condition (4.35) provide two equations while the balance of linear momentum (4.106) and the balance of angular momentum (4.102) each provide three equations. There are therefore eight equations to be satisfied, and this matches the number of unknowns. It is quite common, in simple examples, for the balance of angular momentum to reduce to two equations once the Lagrange multiplier has been eliminated because $\mathbf{n}$, being a unit vector, can be described by two independent parameters. An explicit example involving such a reduction in the number of equations occurs in Section 5.2 below at equations (5.17) and (5.18).

We conclude this Section with some comments on boundary conditions. At a solid boundary or interface it is common to assume that the nematic liquid crystal adheres to the solid surface so that the velocity is subject to the usual no-slip condition; for example, at a solid interface at rest it is assumed that $\mathbf{v} = \mathbf{0}$ on the boundary. (Recall that the no-slip condition is interpreted as indicating that the fluid itself does not 'slip' past a solid boundary.) The corresponding director boundary condition is often the strong anchoring condition discussed in Section 2.6.2, so that the alignment of the director at the surface is insensitive to the flow. Of course, other boundary conditions for the director are possible, but most problems in the literature employ the strong anchoring hypothesis, which we adopt in this text when discussing flow problems. Nevertheless, some workers do occasionally consider the inclusion of a surface viscosity for motions of the director near a solid surface when weak anchoring is employed [141].

4.2.5   Summary of the Ericksen–Leslie Dynamic Equations

It is convenient at this point to summarise the Ericksen–Leslie dynamic equations for nematics in the incompressible isothermal theory when the director inertial term is neglected. These are the most frequently used forms of the equations and we state them in the notation introduced in the previous Sections. They consist of the constraints

$$n_i n_i = 1, \qquad v_{i,i} = 0, \tag{4.118}$$

together with the balance laws which arise from linear and angular momentum, namely,

$$\rho \dot{v}_i = \rho F_i - (p + w_F)_{,i} + \tilde{g}_j n_{j,i} + G_j n_{j,i} + \tilde{t}_{ij,j}, \tag{4.119}$$

$$\left(\frac{\partial w_F}{\partial n_{i,j}} \right)_j - \frac{\partial w_F}{\partial n_i} + \tilde{g}_i + G_i = \lambda n_i, \tag{4.120}$$

where F_i is the external body force per unit mass, G_i is the generalised body force which is related through equation (4.100) to the external body moment K_i per unit mass, ρ is the density, p is the pressure, and w_F is the elastic energy density for nematics. The superposed dot denotes the usual material time derivative (4.5). The scalar function λ is a Lagrange multiplier which can usually be eliminated or evaluated by taking the scalar product of equation (4.120) with **n**. The constitutive equations for the viscous stress $\tilde{t}_{ij}$ and the vector $\tilde{g}_i$ are

$$
\begin{aligned}
\tilde{t}_{ij} &= \alpha_1 n_k A_{kp} n_p n_i n_j + \alpha_2 N_i n_j + \alpha_3 n_i N_j + \alpha_4 A_{ij} \\
&\quad + \alpha_5 n_j A_{ik} n_k + \alpha_6 n_i A_{jk} n_k, \tag{4.121} \\
\tilde{g}_i &= -\gamma_1 N_i - \gamma_2 A_{ip} n_p, \tag{4.122} \\
\gamma_1 &= \alpha_3 - \alpha_2 \geq 0, \tag{4.123} \\
\gamma_2 &= \alpha_3 + \alpha_2 = \alpha_6 - \alpha_5, \tag{4.124} \\
A_{ij} &= \tfrac{1}{2}(v_{i,j} + v_{j,i}), \tag{4.125} \\
N_i &= \dot{n}_i - W_{ij} n_j, \qquad W_{ij} = \tfrac{1}{2}(v_{i,j} - v_{j,i}), \tag{4.126}
\end{aligned}
$$

where $\alpha_1, \alpha_2, ..., \alpha_6$, are the Leslie viscosities, A_{ij} is the rate of strain tensor, W_{ij} is the vorticity tensor, N_i is the co-rotational time flux of the director **n** and a superposed dot again represents the material time derivative. The Parodi relation

$$
\gamma_2 = \alpha_6 - \alpha_5 = \alpha_2 + \alpha_3, \tag{4.127}
$$

is assumed to hold and the Leslie viscosities must additionally satisfy the inequalities (4.91) to (4.95). The stress tensor and couple stress tensor are, by equations (4.56), (4.57) and (4.60),

$$
t_{ij} = -p\,\delta_{ij} - \frac{\partial w_F}{\partial n_{p,j}} n_{p,i} + \tilde{t}_{ij}, \tag{4.128}
$$

$$
l_{ij} = \epsilon_{ipq} n_p \frac{\partial w_F}{\partial n_{q,j}}, \tag{4.129}
$$

respectively, where the elastic energy for nematics is

$$
w_F = \tfrac{1}{2}(K_1 - K_2 - K_4)(n_{i,i})^2 + \tfrac{1}{2}K_2 n_{i,j} n_{i,j} + \tfrac{1}{2}K_4 n_{i,j} n_{j,i} \\
+ \tfrac{1}{2}(K_3 - K_2) n_j n_{i,j} n_k n_{i,k}. \tag{4.130}
$$

It is also worth noting from (2.147) that when partial derivatives commute

$$
\left(\frac{\partial w_F}{\partial n_{i,j}}\right)_{,j} - \frac{\partial w_F}{\partial n_i} = (K_1 - K_2)n_{j,ji} + K_2 n_{i,jj} + (K_3 - K_2)(n_j n_k n_{i,k})_{,j} \\
- (K_3 - K_2)n_j n_{k,j} n_{k,i}. \tag{4.131}
$$

The balance laws (4.119) and (4.120) should be replaced by equations (4.116) and (4.103), respectively, if the director inertial term is to be included.

Simplifications

Some simplifications are possible when the relations (4.112) hold for the forces F_i and G_i. These relations are particularly useful when including the gravitational field and/or electromagnetic fields. In this case the balance of linear momentum (4.119) simplifies to

$$
\rho \dot{v}_i = \tilde{g}_j n_{j,i} - \tilde{p}_{,i} + \tilde{t}_{ij,j}, \tag{4.132}
$$

with
$$\tilde{p} = p + w_F - \Psi, \qquad \Psi = -\rho\Psi_g + \Psi_m, \qquad (4.133)$$

where p is an arbitrary pressure and the forms for F_i and G_i in the balances of linear and angular momentum (4.120) become, respectively,

$$\rho F_i = \frac{\partial\Psi}{\partial x_i}, \qquad G_i = \frac{\partial\Psi}{\partial n_i}, \qquad (4.134)$$

when ρ is constant. The term $\Psi_g = \Psi_g(\mathbf{x})$ is the gravitational potential and $\Psi_m = \Psi_m(\mathbf{n}, \mathbf{x})$ is the potential for a magnetic field $\mathbf{H}$ given by

$$\Psi_m = -w_{mag} = \tfrac{1}{2}\chi_{m_\perp}H^2 + \tfrac{1}{2}\chi_a(\mathbf{n}\cdot\mathbf{H})^2, \qquad (4.135)$$

where w_{mag} is the usual magnetic energy (2.106). There is an analogous expression for electric fields using the electric energy w_{elec} in equation (2.86). If the director inertial term is to be included then (4.132) should be replaced by equation (4.117).

The one-constant approximation (2.62) for the elastic energy when strong anchoring is assumed is also useful for reducing the elastic energy to the form given by (2.67). In this case the simplified elastic energy approximation is

$$w_F = \tfrac{1}{2}K n_{i,j} n_{i,j}, \qquad (4.136)$$

and equation (4.131) becomes

$$\left(\frac{\partial w_F}{\partial n_{i,j}}\right)_{,j} - \frac{\partial w_F}{\partial n_i} = K n_{i,jj}, \qquad (4.137)$$

which then reduces equation (4.120) to

$$K n_{i,jj} + \tilde{g}_i + G_i = \lambda n_i. \qquad (4.138)$$

Remark (i)

The linearised version of the Ericksen–Leslie dynamic equations coincides with that discussed by Martin, Parodi and Pershan [192], as mentioned by Leslie [168], with suitable reorganisation of the terms in the equations.

Remark (ii)

When the director gradients $n_{i,j}$ are neglected (bearing in mind that $\mathbf{n}$ may still depend upon time t) then the dynamic equations derived by Leslie [162, 174] for an anisotropic fluid are recovered (recall that elastic effects are often neglected in simplified versions of anisotropic fluids, but are included in general liquid crystal materials). For example, in the simple case when F_i and G_i are given by equations (4.134), the constraints (4.118) remain as stated while the governing dynamic equations (4.132) and (4.120) reduce to the simplified forms (cf. [162, p.366])

$$\rho\dot{v}_i = -\tilde{p}_{,i} + \tilde{t}_{ij,j}, \qquad (4.139)$$
$$\tilde{g}_i + G_i = \lambda n_i, \qquad (4.140)$$

where

$$\tilde{p} = p - \Psi, \qquad \Psi = -\rho\Psi_g + \Psi_m. \qquad (4.141)$$

4.3   Reformulation of the Dynamic Equations

In this Section we derive a particularly convenient and useful reformulation of the dynamic equations for nematic liquid crystals. The Ericksen–Leslie equations summarised in Section 4.2.5 can be reformulated in a manner similar to the reformulation of the equilibrium equations in Section 2.7 when it is supposed that

$$\mathbf{n} = \mathbf{f}(\theta_1, \theta_2), \quad \mathbf{f} \cdot \mathbf{f} = 1, \quad w_F = \widehat{w}_F(\theta_\alpha, \theta_{\alpha,i}), \quad \alpha = 1, 2, \tag{4.142}$$

where, for example, $\mathbf{n}$ could be of the form

$$\mathbf{n} = (\cos\theta_1 \cos\theta_2, \cos\theta_1 \sin\theta_2, \sin\theta_1), \quad \theta_1 = \theta(\mathbf{x}, t), \quad \theta_2 = \phi(\mathbf{x}, t). \tag{4.143}$$

Equations (2.211) to (2.216) remain valid for $w_F = \widehat{w}_F(\theta_\alpha, \theta_{\alpha,i})$, but to make progress in the dynamic equations some further properties of the dissipation function must be investigated. It will be assumed throughout this Section that the inertial term involving $\sigma\ddot{\mathbf{n}}$ is absent since, as mentioned above, it is considered to be negligible in most applications.

As pointed out by Ericksen [80], when the Parodi relation (4.96) applies the vector $\tilde{g}_i$ and the viscous stress $\tilde{t}_{ij}$ in equations (4.77) and (4.86) can be obtained directly from the dissipation function $\mathcal{D}$ (which we can accept for our purposes as being defined by equation (4.97)), through the properties

$$\tilde{g}_i = -\frac{1}{2}\frac{\partial\mathcal{D}}{\partial\dot{n}_i}, \quad \tilde{t}_{ij} = \frac{1}{2}\frac{\partial\mathcal{D}}{\partial v_{i,j}}. \tag{4.144}$$

These properties can be verified directly by simple calculations. For example, since $N_i = \dot{n}_i - W_{ij}n_j$ by (4.9),

$$
\begin{aligned}
\frac{\partial\mathcal{D}}{\partial\dot{n}_i} &= \frac{\partial}{\partial\dot{n}_i}\left(\tilde{t}_{pq}A_{pq} - N_p\tilde{g}_p\right) \\
&= \frac{\partial}{\partial\dot{n}_i}\left(A_{pq}\left(\alpha_2 N_p n_q + \alpha_3 n_p N_q\right)\right) - \tilde{g}_p\delta_{pi} - N_p\frac{\partial\tilde{g}_p}{\partial\dot{n}_i} \\
&= (\alpha_2 + \alpha_3)A_{iq}n_q - \tilde{g}_i + \gamma_1 N_i \\
&= -2\tilde{g}_i,
\end{aligned}
\tag{4.145}
$$

provided the Parodi relation $\gamma_2 = \alpha_2 + \alpha_3$ holds, resulting in property $(4.144)_1$. Property $(4.144)_2$ can similarly be verified, although the calculations are straightforward but somewhat lengthier.

We introduce reformulated versions of the potential Ψ in $(4.112)_3$ and dissipation function $\mathcal{D}$ in (4.97) by

$$\Psi(n_i, x_i) = \widehat{\Psi}(\theta_\alpha, x_i), \quad \mathcal{D}(A_{ij}, N_i, n_i) = 2\widehat{\mathcal{D}}(v_{i,j}, \dot{\theta}_\alpha, \theta_\alpha), \quad \alpha = 1, 2. \tag{4.146}$$

In addition to the properties (2.212) to (2.216) there is also the property

$$\dot{n}_k = \frac{\partial f_k}{\partial\theta_\alpha}\dot{\theta}_\alpha, \tag{4.147}$$

(summing over α) so that by the result $(4.144)_1$

$$\frac{\partial \widehat{\mathcal{D}}}{\partial \dot{\theta}_\alpha} = \frac{1}{2} \frac{\partial \mathcal{D}}{\partial \dot{n}_k} \frac{\partial \dot{n}_k}{\partial \dot{\theta}_\alpha} = -\tilde{g}_k \frac{\partial f_k}{\partial \theta_\alpha}. \tag{4.148}$$

From this result and $(2.212)_1$ we also have

$$\tilde{g}_j n_{j,i} = \tilde{g}_j \frac{\partial f_j}{\partial \theta_\alpha} \theta_{\alpha,i} = -\frac{\partial \widehat{\mathcal{D}}}{\partial \dot{\theta}_\alpha} \theta_{\alpha,i}. \tag{4.149}$$

Further, using equation $(4.112)_2$,

$$\frac{\partial \widehat{\Psi}}{\partial \theta_\alpha} = \frac{\partial \Psi}{\partial n_i} \frac{\partial f_i}{\partial \theta_\alpha} = G_i \frac{\partial f_i}{\partial \theta_\alpha}. \tag{4.150}$$

Combining the results (2.216), (4.148) and (4.150) we see that

$$\left(\frac{\partial \widehat{w}_F}{\partial \theta_{\alpha,i}} \right)_{,i} - \frac{\partial \widehat{w}_F}{\partial \theta_\alpha} - \frac{\partial \widehat{\mathcal{D}}}{\partial \dot{\theta}_\alpha} + \frac{\partial \widehat{\Psi}}{\partial \theta_\alpha} = \left[\left(\frac{\partial w_F}{\partial n_{k,i}} \right)_{,i} - \frac{\partial w_F}{\partial n_k} + \tilde{g}_k + G_k \right] \frac{\partial f_k}{\partial \theta_\alpha}, \tag{4.151}$$

and therefore by equations (2.209) and (4.102) the balance of angular momentum becomes

$$\left(\frac{\partial \widehat{w}_F}{\partial \theta_{\alpha,i}} \right)_{,i} - \frac{\partial \widehat{w}_F}{\partial \theta_\alpha} - \frac{\partial \widehat{\mathcal{D}}}{\partial \dot{\theta}_\alpha} + \frac{\partial \widehat{\Psi}}{\partial \theta_\alpha} = 0, \qquad \alpha = 1, 2, \tag{4.152}$$

while with the aid of the relations (4.115), $(4.144)_2$, $(4.146)_2$ and (4.149) the balance of linear momentum (4.114) becomes

$$\rho \dot{v}_i = \left(\frac{\partial \widehat{\mathcal{D}}}{\partial v_{i,j}} \right)_{,j} - \frac{\partial \widehat{\mathcal{D}}}{\partial \dot{\theta}_\alpha} \theta_{\alpha,i} - \tilde{p}_{,i}, \qquad i = 1, 2, 3, \tag{4.153}$$

summing over the index α, with

$$\tilde{p} = p + \widehat{w}_F - \widehat{\Psi}. \tag{4.154}$$

These alternative forms given in (4.152) and (4.153) for the main balance equations allow the complete dynamic equations to be derived more easily and rapidly, remembering, of course, that under the assumption of incompressibility the constraint $(4.118)_2$ (div $\mathbf{v} = 0$) must also hold. Applications of these equations for one or two orientation angles can be found in Sections 5.6.2, 5.9.1 and 5.9.2.

4.4   The Nematic Viscosities

It is appropriate at this point to discuss some of the properties of the nematic viscosity coefficients in relation to basic experimental ideas and give some insight to their physical interpretation. It is also important to give the relationship between the notation used for the Leslie viscosities and the notation used by other workers in the field, especially experimentalists. From the derivation of the dynamic equations in Section 4.2, it is evident that five independent viscosities and four elastic constants are necessary for a full description of the hydrodynamics of nematic liquid crystals. As we shall show, with the exception of α_4, the Leslie viscosities cannot in general be identified individually, but certain linear combinations of them can be identified experimentally via measurements of the type first proposed by Miesowicz, who introduced the Miesowicz viscosities for nematics. It appears that the first accurate determination of anisotropic viscosities in nematics was carried out by Miesowicz [201, 202] in the 1930s and reported later in 1946, although Zwetkoff [288] also reported some early viscometry measurements, as we shall discuss later. Miesowicz used a strong magnetic field to align a nematic liquid crystal sample and measured viscosities using an oscillating plate viscometer; in strong fields it is possible to neglect boundary effects and gradients $\nabla \mathbf{n}$ of the director. An expert first-hand historical review on the Miesowicz viscosities and their measurement, together with a detailed description of the mechanical apparatus used in the experiments, was written by Miesowicz in 1983 [203]. Useful and concise information on the viscosities can be found in the recent short review articles by Moscicki [205], Martins [194], Belyaev [13] and Schneider and Kneppe [247]; it is also worth consulting the book by de Jeu [137]. We shall begin by introducing the Miesowicz viscosities and investigate their relationship to the Leslie viscosities.

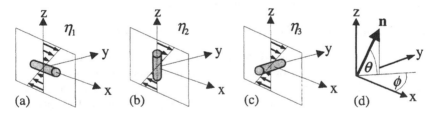

Figure 4.1: In each case the shear is in the xz-plane with the velocity $\mathbf{v}$ indicated by the arrows lying within the plane of shear, while $\nabla \mathbf{v}$ is always along the z-direction. The director $\mathbf{n}$ is represented by the cylinder where (a) $\mathbf{n} \| \mathbf{v}$, (b) $\mathbf{n} \| \nabla \mathbf{v}$ and (c) $\mathbf{n} \perp \mathbf{v}$ and $\mathbf{n} \perp \nabla \mathbf{v}$. The apparent viscosity $\eta(\theta, \phi)$ is defined in equation (4.160) using the angles specified in (d).

Miesowicz distinguished three principal viscosity coefficients, η_1, η_2 and η_3, that could be independently measured experimentally by considering the orientation of the director $\mathbf{n}$ in relation to the flow velocity $\mathbf{v}$. The three basic flow geometries considered by Miesowicz are depicted in Fig. 4.1 and allow the measurement of the viscosities:

(a) η_1 when $\mathbf{n}$ is parallel to $\mathbf{v}$,
(b) η_2 when $\mathbf{n}$ is parallel to $\nabla \mathbf{v}$,
(c) η_3 when $\mathbf{n}$ is orthogonal to both $\mathbf{v}$ and $\nabla \mathbf{v}$.

We have adopted the notation used by Miesowicz, but it should be pointed out that some workers occasionally use what is commonly called the Helfrich notation [124], which interchanges the above definitions for η_1 and η_2. If $\mathbf{n}$ is fixed in an arbitrary direction with respect to $\mathbf{v}$ and $\nabla\mathbf{v}$ then the effective viscosity is a linear combination of the three principal Miesowicz viscosities and a fourth viscosity, η_{12}, which cannot be visualised in the usual way. Nevertheless, it has been suggested that η_{12} is related to a stretch type of deformation, as mentioned by de Jeu [137, p.101].

For the geometry depicted in Fig. 4.1 we have

$$\mathbf{n} = (\cos\theta\cos\phi, \cos\theta\sin\phi, \sin\theta), \qquad \mathbf{v} = (v_x(z), 0, 0), \qquad (4.155)$$

where ϕ and $\frac{\pi}{2} - \theta$ are the angles $\mathbf{n}$ makes with $\mathbf{v}$ and $\nabla\mathbf{v}$, respectively. In this notation, the angles θ and ϕ coincide with the angles used in many of the problems discussed in this text, including the tilt and twist static solutions described earlier in Chapter 3 via Fig. 3.3. It is also supposed for the present that θ and ϕ are constant angles, an assumption made possible by the presence of a sufficiently strong magnetic field which aligns the director and enables the director gradients to be neglected. The only non-zero terms in the rate of strain tensor $\mathbf{A}$ and vorticity tensor $\mathbf{W}$ are

$$A_{13} = A_{31} = \tfrac{1}{2}v_{x,z}, \qquad W_{13} = -W_{31} = \tfrac{1}{2}v_{x,z}, \qquad (4.156)$$

leading to

$$\mathbf{N} = \dot{\mathbf{n}} - \mathbf{W}\mathbf{n} = \tfrac{1}{2}(-v_{x,z}n_3, 0, v_{x,z}n_1). \qquad (4.157)$$

Substituting equation $(4.156)_1$ into the viscous stress given by (4.86) then delivers

$$\tilde{t}_{ij} = \alpha_1 v_{x,z} n_1 n_3 n_i n_j + \alpha_2 N_i n_j + \alpha_3 n_i N_j + \alpha_4 A_{ij} + \alpha_5 n_j A_{ik} n_k + \alpha_6 n_i A_{jk} n_k. \quad (4.158)$$

When elastic terms are neglected the shear stress relevant to Fig. 4.1 is the viscous force along the x-direction per unit area in a plane normal to the z-axis; this is simply $\tilde{t}_{13}$. The apparent viscosity $\eta(\theta, \phi)$ is defined in the usual way for such a geometry by (cf. de Gennes and Prost [110, p.211], who use the transpose of the viscous stress defined in this text)

$$\eta = \frac{\text{shear stress}}{2A_{13}} = \frac{t_{13}}{v_{x,z}} = \frac{\tilde{t}_{13}}{v_{x,z}}, \qquad (4.159)$$

since $t_{13} = \tilde{t}_{13}$ by equation (4.56) when director gradients are absent. Inserting the appropriate terms for $\mathbf{A}$, $\mathbf{n}$ and $\mathbf{N}$ into equation (4.158) then gives the apparent viscosity

$$\eta(\theta, \phi) = \eta_1 \cos^2\theta \cos^2\phi + \eta_2 \sin^2\theta + \eta_3 \sin^2\phi \cos^2\theta + \tfrac{1}{4}\eta_{12} \sin^2(2\theta) \cos^2\phi, \quad (4.160)$$

where we have defined the following Miesowicz viscosities in terms of the Leslie viscosities:

$$\eta_1 = \tfrac{1}{2}(\alpha_3 + \alpha_4 + \alpha_6) = \tfrac{1}{2}(\alpha_2 + 2\alpha_3 + \alpha_4 + \alpha_5), \qquad (4.161)$$

$$\eta_2 = \tfrac{1}{2}(-\alpha_2 + \alpha_4 + \alpha_5), \qquad (4.162)$$

$$\eta_3 = \tfrac{1}{2}\alpha_4, \qquad (4.163)$$

$$\eta_{12} = \alpha_1. \qquad (4.164)$$

The second equality in (4.161) was obtained using the Parodi relation (4.96). Similar equivalent descriptions of $\eta(\theta, \phi)$ can be found in the literature: see, for example, Moscicki [205], who has an equally valid description using exactly these viscosities but with different definitions of θ and ϕ. Further examples include those mentioned by Gähwiller [99, Eqn.(7)] and Schneider and Kneppe [247, p.456] (note that in both References [99, 247] the Authors interchange the definitions of the above η_1 and η_2). The three cases in Fig. 4.1 then correspond to the three types of experiment which allow the measurements of (a) η_1 when $\theta = 0$ and $\phi = 0$, (b) η_2 when $\theta = \frac{\pi}{2}$, and (c) η_3 when $\theta = 0$ and $\phi = \frac{\pi}{2}$. The contribution of η_{12} to the apparent viscosity η is clearly maximised when the director is immobilised in the shear plane at an angle of $\frac{\pi}{4}$ to $\mathbf{v}$, in which event $\theta = \frac{\pi}{4}$ and $\phi = 0$, so that

$$\eta_{12} = 4\eta(\tfrac{\pi}{4}, 0) - 2(\eta_1 + \eta_2). \tag{4.165}$$

When this is the case, $\eta(\frac{\pi}{4}, 0)$ can be measured and then η_{12} can be calculated from the relation (4.165) once η_1 and η_2 have been determined.

So far we have introduced four Miesowicz viscosities. Two other viscosities can be proposed by considering the following. The director $\mathbf{n}$ in Fig. 4.1(a), if free to move, will rotate due to a viscous torque; the viscosity coefficient γ_1 is introduced to describe this situation and characterises the torque associated with a rotation of $\mathbf{n}$. For this reason γ_1 is often called the *rotational viscosity* or *twist viscosity*. The coefficient γ_1 generally determines the rate of relaxation of the director. Also, a rotation of $\mathbf{n}$ due to body forces will induce a flow. The viscosity coefficient γ_2 characterises the contribution to the torque due to a shear velocity gradient in the nematic and is sometimes referred to as the *torsion coefficient* in the velocity gradient: it leads to a coupling between the orientation of the director and shear flow. The two viscosities γ_1 and γ_2 have no counterpart in isotropic fluids. We therefore have a total of six viscosities: four Miesowicz viscosities plus γ_1 and γ_2. It turns out, as will be seen in the problems to be discussed in later Sections, that γ_1 and γ_2 are precisely the viscosities introduced in the constitutive theory at equations (4.78) and (4.79), namely,

$$\gamma_1 = \alpha_3 - \alpha_2, \qquad \gamma_2 = \alpha_2 + \alpha_3 = \alpha_6 - \alpha_5, \tag{4.166}$$

the latter expression being a statement of the Parodi relation. Notice that γ_1 is necessarily non-negative by (4.91). The rotational viscosity γ_1 may be determined experimentally by the method of Zwetkoff, to be discussed in Section 5.4 below.

From the form of the viscous stress in equation (4.158) it is also worth remarking that $\eta_3 = \frac{1}{2}\alpha_4$ corresponds to the usual viscosity that arises in standard isotropic Newtonian fluids where, of course, the director $\mathbf{n}$ is absent (see also Section 5.5.1 on page 176).

The six viscosities in equation (4.161) to (4.164) and (4.166) can now all be given in terms of the five independent Leslie viscosities arising in the Ericksen–Leslie dynamic theory, assuming the widely accepted Parodi relation (4.96) holds. Hence only five of these viscosities are needed to form a canonical set of viscosities.

There is also experimental interest in modelling flows related to the elastic splay, twist and bend deformations (see Fig. 2.1 and Section 3.4.1). Flows related to these deformations occur in the dynamics of the Freedericksz transitions and are characterised by the viscosity coefficients η_{splay}, η_{twist} and η_{bend}, respectively, where we define [68, p.390]

$$\eta_{splay} = \gamma_1 - (\eta_1 - \eta_2 + \gamma_1)^2/4\eta_1 = \gamma_1 - \alpha_3^2/\eta_1, \tag{4.167}$$

$$\eta_{twist} = \gamma_1, \tag{4.168}$$

$$\eta_{bend} = \gamma_1 - (\eta_1 - \eta_2 - \gamma_1)^2/4\eta_2 = \gamma_1 - \alpha_2^2/\eta_2. \tag{4.169}$$

We do not pursue a derivation of these particular viscosities here, but introduce them for reference; they originate in the dynamics of Freedericksz transitions which are discussed later in Section 5.9, where it is also shown that they are necessarily non-negative (see pages 228 and 235). Table 4.1 gives a summary of the viscosities discussed in this Section in terms of the Leslie viscosities. The viscosities η_1, η_2, η_3, η_{12} and γ_1 constitute a canonical set of five independent viscosities for nematics.

Miesowicz viscosities	$\eta_1 = \frac{1}{2}(\alpha_3 + \alpha_4 + \alpha_6) = \frac{1}{2}(\alpha_2 + 2\alpha_3 + \alpha_4 + \alpha_5)$
	$\eta_2 = \frac{1}{2}(-\alpha_2 + \alpha_4 + \alpha_5)$
	$\eta_3 = \frac{1}{2}\alpha_4$
	$\eta_{12} = \alpha_1$
rotational viscosity	$\gamma_1 = \alpha_3 - \alpha_2$
torsion coefficient	$\gamma_2 = \alpha_2 + \alpha_3 = \alpha_6 - \alpha_5 = \eta_1 - \eta_2$
splay viscosity	$\eta_{splay} = \gamma_1 - (\eta_1 - \eta_2 + \gamma_1)^2/4\eta_1 = \gamma_1 - \alpha_3^2/\eta_1$
twist viscosity	$\eta_{twist} = \gamma_1 = \alpha_3 - \alpha_2$
bend viscosity	$\eta_{bend} = \gamma_1 - (\eta_1 - \eta_2 - \gamma_1)^2/4\eta_2 = \gamma_1 - \alpha_2^2/\eta_2$

Table 4.1: The principal viscosities in nematics in terms of the Leslie viscosities. The Miesowicz viscosities and the rotational viscosity make up a canonical set of five independent viscosities.

The Leslie viscosities can also be expressed in terms of the Miesowicz viscosities through the relations

$$\alpha_1 = \eta_{12}, \tag{4.170}$$

$$\alpha_2 = \frac{1}{2}(\eta_1 - \eta_2 - \gamma_1), \tag{4.171}$$

$$\alpha_3 = \frac{1}{2}(\eta_1 - \eta_2 + \gamma_1), \tag{4.172}$$

$$\alpha_4 = 2\eta_3, \tag{4.173}$$

$$\alpha_5 = \frac{1}{2}(\eta_1 + 3\eta_2 - 4\eta_3 - \gamma_1), \tag{4.174}$$

$$\alpha_6 = \alpha_2 + \alpha_3 + \alpha_5 = \frac{1}{2}(3\eta_1 + \eta_2 - 4\eta_3 - \gamma_1), \tag{4.175}$$

obtained by simple manipulations. It is also worth noting for future reference that

$$\alpha_2 = \frac{1}{2}(\gamma_2 - \gamma_1), \qquad \alpha_3 = \frac{1}{2}(\gamma_1 + \gamma_2). \tag{4.176}$$

We emphasise again that these rely on the Parodi relation (4.96) being valid. The coefficients α_1 to α_5 can also be considered as a canonical set of five viscosities. This is known to be experimentally viable for a large majority of nematics when they are treated as incompressible.

Table D.3 on page 330 gives a collection of viscosities for the nematic phases of the materials MBBA, PAA and 5CB. (Note that the coefficient α_1 quoted in de Gennes and Prost [110, p.231] for MBBA near 25°C (based upon the work of Gäwiller [97]) is positive, whereas a more recent report by Kneppe, Schneider and Sharma [155] has revealed α_1 to be negative.) For a brief introduction to the major experimental techniques used for evaluating the nematic viscosities the reader is referred to the review by Moscicki [205].

Chapter 5

Applications of Dynamic Theory of Nematics

5.1 Introduction

The applications of the dynamic theory for nematic liquid crystals are vast in number and it is impossible in an introductory book of this scope to cover all the major topics. The aim here is to equip readers with the essential tools required to access the literature and the major reviews, many of which are mentioned below.

An elementary example of flow in nematic liquid crystals is introduced in Section 5.2. This example motivates the basic ideas of flow stability and instability in nematics and also introduces the concepts of flow-aligning and non-flow-aligning nematic liquid crystals. Section 5.3 presents an example of flow where it is possible to have a transverse component of flow, sometimes called a secondary component of flow. This secondary component of flow is in addition to the component of flow that is present in the direction of an applied pressure gradient, and is in a direction perpendicular to that of the pressure gradient.

The Zwetkoff experiment is described in Section 5.4. This remarkable experiment, involving a magnetic field which rotates relative to a liquid crystal sample, allows the direct measurement of the rotational viscosity γ_1 in nematics. It also allows a straightforward measurement of the magnetic anisotropy χ_a if it is positive, via the observation of a critical angular frequency ω_c.

Section 5.5 describes two simple shear flows: shear flow near a boundary and shear flow between two parallel plates. This leads on to a discussion of scaling properties, and it is pointed out in Section 5.5.5 that the apparent viscosity, defined by equation (5.146) below, scales differently from that of an isotropic fluid. Stability and instability of oscillatory shear flow are discussed in Section 5.6; this Section contains what is perhaps the most advanced analysis in this Chapter.

Couette flow is considered in Section 5.7 and Poiseuille flow is discussed in Section 5.8. Some comments on a scaling analysis for Poiseuille flow of a nematic liquid crystal are contained in Section 5.8.3. These scaling results are important in establishing and validating the Ericksen–Leslie dynamic theory of nematics since they make predictions about flow which were confirmed experimentally.

Section 5.9 introduces the dynamics of the Freedericksz transition in the classical geometries described in Chapter 3 in Section 3.4.1. The 'switch-on' and 'switch-off' times will be defined when flow is considered to be negligible in the usual 'twist' geometry, as detailed in Section 5.9.1. In some instances, however, flow turns out to be quite influential and leads to the phenomena of *backflow* and *kickback*, as to be discussed in detail in the case of the 'splay' geometry in Section 5.9.2. Backflow in the 'bend' geometry is discussed in Section 5.9.3.

Light scattering is introduced and modelled for nematic liquid crystals in Section 5.10. Suitable introductory references to the literature on light scattering are also given.

Further aspects of the dynamics of nematic liquid crystals can be found in the books by Chandrasekhar [38], de Gennes and Prost [110] and, for those interested in more detailed and physically motivated results in general liquid crystals, the book on the rheology of liquid crystals by Khabibullaev, Gevorkyan and Lagunov [145].

5.2   A Simple Flow Alignment

To grasp a basic understanding of the response of the director **n** to flow, consider a sample of nematic liquid crystal where the effects of boundaries and external forces and fields are ignored. In the simplest possible scenario the director gradients $n_{i,j}$ and the elastic energy w_F are neglected and attention can be focused on the possible influence of shear flow upon the director. We present an extension of the simple shear flow example given by Leslie [162] for anisotropic fluids, and follow the case contained in the short reviews by Leslie [171, 174]. Consider an arbitrary time dependent director as depicted in Fig. 5.1(c) of the form

$$\mathbf{n} = (\cos\theta\cos\phi, \cos\theta\sin\phi, \sin\theta), \qquad \theta = \theta(t), \quad \phi = \phi(t), \qquad (5.1)$$

where the velocity is given by

$$\mathbf{v} = (kz, 0, 0), \qquad k \text{ a positive constant.} \qquad (5.2)$$

The constraints in equation (4.118) are clearly satisfied. The governing Ericksen–Leslie dynamic equations (4.119) and (4.120) become, respectively,

$$\rho\dot{v}_i = -p_{,i} + \tilde{t}_{ij,j}, \qquad (5.3)$$

$$\tilde{g}_i = \lambda n_i, \qquad (5.4)$$

with $\tilde{t}_{ij}$ and $\tilde{g}_i$ given by equations (4.121) and (4.122). Taking the scalar product of equation (5.4) with **n** and using the property $N_i n_i = 0$ from equation (4.12), shows that the Lagrange multiplier λ is given by

$$\lambda = -\gamma_2 n_p A_{pq} n_q, \qquad (5.5)$$

which in turn allows equation (5.4) to be formulated as

$$\gamma_1 N_i + \gamma_2 (A_{ip} n_p - n_i n_p A_{pq} n_q) = 0. \qquad (5.6)$$

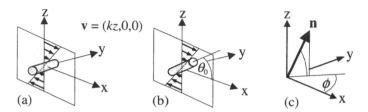

Figure 5.1: The two types of steady solutions. The velocity $\mathbf{v}$ is represented by the arrows lying within the plane of shear. (a) $\theta = 0$, $\phi = \frac{\pi}{2}$, and (b) $\theta = \pm\theta_0$, $\phi = 0$, only the $+\theta_0$ case being shown since the $-\theta_0$ case is simply obtained by a reflection in the xy-plane. The angles θ and ϕ are defined in (c) and the Leslie angle θ_0 in the plane of shear is defined by equation (5.23).

This expression allows N_i to be eliminated from the viscous stress $\tilde{t}_{ij}$ so that it becomes

$$\tilde{t}_{ij} = \alpha n_k A_{kp} n_p n_i n_j + \alpha_4 A_{ij} + \beta(n_j A_{ik} n_k + n_i A_{jk} n_k), \tag{5.7}$$

where, upon using (4.124) to substitute for α_6,

$$\alpha = \alpha_1 + (\alpha_2 + \alpha_3)\gamma_2\gamma_1^{-1}, \qquad \beta = \alpha_5 - \alpha_2\gamma_2\gamma_1^{-1}, \tag{5.8}$$

assuming for the present that $\gamma_1 \equiv \alpha_3 - \alpha_2 \neq 0$ so that $\gamma_1 > 0$, using (4.123). Furthermore, the only non-zero terms in the rate of strain tensor $\mathbf{A}$ and vorticity tensor $\mathbf{W}$, defined in equations (4.125) and (4.126), are

$$A_{13} = A_{31} = \tfrac{1}{2}k, \qquad W_{13} = -W_{31} = \tfrac{1}{2}k. \tag{5.9}$$

It then follows from (5.1), (5.2), (5.7)and (5.9)$_1$ that

$$\tilde{t}_{ij,j} = 0 \quad \text{and} \quad \dot{v}_i = 0, \tag{5.10}$$

because $\tilde{t}_{ij}$ does not depend upon any spatial variables, and hence the balance of linear momentum equation (5.3) is satisfied provided the pressure p is constant. This means that the governing dynamic equations reduce to the reformulated angular momentum equations in (5.6). Using the results in (5.9), it is found that

$$\begin{aligned}
A_{1j}n_j &= \tfrac{1}{2}k\sin\theta, & A_{2j}n_j &= 0, & A_{3j}n_j &= \tfrac{1}{2}k\cos\theta\cos\phi, \\
W_{1j}n_j &= \tfrac{1}{2}k\sin\theta, & W_{2j}n_j &= 0, & W_{3j}n_j &= -\tfrac{1}{2}k\cos\theta\cos\phi, \\
&& n_i A_{ij}n_j &= k\sin\theta\cos\theta\cos\phi,
\end{aligned} \tag{5.11}$$

while the components of $\mathbf{N} = \dot{\mathbf{n}} - \mathbf{W}\mathbf{n}$ are given by

$$\begin{aligned}
N_1 &= -\sin\theta\cos\phi\,\dot{\theta} - \cos\theta\sin\phi\,\dot{\phi} - \tfrac{1}{2}k\sin\theta, \\
N_2 &= -\sin\theta\sin\phi\,\dot{\theta} + \cos\theta\cos\phi\,\dot{\phi}, \\
N_3 &= \cos\theta\,\dot{\theta} + \tfrac{1}{2}k\cos\theta\cos\phi,
\end{aligned} \tag{5.12}$$

where the superposed dot representing the material time derivative has now been reduced to denoting ordinary differentiation with respect to time because the spatial derivatives of the director are zero. Substituting this information into equation (5.6) yields the equations

$$-\gamma_1(\sin\theta\cos\phi\,\dot\theta + \cos\theta\sin\phi\,\dot\phi + \tfrac{1}{2}k\sin\theta) + \tfrac{1}{2}\gamma_2 k\sin\theta(1-2\cos^2\theta\cos^2\phi) \;=\; 0, \quad(5.13)$$

$$\gamma_1(-\sin\theta\sin\phi\,\dot\theta + \cos\theta\cos\phi\,\dot\phi) - \gamma_2 k\cos^2\theta\sin\theta\sin\phi\cos\phi \;=\; 0, \quad(5.14)$$

$$\cos\theta[\gamma_1(\dot\theta + \tfrac{1}{2}k\cos\phi) + \tfrac{1}{2}\gamma_2 k\cos\phi\cos(2\theta)] \;=\; 0. \quad(5.15)$$

Multiplying (5.13) by $\cos\phi$, (5.14) by $\sin\phi$ and adding the resulting equations eliminates $\dot\phi$ to give

$$\gamma_1\sin\theta(\dot\theta + \tfrac{1}{2}k\cos\phi) + \tfrac{1}{2}\gamma_2 k\sin\theta\cos\phi\cos(2\theta) = 0, \qquad(5.16)$$

which, when multiplied by $\sin\theta$ and added to equation (5.15) multiplied by $\cos\theta$, yields the equation for $\dot\theta$, namely,

$$\gamma_1\dot\theta = -\tfrac{1}{2}k(\gamma_1 + \gamma_2\cos(2\theta))\cos\phi. \qquad(5.17)$$

A similar process can be applied to equations (5.13) and (5.14) to eliminate $\dot\theta$ which then gives the equation for $\dot\phi$ as

$$\gamma_1\cos\theta\,\dot\phi = \tfrac{1}{2}k(\gamma_2 - \gamma_1)\sin\theta\sin\phi. \qquad(5.18)$$

Employing the relations (4.123) and (4.124), equations (5.17) and (5.18) reduce to

$$\gamma_1\dot\theta \;=\; k(\alpha_2\sin^2\theta - \alpha_3\cos^2\theta)\cos\phi, \qquad(5.19)$$

$$\gamma_1\cos\theta\,\dot\phi \;=\; k\alpha_2\sin\theta\sin\phi. \qquad(5.20)$$

The Ericksen–Leslie dynamic equations therefore finally reduce to the two equations (5.19) and (5.20) for $\dot\theta$ and $\dot\phi$: these are the equations that describe the orientation of **n** in the simplest shear flow when boundary effects and elastic constants are ignored and external forces and fields are absent.

It is natural to proceed by seeking steady solutions ($\dot\theta = \dot\phi = 0$) to equations (5.19) and (5.20) because these will correspond to the uniform alignments of **n** that are possible in shear flow, bearing in mind that **n** and $-$**n** are indistinguishable in nematic theory. For any values of the viscosities there is always the steady solution provided by

$$\theta = 0, \quad \phi = \tfrac{\pi}{2}. \qquad(5.21)$$

This solution corresponds to a uniform alignment of the director **n** normal to the plane of shear, as indicated in Fig. 5.1(a): Pieranski and Guyon [221] observed such an orientation for MBBA for low shear rates (k small). However, if α_2 and α_3 are non-zero and have the same sign then $\alpha_2\alpha_3 > 0$ and other possible steady solutions will be provided by

$$\theta = \pm\theta_0, \quad \phi = 0, \qquad(5.22)$$

where the acute angle θ_0, referred to as the *flow alignment angle* or *Leslie angle*, is defined by

$$\tan^2\theta_0 = \frac{\alpha_3}{\alpha_2}. \qquad(5.23)$$

(See equations (5.252) and (5.253) below for an equivalent definition in terms of γ_1 and γ_2). Notice that if α_2 and α_3 have different signs then the only available steady solution is given by (5.21). There are then two possibilities, given that $\gamma_1 = \alpha_3 - \alpha_2 > 0$, namely,

$$
\begin{array}{llll}
\text{(i)} & \alpha_2 < \alpha_3 < 0 & \text{and} & 0 < \theta_0 < \frac{\pi}{4}, \quad \phi = 0, \\
\text{(ii)} & \alpha_3 > \alpha_2 > 0 & \text{and} & \frac{\pi}{4} < \theta_0 < \frac{\pi}{2}, \quad \phi = 0.
\end{array}
\tag{5.24}
$$

This indicates that the steady solutions result in the director being in the plane of shear at an acute angle θ_0 to the direction of flow, the $+\theta_0$ case being depicted in Fig. 5.1(b) as an example. This has been observed by Gähwiller [98] for MBBA where θ_0 was seen to be a relatively small angle, in which case possibility (i) is physically relevant; for $0 < \theta_0 < \frac{\pi}{4}$ this would also imply that $\gamma_2 = \alpha_3 + \alpha_2 < 0$ (because both α_2 and α_3 would be negative) and $0 < \gamma_1 < -\gamma_2$ (cf. Table D.3 for the experimental values of MBBA, PAA and 5CB). In possibility (ii) we always have $0 < \gamma_1 < \gamma_2$ for $\frac{\pi}{4} < \theta_0 < \frac{\pi}{2}$.

Stability and Instability

The question remains as to the stability of each type of steady solution, especially given that some materials could exhibit either forms of the solutions (5.21) or (5.24). Notice that $\alpha_2 \neq \alpha_3$ when $\gamma_1 > 0$, which is always the case in this Section.

Firstly, consider small perturbations to the steady solution (5.21) and let

$$
\theta = u, \quad \phi = \frac{\pi}{2} + v, \quad |u|, |v| \ll 1,
\tag{5.25}
$$

where u and v are functions of time t. Inserting these forms into the governing equations (5.19) and (5.20) and linearising in terms of u and v gives the perturbation equations

$$
\gamma_1 \dot{u} = k\alpha_3 v, \quad \gamma_1 \dot{v} = k\alpha_2 u.
\tag{5.26}
$$

If either α_2 or α_3 is zero, then solutions are clearly linearly unstable; for example, if $\alpha_2 = 0$ then $v \equiv c$, c being a non-zero constant for non-zero perturbations, and therefore $|u|$ grows linearly with time t. In all other feasible instances the equations (5.26) can be written as the system

$$
\begin{bmatrix} \dot{u} \\ \dot{v} \end{bmatrix} = \begin{bmatrix} 0 & k\alpha_3 \gamma_1^{-1} \\ k\alpha_2 \gamma_1^{-1} & 0 \end{bmatrix} \begin{bmatrix} u \\ v \end{bmatrix},
\tag{5.27}
$$

which has the eigenvalues $\lambda = \pm k\sqrt{\alpha_2 \alpha_3}\gamma_1^{-1}$. The steady solution (5.21) is therefore linearly unstable if $\alpha_2\alpha_3 > 0$ and linearly stable when $\alpha_2\alpha_3 < 0$ (in the sense that the time-dependent solutions to the perturbations u and v are periodic of period $2\pi/|\lambda|$); in particular, the steady solution (5.21) cannot be asymptotically stable for $\alpha_2\alpha_3 < 0$.

Secondly, consider perturbations to the steady solutions (5.24) of the form

$$
\theta = \pm\theta_0 + u, \quad \phi = v, \quad |u|, |v| \ll 1,
\tag{5.28}
$$

recalling that these steady solutions are only available when $\alpha_2\alpha_3 > 0$. Inserting the above expressions into equations (5.19) and (5.20) and using the definition of θ_0 in equation (5.23) yields the linearised perturbation equations

$$\gamma_1 \dot{u} = k(\alpha_2 + \alpha_3)\sin(\pm 2\theta_0)u, \qquad (5.29)$$
$$\gamma_1 \cos\theta_0 \dot{v} = k\alpha_2 \sin(\pm\theta_0)v. \qquad (5.30)$$

Hence, if $\alpha_2 < \alpha_3 < 0$ then the steady solution $\theta = \theta_0$, $\phi = 0$, is clearly linearly stable, while $\theta = -\theta_0$, $\phi = 0$, is unstable. Similarly, if $\alpha_3 > \alpha_2 > 0$ then $\theta = -\theta_0$, $\phi = 0$, is stable and $\theta = \theta_0$, $\phi = 0$, is unstable.

To summarise, we have now demonstrated that if $\alpha_2\alpha_3 > 0$ (equivalent to the condition $\gamma_1 < |\gamma_2|$ when $\gamma_1 > 0$, by (4.176)) then the steady solution (5.21) pictured in Fig. 5.1(a) is linearly unstable while the steady solution $\theta = \theta_0$, $\phi = 0$, from (5.22), pictured in Fig. 5.1(b), is stable if additionally $\alpha_2 < \alpha_3 < 0$. If $\alpha_3 > \alpha_2 > 0$ then the steady solution $\theta = -\theta_0$, $\phi = 0$, in (5.22) is stable. When $\alpha_2\alpha_3 < 0$ then solutions (5.22) are not possible and in such cases the only steady solution is given by (5.21), which is never asymptotically stable, but can be linearly stable, in the sense indicated above. These results coincide with the results obtained by a nonlinear analysis of the above problem, and the reader is referred to the article by Leslie [171] for details. Results in a similar style in the context of an earlier theory for anisotropic fluids have been discussed by Ericksen [72].

Flow-Aligning and Non-Flow-Aligning Nematics

The flow alignments discussed above may be considered as being obtained by very crude and basic assumptions on the director, but they do nevertheless give a fair indication of the possible alignments for the director $\mathbf{n}$ well away from boundaries for high shear rates. It is also evident from this investigation that the two viscosities α_2 and α_3 (or, equivalently by (4.176), γ_1 and γ_2) play crucial rôles in determining the flow alignment of the director. Nematic liquid crystals are called *flow-aligning* (cf. Fig. 5.1(b)) whenever

$$\alpha_2\alpha_3 > 0 \quad \text{and} \quad \alpha_2 < \alpha_3 < 0, \qquad (5.31)$$

and are called *non-flow-aligning* whenever $\alpha_2\alpha_3 < 0$. For the viscosities at the given temperatures indicated in Table D.3 it follows that MBBA, PAA and 5CB can be flow-aligning. In flow-aligning materials in the absence of applied fields, the velocity gradient aligns the director at the Leslie angle θ_0 to the direction of flow, this flow alignment angle being defined as the acute angle provided by the solution of equation (5.23).

One essential feature is that an appropriate choice of shear direction can produce a flow instability, such an instability often being called a Pieranski–Guyon instability. In their original work on MBBA, Pieranski and Guyon [221] incorporated the effects of elasticity and found that the orientation (5.21) shown in Fig. 5.1(a) was actually stable for low shear rates, but as the speed of the moving plate inducing the shear increased there was a critical value of the shear rate at which the director became unstable and began to reorient into the plane of the shear, in agreement

with the elementary instability result predicted for the solution (5.21) via the perturbation (5.25). Further, the director eventually began to turn into the plane of
shear and align at an angle to the direction of flow, as can be anticipated by the
above elementary stability analysis. For more details on experimental and theoretical work related to this phenomenon the reader may also consult the articles by
Dubois-Violette *et al.* [65, 66]. A short review of such instabilities has been given
by Leslie [168] while a more recent brief review has been compiled by Atkin and
Leslie [68, §8.5].

5.3   A Transverse Flow Effect

We now examine how flow is affected when the director is fixed by a very strong
magnetic field. Consider the situation pictured in Fig. 5.2 where a strong magnetic

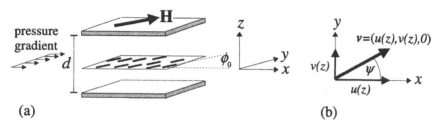

(a)   (b)

Figure 5.2: (a) A very strong magnetic field is applied parallel to the bounding
plates at an angle ϕ_0 to the direction of an external pressure gradient parallel to the
x-axis as shown. The director orientation is uniform across the sample depth d in
the z-direction, a typical cross-section showing the alignment for n when $\chi_a > 0$.
(b) The flow profile for v with components u and v as shown in the xy-plane. The
transverse flow effect is clear because the flow makes an angle ψ, defined by equation
(5.51), with the direction of the pressure gradient in the x-direction.

field **H** parallel to the bounding plates of a sample is inclined at an angle $\phi_0 > 0$
to the direction of an external pressure gradient parallel to the x-axis. It will be
assumed that the director is parallel to **H** everywhere except possibly in very thin
layers near the boundaries; Fig. 5.2 shows the case for $\chi_a > 0$ when the director
is attracted to be parallel to the magnetic field (cf. Section 2.3.2). It is therefore
conceivable, to a good approximation, that the director and the magnetic field take
the forms

$$\mathbf{n} = (\cos \phi_0, \sin \phi_0, 0), \qquad \mathbf{H} = H(\cos \phi_0, \sin \phi_0, 0), \qquad (5.32)$$

uniformly across the bulk of the sample. It is necessary for the velocity to take the
form

$$\mathbf{v} = (u(z), v(z), 0), \qquad (5.33)$$

u and v satisfying the no-slip boundary condition so that they are both equated
to zero on the boundaries at $z = 0$ and $z = d$. Since the director is fixed by
a very strong field, director gradients can be ignored in this basic set-up so that

ϕ_0 is a constant. The effects of gravity will also be neglected. The constraints (4.118) are clearly satisfied for the above choices of **n** and **v**. Given that ϕ_0 is a fixed constant, the balance of angular momentum law for the director **n** may be disregarded in a first approximation to the problem considered here, as suggested by Leslie [168, p.32]. This means that equations (4.140) need not concern us here and that the relevant governing dynamic equations are (4.139) for the balance of linear momentum, which are

$$\tilde{t}_{ij,j} = p_{,i}, \quad i = 1, 2, 3. \tag{5.34}$$

Notice, by equation (4.141), that $\tilde{p}_{,i} = p_{,i}$ because gravity has been neglected and the magnetic potential Ψ_m does not depend explicitly on **x**. Our main concern is then reduced to solving equations (5.34) subject to the no-slip boundary conditions on u and v. It therefore only remains to solve

$$\tilde{t}_{i3,3} = p_{,i}, \quad i = 1, 2, 3, \tag{5.35}$$

subject to the no slip boundary conditions

$$u(0) = u(d) = v(0) = v(d) = 0. \tag{5.36}$$

The non-zero contributions to the rate of strain tensor **A** and the vorticity tensor **W** are

$$A_{13} = A_{31} = \tfrac{1}{2}u'(z), \quad A_{23} = A_{32} = \tfrac{1}{2}v'(z), \tag{5.37}$$
$$W_{13} = -W_{31} = \tfrac{1}{2}u'(z), \quad W_{23} = -W_{32} = \tfrac{1}{2}v'(z), \tag{5.38}$$

where a prime represents differentiation with respect to z. It is obvious that $\dot{\mathbf{n}} \equiv 0$ and therefore the co-rotational time flux **N** is given via (4.126), (5.32) and (5.38) by

$$\mathbf{N} = (0, 0, \tfrac{1}{2}[u'\cos\phi_0 + v'\sin\phi_0]). \tag{5.39}$$

The non-zero components of the dynamic viscous stress $\tilde{t}_{ij}$ in equation (4.121) can be calculated using equations $(5.32)_1$, (5.37), (5.38) and (5.39) to reveal

$$\begin{aligned}
\tilde{t}_{13} &= \tfrac{1}{2}u'\left[\alpha_4 + (\alpha_3 + \alpha_6)\cos^2\phi_0\right] + \tfrac{1}{2}v'(\alpha_3 + \alpha_6)\sin\phi_0\cos\phi_0, & (5.40) \\
\tilde{t}_{23} &= \tfrac{1}{2}v'\left[\alpha_4 + (\alpha_3 + \alpha_6)\sin^2\phi_0\right] + \tfrac{1}{2}u'(\alpha_3 + \alpha_6)\sin\phi_0\cos\phi_0, & (5.41) \\
\tilde{t}_{31} &= \tfrac{1}{2}u'\left[\alpha_4 + (\alpha_2 + \alpha_5)\cos^2\phi_0\right] + \tfrac{1}{2}v'(\alpha_2 + \alpha_5)\sin\phi_0\cos\phi_0, & (5.42) \\
\tilde{t}_{32} &= \tfrac{1}{2}v'\left[\alpha_4 + (\alpha_2 + \alpha_5)\sin^2\phi_0\right] + \tfrac{1}{2}u'(\alpha_2 + \alpha_5)\sin\phi_0\cos\phi_0. & (5.43)
\end{aligned}$$

Equations (5.35) are

$$\tilde{t}_{13,3} = \frac{\partial p}{\partial x}, \tag{5.44}$$

$$\tilde{t}_{23,3} = \frac{\partial p}{\partial y}, \tag{5.45}$$

$$0 = \frac{\partial p}{\partial z}. \tag{5.46}$$

Equation (5.46) shows that the pressure p is restricted to be a function of x and y. However, we know from equations (5.40) and (5.41) that the left-hand sides of (5.44) and (5.45) must be functions of z only, indicating that $p_{,x} = a$ and $p_{,y} = b$ for some constants a and b. Hence we can take

$$p(x, y) = ax + by + p_0, \tag{5.47}$$

where p_0 is an arbitrary constant; but we are considering a gradient flow in the x-direction and therefore $a \neq 0$ while b is arbitrary. Equations (5.44) and (5.45) can now be integrated to give

$$\tilde{t}_{13} = \tfrac{1}{2}u' \left[\alpha_4 + (\alpha_3 + \alpha_6) \cos^2\phi_0 \right] + \tfrac{1}{2}v'(\alpha_3 + \alpha_6) \sin\phi_0 \cos\phi_0 = az + c, \tag{5.48}$$
$$\tilde{t}_{23} = \tfrac{1}{2}v' \left[\alpha_4 + (\alpha_3 + \alpha_6) \sin^2\phi_0 \right] + \tfrac{1}{2}u'(\alpha_3 + \alpha_6) \sin\phi_0 \cos\phi_0 = bz + e, \tag{5.49}$$

for some arbitrary constants c and e.

Now suppose that the transverse component of the shear stress is zero, that is, $\tilde{t}_{23} \equiv 0$, resulting in $b = e = 0$, the longitudinal shear stress being given by $\tilde{t}_{13}$. Taking into account the boundary conditions (5.36), equation (5.49) then integrates to reveal [168]

$$v[\alpha_4 + (\alpha_3 + \alpha_6) \sin^2\phi_0] + u(\alpha_3 + \alpha_6) \sin\phi_0 \cos\phi_0 = 0. \tag{5.50}$$

The flow is therefore diverted by an angle ψ to the direction of the imposed pressure gradient, this angle of deflection being given via the relation

$$\tan\psi = \frac{v}{u} = -\frac{(\alpha_3 + \alpha_6) \sin\phi_0 \cos\phi_0}{\alpha_4 + (\alpha_3 + \alpha_6) \sin^2\phi_0} = -\frac{(\eta_3 - \eta_1) \sin\phi_0 \cos\phi_0}{\eta_1 \sin^2\phi_0 + \eta_3 \cos^2\phi_0}, \tag{5.51}$$

as depicted in Fig. 5.2(b); the result is also stated in terms of the two main Miesowicz viscosities η_1 and η_3 as defined in Table 4.1 on page 158. Integration of equation (5.48) and an application of the boundary conditions (5.36) provides a second linear equation in u and v; solving this equation and (5.50) simultaneously delivers the solutions $u(z)$ and $v(z)$ in terms of z and the non-zero constant a, this constant being related to the applied pressure gradient. The fluid motion is driven by the pressure gradient $\nabla p = (a, 0, 0)$. It is well known that $a < 0$ for a flow which is expected to have a component in the positive x-direction as shown in Fig. 5.2, and $a > 0$ for flow in the opposite direction. The solutions can be obtained relatively easily and are expressed most succinctly in terms of the Miesowicz viscosities as

$$u(z) = -\frac{\left[\eta_1 \sin^2\phi_0 + \eta_3 \cos^2\phi_0 \right]}{2\eta_1 \eta_3} az(d - z), \tag{5.52}$$

$$v(z) = -\frac{(\eta_3 - \eta_1)}{2\eta_1 \eta_3} \sin\phi_0 \cos\phi_0 \, az(d - z). \tag{5.53}$$

Nevertheless, the main result is equation (5.51). This predicted theoretical result is entirely consistent with that obtained by Pieranski and Guyon [222] who carried out an experiment that matched and confirmed the presence of the angle ψ for the nematic MBBA. These authors adopted the above geometry, except that the pressure gradient was applied parallel to the y-axis and, in their notation, they set $\eta_a = \eta_3$ and $\eta_b = \eta_1$. The above analysis can be adapted from equations (5.48) and (5.49) onwards to replicate their results, bearing in mind that the angle ψ defined in their work is measured relative to the y-axis rather than the x-axis and is equivalent to $\frac{\pi}{2} - \psi$ where ψ is defined by (5.51). Fig. 5.3, obtained using equation (5.51),

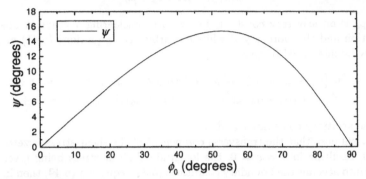

Figure 5.3: The flow is diverted by the angle ψ depicted in Fig. 5.2(b). The director is held fixed by a strong magnetic field at the angle ϕ_0. The angle ψ has been calculated here for MBBA near 25°C using the relation (5.51) for choices of ϕ_0 between 0° and 90°.

shows how the angle ψ varies with ϕ_0 for the viscosities given for MBBA near 25°C in Table D.3 on page 330. Observe that for isotropic fluids, where $\frac{1}{2}\alpha_4$ is the usual viscosity that arises in Newtonian fluids, we have $\eta_1 = \eta_3 = \frac{1}{2}\alpha_4$ and the results (5.52) and (5.53) reduce to

$$u = -\frac{a}{\alpha_4} z(d - z), \qquad v \equiv 0, \tag{5.54}$$

which is the classical situation when the flow is Newtonian [159, p.56] (see also Section 5.5.1 on page 176).

The results of this Section have shown that if the director alignment is fixed by a strong magnetic field not to be in the plane of shear or perpendicular to this plane, that is, $\phi_0 \neq 0$ or $\frac{\pi}{2}$, then a transverse secondary flow must occur. The component of flow in the direction of the pressure gradient is $u(z)$ while $v(z)$ is the component of flow in the y-direction, sometimes called the transverse flow component, or secondary flow component. The reader is referred to the paper by Pieranski and Guyon [222] and the short summaries contained in the texts by Chandrasekhar [38, p.157] and de Gennes and Prost [110, p.212] for a discussion in terms of a transverse pressure gradient. More general time dependent solutions incorporating transverse flow have been discussed by Leslie in his reviews [168, 169].

5.4   The Zwetkoff Experiment

As mentioned in Section 4.4, the rotational viscosity γ_1 can be determined by a method due to Zwetkoff [288]. This method was also used by Gasparoux and

Prost [101] to measure the magnetic anisotropy of MBBA and PAA as well as the viscosity γ_1 for these nematics [226]. In this experiment a circular cylinder of radius R and height L containing a nematic liquid crystal is suspended vertically by a torsion wire in the presence of a uniform magnetic field acting in a horizontal plane. The magnetic field is then rotated at a constant angular velocity ω about the vertical axis; from an experimental point of view, it can sometimes be more convenient to rotate the sample in a fixed magnetic field, this situation being equivalent. If the diameter of the cylinder is large, then the boundary effects and spatial gradients of the director can be ignored. The fluid itself can also be considered at rest, in the sense that only the director may vary with time, and so the velocity of the fluid is identically zero. We shall show that for low angular velocities the rotational viscosity γ_1 can be measured directly via the torque transmitted to the torsion wire, this being a consequence of the relation (5.86) which we derive below.

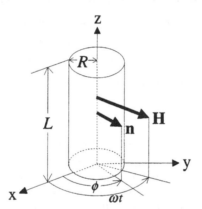

Figure 5.4: The coordinates and notation introduced to describe the alignment of the director **n** and magnetic field **H** in the Zwetkoff experiment. The nematic liquid crystal is contained in a cylinder of radius R and height L as shown. The magnetic field rotates with angular velocity ω around the z-axis, with the phase angles for **n** and **H** being given by ϕ and ωt, respectively.

Under these aforementioned circumstances we suppose that the director and magnetic field take the forms depicted in Fig. 5.4:

$$\mathbf{n} = (\cos\phi(t), \sin\phi(t), 0), \qquad (5.55)$$
$$\mathbf{H} = H(\cos(\omega t), \sin(\omega t), 0), \qquad (5.56)$$

where ω is the constant angular velocity about the z-axis, H is the magnitude of the magnetic field and $\phi(t)$ is the orientation angle of the director as shown in Fig. 5.4. The rate of strain tensor **A** and vorticity tensor **W** are zero because $\mathbf{v} \equiv \mathbf{0}$. The generalised body force **G** and co-rotational time flux **N** are given via equations (4.133), (4.134), (4.135) and (4.126) as

$$G_i = \chi_a(\mathbf{n} \cdot \mathbf{H})H_i, \qquad N_i = \dot{n}_i, \qquad (5.57)$$

when the effects of gravity are neglected, with the superposed dot now denoting the ordinary derivative d/dt. In this example the viscous stress (4.121) is

$$\tilde{t}_{ij} = \alpha_2 N_i n_j + \alpha_3 n_i N_j = \alpha_2 \dot{n}_i n_j + \alpha_3 n_i \dot{n}_j. \tag{5.58}$$

Under the further assumption that there is no other external body force $\mathbf{F}$, the governing Ericksen–Leslie equations (4.119) and (4.120) in this case are

$$p_{,i} = 0, \tag{5.59}$$

$$\lambda n_i = -\gamma_1 \dot{n}_i + \chi_a (\mathbf{n} \cdot \mathbf{H}) H_i. \tag{5.60}$$

The balance of linear momentum (5.59) integrates at once to reveal that p may take the form of an arbitrary constant, as could be expected given the set-up of this particular problem and the incompressibility constraint. The scalar Lagrange multiplier in (5.60) is evaluated in the usual way by taking the scalar product with $\mathbf{n}$. This gives $\lambda = \chi_a (\mathbf{n} \cdot \mathbf{H})^2$ so that the balance of angular momentum becomes

$$\chi_a (\mathbf{n} \cdot \mathbf{H})^2 n_i = -\gamma_1 \dot{n}_i + \chi_a (\mathbf{n} \cdot \mathbf{H}) H_i. \tag{5.61}$$

Inserting the appropriate forms for $\mathbf{n}$ and $\mathbf{H}$ gives the two equations

$$-\gamma_1 \dot{\phi} \sin\phi + \chi_a H^2 \cos^2(\omega t - \phi) \cos\phi = \chi_a H^2 \cos(\omega t - \phi) \cos(\omega t), \tag{5.62}$$

$$\gamma_1 \dot{\phi} \cos\phi + \chi_a H^2 \cos^2(\omega t - \phi) \sin\phi = \chi_a H^2 \cos(\omega t - \phi) \sin(\omega t). \tag{5.63}$$

Multiplying equation (5.63) by $\cos\phi$ and subtracting equation (5.62) multiplied by $\sin\phi$ finally delivers the governing dynamic equation

$$\gamma_1 \dot{\phi} = \tfrac{1}{2} \chi_a H^2 \sin[2(\omega t - \phi)]. \tag{5.64}$$

It is perhaps appropriate to mention here that the method of Zwetkoff outlined above will only be of substantial use when $\chi_a > 0$. From Fig. 5.4 it is seen that the director $\mathbf{n}$ will attempt to remain in the xy-plane parallel to $\mathbf{H}$ if it is attracted to the field, which is the case when $\chi_a > 0$. We shall demonstrate that the director $\mathbf{n}$ will eventually rotate and follow $\mathbf{H}$ if $\chi_a > 0$, but with a certain phase lag. If $\chi_a < 0$ then the director is repelled by the field and will attempt to be perpendicular to it at all times; in this case the director will not rotate in a similar way with the field, but will always attempt to remain simply perpendicular to $\mathbf{H}$ by trying to align itself parallel to the z-axis. This situation cannot be taken into account by the ansatz (5.55) and for this reason we shall simplify matters by seeking solutions only for $\chi_a > 0$. These physical considerations, which assume that the sample is large enough so that boundary effects can be neglected, were emphasised by Gasparoux and Prost [101].

For convenience we define a critical frequency ω_c by

$$\omega_c = \frac{1}{2} \frac{\chi_a}{\gamma_1} H^2. \tag{5.65}$$

The solution of (5.64) is easily obtained by setting $u = 2(\omega t - \phi(t))$ and employing the separation of variables method to find that

$$\frac{1}{2} \int \frac{du}{\omega - \omega_c \sin u} = \int dt. \tag{5.66}$$

The solution splits into three separate cases [101]:

Case (i) $\omega < \omega_c$

The solution in this case may be written as [116, 2.551.3]

$$\tan(\omega t - \phi) = \frac{1}{\omega} \frac{A_0 \left(\omega_c - \sqrt{\omega_c^2 - \omega^2}\right) + \left(\omega_c + \sqrt{\omega_c^2 - \omega^2}\right) \exp\left(-2t\sqrt{\omega_c^2 - \omega^2}\right)}{A_0 + \exp\left(-2t\sqrt{\omega_c^2 - \omega^2}\right)},$$

(5.67)

where A_0 is a (negative) constant determined by the initial condition. (Note: there appears to be a minor misprint in [101].) The important points to note are that

$$\omega t - \phi \to \tan^{-1}\left[\frac{1}{\omega}\left(\omega_c - \sqrt{\omega_c^2 - \omega^2}\right)\right] \quad \text{as } t \to \infty,$$

(5.68)

and

$$\dot{\phi} \to \omega \quad \text{as } t \to \infty.$$

(5.69)

These results show that when $\omega < \omega_c$ the director eventually follows the orientation of the magnetic field with the same angular velocity, but with a constant phase lag given by the limit in equation (5.68).

Case (ii) $\omega = \omega_c$

The solution here is provided by [116, 2.555.3]

$$\tan(\omega_c t - \phi) = 1 - \frac{[1 + \tan \phi(0)]}{1 + \omega_c t \left[1 + \tan \phi(0)\right]},$$

(5.70)

from which it is clear that

$$\omega_c t - \phi \to \frac{\pi}{4} \quad \text{as } t \to \infty,$$

(5.71)

and

$$\dot{\phi} \to \omega_c \quad \text{as } t \to \infty.$$

(5.72)

This shows that when $\omega = \omega_c$ the director eventually follows the magnetic field with the same angular velocity, but with a phase lag of $\frac{\pi}{4}$. Also, if $\phi(0) = -\frac{\pi}{4}$ then $\phi(t) = \omega_c t - \frac{\pi}{4}$ for all times $t \geq 0$.

Case (iii) $\omega > \omega_c$

In this case the solution is given via [116, 2.551.3]

$$\tan(\omega t - \phi) = \frac{1}{\omega}\left(\omega_c + \sqrt{\omega^2 - \omega_c^2} \tan\left[t\sqrt{\omega^2 - \omega_c^2} + B_0\right]\right),$$

(5.73)

where the constant B_0 is determined by the initial condition. This result shows that the director again follows the magnetic field, but with a more complicated time-dependent phase lag.

To exploit these solutions and complete the application it now remains to evaluate the torque transmitted to the torsion wire. It will be seen that the torque exerted by the rotating magnetic field on the director will be transferred to the curved surface of the cylindrical container and therefore to the torsion wire, with no resultant torques arising from the upper and lower ends of the container, as we shall see below. The stress tensor is given via (4.56) and (5.58) by

$$t_{ij} = -p\delta_{ij} + \alpha_2 N_i n_j + \alpha_3 n_i N_j, \tag{5.74}$$

where, by (5.55) and (5.57)$_2$,

$$\mathbf{N} = \dot{\phi}(-\sin\phi, \cos\phi, 0). \tag{5.75}$$

Straightforward calculations reveal that the non-zero components of t_{ij} are

$$
\begin{aligned}
t_{11} &= -p - (\alpha_2 + \alpha_3)\dot{\phi}\sin\phi\cos\phi, \\
t_{22} &= -p + (\alpha_2 + \alpha_3)\dot{\phi}\sin\phi\cos\phi, \\
t_{12} &= \dot{\phi}(\alpha_3\cos^2\phi - \alpha_2\sin^2\phi), \\
t_{21} &= \dot{\phi}(\alpha_2\cos^2\phi - \alpha_3\sin^2\phi), \\
t_{33} &= -p.
\end{aligned}
\tag{5.76}
$$

It is well known (cf. Spencer [256, p.47]) that the traction τ exerted across a boundary surface by material inside a given volume, onto material on the outside of the volume, is given by

$$\tau_i = t_{ij}\nu_j, \quad i = 1, 2, 3, \tag{5.77}$$

where ν is the unit normal to the surface that is directed *into* the interior of the volume. The total torque on the boundary S about the origin due to surface tractions τ is given by

$$\mathbf{T} = \int_S \mathbf{r} \times \tau \, dS, \tag{5.78}$$

where $\mathbf{r}$ is the position vector with respect to the origin. For this cylindrical example, $S = S_0 \cup S_+ \cup S_-$, where S_0 is the curved surface of the cylinder and S_+ and S_- are the upper and lower surfaces of the cylinder, respectively.

In this Zwetkoff experiment, the inward unit normal to the curved surface S_0 of the cylinder is

$$\nu = -(\cos\alpha, \sin\alpha, 0), \qquad \tan\alpha = \frac{y}{x}, \tag{5.79}$$

in the negative r-direction, where α is the usual plane polar angle and r is the distance from the z-axis. The torque per unit area about the origin at a point on S_0 is

$$
\begin{aligned}
\mathbf{M}_0 &= (R\cos\alpha, R\sin\alpha, z) \times \tau \\
&= (-zt_{2j}\nu_j, \ zt_{1j}\nu_j, \ \alpha_3 R\dot{\phi}\sin^2(\alpha - \phi) - \alpha_2 R\dot{\phi}\cos^2(\alpha - \phi)), \tag{5.80}
\end{aligned}
$$

with the total torque, $\mathbf{T}_0$, on the curved surface of the cylinder being the integral of $\mathbf{M}_0$ over S_0. Hence the torque on S_0 due to the liquid crystal can be calculated via cylindrical polar coordinates to yield

$$
\begin{aligned}
\mathbf{T}_0 &= \int_{S_0} \mathbf{M}_0 \, dS \\
&= \int_0^L \int_0^{2\pi} \mathbf{M}_0 \, R \, d\alpha dz \\
&= (0, 0, \mathcal{M}),
\end{aligned}
\tag{5.81}
$$

where

$$
\mathcal{M} = \int_0^L \int_0^{2\pi} M_{03} \, R \, d\alpha dz,
\tag{5.82}
$$

is the only non-zero component of the total torque on S_0, with M_{03} denoting the third component of $\mathbf{M}_0$. This expression is easily evaluated to find that

$$
\mathcal{M} = \gamma_1 V \dot{\phi},
\tag{5.83}
$$

where $V = \pi R^2 L$ is the volume of the cylinder and $\gamma_1 = \alpha_3 - \alpha_2 \geq 0$.

It can be verified that the upper and lower ends of the cylindrical container contribute zero total torques to the system. For example, if $\mathbf{M}_+$ and $\mathbf{M}_-$ represent the torques per unit area about the origin at points on the upper and lower ends S_+ and S_-, respectively, it is found that at any point on S_+

$$
\mathbf{M}_+ = pr(\sin\alpha, -\cos\alpha, 0), \qquad 0 \leq r \leq R, \quad 0 \leq \alpha \leq 2\pi,
\tag{5.84}
$$

and therefore the total torque on S_+ is

$$
\mathbf{T}_+ = \int_{S_+} \mathbf{M}_+ dS = \int_0^{2\pi} \int_0^R \mathbf{M}_+ r dr d\alpha = 0.
\tag{5.85}
$$

By similar reasoning, $\mathbf{T}_- = 0$ also. Thus, in effect, $\mathbf{M}_0$ given in equation (5.80) gives rise to the total torque for the complete system when it is integrated over S_0, so that the total torque due to the liquid crystal is actually $\mathbf{T} = \mathbf{T}_0 + \mathbf{T}_+ + \mathbf{T}_- = \mathbf{T}_0$, given in this example by (5.81) and (5.82). This leads in turn to showing that the total torsion is $\mathcal{M}$ given by equation (5.83).

For $\omega \leq \omega_c$ we see from the results in (5.69), (5.72) and (5.83) that the torsion $\mathcal{M}$ at long times becomes

$$
\mathcal{M} = \gamma_1 V \omega.
\tag{5.86}
$$

This result allows the rotational viscosity γ_1 to be determined by measurement of the torsion in the wire for $\omega \leq \omega_c$. In this situation neither the phase lag nor the magnetic anisotropy need to be known for the measurement of γ_1 to be made via small angular velocities. Prost and Gasparoux [226] used the result (5.86) to measure the rotational viscosity γ_1 for the nematics MBBA and PAA.

Boundary conditions have been neglected in the above exposition. For a short discussion on the possible effects of boundary conditions on the Zwetkoff experiment the reader should consult the review by Leslie [168] and the references cited therein. Brief comments on anchoring conditions and the behaviour of possible defects can be found in de Gennes and Prost [110, pp.223–225], together with a description of another similar experiment involving controlled boundary conditions for a sample in a rotating field. For example, Brochard, Léger and Meyer [27] considered a sample of homeotropically aligned nematic liquid crystal between two flat plates with the field rotating in the plane of the plates. The interested reader is referred to [27, 110] for further details and references.

5.5 Shear Flow

In this Section we describe in detail the shear flow examples discussed by Leslie [163] for nematic liquid crystals. We first make some comments on Newtonian and non-Newtonian behaviour of fluids in Section 5.5.1 before going on to derive the general explicit governing equations for shear flow, equations (5.121) and (5.122), in Section 5.5.2. We then specialise in Sections 5.5.3 and 5.5.4 to the specific problems of shear flow near a boundary and shear flow between parallel plates. Section 5.5.5 discusses some scaling properties for nematics.

5.5.1 Newtonian and Non-Newtonian Behaviour

Before proceeding to investigate typical shear flow behaviour in the following subsections, we examine and comment upon a particularly simple shear flow solution to the usual Ericksen–Leslie dynamic equations. Consider a shear flow in the x-direction between two flat plates in which the director is uniformly aligned perpendicular to the plane of shear, with the director fixed parallel to the bounding plates at $y = 0$ and $y = d$. Suppose also that there is no external body force $\mathbf{F}$ or generalised body force $\mathbf{G}$. Seeking solutions in Cartesian coordinates of the form

$$\mathbf{n} = (0, 0, 1), \qquad \mathbf{v} = (v(y), 0, 0), \tag{5.87}$$

we see that the constraints (4.118) are satisfied, as are the equations for angular momentum (4.120) upon setting $\lambda = 0$. The non-zero components of the resultant viscous stress (4.121) in this example are the shear stresses

$$\tilde{t}_{12} = \tilde{t}_{21} = \frac{1}{2}\alpha_4 \frac{dv}{dy}, \tag{5.88}$$

and hence, by (4.128), the non-zero components of the stress tensor are

$$t_{11} = t_{22} = t_{33} - p, \qquad t_{12} = t_{21} = \frac{1}{2}\alpha_4 \frac{dv}{dy}. \tag{5.89}$$

The linear momentum equations (4.119) then reduce, via (4.121) and (5.87), to solving

$$\frac{\partial p}{\partial x} = \frac{1}{2}\alpha_4 \frac{\partial^2 v}{\partial y^2}, \qquad \frac{\partial p}{\partial y} = \frac{\partial p}{\partial z} = 0. \tag{5.90}$$

The Newtonian viscosity μ of a Newtonian fluid is constant at a fixed temperature and pressure. It is known that during the shear flow of a Newtonian fluid the shear stress σ obeys the relation [86]

$$\sigma = \mu \frac{dv}{dy}, \tag{5.91}$$

and the graph of σ versus dv/dy is a straight line passing through the origin. This means that in this special example of shear flow of a liquid crystal, given by equations (5.87) to (5.90), the flow is evidently Newtonian with Newtonian viscosity $\mu = \frac{1}{2}\alpha_4$, by (5.89), and the uniform alignment of the director, given by $(5.87)_1$, imposed by the boundaries is unaffected by the flow. However, it is known that for liquid crystals under high shear rates this configuration becomes unstable. We do not pursue this topic here and refer the reader to the review by Leslie [168] for further comments and a list of relevant references.

All fluids for which the shear stress σ is of the form

$$\sigma = f\left(\frac{dv}{dy}\right), \tag{5.92}$$

for some function f, are called *non-Newtonian* [86] if the graph of σ versus dv/dy is *not* a straight line through the origin at a given fixed temperature and pressure. More generally, a fluid at a given temperature and pressure is also called non-Newtonian [86] if at least one of the following requirements is *not* fulfilled:

(i) σ as a function of dv/dy is a straight line through the origin,

(ii) the fluid does not exhibit memory effects resulting from elasticity and/or thixotropy,

(iii) the first and second normal stress differences $\sigma_1 = t_{11} - t_{22}$ and $\sigma_2 = t_{22} - t_{33}$ are zero,

(iv) the ratio of extensional to shear viscosity is constant.

The non-Newtonian behaviour of nematic liquid crystals will become evident from the form of the stress tensor given via equation (4.128) and the viscous stress derived in, for example, equations (5.100) to (5.103). We remark that, in general, it is necessary but not sufficient for the graph of σ to be a straight line through the origin for a fluid to be Newtonian; however, it is sufficient for any one of the above four requirements to be violated for a fluid to be regarded as non-Newtonian [86, pp.9–13].

5.5.2 Governing Equations for Shear Flow

To illustrate flow behaviour that is often considered to be more typical than the Newtonian behaviour indicated by (5.88), it will be assumed that the director and the velocity take the forms

$$\mathbf{n} = (\cos\phi(y), \sin\phi(y), 0), \tag{5.93}$$
$$\mathbf{v} = (v(y), 0, 0). \tag{5.94}$$

Clearly, the constraints (4.118) are satisfied and, because of the dependence of $\mathbf{n}$ and $\mathbf{v}$ upon y only, the governing Ericksen–Leslie dynamic equations (4.119) and

(4.120) in the absence of any forces $\mathbf{F}$ and $\mathbf{G}$ become

$$(p + w_F)_{,i} = \tilde{g}_j n_{j,i} + \tilde{t}_{i2,2}, \tag{5.95}$$

$$\left(\frac{\partial w_F}{\partial n_{i,2}}\right)_{,2} - \frac{\partial w_F}{\partial n_i} + \tilde{g}_i = \lambda n_i, \tag{5.96}$$

where w_F is the usual nematic free energy density given by (4.130). The non-zero contributions to the rate of strain tensor A and vorticity tensor W are

$$A_{12} = A_{21} = \tfrac{1}{2}v'(y), \qquad W_{12} = -W_{21} = \tfrac{1}{2}v'(y), \tag{5.97}$$

where a prime denotes differentiation with respect to y. Making use of (5.97), the co-rotational time flux (4.126) is then given by

$$\mathbf{N} = (-\tfrac{1}{2}v'\sin\phi, \tfrac{1}{2}v'\cos\phi, 0), \tag{5.98}$$

while, from equation (4.122),

$$\tilde{\mathbf{g}} = (\tfrac{1}{2}(\gamma_1 - \gamma_2)v'\sin\phi, -\tfrac{1}{2}(\gamma_1 + \gamma_2)v'\cos\phi, 0), \tag{5.99}$$

and the non-zero components of the dynamic viscous stress in (4.121) are

$$\tilde{t}_{11} = \tfrac{1}{2}v'\sin\phi\cos\phi\left[2\alpha_1\cos^2\phi - \alpha_2 - \alpha_3 + \alpha_5 + \alpha_6\right], \tag{5.100}$$
$$\tilde{t}_{12} = \tfrac{1}{2}v'\left[2\alpha_1\sin^2\phi\cos^2\phi + (\alpha_5 - \alpha_2)\sin^2\phi + (\alpha_3 + \alpha_6)\cos^2\phi + \alpha_4\right], \tag{5.101}$$
$$\tilde{t}_{21} = \tfrac{1}{2}v'\left[2\alpha_1\sin^2\phi\cos^2\phi + (\alpha_2 + \alpha_5)\cos^2\phi + (\alpha_6 - \alpha_3)\sin^2\phi + \alpha_4\right], \tag{5.102}$$
$$\tilde{t}_{22} = \tfrac{1}{2}v'\sin\phi\cos\phi\left[2\alpha_1\sin^2\phi + \alpha_2 + \alpha_3 + \alpha_5 + \alpha_6\right]. \tag{5.103}$$

Equations (5.95) are

$$\frac{\partial}{\partial x}(p + w_F) = \frac{\partial \tilde{t}_{12}}{\partial y}, \tag{5.104}$$

$$\frac{\partial}{\partial y}(p + w_F) = \tilde{g}_i n_{i,y} + \frac{\partial \tilde{t}_{22}}{\partial y}, \tag{5.105}$$

$$\frac{\partial}{\partial z}(p + w_F) = 0. \tag{5.106}$$

It is clear from equations (5.104) and (5.106), and the dependence of $\tilde{t}_{12}$ upon y, that

$$p + w_F = x\frac{\partial \tilde{t}_{12}}{\partial y} + f_1(y), \tag{5.107}$$

Inserting (5.107) into (5.105) shows that

$$\tilde{t}_{12} = ay + c, \tag{5.108}$$

for some constants a and c because the right-hand side of (5.105) is a function of y only. (Taking the derivative of (5.104) with respect to y and the derivative of (5.105) with respect x will also lead to the same result.) Therefore

$$p + w_F = ax + f_1(y). \tag{5.109}$$

It is now seen from (5.105) and (5.109) that

$$f_1(y) = p_0 + \tilde{t}_{22} + \int \tilde{g}_i n_{i,y} \, dy, \tag{5.110}$$

for some constant p_0, and hence, by (5.109), the pressure takes the form

$$p = -w_F + ax + p_0 + \tilde{t}_{22} + \int \tilde{g}_i n_{i,y} \, dy. \tag{5.111}$$

Using the form for $\tilde{t}_{12}$ given by equation (5.101), the result (5.108) may be formulated conveniently as [163]

$$g(\phi)\frac{dv}{dy} = ay + c, \tag{5.112}$$

where

$$g(\phi) = \tfrac{1}{2} \left[2\alpha_1 \sin^2\phi \cos^2\phi + (\alpha_5 - \alpha_2)\sin^2\phi + (\alpha_3 + \alpha_6)\cos^2\phi + \alpha_4 \right]. \tag{5.113}$$

Equations (5.95) have now been reduced to (5.112), the pressure being available via (5.111) to allow this reduction.

Our attention is now turned to the remaining equations (5.96). These equations, using (4.131), reduce to

$$K_2 n_{1,22} + (K_3 - K_2)(n_2^2 n_{1,2})_{,2} + \tilde{g}_1 = \lambda n_1, \tag{5.114}$$

$$K_1 n_{2,22} + (K_3 - K_2)\left[(n_2^2 n_{2,2})_{,2} - n_2 n_{p,2} n_{p,2} \right] + \tilde{g}_2 = \lambda n_2. \tag{5.115}$$

The Lagrange multiplier λ can be eliminated from these two equations by multiplying (5.114) by n_2, (5.115) by n_1 and then subtracting the resulting equations to produce, after much tedious algebra, the single equation

$$K_3 n_2 n_{1,22} - K_1 n_1 n_{2,22} + \tilde{g}_1 n_2 - \tilde{g}_2 n_1 = 0. \tag{5.116}$$

Further substitutions for $\tilde{g}_i$ and n_i using equations (5.93) and (5.99) finally reduce this equation to

$$2f(\phi)\frac{d^2\phi}{dy^2} + \frac{df}{d\phi}\left(\frac{d\phi}{dy}\right)^2 - \frac{dv}{dy}\left[\gamma_1 + \gamma_2 \cos(2\phi)\right] = 0, \tag{5.117}$$

where

$$f(\phi) = K_1 \cos^2\phi + K_3 \sin^2\phi \geq 0, \tag{5.118}$$

this last inequality being valid because K_1 and K_3 are necessarily non-negative. (Recall that $\gamma_1 = -\lambda_1$ and $\gamma_2 = -\lambda_2$ in the original notation of Leslie [163].)

The inequality (4.82) (or its equivalent in the form of (4.85)) must be satisfied also. For the type of problem considered here, it becomes

$$0 \leq \mathcal{D} = \tilde{t}_{ij} A_{ij} - N_i \tilde{g}_i = g(\phi)\left(\frac{dv}{dy}\right)^2, \tag{5.119}$$

where (4.176) and the Parodi relation (4.96) in the form $\alpha_6 - \alpha_3 = \alpha_2 + \alpha_5$ have been used when inserting $\tilde{t}_{21}$, given by (5.102), into the inequality (5.119). It is therefore seen that

$$g(\phi) \geq 0, \tag{5.120}$$

in the shear flows examined here. Following Leslie [163], for tractability we shall consider both $f(\phi)$ and $g(\phi)$ to be strictly positive.

The flow $v(y)$ in the x-direction may be induced in a sample of nematic liquid crystal confined between two parallel plates by the horizontal motion of the upper plate. In this case the pressure p given by equation (5.111) may be independent of x so that the constant a can be set to zero. Equations (5.112) and (5.117) then become

$$\frac{dv}{dy} = \frac{c}{g(\phi)}, \tag{5.121}$$

$$2f(\phi)\frac{d^2\phi}{dy^2} + \frac{df}{d\phi}\left(\frac{d\phi}{dy}\right)^2 = c\frac{[\gamma_1 + \gamma_2\cos(2\phi)]}{g(\phi)}, \tag{5.122}$$

where v' in equation (5.117) has been replaced by (5.121). These two equations form the key starting point for the two shear problems in the following subsections. Notice that $\phi(y)$ has been uncoupled form $v(y)$ and that if the upper plate has a positive velocity in the x-direction then the constant c must necessarily be positive, by equation (5.121). Also, as remarked upon by Müller [207], observe that if the fluid were isotropic then $g(\phi)$ would become $\frac{1}{2}\alpha_4$, the usual Newtonian fluid viscosity, and then $v(y)$ would increase linearly between the two plates. However, in nematic liquid crystals the flow $v(y)$ is influenced by the orientation of the director $\mathbf{n}$, as can be seen through the effect of $g(\phi)$ in equation (5.121).

We now proceed to analyse equations (5.121) and (5.122).

5.5.3   Shear Flow Near a Boundary

Consider flow near a stationary solid boundary at $y = 0$ as shown in Fig. 5.5. Following Leslie [163] and the exposition by Müller [207], we expect solutions to equations (5.121) and (5.122) to satisfy the boundary conditions

$$v(0) = 0, \quad \phi(0) = \phi_0, \quad \text{and} \quad \phi(y) \to \phi_\infty \quad \text{as} \quad y \to \infty, \tag{5.123}$$

where ϕ_0 and ϕ_∞ are constants having values between zero and 2π; the director is assumed to be strongly anchored at an angle ϕ_0 on the lower plate. However, since the theory does not distinguish between $\mathbf{n}$ and $-\mathbf{n}$, these values are equivalent to those obtained by adding or subtracting a multiple of π, and so we suppose that they lie between zero and π. The interpretation of such a solution is that the director $\mathbf{n}$ is expected to change rapidly near the boundary and will effectively be constant very much further away from the boundary where the upper plate is located. We may therefore assume that both the first and second derivatives of $\phi(y)$ decay to zero as $y \to \infty$. It follows from equations (5.122) and (5.123)$_3$ that

$$\cos(2\phi_\infty) = -\frac{\gamma_1}{\gamma_2}, \tag{5.124}$$

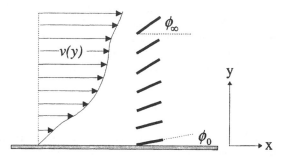

Figure 5.5: The director is strongly anchored on the lower plate at an angle ϕ_0. The expected flow profile is as shown for the velocity $v(y)$ in the x-direction. As $y \to \infty$ the orientation angle $\phi(y)$ tends to the constant angle ϕ_∞, defined in equation (5.124).

and it is therefore assumed that $\gamma_1 \leq |\gamma_2|$, recalling that $\gamma_1 \geq 0$ by equation (4.123). For simplicity we shall forthwith suppose that $\gamma_2 < 0$, the case for $\gamma_2 > 0$ being similar in analysis [163]. For $\gamma_1 \geq 0$ and $\gamma_2 < 0$ we must have the restrictions

$$0 \leq \phi_\infty \leq \tfrac{\pi}{4} \qquad \text{or} \qquad \tfrac{3}{4}\pi \leq \phi_\infty \leq \pi, \tag{5.125}$$

by (5.124). Further use of (5.124) to substitute for γ_1 in equation (5.122), and then multiplying throughout by $d\phi/dy$, gives

$$\frac{d}{dy}\left[f(\phi) \left(\frac{d\phi}{dy}\right)^2 \right] = c\gamma_2 \frac{\cos(2\phi) - \cos(2\phi_\infty)}{g(\phi)} \frac{d\phi}{dy}. \tag{5.126}$$

Integrating (5.126) between y and infinity and employing the boundary condition $(5.123)_3$ results in the relation

$$f(\phi)\left(\frac{d\phi}{dy}\right)^2 = -c\gamma_2 \int_\phi^{\phi_\infty} \frac{\cos(2\psi) - \cos(2\phi_\infty)}{g(\psi)} d\psi, \tag{5.127}$$

given that the derivative of ϕ ought to vanish as $y \to \infty$. Since $c > 0$ and $\gamma_2 < 0$, the above integral must be positive because $f(\phi) > 0$. This shows that ϕ_∞ has to be particularly restricted to the range $0 \leq \phi_\infty \leq \tfrac{\pi}{4}$, given the earlier restrictions (5.125) and the assumption $g(\phi) > 0$. Equation (5.127) may now be written as

$$\frac{d\phi}{dy} = \left[-\frac{c\gamma_2}{f(\phi)} \int_\phi^{\phi_\infty} \frac{\cos(2\psi) - \cos(2\phi_\infty)}{g(\psi)} d\psi \right]^{\frac{1}{2}}, \tag{5.128}$$

and a further integration using the boundary condition $(5.123)_2$ finally yields the solution

$$y = \int_{\phi_0}^\phi \left[-\frac{c\gamma_2}{f(\zeta)} \int_\zeta^{\phi_\infty} \frac{\cos(2\psi) - \cos(2\phi_\infty)}{g(\psi)} d\psi \right]^{-\frac{1}{2}} d\zeta. \tag{5.129}$$

This provides y as a function of ϕ, that is, the solution ϕ is implicitly given as a function of y in (5.129); clearly, from the form of (5.129), ϕ increases monotonically with y. Once the solution ϕ has been determined from equation (5.129), an integration of equation (5.121) combined with the boundary condition (5.123)$_1$ concludes the solution to the problem by determining the velocity

$$v(y) = \int_0^y \frac{c}{g(\phi(\zeta))} \, d\zeta. \tag{5.130}$$

As commented upon by Müller [207], the constant c can be determined by the shear stress required to keep the lower plate in place, since, by (4.56), (5.108) and the given dependence of $\mathbf{n}$ and v upon y,

$$t_{12} = \tilde{t}_{12} = c, \tag{5.131}$$

because a has been set to zero in (5.112), as discussed in the derivation of equation (5.121), the first of the governing equations.

The solutions for $\phi(y)$ and $v(y)$ are given via equations (5.129) and (5.130). In general, the evaluations of the resultant integrals have to be carried out numerically for given values of the elastic constants and viscosities once c is known.

A 'boundary layer' type of behaviour for the above solution has also been mentioned briefly by Leslie [163]. For example, if we set

$$\xi = (|K_1| + |K_3|) \, |c|^{-1}, \tag{5.132}$$

then, by taking basic inequalities, it is easily seen from (5.118) and the solution (5.129) that

$$\lim_{\xi \to 0} y \le \lim_{\xi \to 0} \sqrt{\xi} \int_{\phi_0}^{\phi} \left[-\gamma_2 \int_{\zeta}^{\phi_\infty} \frac{\cos(2\psi) - \cos(2\phi_\infty)}{g(\psi)} \, d\psi \right]^{-\frac{1}{2}} d\zeta = 0, \tag{5.133}$$

provided $\phi \ne \phi_\infty$. This demonstrates the existence of a boundary layer phenomenon when the length $\sqrt{\xi}$ is sufficiently small.

5.5.4   Shear Flow between Parallel Plates

We now investigate the flow between two parallel plates at a constant distance $d = 2h$ apart where the lower plate is at rest and the upper plate is moving at a constant velocity V along a straight line parallel to the x-axis, as shown in Fig. 5.6. The governing equations are again provided by (5.121) and (5.122) and for this example we follow Leslie [163] by assuming that the director is strongly anchored parallel to the plates at the boundaries, and that the solution for ϕ is symmetric in y. We therefore set

$$v(-h) = 0, \quad v(h) = V, \tag{5.134}$$

$$\phi(-h) = \phi(h) = 0, \quad \phi(-y) = \phi(y), \quad -h \le y \le h, \tag{5.135}$$

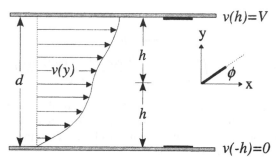

Figure 5.6: The director is strongly anchored in the x-direction on the boundaries, parallel with the bounding plates which are placed at a distance $d = 2h$ apart. The lower plate is fixed with velocity set to zero while the upper plate has a relative shear velocity of V. The orientation angle $\phi(y)$ of the director is defined as shown for the director given by equation (5.93)

and take

$$\left.\frac{d\phi}{dy}\right|_{y=0} = 0, \quad \phi(0) = \phi_0, \tag{5.136}$$

as conditions in the centre of the sample, with ϕ_0 being a constant. As before, the constant c appearing in equations (5.121) and (5.122) must be positive for the geometry described in Fig. 5.6. It will also be assumed that $\gamma_2 < 0$, as in the previous Section, so that it is possible to introduce the constant ϕ_1 satisfying

$$\cos(2\phi_1) = -\frac{\gamma_1}{\gamma_2}, \tag{5.137}$$

analogous to equation (5.124).

Using (5.137) to substitute for γ_1 in (5.122) and then multiplying throughout by $d\phi/dy$ gives an equation analogous to (5.126) with ϕ_∞ replaced by ϕ_1. Integrating this resultant equation from y to zero and applying the conditions (5.136) leads to a result similar to equation (5.128), namely,

$$\frac{d\phi}{dy} = \left[-\frac{c\gamma_2}{f(\phi)} \int_\phi^{\phi_0} \frac{\cos(2\psi) - \cos(2\phi_1)}{g(\psi)} d\psi \right]^{\frac{1}{2}}, \quad 0 \le \phi \le \phi_0, \tag{5.138}$$

where, without loss of generality, it has been supposed that $\phi_0 \ge 0$ so that the positive square root can be taken to arrive at (5.138), for reasons similar to those given after equation (3.117) on page 74 when the Freedericksz transitions were investigated. It can now be concluded from the above that $0 \le \phi_0 \le \phi_1$ and that $0 \le \phi_1 \le \frac{\pi}{4}$, because, as before, for $c > 0$ and $\gamma_2 < 0$ the integral in (5.138) must always remain non-negative. The required solution for ϕ in the range $-h \le y \le 0$ is now obtained by integrating (5.138) and using the boundary condition (5.135)$_1$

to find that

$$(h+y)\sqrt{c} \;=\; F_{\phi_0}(\phi), \quad -h \leq y \leq 0, \tag{5.139}$$

$$F_{\phi_0}(\phi) \;=\; \int_0^\phi \left[-\frac{\gamma_2}{f(\zeta)} \int_\zeta^{\phi_0} \frac{\cos(2\psi) - \cos(2\phi_1)}{g(\psi)} d\psi \right]^{-\frac{1}{2}} d\zeta. \tag{5.140}$$

By symmetry, the solution for the range $0 \leq y \leq h$ obeys the relation (5.138) with the negative square root being chosen, and this ultimately leads, by the same reasoning, to the solution on this interval being obtained by simply replacing y with $-y$ in the solution (5.139). Therefore

$$(h-y)\sqrt{c} = F_{\phi_0}(\phi), \quad 0 \leq y \leq h. \tag{5.141}$$

As in the previous subsection, the constant c is essentially the shear stress $\tilde{t}_{12}$ because $a = 0$ in (5.108), so that equation (5.131) holds as before. When c is known, then the angle ϕ_0 in the middle of the sample at $y = 0$ is determined via (5.139) and (5.140) by the relation

$$h\sqrt{c} = F_{\phi_0}(\phi_0). \tag{5.142}$$

This means that the maximum angle ϕ_0 of orientation of the director within the sample depends upon the quantity $h\sqrt{c} = h\sqrt{\tilde{t}_{12}}$.

After the solution ϕ has been determined by equations (5.139) and (5.141), the solution for the velocity can be evaluated by integrating equation (5.121) to give

$$v(y) = \int_{-h}^y \frac{c}{g(\phi(\zeta))} d\zeta. \tag{5.143}$$

Using the boundary condition $(5.134)_2$ on v in equation (5.143) and the symmetry requirement $(5.135)_2$ shows that the constants V and $c = \tilde{t}_{12}$ must additionally satisfy the relation

$$V = c \int_{-h}^h \frac{d\zeta}{g(\phi(\zeta))} = 2c \int_0^h \frac{d\zeta}{g(\phi(\zeta))}. \tag{5.144}$$

The solutions for $\phi(y)$ and $v(y)$ are given by equations (5.139), (5.140), (5.141) and (5.143). In general, the integrals appearing have to be evaluated numerically for given values of the elastic constants and viscosities once the constant c has been given. Further, as in the previous example, the existence of a boundary layer phenomenon can be demonstrated and the interested reader is referred to Leslie [163] for more details.

As pointed out by Currie [59], the symmetric solution to the above problem need not be the only solution, as he discussed via a phase-plane analysis of the general equations (5.121) and (5.122), motivated by his earlier observations for Couette flow [57]. Further studies of the apparent viscosity were also carried out by Currie [60] for various viscometric flows. Under some simplifying assumptions, for example, setting $K_1 = K_2 = K_3$ and $\alpha_1 = 0$, Currie and MacSithigh [61] managed to distinguish between various solutions and select several candidates as possible stable solutions. McIntosh, Leslie and Sloan [185] returned to the problem of non-uniqueness and investigated simple shear flow, in a slightly more general setting than that used here, by numerical methods making use of similar approximations to those employed by Currie and MacSithigh. Their stability arguments help eliminate most solutions, indicating that at most two types of solution may be stable. For other more general comments the reader is referred to the review by Leslie [168].

5.5.5   Scaling Properties

It is well known from elementary fluid dynamics that an isotropic fluid with constant viscosity η undergoing a shear between two parallel plates, separated by the distance $d = 2h$ as described above in Fig. 5.6, is characterised by (cf. Landau and Lifshitz [159, p.55])

$$\sigma_{12} = \eta \frac{V}{2h}, \qquad v(y) = \frac{V}{2}\left(1 + \frac{y}{h}\right), \quad -h \leq y \leq h, \tag{5.145}$$

where σ_{ij} is the usual stress tensor and v is the velocity, the boundary conditions on v given by (5.134) clearly being satisfied. Motivated by this classical result, we now define the apparent viscosity η of a nematic liquid crystal in this situation by [163]

$$\eta = t_{12}\frac{2h}{V}, \tag{5.146}$$

and look at the implications and consequences of this definition upon the results contained in the previous subsection. This coincides with the definition in equation (4.159) when $v(y)$ is given by $(5.145)_2$, upon noting the obvious change of coordinates. In the context of shear flow of a nematic between parallel plates in the geometry of Fig. 5.6, $t_{12} = \tilde{t}_{12}$ by equations (4.56), given the dependence of $\mathbf{n}$ and v upon y, and therefore by (5.131)

$$\eta = \tilde{t}_{12}\frac{2h}{V} = 2\frac{hc}{V}. \tag{5.147}$$

It is clear from (5.139) and (5.141) that the director orientation angle $\phi(y)$ may be written implicitly as the function $\hat{\phi}(\sqrt{c}y)$. Therefore, from the relation (5.144) it is seen that

$$V = 2\sqrt{c}\int_0^{h\sqrt{c}} \frac{ds}{g(\hat{\phi}(s))}. \tag{5.148}$$

Inserting this result into the apparent viscosity (5.147) gives

$$\eta = h\sqrt{c}\left[\int_0^{h\sqrt{c}} \frac{ds}{g(\hat{\phi}(s))}\right]^{-1}, \tag{5.149}$$

which shows that η depends upon the quantity $h\sqrt{c}$; notice also that $\hat{\phi}(s)$ depends on ϕ_0, which in turn depends implicitly on $h\sqrt{c}$ by equation (5.142). Moreover,

$Vd = 2Vh$ is a function of $h\sqrt{c}$ by equation (5.148), which means that any function of $h\sqrt{c}$ can be expressed as an equivalent function of Vd. Therefore η can also be expressed as a function of Vd and so we have

$$\eta = \mathcal{F}(h\sqrt{c}) = \mathcal{G}(Vd), \qquad (5.150)$$

for some functions $\mathcal{F}$ and $\mathcal{G}$. This result is important because it demonstrates that if the distance d between the plates is decreased, then the velocity V of the upper plate must be increased if the same apparent viscosity is to be maintained.

The dependence of the viscosity on Vd is evidence of a particular quantity being 'scaled.' Ericksen [79] considered the more general case to see if any other quantities may be scaled in a way that the form of the quantities remains the same. To this end, he scaled space and time via the transformations

$$\mathbf{x} = d\mathbf{x}^*, \quad t = qt^*. \qquad (5.151)$$

For the quantities appearing in Section 4.2.5 at equations (4.118) to (4.131) it is easy to establish the following collection of results under the above transformation:

$$n_i \to n_i^*, \quad n_{i,j} \to d^{-1}(n_{i,j})^*, \quad v_i \to dq^{-1}v_i^*, \quad v_{i,j} \to q^{-1}(v_{i,j})^*,$$

$$\dot{n}_i \to q^{-1}(\dot{n}_i)^*, \quad \dot{v}_i \to dq^{-2}(\dot{v}_i)^*, \quad A_{i,j} \to q^{-1}A_{i,j}^*, \quad W_{ij} \to q^{-1}W_{i,j}^*,$$

$$N_i \to q^{-1}N_i^*, \quad \tilde{g}_i \to q^{-1}\tilde{g}_i^*, \quad \tilde{t}_{ij} \to q^{-1}(\tilde{t}_{ij})^*, \quad \tilde{t}_{ij,j} \to (dq)^{-1}(\tilde{t}_{ij,j})^*, \qquad (5.152)$$

$$w_F \to d^{-2}w_F^*, \quad w_{F,i} \to d^{-3}(w_{F,i})^*, \quad \frac{\partial w_F}{\partial n_{i,j}} \to d^{-1}\left(\frac{\partial w_F}{\partial n_{i,j}}\right)^*.$$

Under these transformations, the constraints (4.118) are simply replaced by their equivalent transformed versions, namely,

$$n_i^* n_i^* = 1, \quad (v_{i,i})^* = 0, \qquad (5.153)$$

while the key dynamic equations (4.119) and (4.120) arising from the balances of linear and angular momentum become, respectively,

$$\rho\frac{d}{q^2}(\dot{v}_i)^* = \rho F_i - p_{,i} - \frac{1}{d^3}(w_{F,i})^* + \frac{1}{qd}(\tilde{g}_j n_{j,i})^* + \frac{1}{d}G_j(n_{j,i})^* + \frac{1}{dq}(\tilde{t}_{ij,j})^*, \quad (5.154)$$

$$\frac{1}{d^2}\left[\left(\frac{\partial w_F}{\partial n_{i,j}}\right)_j\right]^* - \frac{1}{d^2}\left[\frac{\partial w_F}{\partial n_i}\right]^* + \frac{1}{q}\tilde{g}_i + G_i = \lambda n_i. \qquad (5.155)$$

Therefore if we set

$$q = d^2, \quad F_i = d^{-3}F_i^*, \quad p = d^{-2}p^* \quad G_i = d^{-2}G_i^*, \quad \lambda = d^{-2}\lambda^*, \qquad (5.156)$$

then the transformed dynamic equations have the same forms as the original equations. Adopting the relations (5.156), the velocity $\mathbf{v}$ and the stress tensor (4.56) transform in this scaling according to

$$v_i^* = dv_i, \quad t_{ij}^* = d^2 t_{ij}, \qquad (5.157)$$

while the director, being a unit vector, is unaltered.

As a result of the above scaling properties, the shear flow problem outlined previously in subsection 5.5.4 can be reduced from one involving the gap width d and the relative shear velocity V to a problem with unit gap width and relative velocity V^* where

$$V^* = Vd, \tag{5.158}$$

with the strong anchoring conditions $(5.135)_1$ on the director being unchanged. In the new variables the boundary conditions (5.134) can be rewritten as

$$v^*(-\tfrac{1}{2}) = 0, \qquad v^*(\tfrac{1}{2}) = V^* = Vd. \tag{5.159}$$

As mentioned by Leslie [168, 171], if there is a unique solution $\mathbf{n}$ to this problem (via the solution for ϕ) then it follows that

$$\mathbf{n} = \mathbf{n}^* = \mathbf{n}^*(\mathbf{x}^*, V^*) = \mathbf{n}^*(d^{-1}\mathbf{x}, Vd), \tag{5.160}$$

which shows that in simple shear flow an optical property must depend upon the relative speed V through the quantity Vd. Evidence of this scaling occurring in optical properties has been confirmed by Wahl and Fischer [276] for an experiment similar to simple shear flow. These authors looked at the behaviour of the nematic MBBA when it was sheared between two parallel plates, one at rest and the other rotating about its normal. In contrast to the problem considered in the previous subsection, the surface alignment of the director was normal to the plates, but at small speeds of rotation of the upper plate their set-up approximated the simple shear experiment outlined above and their results on optical measurements exhibited dependence on the quantity Vd.

We know from the result in (5.150) that the apparent viscosity η depends on the product Vd. It is also the case, following on from (5.146), (5.147), (5.152) and (5.156), that with $d = 2h$ and $q = d^2$, as selected above, we have

$$\eta = \tilde{t}_{12}\frac{d^2}{Vd} = \frac{\tilde{t}_{12}^*}{V^*} = \eta^*, \tag{5.161}$$

where η^* is the apparent viscosity in the reduced problem with unit gap width. It now follows from equation (5.150) that the apparent viscosity in the transformed variables according to the above scaling coincides with the viscosity in the original variables and that the dependence of this apparent viscosity upon the product Vd remains as before because $\eta = \mathcal{G}(Vd) = \mathcal{G}(V^*)$. This result is in stark contrast to that for an isotropic viscoelastic fluid for which the apparent viscosity, similarly defined, is generally a function of the quantity V/d (see, for example, Coleman and Noll [50]). Nematic liquid crystals are said to scale differently from such fluids. See also Section 5.8.3 below for other scaling properties.

5.6 Oscillatory Shear Flow

In this Section an investigation will be made of the response of a nematic liquid crystal to an induced oscillatory shear flow. After deriving some possible solutions in Section 5.6.1, the onset of an instability which is relevant to devices will also be demonstrated in Section 5.6.2. This particular instability induced by the oscillatory shearing of a homogeneously aligned sample of nematic liquid crystal is often taken to be indicative of more complex flow behaviour such as 'bands' and 'rolls' sometimes called mechanical Williams domains: see Clark, Saunders, Shanks and Leslie [46]. We shall follow part of the work contained in the experimental and theoretical work by Clark *et al.* [46] and Leslie [169] and consider a sample of nematic liquid crystal confined between two parallel glass plates a distance d apart as shown in Fig. 5.7.

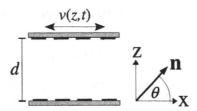

Figure 5.7: The director at the surfaces of a cell of depth d is strongly anchored parallel to the plates. The lower plate is fixed and the upper plate is oscillating with an amplitude a and frequency ω in the x-direction, with velocity $v(d,t) = a\omega\cos(\omega t)$.

5.6.1   Oscillatory Shear Flow Solutions

It is supposed that the director is strongly anchored parallel to the bounding plates and that the lower plate is fixed while the upper plate is subject to sinusoidal oscillations parallel to the initial alignment, as indicated in Fig. 5.7. As pointed out by Leslie [169, 171, 174], in a preliminary investigation it appears reasonable to suppose that both the surface and inertial effects are of secondary importance and concentrate on the influence of the oscillatory flow induced by the oscillating upper plate. This is equivalent to supposing that the associated Reynolds number for this problem, defined by

$$\mathcal{R}_e = dv\frac{\rho}{\eta}, \tag{5.162}$$

is small, and that the Ericksen number, defined by

$$\mathcal{E}_r = dv\frac{\eta}{K}, \tag{5.163}$$

is large. In these formulae d denotes a typical length, v a typical velocity, ρ the density, η a representative dynamic viscosity and K a typical elastic constant. Flows of the same type with the same Reynolds number are called similar [159, p.62]. The Ericksen number was first introduced by de Gennes [109] and is based upon an earlier dimensional argument by Ericksen [79]; it is a measure of the ratio of viscous to elastic forces and moments. The Reynolds number and Ericksen number are both dimensionless. We can write [109]

$$\mathcal{E}_r = \frac{1}{\mu}\mathcal{R}_e, \qquad \mu = \frac{K\rho}{\eta^2}, \tag{5.164}$$

and realise that for the typical values $K = 10^{-11}$ N, $\rho = 1000$ kg m^{-3} and $\eta = 0.01$ Pa s, we have $\mu \sim 10^{-4}$. This indicates that the dimensionless parameter μ is, in practice, generally small, and that the assumptions made at the beginning of this paragraph are reasonable, namely, that $\mathcal{R}_e$ can be considered small while $\mathcal{E}_r$ may be relatively large. (Recall that a small Reynolds number is often desirable for common problems in fluids in order to avoid the usual hydrodynamic instabilities that may arise when the Reynolds number is large: cf. Landau and Lifshitz [159, p.103].)

In these circumstances, motivated by the set-up in Fig. 5.7, it is appropriate in this basic investigation to seek solutions to the Ericksen–Leslie dynamic equations of the form

$$\mathbf{v} = (v(z,t), 0, 0), \qquad v(z,t) = aw\frac{z}{d}\cos(\omega t); \tag{5.165}$$

$$\mathbf{n} = (\cos\theta(t), 0, \sin\theta(t)), \tag{5.166}$$

where the oscillations are of amplitude a and frequency w. The displacement at the upper plate is $x(t) = a\sin(\omega t)$, which implies that the velocity there is $dx/dt = aw\cos(\omega t)$.

The constraints (4.118) are obviously satisfied. The governing dynamic equations (4.139) and (4.140), in the absence of any external body force F_i or generalised body force G_i, are

$$\rho\dot{v}_i = -p_{,i} + \tilde{t}_{ij,j}, \tag{5.167}$$

$$\tilde{g}_i = \lambda n_i, \tag{5.168}$$

where we have neglected the effects of gravity and $\tilde{t}_{ij}$ and $\tilde{g}_i$ are defined by (4.121) and (4.122), respectively. A simple calculation involving (4.5) shows

$$\dot{\mathbf{v}} = (\frac{\partial v}{\partial t}, 0, 0) = -(aw^2\frac{z}{d}\sin(\omega t), 0, 0), \tag{5.169}$$

while the non-zero contributions to the rate of strain tensor $\mathbf{A}$ and vorticity tensor $\mathbf{W}$ are

$$A_{13} = A_{31} = \frac{aw}{2d}\cos(\omega t), \qquad W_{13} = -W_{31} = \frac{aw}{2d}\cos(\omega t), \tag{5.170}$$

thereby showing that the components of the co-rotational time flux $\mathbf{N}$ defined in equation (4.126) are

$$N_1 = -\dot{\theta}\sin\theta - \frac{aw}{2d}\cos(\omega t)\sin\theta, \tag{5.171}$$

$$N_2 = 0, \tag{5.172}$$

$$N_3 = \dot{\theta}\cos\theta + \frac{aw}{2d}\cos(\omega t)\cos\theta, \tag{5.173}$$

where the dot, which represents the material time derivative, has now effectively been reduced to representing ordinary differentiation with respect to time. Hence, because $\tilde{t}_{ij}$ does not depend on any spatial variables,

$$\tilde{t}_{ij,j} = 0, \tag{5.174}$$

and therefore, by (5.169), equations (5.167) reduce to

$$p_{,x} = -\rho\frac{\partial v}{\partial t}, \quad p_{,y} = 0, \quad p_{,z} = 0. \tag{5.175}$$

Following Clark *et al.* [46], the fluid inertial contribution $\rho\dot{v}_1 = \rho\partial v/\partial t$ may be neglected because experimental sample depths are often very thin, or it may be omitted for the reasons suggested by Pieranski *et al.* [220]. Therefore we can solve for the balance of linear momentum by setting the pressure p to be

$$p = p_0(t), \tag{5.176}$$

where $p_0(t)$ is an arbitrary function of time, in which case the equations (5.167) are then satisfied on account of the result in (5.174). It now only remains to solve equations (5.168).

We have from equation (5.166), the definition (4.122) and the results $(5.170)_1$ and (5.171) to (5.173) that equations (5.168) can be written as the two equations

$$-\gamma_1 N_1 - \gamma_2 A_{13}\sin\theta = \lambda\cos\theta, \tag{5.177}$$
$$-\gamma_1 N_3 - \gamma_2 A_{31}\cos\theta = \lambda\sin\theta. \tag{5.178}$$

Multiplying equation (5.177) by $\sin\theta$ and equation (5.178) by $\cos\theta$ and subtracting eliminates the Lagrange multiplier λ, finally leaving the single governing dynamic equation

$$\gamma_1\frac{d\theta}{dt} + [\gamma_1 + \gamma_2\cos(2\theta)]\frac{a\omega}{2d}\cos(\omega t) = 0. \tag{5.179}$$

If required, λ can be obtained by taking the scalar product of equation (5.168) with **n** to find

$$\lambda = -\gamma_2\frac{a\omega}{2d}\cos(\omega t)\sin(2\theta). \tag{5.180}$$

By the definitions of the viscosity coefficients γ_1 and γ_2 in equations (4.123) and (4.124), the governing equation (5.179) can be written in the equivalent convenient form

$$\frac{d\theta}{dt} = -\left[\sin^2\theta - \epsilon\cos^2\theta\right]A\omega\cos(\omega t), \tag{5.181}$$

where

$$\epsilon = \frac{\alpha_3}{\alpha_2}, \quad\text{and}\quad A = \frac{|\alpha_2|a}{\gamma_1 d}, \tag{5.182}$$

provided $\gamma_1 > 0$ and $\alpha_2 < 0$ (cf. Table D.3). This equation can be solved by the separation of variables method to find that there are two easily verified solutions depending upon the sign of ϵ (cf. [116, 2.562.1], Leslie [169]), namely,

$$\theta(t) = \tan^{-1}\{\delta\tanh[A\delta\sin(\omega t)]\}, \quad \delta = \sqrt{\epsilon}, \quad\text{when}\quad \epsilon > 0, \tag{5.183}$$

and

$$\theta(t) = -\tan^{-1}\{\delta\tan[A\delta\sin(\omega t)]\}, \quad \delta = \sqrt{|\epsilon|}, \quad\text{when}\quad \epsilon < 0. \tag{5.184}$$

(Note that ϵ has a different sign in its definition in [169].) In both these solutions $\theta(0)$ has been set to zero, consistent with the assumed homogeneously aligned sample at $t = 0$ when the director is initially aligned everywhere parallel to the plates.

The solutions to the full dynamic equations (5.167) and (5.168) are therefore given by (5.165), (5.166), (5.183) and (5.184) with the pressure p and Lagrange multiplier λ provided by equations (5.176) and (5.180), respectively. We now comment on these solutions for flow-aligning and non-flow-aligning materials.

Flow-Aligning Nematics

For a flow-aligning nematic, that is, for a nematic where $\alpha_2\alpha_3 > 0$ and $\alpha_2 < \alpha_3 < 0$ (see equation (5.31)) we have that $\epsilon > 0$ and therefore the solution for $\theta(t)$ is given by (5.183). This solution oscillates around its initial value of zero with an amplitude that is clearly independent of the frequency ω. The amplitude increases as the amplitude a of the oscillation on the upper plate increases, as can be seen from (5.182)$_2$; it is also evident from (5.183) that the amplitude is uniformly bounded, since it is never greater than $\tan^{-1}(\delta)$ for any value of a.

Non-Flow-Aligning Nematics

For a non-flow-aligning nematic, that is, a nematic where $\alpha_2\alpha_3 < 0$, we have that $\epsilon < 0$ and the solution $\theta(t)$ is provided by equation (5.184). The amplitude of this oscillating solution is independent of ω, as before. However, in this case the amplitude of the oscillation in the solution can increase without bound as the amplitude a of the oscillation on the upper plates increases. Further, if $\delta < 1$, then the amplitude of the solution remains relatively small until the value a increases to a critical value a_c defined by

$$a_c = \frac{\gamma_1 d\pi}{2\delta|\alpha_2|}, \tag{5.185}$$

as is seen from equations (5.182)$_2$ and (5.184). The amplitude of the solution increases sharply as a approaches a_c.

5.6.2   Stability and Instability

To examine the stability of the solutions (5.183) and (5.184), consider solutions of the form

$$\mathbf{n} = (\cos\theta\cos\phi, \cos\theta\sin\phi, \sin\theta), \quad \theta = \theta(t), \quad \phi = \phi(t), \tag{5.186}$$

$$\mathbf{v} = (\kappa(t)z, \tau(t)z, 0), \quad \kappa(t) = a\frac{\omega}{d}\cos(\omega t), \tag{5.187}$$

where a velocity component in the y-direction has been included to allow for the possibility of transverse flow. These forms mentioned by Leslie [174] are slightly easier to handle than the forms introduced by Clark *et al.* [46] which had $\theta = \theta(z, t)$ and $\phi = \phi(z, t)$, but lead to similar conclusions. The required constraints (4.118) are satisfied. When $\tau \equiv \phi \equiv 0$ these forms represent the solutions already obtained for $\theta(t)$ in the previous Section and therefore the above more general situation can be indicative of the stability of these earlier solutions, in a sense to be made clear below. The derivation and identification of first integrals for solutions will be given in some detail as it is quite common to deal with similar equations occurring in the literature. Once a first integral for ϕ has been obtained in terms of the solution θ, some deductions on the stability of the oscillatory shear flow solutions (5.183) and (5.184) will be made in a special case. The stability argument is perhaps a little unusual and is rather technical, but the details are worth pursuing because the interpretation of the results is known to be in good agreement with the experimental observations contained in [46], where more exhaustive details for all possible cases are given.

We shall use the reformulated version of the dynamic equations given in Section 4.3 once the dissipation function has been determined. This will allow a more swift derivation of the dynamic equations contained in [46], as expounded by Leslie [174]. The non-zero components of the rate of strain tensor $\mathbf{A}$ and vorticity tensor $\mathbf{W}$ are, from (4.125), $(4.126)_2$ and (5.187),

$$A_{13} = A_{31} = \tfrac{1}{2}\kappa, \qquad A_{23} = A_{32} = \tfrac{1}{2}\tau, \tag{5.188}$$
$$W_{13} = -W_{31} = \tfrac{1}{2}\kappa, \qquad W_{23} = -W_{32} = \tfrac{1}{2}\tau. \tag{5.189}$$

By (4.126) we therefore have

$$N_1 = -\dot{\theta}\sin\theta\cos\phi - \dot{\phi}\cos\theta\sin\phi - \tfrac{1}{2}\kappa\sin\theta, \tag{5.190}$$
$$N_2 = -\dot{\theta}\sin\theta\sin\phi + \dot{\phi}\cos\theta\cos\phi - \tfrac{1}{2}\tau\sin\theta, \tag{5.191}$$
$$N_3 = \dot{\theta}\cos\theta + \tfrac{1}{2}\kappa\cos\theta\cos\phi + \tfrac{1}{2}\tau\cos\theta\sin\phi, \tag{5.192}$$

where the superposed dot now represents differentiation with respect to time t, given the above forms for $\mathbf{n}$ and $\mathbf{v}$. The results in equations (5.188) and (5.190) to (5.192) can be used with the definition of $\mathbf{n}$ to evaluate the terms appearing in the dissipation function as defined by equation (4.85). It proves convenient to introduce the time dependent functions

$$\xi(t) = \kappa\cos\phi + \tau\sin\phi, \qquad \zeta(t) = \kappa\sin\phi - \tau\cos\phi, \tag{5.193}$$

which satisfy the relation
$$\xi^2 + \zeta^2 = \kappa^2 + \tau^2. \tag{5.194}$$

The following can then be verified:

$$(n_i A_{ij} n_j)^2 = \xi^2 \sin^2\theta\cos^2\theta, \tag{5.195}$$
$$N_i A_{ij} n_j = \tfrac{1}{2}\xi\dot{\theta}(\cos^2\theta - \sin^2\theta) - \tfrac{1}{4}(\xi^2 + \zeta^2)\sin^2\theta$$
$$\qquad\qquad - \tfrac{1}{2}\dot{\phi}\zeta\sin\theta\cos\theta + \tfrac{1}{4}\xi^2\cos^2\theta, \tag{5.196}$$
$$A_{ij} A_{ij} = \tfrac{1}{2}(\xi^2 + \zeta^2), \tag{5.197}$$
$$n_i A_{ij} A_{jk} n_k = \tfrac{1}{4}(\xi^2 + \zeta^2)\sin^2\theta + \tfrac{1}{4}\xi^2\cos^2\theta, \tag{5.198}$$
$$N_i N_i = \dot{\theta}^2 + \dot{\phi}^2\cos^2\theta + \tfrac{1}{4}(\xi^2 + \zeta^2)\sin^2\theta + \dot{\theta}\xi$$
$$\qquad\qquad + \dot{\phi}\zeta\sin\theta\cos\theta + \tfrac{1}{4}\xi^2\cos^2\theta, \tag{5.199}$$

The above quantities can be inserted into the dissipation function $\widehat{\mathcal{D}} = \frac{1}{2}\mathcal{D}$ defined by equations (4.85) and (4.146)$_2$. After some straightforward rearrangements and computations, and using the Parodi relation (4.127) when appropriate, we can collect terms in powers of ξ and ζ to obtain [174]

$$\widehat{\mathcal{D}} = \frac{1}{2}\left[g(\theta)\xi^2 + h(\theta)\zeta^2\right] + \frac{1}{2}\gamma_1\left(\frac{d\theta}{dt}\right)^2 + \frac{1}{2}\gamma_1\cos^2\theta\left(\frac{d\phi}{dt}\right)^2$$
$$+m(\theta)\xi\frac{d\theta}{dt} - \alpha_2\zeta\sin\theta\cos\theta\frac{d\phi}{dt}, \qquad (5.200)$$

where the functions $g(\theta)$, $h(\theta)$ and $m(\theta)$ are defined by

$$g(\theta) = \frac{1}{2}\left[\alpha_4 + (\alpha_5 - \alpha_2)\sin^2\theta + (\alpha_3 + \alpha_6)\cos^2\theta\right] + \alpha_1\sin^2\theta\cos^2\theta, \quad (5.201)$$
$$h(\theta) = \frac{1}{2}\left[\alpha_4 + (\alpha_5 - \alpha_2)\sin^2\theta\right], \qquad (5.202)$$
$$m(\theta) = \alpha_3\cos^2\theta - \alpha_2\sin^2\theta. \qquad (5.203)$$

The definition for $g(\theta)$ is consistent with that defined earlier at equation (5.113) in Section 5.5 and later in Sections 5.7.2, 5.8.2; both $m(\theta)$ and $g(\theta)$ arise again in Section 5.9.2 at equations (5.438) and (5.439) when discussing the dynamics of the Freedericksz transitions, as they do together with $h(\theta)$ when light scattering is investigated in Section 5.10. The dissipation function is a quadratic form and can be expressed as

$$\widehat{\mathcal{D}} = \frac{1}{2}\begin{bmatrix} \xi & \zeta & \dot\theta & \dot\phi \end{bmatrix}\begin{bmatrix} g(\theta) & 0 & m(\theta) & 0 \\ 0 & h(\theta) & 0 & -\alpha_2\sin\theta\cos\theta \\ m(\theta) & 0 & \gamma_1 & 0 \\ 0 & -\alpha_2\sin\theta\cos\theta & 0 & \gamma_1\cos^2\theta \end{bmatrix}\begin{bmatrix} \xi \\ \zeta \\ \dot\theta \\ \dot\phi \end{bmatrix}. \quad (5.204)$$

Following from the comments after equation (2.56), $\widehat{\mathcal{D}}$ is non-negative if and only if the determinants of all the principal submatrices of the above symmetric matrix are non-negative. Hence non-negativity of the dissipation function requires that

$$g(\theta) \geq 0, \qquad h(\theta) \geq 0, \qquad (5.205)$$
$$\gamma_1 g(\theta) - m^2(\theta) \geq 0, \qquad \gamma_1 h(\theta) - \alpha_2^2\sin^2\theta \geq 0. \qquad (5.206)$$

Interchanging the rôles of ξ and $\dot\theta$ in the matrix representations in (5.204) also delivers the known result from equation (4.123), namely, $\gamma_1 \geq 0$. If we assume that $\gamma_1 > 0$ then we can introduce two modified viscosity functions $G(\theta)$ and $H(\theta)$ which will be of use below, defined by

$$G(\theta) = g(\theta) - m^2(\theta)/\gamma_1 \geq 0, \qquad (5.207)$$
$$H(\theta) = h(\theta) - \alpha_2^2\sin^2\theta/\gamma_1 \geq 0. \qquad (5.208)$$

Since w_F and Ψ are, as above, assumed to be absent, the dynamic equations (4.152) and (4.153) for the balances of angular and linear momentum in this case are

$$\frac{\partial\widehat{\mathcal{D}}}{\partial\dot\theta} = 0, \qquad \frac{\partial\widehat{\mathcal{D}}}{\partial\dot\phi} = 0, \qquad (5.209)$$

and, since θ and ϕ do not depend upon z and $\tilde{p} = p$ (cf. 4.133),

$$\rho \dot{v}_i = \left(\frac{\partial \widehat{D}}{\partial v_{i,j}} \right)_{,j} - p_{,i}, \qquad i = 1, 2, 3. \tag{5.210}$$

Inserting the dissipation function given by (5.200) into equations (5.209) gives the two equations

$$\gamma_1 \frac{d\theta}{dt} + m(\theta)\xi = 0, \tag{5.211}$$

$$\gamma_1 \cos\theta \frac{d\phi}{dt} - \alpha_2 \zeta \sin\theta = 0, \tag{5.212}$$

assuming $\theta \neq \pm\frac{\pi}{2}$. We have that $\widehat{D}$ only depends on time t and, additionally, $\dot{v}_i = \partial v_i / \partial t$. We can neglect the fluid inertia terms in (5.210), as before, and proceed to write equations (5.210) as

$$p_{,x} = p_{,y} = p_{,z} = 0. \tag{5.213}$$

Hence, setting $p = p_0(t)$ for some arbitrary function p_0, equations (5.210) can be integrated and reduce to

$$\tilde{t}_{13} = \frac{\partial \widehat{D}}{\partial v_{1,z}} = \sigma_1(t), \qquad \tilde{t}_{23} = \frac{\partial \widehat{D}}{\partial v_{2,z}} = \sigma_2(t), \tag{5.214}$$

where σ_1 and σ_2 are arbitrary functions of time and use has been made of the relations $(4.144)_2$ and $(4.146)_2$. Following Clark *et al.* [46] and Leslie [169], it seems reasonable in this problem to set the transverse component of the stress to zero, so that $\sigma_2(t) = 0$. Clearly, for the above forms of solutions, $v_{1,z} = \kappa$, $v_{2,z} = \tau$ and so carrying out the differentiations in (5.214), recalling that ξ and ζ appearing in $\widehat{D}$ are functions of these variables, equations (5.214) are

$$\left[g(\theta)\xi + m(\theta)\frac{d\theta}{dt} \right] \cos\phi + \left[h(\theta)\zeta - \alpha_2 \sin\theta\cos\theta\frac{d\phi}{dt} \right] \sin\phi = \sigma_1(t), \tag{5.215}$$

$$\left[g(\theta)\xi + m(\theta)\frac{d\theta}{dt} \right] \sin\phi - \left[h(\theta)\zeta - \alpha_2 \sin\theta\cos\theta\frac{d\phi}{dt} \right] \cos\phi = 0. \tag{5.216}$$

Multiplying (5.215) by $\cos\phi$ and (5.216) by $\sin\phi$ and adding the resulting equations gives

$$g(\theta)\xi + m(\theta)\frac{d\theta}{dt} = \sigma_1 \cos\phi, \tag{5.217}$$

whereas multiplying (5.215) by $\sin\phi$ and (5.216) by $\cos\phi$ and subtracting gives

$$h(\theta)\zeta - \alpha_2 \sin\theta\cos\theta\frac{d\phi}{dt} = \sigma_1 \sin\phi. \tag{5.218}$$

It is now possible to eliminate the velocity gradients ξ and ζ from the angular momentum balance equations (5.211) and (5.212) by using the linear momentum balance equations (5.217) and (5.218) to give the final form of the governing dynamic equations. These two final equations are

$$\gamma_1 G(\theta)\frac{d\theta}{dt} + \sigma_1 m(\theta)\cos\phi = 0, \tag{5.219}$$

$$\gamma_1 H(\theta)\cos\theta\frac{d\phi}{dt} - \sigma_1\alpha_2\sin\theta\sin\phi = 0, \tag{5.220}$$

where $G(\theta)$ and $H(\theta)$ are as defined in (5.207) and (5.208).

We are now in a position to exploit the governing equations (5.219) and (5.220) under the assumption that G and H are strictly positive. These equations easily lead to the equation

$$\frac{d\phi}{d\theta} = -\tan\theta\tan\phi\frac{\alpha_2 G(\theta)}{m(\theta)H(\theta)}. \tag{5.221}$$

Notice that the arbitrary function σ_1 is no longer present. Employing the substitution

$$s = \tan^2\theta, \tag{5.222}$$

the above equation can be written as

$$2\cot\phi\frac{d\phi}{ds} = -\frac{\alpha_2 G(\theta)}{(1+s)m(\theta)H(\theta)}, \tag{5.223}$$

where θ is now a function of s. It is convenient to introduce the parameters

$$\lambda = \alpha_4\left(\alpha_3 + \alpha_4 + \alpha_6 - 2\alpha_3^2/\gamma_1\right)^{-1}, \tag{5.224}$$

$$\beta = \left(\gamma_2^2/\gamma_1 + \alpha_1\right)\left(\alpha_3 + \alpha_4 + \alpha_6 - 2\alpha_3^2/\gamma_1\right)^{-1}, \tag{5.225}$$

with $\gamma_1 = \alpha_3 - \alpha_2$ and $\gamma_2 = \alpha_2 + \alpha_3$, as usual. Notice also that $H(0) = \alpha_4/2$ and $\lambda = \alpha_4/2G(0)$ and therefore by the assumed positivity of G and H it follows that $\alpha_4 > 0$ and $\lambda > 0$. Using these parameters and the substitution (5.222), and observing that

$$\alpha_5 - \alpha_2 - 2\alpha_2^2/\gamma_1 = \alpha_3 + \alpha_6 - 2\alpha_3^2/\gamma_1, \tag{5.226}$$

by the Parodi relation (4.127), the result in equation (5.223) can be written in the form [46, 174]

$$2\cot\phi\frac{d\phi}{ds} = \frac{[(1+s)^2 + 2\beta s]}{(s+1)(s-\epsilon)(s+\lambda)}, \tag{5.227}$$

with $\epsilon = \alpha_3/\alpha_2$ as previously introduced at (5.182). We shall make the assumption that

$$\lambda \neq 1, \quad \lambda \neq -\epsilon, \quad \epsilon \neq -1, \tag{5.228}$$

other possibilities being discussed by Clark *et al.* [46]. Equation (5.227) can be integrated directly by the separation of variables method to find that

$$\sin^2\phi = C(1 + \tan^2\theta)^p \left|\epsilon - \tan^2\theta\right|^q (\lambda + \tan^2\theta)^{1-p-q}, \tag{5.229}$$

where C is a constant of integration and

$$p = \frac{2\beta}{(1+\epsilon)(\lambda-1)}, \qquad q = \frac{(1+\epsilon)^2 + 2\beta\epsilon}{(1+\epsilon)(\lambda+\epsilon)}. \tag{5.230}$$

If at $t = 0$ we have $\theta = 0$ and $\phi = \phi_0 \neq 0$ then the solution (5.229) for ϕ takes the form

$$\sin^2\phi = \sin^2\phi_0 \, \lambda^{p+q-1}|\epsilon|^{-q}(1 + \tan^2\theta)^p \, |\epsilon - \tan^2\theta|^q \, (\lambda + \tan^2\theta)^{1-p-q}, \qquad (5.231)$$

which now allows us to consider the stability of flow-aligning and non-flow-aligning nematics.

Following the method used by Clark et al. [46], consider the ratio of the second component of the director for $t > 0$ divided by its value at $t = 0$. This ratio gives some measure of the growth of any perturbation ϕ out of the plane of shear and can therefore lead to conclusions about the stability or instability of solutions. From (5.186) and (5.231) we have, since $\theta(0) = 0$, $\phi(0) = \phi_0$ and $\cos^2\theta = (1 + \tan^2\theta)^{-1}$,

$$\frac{n_y^2}{(n_y^2)_{t=0}} = \lambda^{p+q-1}|\epsilon|^{-q}(1 + \tan^2\theta)^{p-1} \, |\epsilon - \tan^2\theta|^q \, (\lambda + \tan^2\theta)^{1-p-q}. \qquad (5.232)$$

Stability for Flow-Aligning Nematics

Suppose that $\alpha_2\alpha_3 > 0$ with $\alpha_3 < \alpha_2 < 0$ so that the nematic is flow-aligning with $\epsilon > 0$. Suppose we have the corresponding solution (5.183) for $\theta(t)$ and that we perturb it by a small amount ϕ out of the plane of shear. To simplify the problem, also suppose the restrictions (5.228) hold and that $\epsilon \ll 1$. Substituting the original solution for θ into equation (5.232) shows that

$$\frac{n_y^2}{(n_y^2)_{t=0}} \approx \left\{1 - \tanh^2\left[A\delta\sin(\omega t)\right]\right\}^q, \qquad (5.233)$$

noting that

$$q \sim 1/\lambda > 0 \qquad \text{for } |\epsilon| \ll 1, \qquad (5.234)$$

by (5.230)$_2$. Given the behaviour of the tanh function, it is clear that the perturbation ϕ will remain bounded for any value of q consistent with the assumptions (5.228). We can therefore conclude that in the simple oscillatory shear problem for flow-aligning nematics the solution $\theta(t)$ provided by equation (5.183) may be considered as being stable.

Instability for Non-Flow-Aligning Nematics

Now suppose that $\alpha_2\alpha_3 < 0$ so that the nematic is non-flow-aligning with $\epsilon < 0$ and that we have the corresponding solution (5.184) for $\theta(t)$. Recall that we have set $\alpha_2 < 0$ and so here we have $\alpha_3 > 0$. As above, suppose that this solution is perturbed by a small amount ϕ out of the plane of shear and, for simplicity, suppose the restrictions (5.228) are valid with $|\epsilon| \ll 1$. Substituting for θ in equation (5.232) and noting that $\epsilon = -|\epsilon| = -\delta^2$, we have

$$\frac{n_y^2}{(n_y^2)_{t=0}} = \left(1 + |\epsilon|T^2\right)^{p-1}\left(1 + T^2\right)^q\left(1 + |\epsilon|T^2/\lambda\right)^{1-p-q}, \qquad (5.235)$$

with

$$T = \tan\left[\frac{|\alpha_2|a}{\gamma_1 d}\delta\sin(\omega t)\right]. \qquad (5.236)$$

It is now clear that as the amplitude a of the oscillation at the upper plate approaches the critical value a_c defined by equation (5.185) then T can become arbitrarily large and that near a_c we have

$$\frac{n_y^2}{(n_y^2)_{t=0}} \approx |\epsilon|^{-\frac{1}{\lambda}}\lambda^{p+q-1}, \qquad (5.237)$$

employing the approximation (5.234). This result shows that deviations from the shear plane increase sharply with the angle θ as a approaches a_c. This behaviour is interpreted as a flow instability [46, 169] and is in good agreement with the observations reported in [46].

Other cases for different values of ϵ, for example $\epsilon \gg 1$, have been considered by Clark *et al.* [46]. Further, homeotropic boundary conditions rather than the above planar boundary conditions have also been examined in a similar style in [46].

A great deal of experimental and theoretical work has been carried out recently on oscillatory shear problems in nematics. Readers will find much of interest in the experimental work of Borzsonyi, Buka, Krekhov and Kramer [20] and Mullin and Peacock [208]. More theoretical work can be found in the articles by Krekhov and Kramer [156] and Hogan, Mullin and Woodford [128] (who also consider experimental results) and the references cited therein.

5.7   Couette Flow

In this Section we consider flow of a sample of anisotropic fluid or nematic liquid crystal occupying the annular region $R_1 \leq r \leq R_2$ between two concentric cylinders in relative rotation. The inner and outer concentric cylinders have fixed radii R_1 and R_2 and corresponding constant angular velocities Ω_1 and Ω_2, respectively, as shown in Fig. 5.8. The usual cylindrical coordinate system (r, θ, z) having basis vectors $\mathbf{e}_r$, $\mathbf{e}_\theta$ and $\mathbf{e}_z$ will be used, with the z-axis coinciding with the common axis of the cylinders, perpendicular to the page in Fig. 5.8. We consider two cases: firstly, the solution for an anisotropic fluid first discussed by Leslie [162] where the director gradients $n_{i,j}$ are absent (and therefore the elastic energy is neglected), and, secondly, a symmetric solution for nematic liquid crystals considered by Atkin and Leslie [6] which incorporates the elastic energy and director gradients. Both these problems were influenced by the earlier work of Verma [272], who found an exact solution for a special form of anisotropic fluid.

5.7.1   The Anisotropic Fluid Case

The Ericksen–Leslie theory from Section 4.2.5 will be used with all director gradients and the elastic energy being set to zero, so that we are dealing with an anisotropic fluid. Incorporating the gravitational potential Ψ_g, the relevant dynamic equations

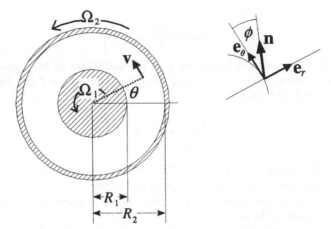

Figure 5.8: The geometrical set-up for one type of Couette flow. The anisotropic
fluid or liquid crystal is placed in the annular region between two concentric cylinders of radii R_1 and R_2 which have corresponding angular velocities Ω_1 and Ω_2,
respectively. The usual cylindrical polar coordinates (r, θ, z) are used as discussed in
the text, with the z-axis perpendicular to the page and coincident with the common
axis of the cylinders. The velocity $\mathbf{v}$ is assumed to have a component only in the
$\mathbf{e}_\theta$ direction. The director $\mathbf{n}$ makes an angle ϕ with respect to the local coordinate
axes $\mathbf{e}_r$ and $\mathbf{e}_\theta$ as shown.

in Cartesian component form are (4.139) and (4.140) with (4.141), which in this
example are

$$\rho \dot{v}_i = -\tilde{p}_{,i} + \tilde{t}_{ij,j}, \qquad \tilde{p} = p + \rho \Psi_g, \tag{5.238}$$

$$\tilde{g}_i = \lambda n_i, \tag{5.239}$$

in the absence of any generalised body force G_i. Motivated by the geometry in
Fig. 5.8, solutions to these equations will be sought in which the velocity $\mathbf{v}$ and
director $\mathbf{n}$, expressed in cylindrical polar coordinates, are of the forms [162]

$$v_r = 0, \quad v_\theta = r\omega(r), \quad v_z = 0, \tag{5.240}$$

$$n_r = \sin\phi, \quad n_\theta = \cos\phi, \quad n_z = 0, \tag{5.241}$$

where ω and ϕ are functions of r with $\omega(r)$ satisfying the boundary conditions

$$\omega(R_1) = \Omega_1, \quad \omega(R_2) = \Omega_2, \tag{5.242}$$

these being consistent with the usual no-slip boundary conditions on $\mathbf{v}$. The constraints (4.118) are clearly satisfied. Equations (5.238) and (5.239) will be transformed into the corresponding relevant physical components in the cylindrical polar
coordinate system. (Recall that all equations involving vector and tensor quantities
given in Cartesian coordinates are equally valid when they are expressed in terms
of curvilinear coordinates provided the physical components are used [115, p.134].)

The velocity is independent of time t and therefore $\dot{\mathbf{v}} = (\mathbf{v} \cdot \nabla)\mathbf{v}$. Using the general formula stated at equation (C.4) in Appendix C (with $\mathbf{n}$ set to $\mathbf{v}$), the physical components of $\dot{\mathbf{v}}$ are therefore given by

$$\dot{v}_r = -r\omega^2, \quad \dot{v}_\theta = 0, \quad \dot{v}_z = 0. \tag{5.243}$$

Similarly, since $\mathbf{n}$ is independent of t, the same formula can be applied again to show that the physical components of $\dot{\mathbf{n}} = (\mathbf{v} \cdot \nabla)\mathbf{n}$ are

$$\dot{n}_r = -\omega \cos\phi, \quad \dot{n}_\theta = \omega \sin\phi, \quad \dot{n}_z = 0. \tag{5.244}$$

The non-zero physical components of the rate of strain tensor A and vorticity tensor W are obtained by applying the formulae (C.11) to (C.17) in Appendix C, giving

$$A_{r\theta} = A_{\theta r} = \tfrac{1}{2}r\omega', \quad W_{r\theta} = -W_{\theta r} = -\tfrac{1}{2}r\omega' - \omega, \tag{5.245}$$

where a prime denotes differentiation with respect to r. From equations (4.126), (5.241), (5.244) and (5.245)$_2$, the co-rotational time flux is given by

$$N_r = \tfrac{1}{2}r\omega' \cos\phi, \quad N_\theta = -\tfrac{1}{2}r\omega' \sin\phi, \quad N_z = 0, \tag{5.246}$$

and this further allows $\tilde{\mathbf{g}}$ in equation (4.122) to be evaluated as

$$\tilde{g}_r = -\tfrac{1}{2}(\gamma_1 + \gamma_2)r\omega' \cos\phi, \quad \tilde{g}_\theta = \tfrac{1}{2}(\gamma_1 - \gamma_2)r\omega' \sin\phi, \quad \tilde{g}_z = 0. \tag{5.247}$$

We are now in a position to solve equations (5.239) before proceeding to solve (5.238).

Inserting the above results for $\tilde{\mathbf{g}}$ into (5.239) reduces them to the two equations

$$-\tfrac{1}{2}(\gamma_1 + \gamma_2)r\omega' \cos\phi = \lambda \sin\phi, \tag{5.248}$$
$$\tfrac{1}{2}(\gamma_1 - \gamma_2)r\omega' \sin\phi = \lambda \cos\phi, \tag{5.249}$$

where λ is a Lagrange multiplier. Multiplying (5.248) by $\cos\phi$, (5.249) by $\sin\phi$ and subtracting allows λ to be eliminated, leaving the single equation

$$r\omega' \left[\gamma_1 + \gamma_2 \cos(2\phi)\right] = 0. \tag{5.250}$$

Alternatively, λ can be evaluated by taking the scalar product of equations (5.239) with $\mathbf{n}$ to find

$$\lambda = -\gamma_2 r\omega' \sin\phi \cos\phi. \tag{5.251}$$

Inserting this value into equations (5.239) yields equation (5.250) again. If we suppose $\omega' \neq 0$, then it is evident that $\phi(r)$ must take a constant value determined by a relation which involves only the rotational viscosity γ_1 and the torsion coefficient γ_2. This value is given by the relation

$$\cos(2\phi) = -\frac{\gamma_1}{\gamma_2}, \tag{5.252}$$

which delivers a real valued constant ϕ provided $\gamma_1 \leq |\gamma_2|$, with $\gamma_2 \neq 0$, recalling that γ_1 is necessarily non-negative. It is convenient to record for later use that ϕ also satisfies the identities

$$\sin^2\phi = \frac{\alpha_3}{\gamma_2}, \qquad \cos^2\phi = \frac{\alpha_2}{\gamma_2}, \qquad \tan^2\phi = \frac{\alpha_3}{\alpha_2}, \tag{5.253}$$

obtained from (5.252) and the definitions in equations (4.123) and (4.124). This demonstrates that the angle ϕ coincides with the flow alignment angle (Leslie angle) θ_0 defined by equation (5.23). When the Parodi relation (4.127), in conjunction with (4.123) and (4.124), is assumed, then we remark here for later use that the following general identity is valid:

$$(\alpha_6 - \alpha_3)\sin^2\phi + (\alpha_2 + \alpha_5)\cos^2\phi \;=\; (\alpha_5 - \alpha_2)\sin^2\phi + (\alpha_6 + \alpha_3)\cos^2\phi$$
$$-(\gamma_1 + \gamma_2\cos(2\phi)). \tag{5.254}$$

The relation (5.252) which is available in this particular example shows that this identity may be reduced to

$$(\alpha_6 - \alpha_3)\sin^2\phi + (\alpha_2 + \alpha_5)\cos^2\phi = (\alpha_5 - \alpha_2)\sin^2\phi + (\alpha_6 + \alpha_3)\cos^2\phi. \tag{5.255}$$

Equations (5.239) have now been completely solved, showing that ϕ must be a constant given by (5.252) if $\omega' \neq 0$, and so it only remains to solve equations (5.238) to find p and the solution for $\omega(r)$.

The non-zero components of the viscous stress are found by appropriately inserting the results (5.241), (5.245)$_1$ and (5.246) into equation (4.121), to give

$$\tilde{t}_{rr} = Ar\omega', \quad \tilde{t}_{\theta\theta} = Br\omega', \quad \tilde{t}_{r\theta} = \tilde{t}_{\theta r} = Cr\omega', \tag{5.256}$$

where A, B and C are constants defined by

$$A = \tfrac{1}{2}\sin\phi\cos\phi\left(2\alpha_1\sin^2\phi + \alpha_2 + \alpha_3 + \alpha_5 + \alpha_6\right), \tag{5.257}$$
$$B = \tfrac{1}{2}\sin\phi\cos\phi\left(2\alpha_1\cos^2\phi - \alpha_2 - \alpha_3 + \alpha_5 + \alpha_6\right), \tag{5.258}$$
$$C = \tfrac{1}{2}\left[2\alpha_1\sin^2\phi\cos^2\phi + (\alpha_2 + \alpha_5)\cos^2\phi + (\alpha_6 - \alpha_3)\sin^2\phi + \alpha_4\right]. \tag{5.259}$$

The identity (5.255) allows the verification that $\tilde{t}_{r\theta} = \tilde{t}_{\theta r}$ in equation (5.256). The results in (5.256) can be inserted into the formulae (C.18), (C.19) and (C.20) in Appendix C to find the physical components of $\tilde{t}_{ij,j}$ in cylindrical polar coordinates. These resulting components, in conjunction with (5.243) and equation (C.3), allow equations (5.238) to be given as

$$\frac{\partial \tilde{p}}{\partial r} = \frac{\partial \tilde{t}_{rr}}{\partial r} + \frac{1}{r}\left(\tilde{t}_{rr} - \tilde{t}_{\theta\theta}\right) + \rho r\omega^2, \tag{5.260}$$

$$\frac{1}{r}\frac{\partial \tilde{p}}{\partial \theta} = \frac{\partial \tilde{t}_{\theta r}}{\partial r} + \frac{1}{r}\left(\tilde{t}_{\theta r} + \tilde{t}_{r\theta}\right), \tag{5.261}$$

$$\frac{\partial \tilde{p}}{\partial z} = 0. \tag{5.262}$$

It is clear from equation (5.262) that $\tilde{p} = \tilde{p}(r, \theta)$. Given the dependency of the viscous stress components upon r, we can take the derivatives of equations (5.260) and (5.261) with respect to θ and r, respectively, to show that (cf. Atkin and Leslie [6])

$$\tilde{p} = a\theta + H(r), \tag{5.263}$$

where, on account of the results in (5.256), a is an arbitrary constant and the function $H(r)$ satisfies

$$\frac{dH}{dr} = \frac{d\tilde{t}_{rr}}{dr} + \frac{1}{r}\left(\tilde{t}_{rr} - \tilde{t}_{\theta\theta}\right) + \rho r\omega^2. \tag{5.264}$$

For a single valued solution (for example, we must have $\tilde{p}(r, \theta) = \tilde{p}(r, \theta + 2\pi)$) it is required that a be set to zero, leading to

$$\tilde{p} = H(r), \tag{5.265}$$

which then reduces equation (5.261) to

$$\frac{d\tilde{t}_{\theta r}}{dr} + \frac{1}{r}\left(\tilde{t}_{\theta r} + \tilde{t}_{r\theta}\right) = 0. \tag{5.266}$$

On account of equation $(5.256)_3$, this further reduces, in this example, to

$$\frac{d^2\omega}{dr^2} + \frac{3}{r}\frac{d\omega}{dr} = 0. \tag{5.267}$$

Applying the boundary conditions (5.242), this differential equation is easily solved to find the solution

$$\omega(r) = L + \frac{M}{r^2}, \tag{5.268}$$

where the constants L and M are given by

$$L = \frac{R_2^2\Omega_2 - R_1^2\Omega_1}{R_2^2 - R_1^2}, \quad M = \frac{R_1^2 R_2^2(\Omega_1 - \Omega_2)}{R_2^2 - R_1^2}. \tag{5.269}$$

This result also shows that we can now write, via equations $(5.238)_2$, (5.256), (5.264), (5.265) and (5.268),

$$p = -\rho\Psi_g + \tilde{p} = p_0 - \rho\Psi_g - (A + B)\frac{M}{r^2} + \int_{R_1}^r \rho s\omega^2(s)\,ds, \tag{5.270}$$

where p_0 is an arbitrary constant pressure.

The complete solution to this particular Couette flow problem for an anisotropic fluid is given by the results for ϕ and ω in equations (5.252) and (5.268), respectively, with the Lagrange multiplier λ being given by (5.251) and the pressure p by (5.270).

It is worth commenting on these results. It is known that for an incompressible isotropic viscous fluid in the above form of Couette flow, $v_\theta(r) = r\omega(r)$ is identical to that obtained above [159, p.60]. In particular, for $\Omega_1 = \Omega_2 = \Omega$ we have $v_\theta = r\Omega$, which shows that the entire fluid rotates rigidly with the cylinders. Also, when the outer cylinder is absent, so that $\Omega_2 = 0$ and $R_2 = \infty$, it is seen from equations (5.268) and (5.269) that $v_\theta = \Omega_1 R_1^2/r$. The main difference for an anisotropic fluid is the alignment of the director **n** given via (5.241) and (5.252) and the viscous stress given by (5.256): here it is shown that the director always makes a constant angle ϕ with the tangent lines parallel to the e_θ-direction: see Fig. 5.8. Also, nothing has been said about boundary conditions on the director in this example: the director is everywhere fixed at this angle ϕ to the e_θ-direction. We shall incorporate elastic effects in the next Section when we deal with a nematic liquid crystal in the same type of Couette flow.

There is one simple special case considered by Leslie [162] which we now mention for completeness. If the flow profile remains as supposed in equation (5.240), but the director profile in (5.241) is replaced by

$$n_r = 0, \quad n_\theta = 0, \quad n_z = 1, \tag{5.271}$$

then it is straightforward to find that the only non-zero components of the viscous stress (4.121) are

$$\tilde{t}_{r\theta} = \tilde{t}_{\theta r} = \tfrac{1}{2}\alpha_4 r\omega'. \tag{5.272}$$

In this case, following the same sorts of computations used to obtain the results in equations (5.244) to (5.247), we find that

$$\dot{n}_i = 0, \quad N_i = 0, \quad \tilde{g}_i = 0, \quad i = 1, 2, 3, \tag{5.273}$$

which shows that equations (5.239) are satisfied with $\lambda = 0$. The remaining governing equations (5.260), (5.261) and (5.262), corresponding to equations (5.238), are replaced with

$$\frac{\partial \tilde{p}}{\partial r} = \rho r\omega^2, \tag{5.274}$$

$$\frac{1}{r}\frac{\partial \tilde{p}}{\partial \theta} = \frac{\partial \tilde{t}_{\theta r}}{\partial r} + \frac{2\tilde{t}_{r\theta}}{r}, \tag{5.275}$$

$$\frac{\partial \tilde{p}}{\partial z} = 0. \tag{5.276}$$

By the same procedure used to obtain the earlier results above in equations (5.263) to (5.266), we arrive at equation (5.267) as before, which shows that the solution for $v(r) = r\omega(r)$ is again given via the results stated in equations (5.268) and (5.269). However, in this case the pressure p is given by

$$p = -\rho\Psi_g + \tilde{p} = p_0 - \rho\Psi_g + \int_{R_1}^{r} \rho s\omega^2(s)\, ds, \tag{5.277}$$

p_0 being an arbitrary constant pressure.

5.7.2   The Nematic Liquid Crystal Case

We now investigate the Couette flow of a nematic liquid crystal in precisely the same type of experiment discussed above for an anisotropic fluid described in Fig. 5.8 and in essence follow the work of Atkin and Leslie [6]. The one-constant approximation for the nematic elastic energy will be assumed in order to simplify the presentation:

the reader is referred to Atkin and Leslie [6] for full details when the elastic constants are not all equal. The forms for the velocity and the director remain as given above except that ϕ will turn out to be a non-constant function of r. Therefore we set

$$v_r = 0, \quad v_\theta = r\omega(r), \quad v_z = 0, \tag{5.278}$$
$$n_r = \sin\phi(r), \quad n_\theta = \cos\phi(r), \quad n_z = 0. \tag{5.279}$$

The constraints (4.118) are again satisfied. As before, the boundary conditions on ω are given by

$$\omega(R_1) = \Omega_1, \quad \omega(R_2) = \Omega_2, \tag{5.280}$$

while for the director we assume strong anchoring at the boundaries with

$$\phi(R_1) = \phi(R_2) = 0, \tag{5.281}$$

so that the director is tangential to the cylinder walls at the boundaries and is aligned parallel to the θ-direction. In the absence of any generalised body force G_i, the Cartesian component forms for the dynamic equations analogous to those for an anisotropic fluid, stated at equations (5.238) and (5.239), are given via (4.120), (4.132), (4.133) and (4.137) by

$$\rho\dot{v}_i = \tilde{g}_j n_{j,i} - \tilde{p}_{,i} + \tilde{t}_{ij,j}, \quad \tilde{p} = p + \rho\Psi_g + w_F, \tag{5.282}$$
$$K n_{i,jj} + \tilde{g}_i = \lambda n_i, \tag{5.283}$$

where, by (4.136),

$$w_F = \tfrac{1}{2} K n_{i,j} n_{i,j}, \quad K > 0, \tag{5.284}$$

is the adopted one-constant approximation for the elastic energy when strong anchoring is assumed. (The elastic contribution on the left-hand side of equation (5.283) is a consequence of setting $K \equiv K_1 = K_2 = K_3$ in the expression (4.131).) Solutions for $\phi(r)$ and $\omega(r)$ to equations (5.282) and (5.283) subject to the boundary conditions (5.280) and (5.281) are sought here and the relevant equations will now be expressed in terms of the relevant physical components in the cylindrical polar coordinate system.

Firstly, notice that the expressions for $\dot{v}_i$, $\dot{n}_i$, A_{ij}, W_{ij}, N_i and $\tilde{g}_i$ are exactly the same as those derived earlier in equations (5.243) to (5.247), the only difference being that ϕ will turn out not to be constant. From formulae (C.8) to (C.10) in Appendix C and the form for the director given by (5.279), the non-zero physical components of $[\nabla\mathbf{n}]_{ij} = n_{i,j}$ in cylindrical coordinates are

$$[\nabla\mathbf{n}]_{11} = \phi'\cos\phi, \quad [\nabla\mathbf{n}]_{12} = -\frac{\cos\phi}{r}, \tag{5.285}$$

$$[\nabla\mathbf{n}]_{21} = -\phi'\sin\phi, \quad [\nabla\mathbf{n}]_{22} = \frac{\sin\phi}{r}, \tag{5.286}$$

recalling that ϕ is now a function of r and that a prime denotes differentiation with respect to r. The divergence $n_{i,jj}$, $i = 1, 2, 3$, can be obtained by using the results

(5.285) and (5.286) and setting $\mathsf{T} = (\nabla \mathbf{n})$ in formulae (C.18) to (C.20) to find that

$$[\nabla \cdot (\nabla \mathbf{n})]_1 = \phi'' \cos \phi - (\phi')^2 \sin \phi + \frac{\phi' \cos \phi}{r} - \frac{\sin \phi}{r^2}, \qquad (5.287)$$

$$[\nabla \cdot (\nabla \mathbf{n})]_2 = -\phi'' \sin \phi - (\phi')^2 \cos \phi - \frac{\phi' \sin \phi}{r} - \frac{\cos \phi}{r^2}, \qquad (5.288)$$

$$[\nabla \cdot (\nabla \mathbf{n})]_3 = 0. \qquad (5.289)$$

By the results in equations (5.247) and (5.287) to (5.289), equations (5.283) can be replaced by the two equations

$$K \left(\phi'' \cos \phi - (\phi')^2 \sin \phi + \frac{\phi' \cos \phi}{r} - \frac{\sin \phi}{r^2} \right)$$
$$-\tfrac{1}{2}(\gamma_1 + \gamma_2) r \omega' \cos \phi = \lambda \sin \phi, \qquad (5.290)$$

$$K \left(-\phi'' \sin \phi - (\phi')^2 \cos \phi - \frac{\phi' \sin \phi}{r} - \frac{\cos \phi}{r^2} \right)$$
$$+\tfrac{1}{2}(\gamma_1 - \gamma_2) r \omega' \sin \phi = \lambda \cos \phi. \qquad (5.291)$$

The Lagrange multiplier λ can be eliminated from these two equations in a similar way to that used to obtain equation (5.250) earlier, resulting in the equation

$$K \left(\frac{d^2 \phi}{dr^2} + \frac{1}{r} \frac{d\phi}{dr} \right) - \frac{r}{2} \frac{d\omega}{dr} [\gamma_1 + \gamma_2 \cos(2\phi)] = 0. \qquad (5.292)$$

If desired, the Lagrange multiplier can be evaluated directly by taking the scalar product of equations (5.283) with $\mathbf{n}$ to find, with the aid of equations (5.247) and (5.287) to (5.289),

$$\lambda = -\gamma_2 r \omega' \sin \phi \cos \phi - K(\phi')^2 - Kr^{-2}. \qquad (5.293)$$

Equations (5.283) have now been reduced to the single equation (5.292) and it now remains to tackle the remaining dynamic equations (5.282).

Notice that in the nematic case we no longer have the identities (5.252) and (5.253). Some straightforward calculations show that from equations (4.121) and the expressions (5.245), (5.246), (4.123), (4.124) and the identity (5.254) we have

$$\tilde{t}_{rr} = A(\phi) r \omega', \qquad \tilde{t}_{r\theta} = \left\{ g(\phi) - \tfrac{1}{2} [\gamma_1 + \gamma_2 \cos(2\phi)] \right\} r \omega', \qquad (5.294)$$

$$\tilde{t}_{\theta r} = g(\phi) r \omega', \qquad \tilde{t}_{\theta\theta} = B(\phi) r \omega', \qquad (5.295)$$

where

$$A(\phi) = \tfrac{1}{2} \sin \phi \cos \phi \left(2\alpha_1 \sin^2 \phi + \alpha_2 + \alpha_3 + \alpha_5 + \alpha_6 \right), \qquad (5.296)$$

$$B(\phi) = \tfrac{1}{2} \sin \phi \cos \phi \left(2\alpha_1 \cos^2 \phi - \alpha_2 - \alpha_3 + \alpha_5 + \alpha_6 \right), \qquad (5.297)$$

$$g(\phi) = \tfrac{1}{2} \left[2\alpha_1 \sin^2 \phi \cos^2 \phi + (\alpha_5 - \alpha_2) \sin^2 \phi + (\alpha_6 + \alpha_3) \cos^2 \phi + \alpha_4 \right]. \qquad (5.298)$$

The viscous stress need not be symmetric in this nematic liquid crystal case, in contrast to that in equation (5.256) for the purely anisotropic case. The function

$g(\phi)$ is identical to that introduced earlier at equation (5.113), and it can be verified via equations (4.82), $(5.245)_1$, (5.246), (5.247), (5.294) and (5.295) that the inequality $g(\phi) \geq 0$ in equation (5.120) must hold again when the Parodi relation is assumed. It will henceforth be assumed that $g(\phi) > 0$ in order to simplify the analysis. It also follows from (5.247), (5.285) and (5.286) that (summing over the repeated index i)

$$\tilde{g}_i n_{i,r} = -\tfrac{1}{2} r \omega' \phi' \left[\gamma_1 + \gamma_2 \cos(2\phi) \right], \qquad (5.299)$$

$$\tilde{g}_i n_{i,\theta} = \tfrac{1}{2} \omega' \left[\gamma_1 + \gamma_2 \cos(2\phi) \right], \qquad (5.300)$$

$$\tilde{g}_i n_{i,z} = 0. \qquad (5.301)$$

Employing the results in equations (5.243), (5.299), (5.300) and the formulae in equations (C.3) and (C.18) to (C.20) (with $T_{ij} = \tilde{t}_{ij}$), equations (5.282) become

$$\frac{\partial \tilde{p}}{\partial r} = \frac{\partial \tilde{t}_{rr}}{\partial r} + \frac{1}{r}(\tilde{t}_{rr} - \tilde{t}_{\theta\theta}) + \rho r \omega^2 - \frac{1}{2} r \omega' \phi' \left[\gamma_1 + \gamma_2 \cos(2\phi) \right], \qquad (5.302)$$

$$\frac{1}{r} \frac{\partial \tilde{p}}{\partial \theta} = \frac{\partial \tilde{t}_{\theta r}}{\partial r} + \frac{1}{r}(\tilde{t}_{\theta r} + \tilde{t}_{r\theta}) + \frac{1}{2} \omega' \left[\gamma_1 + \gamma_2 \cos(2\phi) \right], \qquad (5.303)$$

$$\frac{\partial \tilde{p}}{\partial z} = 0. \qquad (5.304)$$

Using an argument similar to that contained in the previous example at equations (5.263) to (5.266), equations (5.302) to (5.304) reduce to the single equation

$$\frac{d\tilde{t}_{\theta r}}{dr} + \frac{1}{r}(\tilde{t}_{\theta r} + \tilde{t}_{r\theta}) + \frac{1}{2} \frac{d\omega}{dr} \left[\gamma_1 + \gamma_2 \cos(2\phi) \right] = 0, \qquad (5.305)$$

and an expression for $\tilde{p}$, namely,

$$\tilde{p} = H(r), \qquad (5.306)$$

where H satisfies the differential equation

$$\frac{dH}{dr} = \frac{d\tilde{t}_{rr}}{dr} + \frac{1}{r}(\tilde{t}_{rr} - \tilde{t}_{\theta\theta}) + \rho r \omega^2 - \frac{1}{2} r \frac{d\omega}{dr} \frac{d\phi}{dr} \left[\gamma_1 + \gamma_2 \cos(2\phi) \right]. \qquad (5.307)$$

Substituting the relevant expressions from (5.294) and (5.295) into (5.305) shows that the dynamic equation (5.305) finally reduces to

$$\frac{d}{dr} \left[g(\phi) r \frac{d\omega}{dr} \right] + 2 g(\phi) \frac{d\omega}{dr} = 0, \qquad (5.308)$$

and therefore

$$r^3 g(\phi) \frac{d\omega}{dr} = c, \qquad (5.309)$$

for some arbitrary constant c. Equation (5.309) then allows ω' to be eliminated from (5.292) and enables it to be written as

$$K \left(r^2 \frac{d^2 \phi}{dr^2} + r \frac{d\phi}{dr} \right) - \frac{1}{2} \frac{c}{g(\phi)} \left[\gamma_1 + \gamma_2 \cos(2\phi) \right] = 0. \qquad (5.310)$$

To determine the full solutions one must first find $\phi(r)$ from (5.310): this then permits an evaluation of $\omega(r)$ from equation (5.309). Equations (5.309) and (5.310) are the final reduced forms of the governing equations for $\phi(r)$ and $\omega(r)$ and are precisely those derived by Atkin and Leslie [6] when the one-constant approximation is made in the elastic energy (note that $\gamma_1 = -\lambda_1$ and $\gamma_2 = -\lambda_2$ in the notation of [6]).

It is instructive at this point to comment upon the constant c. The surface traction exerted by an outer cylinder upon the liquid crystal is given via equations (4.37), (4.128) and (4.129) with $\boldsymbol{\nu} = \mathbf{e}_r$ (notice that $t_{zr} = 0$ by equations (4.121), (5.245), (5.279), (5.284), (5.285) and (5.286)). Recalling that there are surface contributions arising from both the stress tensor (4.128) and couple stress tensor (4.129) (cf. the balance law (4.31)), the corresponding moment about the z-axis of such a cylinder of radius r is then, in cylindrical coordinates, with $t_i = t_{ij}\nu_j$ and $l_i = l_{ij}\nu_j$,

$$\mathbf{M} = (r\,\mathbf{e}_r \times \mathbf{t} + \mathbf{l}) = r\tilde{t}_{\theta r}\mathbf{e}_z, \qquad (5.311)$$

obtained after detailed calculations using (5.279) and the results (5.284) to (5.286). The total torque on such a cylinder of height L is then given by $\mathbf{T} = T_3\mathbf{e}_z$ where

$$T_3 = \int_0^L \int_0^{2\pi} M_3\, r\,d\theta dz = \int_0^L \int_0^{2\pi} r^2 \tilde{t}_{\theta r}\, d\theta dz = 2\pi L r^3 g(\phi)\omega' = 2\pi L c, \qquad (5.312)$$

where use has been made of the results (5.295)$_1$ and (5.309). A similar expression occurs for the inner cylinder. As noted by Atkin and Leslie [6], the constant c is therefore related to the magnitude of the moment per unit length of the cylinder. If c is zero then it follows from (5.309) that ω is a constant so that the motion, at most, corresponds to a uniform rigid body rotation. We shall suppose that $c \neq 0$ and note that $c > 0$ when $\Omega_2 > \Omega_1$ and $c < 0$ when $\Omega_2 < \Omega_1$.

Given the boundary conditions (5.281), it is easiest to search for solutions to (5.310) for which ϕ increases from zero at $r = R_1$ to a maximum value ϕ_m at the point r_m, and then decreases to zero at $r = R_2$, with $R_1 < r_m < R_2$. (An equally valid scenario for ϕ decreasing from zero at R_1 to a minimum at $r = r_m$ and then increasing to zero at $r = R_2$ is also possible, but will not be considered here.) Therefore, in addition to (5.281) we suppose

$$\phi'(r_m) = 0, \quad \text{with} \quad \phi(r_m) = \phi_m > 0. \qquad (5.313)$$

Equation (5.310) can be multiplied by ϕ' and then integrated to reveal, by application of the conditions (5.313),

$$r^2 K\left(\frac{d\phi}{dr}\right)^2 = -c\int_\phi^{\phi_m} \frac{1}{g(\psi)}\left[\gamma_1 + \gamma_2\cos(2\psi)\right] d\psi. \qquad (5.314)$$

Defining the angle ϕ_0 by

$$\cos(2\phi_0) = -\frac{\gamma_1}{\gamma_2}, \qquad (5.315)$$

we can rewrite (5.314) as

$$r^2 K\left(\frac{d\phi}{dr}\right)^2 = -c\gamma_2\int_\phi^{\phi_m} \frac{1}{g(\psi)}\left[\cos(2\psi) - \cos(2\phi_0)\right] d\psi. \qquad (5.316)$$

As at the beginning of Section 5.5.3, we shall assume here that $\gamma_1 \leq |\gamma_2|$ and in particular only consider the case for $\gamma_1 \geq 0$ and $\gamma_2 < 0$. These restrictions mean that ϕ_0 is restricted to the ranges (cf. the comments between equations (5.123) and (5.125) above)

$$0 \leq \phi_0 \leq \tfrac{\pi}{4} \quad \text{or} \quad \tfrac{3}{4}\pi \leq \phi_0 \leq \pi. \tag{5.317}$$

It is convenient to introduce the variable

$$t = \ln r, \tag{5.318}$$

and the constants

$$t_1 = \ln(R_1), \quad t_m = \ln(r_m), \quad t_2 = \ln(R_2). \tag{5.319}$$

Equation (5.316) then simplifies to

$$K \left(\frac{d\phi}{dt} \right)^2 = -c\gamma_2 \int_\phi^{\phi_m} \frac{1}{g(\psi)} \left[\cos(2\psi) - \cos(2\phi_0) \right] d\psi. \tag{5.320}$$

The right-hand side of the above equality must necessarily be non-negative for all positive values of ϕ between zero and ϕ_m, given the form of the left-hand side. Further assuming $c > 0$ forces the integral appearing in (5.320) to be non-negative, under the previously stated assumptions that $g(\phi) > 0$ and $\gamma_2 < 0$. This in turn restricts ϕ_0 to be limited to the range $0 \leq \phi_0 \leq \tfrac{\pi}{4}$ identified in (5.317)$_1$, *provided* $\phi_m \leq \phi_0$, which we shall suppose from now on in order to simplify the discussion (cf. Atkin and Leslie [6]). Equation (5.320) can now be put into the same form as equation (5.138), giving

$$\frac{d\phi}{dt} = \left[-c\frac{\gamma_2}{K} \int_\phi^{\phi_m} \frac{\cos(2\psi) - \cos(2\phi_0)}{g(\psi)} d\psi \right]^{\frac{1}{2}}, \quad 0 \leq \phi \leq \phi_m, \tag{5.321}$$

which allows the construction of a symmetric solution $\phi(t)$. The positive square root is taken in (5.321) since it has been supposed that $\phi_m > 0$ and ϕ is increasing on $t_1 \leq t \leq t_m$: the negative square root will then be applicable on the interval $t_m \leq t \leq t_2$ where ϕ will be decreasing. By application of the boundary condition (5.281)$_1$, the solution for $t_1 \leq t \leq t_m$ is given by

$$t - t_1 = \int_0^\phi \left[-c\frac{\gamma_2}{K} \int_\zeta^{\phi_m} \frac{\cos(2\psi) - \cos(2\phi_0)}{g(\psi)} d\psi \right]^{-\frac{1}{2}} d\zeta. \tag{5.322}$$

By the aforementioned symmetry, the solution for $t_m \leq t \leq t_2$ is given by

$$t_2 - t = \int_0^\phi \left[-c\frac{\gamma_2}{K} \int_\zeta^{\phi_m} \frac{\cos(2\psi) - \cos(2\phi_0)}{g(\psi)} d\psi \right]^{-\frac{1}{2}} d\zeta. \tag{5.323}$$

As observed by Atkin and Leslie [6], setting $t = t_m$ and $\phi(t_m) = \phi_m$ in equations (5.322) and (5.323) and adding these equations shows that ϕ_m is determined by the ratio of the radii R_1 and R_2 through the relationship

$$\ln \left(\frac{R_2}{R_1} \right) = t_2 - t_1 = 2 \int_0^{\phi_m} \left[-c\frac{\gamma_2}{K} \int_\zeta^{\phi_m} \frac{\cos(2\psi) - \cos(2\phi_0)}{g(\psi)} d\psi \right]^{-\frac{1}{2}} d\zeta, \tag{5.324}$$

while subtracting these same equations determines r_m as a function of R_1 and R_2 through the relation

$$r_m^2 = R_1 R_2. \tag{5.325}$$

Once $\phi(r)$ has been determined from equations (5.322) and (5.323), with ϕ_m obtained from (5.324), it is possible to determine the function $\omega(r)$ from equations (5.280) and (5.309) as

$$\omega(r) = \Omega_1 + c \int_{R_1}^r \frac{d\zeta}{\zeta^3 g(\phi(\zeta))}, \tag{5.326}$$

with, in particular,

$$\Omega_2 - \Omega_1 = c \int_{R_1}^{R_2} \frac{d\zeta}{\zeta^3 g(\phi(\zeta))}. \tag{5.327}$$

Equation (5.327) gives the relationship between the angular velocity difference $\Omega_2 - \Omega_1$ between the cylinders, the radii of the cylinders R_1 and R_2 and the magnitude of the applied torque (through the constant c via (5.312)). Clearly, for any difference in angular velocities there is always a suitable constant c which will allow the fulfilment of the relation (5.327) for some previously fixed values of R_1 and R_2. The pressure $p = \tilde{p} - \rho \Psi_g - w_F$ can be found by integrating equation (5.307) and using equations (5.284), (5.285) and (5.286) to find w_F to give

$$p = p_0 - \rho \Psi_g - \tfrac{1}{2} K \left[\left(\frac{d\phi}{dr} \right)^2 + \frac{1}{r^2} \right] + \int_{R_1}^r \frac{d}{ds} H(s) ds, \tag{5.328}$$

where p_0 is an arbitrary constant pressure, arising from the assumed incompressibility. Since we have made use of the formulations (5.299) and (5.300), it is worth mentioning that the result in equation (5.292) can be inserted into (5.302) to allow this form of the pressure to be compared and equated with that originally stated by Atkin and Leslie [6] obtained by an earlier description of the dynamic theory.

To summarise, the full solutions for the velocity and the director are provided by equations (5.278) and (5.279) where $\phi(r)$ is the solution provided via equations (5.318), (5.322), (5.323) and (5.324), and $\omega(r)$ is the solution given by equation (5.326), with the constant c (cf. page 206) appearing in these solutions satisfying the relation (5.327). The Lagrange multiplier λ is given by (5.293) and the pressure p by (5.328). As in the shear flow examples studied in Sections 5.5.3 and 5.5.4, the integrals appearing in these solutions in general have to be evaluated numerically.

A detailed and intricate scaling analysis for the Couette flow of nematics, similar in style to that contained in Section 5.5.5, has been carried out by Atkin and Leslie [6]. We do not pursue this aspect of the analysis in this text and refer the reader to Reference [6] for comments on possible experimentally determined quantities such as an apparent viscosity. As already mentioned, solutions for unequal elastic constants can also be found in [6]. A more extensive analysis of Couette flow incorporating the effects of an applied magnetic field has been provided by Currie [57], who also comments on other types of solutions which may be possible.

Pieranski and Guyon [223] have used a very basic application of dynamic theory to examine Couette flow instabilities for nematics under some more complicated profiles incorporating a perturbation to the flow which is dependent upon z. They found that when the director was anchored parallel to the cylinder walls in the θ-direction, as is the case considered here, Taylor instabilities can occur. It is also worth noting that Cladis and Torza [41] have performed experiments and investigated the stability of Couette flow for the nematics HBAB and CBOOA. In contrast to the homogeneous boundary conditions (5.281) imposed in the example considered in this Section, they considered samples with homeotropic boundary conditions where $\mathbf{n} = \mathbf{e}_r$ at the cylinder walls with $R_1 = 0.25$ mm and $R_2 = 0.5$ mm. Various phenomena were shown to occur experimentally that included tumbling and the appearance of a cellular secondary flow, resembling Taylor vortices in isotropic fluids, as the relative shear rate increased. The experimentally minded reader is referred to their article for more details.

5.8 Poiseuille Flow

Here we consider Poiseuille flow due to an applied pressure gradient down a cylindrical capillary tube of radius R as shown in Fig. 5.9. It is natural to choose the usual cylindrical coordinates as introduced above in the Couette flow examples in Section 5.7 with the z-axis coincident with the axis of the capillary. In Section 5.8.1

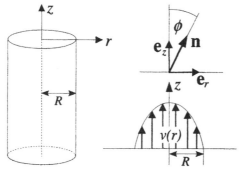

Figure 5.9: The geometrical set-up for Poiseuille flow due to an applied pressure gradient down a circular cylindrical capillary of radius R, described in the usual cylindrical coordinate system. The director makes an angle ϕ with respect to the local coordinate axes in the directions of $\mathbf{e}_r$ and $\mathbf{e}_z$ as shown. For an anisotropic fluid the velocity $v(r)$ in the z-direction is given by (5.351) and has an expected profile as indicated by the bold arrows in the lower diagram with $v(R) = 0$ at the boundary surface of the capillary. The flow profile for a nematic with strong anchoring is much more complex: see equation (5.389).

we shall first look at the anisotropic fluid case where the director gradients $n_{i,j}$ are absent and provide a complete analysis of some solutions discussed by Leslie [162]. Section 5.8.2 goes on to discuss aspects of the more difficult case for a nematic liquid crystal investigated theoretically by Atkin [5] and numerically by Tseng, Silver and Finlayson [271]. We also comment in Section 5.8.3 on an important result from the

work of Atkin [5] on this Poiseuille flow problem where a theoretical scaling analysis
was found to have been confirmed experimentally by Fisher and Fredrickson [89].

5.8.1   The Anisotropic Fluid Case

The Ericksen–Leslie theory from Section 4.2.5 will be used as before with all director
gradients and the elastic energy being set to zero. When the gravitational potential
Ψ_g is included, the resulting Cartesian component forms of the dynamic equations
from (4.139) and (4.140) with (4.141) are

$$\rho \dot{v}_i = -\tilde{p}_{,i} + \tilde{t}_{ij,j}, \qquad \tilde{p} = p + \rho\Psi_g, \tag{5.329}$$
$$\tilde{g}_i = \lambda n_i, \tag{5.330}$$

neglecting any generalised body force G_i. Solutions will be sought for the geometry
given in Fig. 5.9 where the forms for the velocity and the director are taken to be

$$v_r = 0, \quad v_\theta = 0, \quad v_z = v(r), \tag{5.331}$$
$$n_r = \sin\phi, \quad n_\theta = 0, \quad n_z = \cos\phi, \tag{5.332}$$

where ϕ is a function of r and $v(r)$ satisfies the common no-slip boundary condition

$$v(R) = 0. \tag{5.333}$$

Notice that the usual constraints (4.118) are satisfied.

Equations (5.329) and (5.330) will be expressed in the relevant physical components in cylindrical polar coordinates. As in the Couette flow examples above,
the velocity and director are independent of time t and therefore $\dot{\mathbf{v}} = (\mathbf{v} \cdot \nabla)\mathbf{v}$ and
$\dot{\mathbf{n}} = (\mathbf{v} \cdot \nabla)\mathbf{n}$, and by appropriate applications of the formula (C.4) in Appendix C
we have

$$\dot{\mathbf{v}} = \dot{\mathbf{n}} = \mathbf{0}. \tag{5.334}$$

The non-zero components of the rate of strain tensor $\mathbf{A}$ and vorticity tensor $\mathbf{W}$ are
obtained by applying the formulae (C.11) to (C.17) in Appendix C, leading to

$$A_{rz} = A_{zr} = \tfrac{1}{2}v', \qquad W_{rz} = -W_{zr} = -\tfrac{1}{2}v', \tag{5.335}$$

a prime denoting differentiation with respect to r. The co-rotational time flux is
given by equations (4.126), (5.332), (5.334) and (5.335) as

$$N_r = \tfrac{1}{2}v'\cos\phi, \quad N_\theta = 0, \quad N_z = -\tfrac{1}{2}v'\sin\phi, \tag{5.336}$$

which allows $\tilde{\mathbf{g}}$ in (4.122) to be given as

$$\tilde{g}_r = -\tfrac{1}{2}(\gamma_1 + \gamma_2)v'\cos\phi, \quad \tilde{g}_\theta = 0, \quad \tilde{g}_z = \tfrac{1}{2}(\gamma_1 - \gamma_2)v'\sin\phi. \tag{5.337}$$

Insertion of the above results for $\tilde{\mathbf{g}}$ into equation (5.330) reduces them to the
two equations

$$-\tfrac{1}{2}(\gamma_1 + \gamma_2)v'\cos\phi = \lambda\sin\phi, \tag{5.338}$$
$$\tfrac{1}{2}(\gamma_1 - \gamma_2)v'\sin\phi = \lambda\cos\phi. \tag{5.339}$$

Similar to equations (5.248) to (5.251), we can eliminate the Lagrange multiplier
to find that these equations finally reduce to

$$v' \left[\gamma_1 + \gamma_2 \cos(2\phi) \right] = 0, \tag{5.340}$$

with

$$\lambda = -\gamma_2 v' \sin\phi \cos\phi. \tag{5.341}$$

If we suppose that $v' \neq 0$, then the solution ϕ must be the constant value previously
identified in equation (5.252) when we considered Couette flow, namely,

$$\cos(2\phi) = -\frac{\gamma_1}{\gamma_2}. \tag{5.342}$$

The relations (5.253) will also hold in this case if we further suppose $\gamma_1 \leq |\gamma_2|$
with $\gamma_2 \neq 0$ (recall that $\gamma_1 \geq 0$ by (4.91)). Now that the Lagrange multiplier has
been determined and the equations (5.330) have been completely solved to find the
constant solution ϕ, it only remains to solve equations (5.329).

The non-zero components of the viscous stress can be identified from equations
(4.121), (5.332), (5.335) and (5.336), leading to

$$\tilde{t}_{rr} = Av', \quad \tilde{t}_{zz} = Bv', \quad \tilde{t}_{rz} = \tilde{t}_{zr} = Cv', \tag{5.343}$$

where the constants A, B and C are precisely those introduced in equations (5.257)
to (5.259) above (of course, these constants depend upon the constant solution ϕ),
the identity (5.255) being available and used again in this particular case to verify
that $\tilde{t}_{rz} = \tilde{t}_{zr}$. The physical components of $\tilde{t}_{ij,j}$ in cylindrical coordinates can be
determined from inserting the terms in equation (5.343) into formulae (C.18) to
(C.20). These components, together with the result in equation (5.334) for $\dot{\mathbf{v}}$ and
formula (C.3), lead to the remaining dynamic equations (5.329) being formulated
as

$$\frac{\partial \tilde{p}}{\partial r} = \frac{\partial \tilde{t}_{rr}}{\partial r} + \frac{\tilde{t}_{rr}}{r}, \tag{5.344}$$

$$\frac{1}{r}\frac{\partial \tilde{p}}{\partial \theta} = 0, \tag{5.345}$$

$$\frac{\partial \tilde{p}}{\partial z} = \frac{\partial \tilde{t}_{zr}}{\partial r} + \frac{\tilde{t}_{zr}}{r}. \tag{5.346}$$

From equation (5.345) we must have $\tilde{p} = \tilde{p}(r, z)$. Substituting (5.343) for the
viscous stress terms in the above equations, it is straightforward (for example, take
the derivatives of (5.344) and (5.346) with respect to z and r, respectively) to find
that we can set

$$\tilde{p} = -Pz + H(r), \tag{5.347}$$

for some constant P and a function $H(r)$ which satisfies

$$\frac{dH}{dr} = \frac{d\tilde{t}_{rr}}{dr} + \frac{\tilde{t}_{rr}}{r}. \tag{5.348}$$

The constant P is often referred to as the specific driving force of the flow [5]. Putting (5.347) into (5.346) shows that

$$v(r) = -\frac{Pr^2}{4C} + \frac{b}{C}\ln r + d, \qquad (5.349)$$

for some arbitrary constants b and d. An insertion of this result into (5.348) reveals that

$$H(r) = p_0 - \frac{PA}{C}r, \qquad (5.350)$$

for an arbitrary constant pressure contribution p_0. For a finite solution at $r = 0$ we must set $b = 0$. Applying the boundary condition (5.333) to (5.349) gives the constant $d = PR^2/4C$. Hence the solution for the velocity is

$$v(r) = \frac{P}{4C}\left(R^2 - r^2\right), \qquad (5.351)$$

with the pressure p given via $(5.329)_2$, (5.347) and (5.350) as

$$p = \tilde{p} - \rho\Psi_g = p_0 - \rho\Psi_g - Pz - PAr/C. \qquad (5.352)$$

The velocity distribution across the capillary is therefore parabolic.

The complete solution for this Poiseuille flow problem is provided by the solutions for ϕ and $v(r)$ obtained in (5.342) and (5.351). The required Lagrange multiplier λ and the pressure p are given by equations (5.341) and (5.352). The solution for $v(r)$ is of a form reminiscent of the result for Poiseuille flow of an incompressible isotropic Newtonian fluid in a circular capillary [129, p.189], the key difference being the dependence of the constant C on more than one viscosity coefficient.

As noted by Leslie [162], the velocity gradient is zero on the z-axis ($r = 0$), showing that ϕ cannot really be determined there by equations (5.330) (it was necessary to suppose $v' \neq 0$ in order to exploit the consequent equation (5.340) for the determination of ϕ). However, provided the director does not vary at any material point on the z-axis, the constant solution for ϕ is not altered by this indeterminacy.

There is also one special solution for Poiseuille flow mentioned by Leslie [162] when the director takes the form

$$n_r = 0, \quad n_\theta = 1, \quad n_z = 0, \qquad (5.353)$$

except, of course, on the z-axis where the director is again indeterminate. The velocity has the form previously supposed in equation (5.331). It is a simple exercise to verify that

$$\dot{n}_i = 0, \quad N_i = 0, \quad \tilde{g}_i = 0, \qquad (5.354)$$

and to use the results in (5.335) and (5.353) in (4.121) to see that the non-zero components of the viscous stress in this case are

$$\tilde{t}_{rz} = \tilde{t}_{zr} = \tfrac{1}{2}\alpha_4 v'. \qquad (5.355)$$

Equations (5.330) are therefore satisfied by setting $\lambda = 0$ and, using the results (C.3) and (C.18) to (C.20) in Appendix C, equations (5.329) reduce to

$$\frac{\partial \tilde{p}}{\partial r} = 0, \quad \frac{1}{r}\frac{\partial \tilde{p}}{\partial \theta} = 0, \quad \frac{\partial \tilde{p}}{\partial z} = \frac{1}{2}\alpha_4 \left(\frac{d^2 v}{dr^2} + \frac{1}{r}\frac{dv}{dr} \right), \tag{5.356}$$

since equation $(5.334)_1$ remains valid. Hence, by similar reasoning to that used to obtain (5.351) earlier, the boundary condition (5.333) yields

$$v(r) = \frac{P}{2\alpha_4}\left(R^2 - r^2\right), \tag{5.357}$$

where P is the constant specific driving force. The pressure p is given by

$$p = \tilde{p} - \rho\Psi_g = p_0 - \rho\Psi_g - Pz, \tag{5.358}$$

where p_0 is an arbitrary constant contribution to the pressure, arising from the assumed incompressibility. The complete solution for this special Poiseuille flow problem then consists of the solutions for $\mathbf{n}$ and $\mathbf{v}$ provided by equations (5.353), (5.331) and (5.357), with the Lagrange multiplier λ set to zero and the pressure p provided by the result in equation (5.358). The solution for $v(r)$ in (5.357) coincides with the well known result for Poiseuille flow of an incompressible isotropic Newtonian fluid in a circular capillary [129, p.189] where the Newtonian fluid viscosity is $\mu = \frac{1}{2}\alpha_4$.

5.8.2 The Nematic Liquid Crystal Case

Our attention now turns to Poiseuille flow of a nematic liquid crystal in a cylindrical capillary tube in the geometry of Fig. 5.9. For strong anchoring of the director at the cylinder walls the flow distribution is not generally as simple as that depicted in the Figure for an anisotropic fluid. We follow the work of Atkin [5] and simplify the discussion by adopting the one-constant approximation in the elastic energy. The main dynamic equations (5.387) and (5.388) below, special cases taken from [5], will be derived, but further analytical progress appears to be limited in the literature. For this reason, Tseng, Silver and Finlayson [271] solved these equations numerically; we briefly comment on their results and solutions below.

The velocity and director are now assumed to have the forms

$$v_r = 0, \quad v_\theta = 0, \quad v_z = v(r), \tag{5.359}$$
$$n_r = \sin\phi(r), \quad n_\theta = 0, \quad n_z = \cos\phi(r), \tag{5.360}$$

with the no-slip boundary condition

$$v(R) = 0, \tag{5.361}$$

and strong anchoring of the director at the cylindrical wall

$$\phi(R) = \phi_1, \tag{5.362}$$

for some constant angle ϕ_1. As before, $\dot{\mathbf{v}} = (\mathbf{v} \cdot \nabla)\mathbf{v} = \mathbf{0}$, by formula (C.4) in Appendix C with $\mathbf{n}$ set to $\mathbf{v}$. Neglecting any generalised body force G_i and adopting the one-constant approximation for the elastic energy, the Cartesian component forms of the dynamic equations follow from (4.120), (4.132) with (4.133), (4.136)and (4.137) and are

$$\tilde{p}_{,i} = \tilde{g}_j n_{j,i} + \tilde{t}_{ij,j}, \qquad \tilde{p} = p + \rho\Psi_g + w_F, \qquad (5.363)$$
$$Kn_{i,jj} + \tilde{g}_i = \lambda n_i, \qquad (5.364)$$

where

$$w_F = \tfrac{1}{2}Kn_{i,j}n_{i,j}, \quad K > 0. \qquad (5.365)$$

In a similar way to the results for Couette flow of a nematic in Section 5.7.2, it is easily verified that the expressions for $\dot{v}_i$, $\dot{n}_i$, A_{ij}, W_{ij}, N_i and $\tilde{g}_i$ are identical to those stated above in equations (5.334) to (5.337), except that ϕ will now turn out to be a function of r. The formulae (C.8) to (C.10) in Appendix C and the form for the director in (5.360) lead to the physical components of $[\nabla\mathbf{n}]_{ij} = n_{i,j}$ in cylindrical coordinates. The non-zero components are

$$[\nabla\mathbf{n}]_{11} = \phi'\cos\phi, \qquad [\nabla\mathbf{n}]_{22} = \frac{\sin\phi}{r}, \qquad [\nabla\mathbf{n}]_{31} = -\phi'\sin\phi, \qquad (5.366)$$

where a prime denotes differentiation with respect to r, as before. The divergence $n_{i,jj}$ is obtained by inserting $T = (\nabla\mathbf{n})$ into formulae (C.18) to (C.20), giving

$$[\nabla \cdot (\nabla\mathbf{n})]_1 = \phi''\cos\phi - (\phi')^2\sin\phi + \frac{\phi'\cos\phi}{r} - \frac{\sin\phi}{r^2}, \qquad (5.367)$$
$$[\nabla \cdot (\nabla\mathbf{n})]_2 = 0, \qquad (5.368)$$
$$[\nabla \cdot (\nabla\mathbf{n})]_3 = -\phi''\sin\phi - (\phi')^2\cos\phi - \frac{\phi'\sin\phi}{r}. \qquad (5.369)$$

Use of the results from (5.337) and (5.367) to (5.369) allows equations (5.364) to be replaced by the two equations

$$K\left(\phi''\cos\phi - (\phi')^2\sin\phi + \frac{\phi'\cos\phi}{r} - \frac{\sin\phi}{r^2}\right)$$
$$-\tfrac{1}{2}(\gamma_1 + \gamma_2)v'\cos\phi = \lambda\sin\phi, \qquad (5.370)$$

$$K\left(-\phi''\sin\phi - (\phi')^2\cos\phi - \frac{\phi'\sin\phi}{r}\right) + \tfrac{1}{2}(\gamma_1 - \gamma_2)v'\sin\phi = \lambda\cos\phi. \qquad (5.371)$$

Eliminating the Lagrange multiplier λ in a way similar to that used to obtain (5.250) reduces these equations to the single equation

$$K\left(\phi'' + \frac{\phi'}{r} - \frac{1}{r^2}\sin\phi\cos\phi\right) - \tfrac{1}{2}v'\left[\gamma_1 + \gamma_2\cos(2\phi)\right] = 0, \qquad (5.372)$$

which finally leads to

$$K\left(\frac{d^2\phi}{dr^2} + \frac{1}{r}\frac{d\phi}{dr} - \frac{1}{r^2}\sin\phi\cos\phi\right) - \frac{1}{2}\gamma_2\frac{dv}{dr}\left[\cos(2\phi) - \cos(2\phi_0)\right] = 0, \qquad (5.373)$$

where, as for Couette flow, we have introduced the constant angle ϕ_0 given by

$$\cos(2\phi_0) = -\frac{\gamma_1}{\gamma_2}, \tag{5.374}$$

under the previous assumption stated after equation (5.342), that $\gamma_1 \leq |\gamma_2| \neq 0$. If needed, the Lagrange multiplier is easily found by taking the scalar product of equation (5.364) with $\mathbf{n}$, to obtain

$$\lambda = -K(\phi')^2 - \frac{K}{r^2}\sin^2\phi - \gamma_2 v' \sin\phi\cos\phi. \tag{5.375}$$

We now return to the remaining dynamic equations (5.363). Straightforward computations show that the physical components of the viscous stress are given via equations (4.121), (4.123), (4.124), (5.335), (5.336), (5.360) and the identity (5.254) as

$$\tilde{t}_{rr} = A(\phi)v', \qquad \tilde{t}_{rz} = \left\{g(\phi) - \tfrac{1}{2}\left[\gamma_1 + \gamma_2\cos(2\phi)\right]\right\}v', \tag{5.376}$$
$$\tilde{t}_{zr} = g(\phi)v', \qquad \tilde{t}_{zz} = B(\phi)v', \tag{5.377}$$

where $A(\phi)$, $B(\phi)$ and $g(\phi)$ are as defined above in equations (5.296), (5.297) and (5.298). It will also be assumed, as for Couette flow, that $g(\phi) > 0$. (It can be shown that $g(\phi) \geq 0$ in a similar way to that stated after (5.298): see Atkin [5].) Using (5.337) and (5.366) we have, summing over the repeated index i,

$$\tilde{g}_i n_{i,r} = -\tfrac{1}{2}v'\phi'\left[\gamma_1 + \gamma_2\cos(2\phi)\right], \tag{5.378}$$
$$\tilde{g}_i n_{i,\theta} = 0, \qquad \tilde{g}_i n_{i,z} = 0. \tag{5.379}$$

The results from equations (5.376) to (5.379) can be used with the formulae (C.3) and (C.18) to (C.20) to show that in cylindrical coordinates the dynamic equations (5.363) are

$$\frac{\partial \tilde{p}}{\partial r} = \frac{\partial \tilde{t}_{rr}}{\partial r} + \frac{\tilde{t}_{rr}}{r} - \tfrac{1}{2}v'\phi'\left[\gamma_1 + \gamma_2\cos(2\phi)\right], \tag{5.380}$$

$$\frac{1}{r}\frac{\partial \tilde{p}}{\partial \theta} = 0, \tag{5.381}$$

$$\frac{\partial \tilde{p}}{\partial z} = \frac{\partial \tilde{t}_{zr}}{\partial r} + \frac{\tilde{t}_{zr}}{r}. \tag{5.382}$$

By the analogues of the results in equations (5.344) to (5.348) we have

$$\tilde{p} = -Pz + H(r), \tag{5.383}$$

where

$$\frac{dH}{dr} = \frac{d\tilde{t}_{rr}}{dr} + \frac{\tilde{t}_{rr}}{r} - \tfrac{1}{2}v'\phi'\left[\gamma_1 + \gamma_2\cos(2\phi)\right]. \tag{5.384}$$

The constant P is the specific driving force of the flow [5]. The pressure $p = \tilde{p} - w_F - \rho\Psi_g$ can be determined via the integral of equation (5.384). By making use of the quantities (5.366) in the expression (5.365) for w_F we can identify the pressure as

$$p = p_0 - Pz - \rho\Psi_g - \tfrac{1}{2}K\left[\left(\frac{d\phi}{dr}\right)^2 + \frac{\sin^2\phi}{r^2}\right] + \int^r \frac{d}{ds}H(s)\,ds, \qquad (5.385)$$

where p_0 is an arbitrary constant pressure. This expression for the pressure coincides with that stated by Atkin [5] when (5.372) is used to substitute for the last term in brackets on the right-hand side (5.384), the resulting form for H then being inserted into (5.385). Equations (5.380) to (5.382) can now be reduced to the above expression for the pressure p and the consequent equation arising from (5.382), namely,

$$-P = \frac{d}{dr}\left[g(\phi)\frac{dv}{dr}\right] + \frac{1}{r}g(\phi)\frac{dv}{dr}, \qquad (5.386)$$

which integrates to give

$$g(\phi)\frac{dv}{dr} = \frac{b}{r} - \frac{1}{2}Pr, \qquad (5.387)$$

where b is a constant of integration. It is required that dv/dr be finite at $r = 0$ and therefore we set $b = 0$. The result (5.387) can then be used to substitute for dv/dr in equation (5.373) to arrive finally at the governing dynamic equation for ϕ

$$K\left(\frac{d^2\phi}{dr^2} + \frac{1}{r}\frac{d\phi}{dr} - \frac{1}{r^2}\sin\phi\cos\phi\right) + \gamma_2\frac{Pr}{4g(\phi)}\left[\cos(2\phi) - \cos(2\phi_0)\right] = 0, \qquad (5.388)$$

this being the one-constant elastic energy approximation version of the equation derived by Atkin [5] (where $\lambda_2 = -\gamma_2$) when b is set to zero.

The full solution to this Poiseuille flow problem for a nematic is provided by equations (5.387) and (5.388). Once $\phi(r)$ has been determined from (5.388) it can be inserted into (5.387) with $b = 0$ to obtain $v(r)$ by the integral

$$v(r) = \frac{1}{2}\int_r^R \frac{Ps}{g(\phi(s))}\,ds, \qquad (5.389)$$

where the boundary condition (5.361) has been applied.

Further theoretical analysis of the main equations (5.387) and (5.388) appears to be limited in scope. Atkin [5] went on to discuss the existence properties for solutions $\phi(r)$ to (5.388) for this problem. Tseng, Silver and Finlayson [271] continued the investigation of these equations numerically using the boundary conditions

$$\phi(0) = 0, \qquad \phi(R) = -\frac{\pi}{2}, \qquad (5.390)$$

where the requirement $\phi(0) = 0$ has to be used in order to avoid singularities and the director is strongly anchored perpendicular to the cylinder wall (homeotropic boundary alignment). They produced plots of $v(r)$ and $\phi(r)$ for the nematic PAA at 122°C which show that in general most of the reorientation of ϕ occurs near the cylindrical boundary.

Atkin [5] also discussed Poiseuille flow between two coaxial cylinders of radii R_1, R_2 with $0 < R_1 < R_2$. In this case equations (5.387) and (5.388) remain valid, the only difference being that, in general, the constant b appearing in (5.387) need not be zero. Existence and uniqueness properties were discussed by Atkin for this problem also, due to the apparent analytic intractability of these equations to yield an explicit solution.

5.8.3   Results from a Scaling Analysis

Atkin [5] adapted the techniques for finding scaling properties presented in Section 5.5.5 to Poiseuille flow of a nematic through a capillary of radius R under a constant pressure gradient P. For brevity, we do not derive these results here, but rather state some conclusions for reference and refer the reader to [5] for details. Let Q denote the efflux per unit time, that is, the volume of fluid traversing a cross-section of the capillary per unit time, also called the volume flow rate of fluid discharge. (Note that the mass flow rate is generally the volume flow rate multiplied by the density ρ.) The apparent viscosity is defined by

$$\eta = \pi \frac{PR^4}{8Q}, \qquad (5.391)$$

motivated by the Hagen–Poiseuille law of classical viscous fluids [129, p.189], [168, p.39]. Atkin found that Q and η must satisfy the functional relationships

$$Q = R\mathcal{G}(PR^3), \qquad \eta = \mathcal{H}(QR^{-1}), \qquad (5.392)$$

where the functions $\mathcal{G}$ and $\mathcal{H}$ are unknown, the result in $(5.392)_2$ being a consequence of (5.391) and $(5.392)_1$. The results in (5.392) are commonly called the Atkin scaling laws. The properties (5.392) are in contrast to similar relationships for general isotropic viscoelastic fluids where it is known that for Poiseuille flow the efflux and viscosity obey the relations [50]

$$Q = R^3\widehat{\mathcal{G}}(PR), \qquad \eta = \widehat{\mathcal{H}}(QR^{-3}), \qquad (5.393)$$

where $\widehat{\mathcal{G}}$ and $\widehat{\mathcal{H}}$ are again unknown functions.

The theoretical predictions given by (5.392) were confirmed by the experiments of Fisher and Fredrickson [89], who, motivated by the work of Atkin, replotted their data for η as a function of $4Q/\pi R$ for perpendicular (homeotropic) boundary alignment of the director in samples of the nematic PAA. They found that their data for η for different diameters of capillary fitted a single curve [89, Fig.7] in excellent agreement with the above predictions resulting from a scaling analysis. This was the first experimental confirmation of the continuum theory developed by Ericksen and Leslie since it demonstrated that the scaling predicted by equation (5.392) justifies many of the special assumptions made in deriving the constitutive relations of the theory. Also, the usual strong anchoring assumption for the director at the cylindrical surface of the capillary was shown to be justified. The implications and importance of these results are discussed in finer detail in the review by Leslie [168].

5.9   Dynamics of the Freedericksz Transition

The Freedericksz transitions discussed in Section 3.4.1 are simplified from the point of view of statics. We now turn our attention to the dynamics of these situations. As the magnitude of the applied magnetic field is changed, fluid flow is known to be less important in the twist geometry Freedericksz transition, but it is important in the splay and bend geometries. The 'switch-on' and 'switch-off' times, when available, will be defined as appropriate for Freedericksz transitions in Section 5.9.1. These characteristic times are easily identified in the twist geometry when flow is absent, but are more complex in the splay and bend geometries where the flow behaviour leads to what are commonly called the backflow and kickback effects; backflow turns out to be particularly influential in the bend geometry. The phenomena of backflow and kickback will be introduced and discussed in Sections 5.9.2 and 5.9.3 below.

5.9.1   Dynamics in the Twist Geometry

We now consider the dynamics of a simple Freedericksz transition in the twist geometry of Fig. 3.7 on page 77, in the absence of fluid flow. From the physical point of view, there is no hydrodynamical flow because the director, representing the average molecular alignment, can rotate within the xy-plane without any translational motion in this pure twist geometry [137, p.110]. The same form for the director given by (3.134) is adopted again, except that ϕ now depends on time t also. Thus we suppose that $\mathbf{n}$ and $\mathbf{H}$ are given by

$$\mathbf{n} = (\cos\phi(z,t), \sin\phi(z,t), 0), \tag{5.394}$$

$$\mathbf{H} = H(0,1,0), \quad H = |\mathbf{H}|, \tag{5.395}$$

with the strong anchoring boundary conditions matching the analogues of the static boundary conditions (3.105), namely,

$$\phi(0,t) = \phi(d,t) = 0, \quad t \geq 0, \tag{5.396}$$

other notation being as introduced when discussing Freedericksz transitions in Section 3.4.1. For illustrative purposes the aim is to use the reformulated dynamic equations from Section 4.3, as these are actually easier for calculations when $\mathbf{n}$ takes the form in (5.394). In the notation of Section 4.3, the nematic elastic energy density (2.49) in terms of ϕ is simply given by

$$\widehat{w}_F = \tfrac{1}{2}K_2\left(\frac{\partial\phi}{\partial z}\right)^2 \tag{5.397}$$

Since the fluid velocity is absent, $\nabla\cdot\mathbf{v} = 0$ and the constraints on the director and the velocity in (4.118) are satisfied. The rate of strain tensor A_{ij} and the vorticity tensor W_{ij} are both identically zero. The dissipation function in equation (4.85) is therefore given via (4.126) by

$$\mathcal{D} = \gamma_1 N_i N_i = \gamma_1 \frac{\partial n_i}{\partial t}\frac{\partial n_i}{\partial t} = \gamma_1\left(\frac{\partial\phi}{\partial t}\right)^2, \tag{5.398}$$

which leads by $(4.146)_2$ to

$$\widehat{\mathcal{D}} = \tfrac{1}{2}\gamma_1 \left(\frac{\partial \phi}{\partial t}\right)^2, \qquad (5.399)$$

while the magnetic potential, in a cgs units description obtained from equations (2.106) and (2.175), is given by

$$\widehat{\Psi}_m = -w_{mag} = \tfrac{1}{2}\chi_{m\perp}H^2 + \tfrac{1}{2}\chi_a(\mathbf{n} \cdot \mathbf{H})^2 = \tfrac{1}{2}\chi_{m\perp}H^2 + \tfrac{1}{2}\chi_a H^2 \sin^2\phi, \qquad (5.400)$$

where it is assumed that $\chi_a > 0$. We shall neglect the gravitational potential and set $\widehat{\Psi} = \widehat{\Psi}_m$. For the single orientation angle case of equations (4.152), (4.153) and (4.154) we have the balance of linear momentum and angular momentum given by, respectively,

$$\tilde{p}_{,i} = -\frac{\partial \widehat{\mathcal{D}}}{\partial \dot{\phi}}\,\phi_{,i}, \qquad \tilde{p} = p + \widehat{w}_F - \widehat{\Psi}_m, \qquad (5.401)$$

$$\left(\frac{\partial \widehat{w}_F}{\partial \phi_{,z}}\right)_{,z} - \frac{\partial \widehat{w}_F}{\partial \phi} - \frac{\partial \widehat{\mathcal{D}}}{\partial \dot{\phi}} + \frac{\partial \widehat{\Psi}_m}{\partial \phi} = 0, \qquad (5.402)$$

where the superposed dot in this present example now represents the partial derivative with respect to time (in the absence of flow the material time derivative coincides with the partial time derivative) and the circumflexes denote quantities in terms of ϕ and its derivatives. Equations $(5.401)_1$ are simply

$$\tilde{p}_{,x} = \tilde{p}_{,y} = 0, \quad \tilde{p}_{,z} = -\gamma_1 \frac{\partial \phi}{\partial t}\frac{\partial \phi}{\partial z}. \qquad (5.403)$$

The arbitrary pressure p may then take the form, using $(5.401)_2$,

$$p(z,t) = \widehat{\Psi}_m - \widehat{w}_F - \gamma_1 \int \frac{\partial \phi}{\partial t}\frac{\partial \phi}{\partial z}\,dz, \qquad (5.404)$$

where gravity has been neglected, as mentioned above. This form for the pressure p allows a complete solution of the balance of linear momentum equations (5.401) once $\phi(z,t)$ has been obtained: it only remains to find $\phi(z,t)$ as a solution to the balance of angular momentum equation (5.402).

Inserting the expressions (5.397), (5.399) and (5.400) into (5.402) then gives the main governing dynamic equation

$$\gamma_1 \frac{\partial \phi}{\partial t} = K_2 \frac{\partial^2 \phi}{\partial z^2} + \chi_a H^2 \sin\phi \cos\phi, \qquad (5.405)$$

where we shall assume that $\gamma_1 > 0$ (cf. the general requirement (4.91)). Recall that the Freedericksz threshold for the onset of a Freedericksz transition in this geometry is given by

$$H_c = \frac{\pi}{d}\sqrt{\frac{K_2}{\chi_a}}, \qquad (5.406)$$

obtained earlier at equation (3.140) by an application of static theory in Section 3.4.1.

Now imagine a situation where the magnetic field is abruptly switched on with a magnitude H greater than the critical threshold H_c, the initial uniform alignment of $\phi \equiv 0$ becoming unstable. To find the switch-on time, to be defined below, we proceed as follows. Suppose, in the initial stage of the dynamics, that ϕ is small and satisfies the non-zero initial condition

$$\phi(z,0) = \phi_0(z), \quad |\phi_0(z)| \ll 1, \quad \phi_0(0) = \phi_0(d) = 0, \tag{5.407}$$

in addition to the boundary conditions (5.396). Linearising equation (5.405) around the zero state gives the perturbation equation for $\phi(z,t)$ as

$$\gamma_1 \frac{\partial \phi}{\partial t} = K_2 \frac{\partial^2 \phi}{\partial z^2} + \chi_a H^2 \phi. \tag{5.408}$$

This equation can be put into the more convenient form

$$\frac{\partial \phi}{\partial \tau} = \frac{\partial^2 \phi}{\partial z^2} + c\phi, \tag{5.409}$$

where

$$\tau = \frac{K_2}{\gamma_1} t \quad \text{and} \quad c = \chi_a \frac{H^2}{K_2}. \tag{5.410}$$

Using the canonical transformation

$$\phi(z,\tau) = \Phi(z,\tau)e^{c\tau}, \tag{5.411}$$

equation (5.409) and its associated boundary and initial conditions (5.396) and (5.407) become

$$\frac{\partial \Phi}{\partial \tau} = \frac{\partial^2 \Phi}{\partial z^2}, \tag{5.412}$$

$$\Phi(z,0) = \phi_0(z), \quad 0 \le z \le d, \tag{5.413}$$

$$\Phi(0,\tau) = \Phi(d,\tau) = 0, \quad \tau \ge 0. \tag{5.414}$$

By the standard separation of variables technique the solution for $\Phi(z,\tau)$ is given by

$$\Phi(z,\tau) = \sum_{n=1}^{\infty} A_n \sin\left(\frac{n\pi z}{d}\right) \exp\left[-\left(\frac{n\pi}{d}\right)^2 \tau\right], \tag{5.415}$$

where the constant coefficients A_n are the usual half-range Fourier sine series coefficients for ϕ_0, since setting $\tau = 0$ in the above series should deliver $\phi_0(z)$. Therefore

$$A_n = \frac{2}{d} \int_0^d \phi_0(z) \sin\left(\frac{n\pi z}{d}\right) dz. \tag{5.416}$$

Returning to the original variables via (5.410) and (5.411), and using the definition of H_c in (5.406), provides the solution

$$\phi(z,t) = \sum_{n=1}^{\infty} A_n \sin\left(\frac{n\pi z}{d}\right) \exp\left(-\frac{t}{\tau_n}\right), \tag{5.417}$$

where

$$\tau_n = \frac{\gamma_1}{\chi_a(n^2 H_c^2 - H^2)}. \tag{5.418}$$

The first point to notice is that $\tau_n > 0$ for all n when $H < H_c$, indicating that all modes in the series solution decay exponentially with time so that the solution $\phi \equiv 0$ is linearly stable. This result coincides with the conclusion that would be reached by a stability result involving an energy argument for the occurrence of a Freedericksz transition that measures the twist constant K_2: from the comments after equation (3.135), it is analogous to the case discussed in Section 3.4.1 at equations (3.128) to (3.133) for the splay constant K_1. However, τ_1 is certainly negative for H above H_c, indicating the onset of an instability: the first mode in the series solution (5.417) will experience exponential growth before the other modes if H were to be gradually increased above H_c. It is for this reason that for any $H > H_c$ we call $-\tau_1$ the *switch-on time* and denote it by τ_{on}. Thus

$$\tau_{on} = \frac{\gamma_1}{\chi_a(H^2 - H_c^2)} = \frac{\gamma_1}{\chi_a H^2 - K_2 \pi^2/d^2}. \tag{5.419}$$

This switch-on time is also called the reaction time or rise time by some authors and is one measure of the time taken for the reorientation of the director $\mathbf{n}$ to take place. Of course, for higher magnitudes of $\mathbf{H}$ other modes will become unstable. Actually, if $nH_c < H < (n+1)H_c$ then the first n modes are unstable and the sum of these modes approximates the growth of the perturbation with time after the magnetic field is switched on at a given magnitude greater than H_c.

There is an obvious analogous result for electric fields using the simple identification $\chi_a H^2 \leftrightarrow \epsilon_0 \epsilon_a E^2$ introduced earlier with H_c replaced by $E_c = \frac{\pi}{d}\sqrt{\frac{K_2}{\epsilon_0 \epsilon_a}}$. The voltage is $V = Ed$ and we have for $E > E_c$

$$\tau_{on} = \frac{\gamma_1}{\epsilon_0 \epsilon_a(E^2 - E_c^2)} = \frac{\gamma_1 d^2}{\epsilon_0 \epsilon_a(V^2 - V_c^2)} = \frac{\gamma_1 d^2}{K_2 \pi^2(V^2/V_c^2 - 1)}, \tag{5.420}$$

where $V_c = E_c d$ and $\epsilon_a > 0$. Equivalent forms for τ_{on} are presented in (5.420) for ease of comparison with the results of various authors. This result for τ_{on} may be used as a good approximation to the switch-on time in this type of problem, bearing in mind the comments on the more complicated behaviour of electric field effects in liquid crystals mentioned in Section 3.5 above.

The switch-off time can be determined by similar reasoning. Initially, it is assumed that $H > H_c$ and that the post-Freedericksz transition static solution $\phi_0(z)$ is known, for example via the results in Section 3.4.1. From equation (5.405) (or (3.136)) this post-threshold static solution for $H > H_c$ satisfies

$$K_2 \frac{d^2\phi_0}{dz^2} + \chi_a H^2 \sin\phi_0 \cos\phi_0 = 0, \qquad \phi_0(0) = \phi_0(d) = 0, \qquad \phi_0(\tfrac{d}{2}) = \phi_m > 0, \tag{5.421}$$

and clearly depends on the fixed value of H. It is then imagined that the field is suddenly switched off to zero so that $H = 0$ for $t > 0$. The consequent dynamic equation for $\phi(z, t)$ for $t > 0$ is then obtained from equation (5.405) by setting

$H = 0$ for $t > 0$, giving

$$\gamma_1 \frac{\partial \phi}{\partial t} = K_2 \frac{\partial^2 \phi}{\partial z^2}, \tag{5.422}$$

with the boundary conditions (5.396) holding as before, but now with the initial condition $\phi(z, 0) = \phi_0(z)$ which, of course, depends on the initial fixed value of H at $t = 0$ before the field is switched off. The full solution to (5.422) is obtained by the same method used to obtain the solution (5.417) above. It is exactly the same as (5.417) except that the half-range Fourier coefficients A_n are now given by (5.416) with $\phi_0(z)$ being the static solution for $H > H_c$, while, with the aid of (5.406) and (5.418) (with $H = 0$), the constants τ_n are replaced by

$$\tau_n = \frac{\gamma_1}{K_2} \left(\frac{d}{n\pi} \right)^2. \tag{5.423}$$

Clearly, the largest valued τ_n is linked to the slowest exponentially decaying mode in the solution (5.417). The *switch-off time* τ_{off} is defined to be the largest valued τ_n and is equal to τ_1. Therefore

$$\tau_{off} = \frac{\gamma_1}{K_2} \left(\frac{d}{\pi} \right)^2. \tag{5.424}$$

The switch-off time is also called the decay time or relaxation time by some authors. The analogous result for electric fields yields the same value for τ_{off}; the full solution $\phi(z, t)$ for $t > 0$ is of course modified and is obtained via (5.416), (5.417) and (5.423) by replacing ϕ_0 with the analogous static solution for the electric field. In many practical situations, such as that to be encountered below when we discuss the kickback effect, it is often convenient and appropriate to approximate $\phi(z)$ by a fixed constant when $H \gg H_c$. The form in equation (5.424) for τ_{off} is common for many relaxation processes in nematic liquid crystals when electromagnetic fields are set to zero. For example, the following values

$$d = 10 \ \mu m, \quad \gamma_1 = 10^{-1} \ Pa \ s, \quad K_2 = 10^{-11} \ N, \tag{5.425}$$

give $\tau_{off} \approx 10^{-1}$ s, this result being valid when either an electric or magnetic field is switched off. The switch-on time for suitable material parameters can also be calculated by using the expressions (5.419) and (5.420) and data from Table D.3 as appropriate, provided of course that either $\chi_a > 0$ in the magnetic field problem or $\epsilon_a > 0$ for the electric field problem.

The results derived here have been obtained by a brief linear analysis. A full nonlinear analysis would generally require a numerical approach or nonlinear approximations for the sinusoidal terms in the governing equation (5.405), especially for the long-time behaviour: see Pieranski, Brochard and Guyon [220], de Gennes and Prost [110] and Chandrasekhar [38] for more details. Problems involving negative magnetic or dielectric anisotropy can be tackled by a similar mathematical analysis for suitable geometries, but we do not pursue such aspects here.

It is not proposed to derive the switch-on time for a twisted nematic device in this text. However, it should be mentioned that in more complex geometries, such as the twisted nematic device discussed in Section 3.7, results analogous to (5.420) and (5.424) can be derived: see the review by Tarumi and Heckmeier [268, p.593] where it may be found that for the twisted nematic device

$$\tau_{on} \propto \frac{\gamma_1 d^2}{K_{eff}(V^2/V_c^2 - 1)}, \quad \tau_{off} \propto \frac{\gamma_1 d^2}{K_{eff}}, \quad K_{eff} = K_1 + (K_3 - 2K_2)/4. \quad (5.426)$$

Other definitions of switch-on and switch-off times are possible and are often quoted in the literature on measurements from experimental data. However, they are generally related to the above definitions.

5.9.2 Backflow and Kickback in the Splay Geometry

It has been shown in earlier Sections that flow can have a dramatic effect upon the director orientation in liquid crystals. Another important feature of liquid crystals is that the reorientation of the director itself can induce flow. In this Section we attempt to investigate some of the more intricate effects of the coupling between flow and the director orientation and follow the work of Pieranski, Brochard and Guyon [220] and Clark and Leslie [45]. Pieranski *et al.* showed that in both the bend and splay geometries the Freedericksz transition induces flow which influences the switching process. (As mentioned in the opening paragraph of Section 5.9.1, the influence of flow in the twist geometry is often considered negligible.) In general, the induced flow at the onset of the Freedericksz transition enhances the switch-on time by reducing the effective viscosity and thereby reducing the value of τ_{on}, such a flow being called *backflow*. It will be shown that the value of τ_{on} is similar in form to that given by (5.419) with γ_1 replaced by an appropriate reduced viscosity γ_1^*.

We first look at the effect of backflow when the Freedericksz transition occurs in the 'splay' planar to homeotropic transition geometry of Fig. 3.6 on page 73, where a magnetic field is abruptly switched on with a magnitude greater than H_c, employing the notation introduced at equations (3.103), (3.104) and (3.126). We then proceed to examine the effect in the same geometry when a field much greater than H_c is suddenly removed. In this case it will be shown that the director does not immediately relax with a decay rate governed by a result similar to the switch-off time discussed in the twist geometry above: the relaxation is much more complex because the removal of such a field leads to the fluid moving one way and then reversing direction before decreasing to zero as time progresses. This form of backflow is accompanied by what is called the *kickback effect*, which has an unusual and novel influence upon the orientation of the director. In the kickback effect the coupled interplay between the backflow and the director orientation causes the director, which for $H \gg H_c$ is largely oriented parallel to the applied field at $\theta = \frac{\pi}{2}$, to overshoot to an angle greater than $\frac{\pi}{2}$ upon removal of the field before gradually relaxing and returning to the zero field planar alignment of $\theta \equiv 0$ depicted in Fig. 3.6(a): this overshooting phenomenon is called the kickback effect.

Backflow in the Planar to Homeotropic Transition Splay Geometry

It should first be noted that if flow is considered unimportant (for example, this may be the case for some materials when H is very near H_c) then the switch-on time can be derived analogously to that for the twist geometry given by (5.419), the result being

$$\tau_{on} = \frac{\gamma_1}{\chi_a(H^2 - H_c^2)} = \frac{\gamma_1}{\chi_a H^2 - K_1\pi^2/d^2}, \tag{5.427}$$

where $\chi_a > 0$ and H_c is given by

$$H_c = \frac{\pi}{d}\sqrt{\frac{K_1}{\chi_a}}. \tag{5.428}$$

This result for τ_{on} is reasonably accurate on physical grounds, as pointed out by Chandrasekhar [38, pp.163–165]. However, to illustrate the general procedure, we incorporate the effects of flow and follow Pieranski *et al.* [220]. This procedure allows us to investigate possible effects when fields are switched on or off. The method is virtually identical to that needed for the alternative 'bend' homeotropic to planar transition considered in the next Section, where it is known that backflow has a greater influence on the switch-on time.

Motivated by the set-up in Fig. 3.6 and following Pieranski *et al.*, consider

$$\mathbf{n} = (\cos\theta(z,t), 0, \sin\theta(z,t)), \tag{5.429}$$

$$\mathbf{v} = (v(z,t), 0, 0), \tag{5.430}$$

$$\mathbf{H} = H(0,0,1). \tag{5.431}$$

The nematic elastic energy (2.49) is given by

$$\widehat{w}_F = \frac{1}{2}(K_1\cos^2\theta + K_3\sin^2\theta)\left(\frac{\partial\theta}{\partial z}\right)^2. \tag{5.432}$$

Routine calculations reveal that the non-zero components of the rate of strain tensor $\mathbf{A}$ in (4.125) and vorticity tensor $\mathbf{W}$ in (4.126)$_2$ are

$$A_{13} = A_{31} = \frac{1}{2}\frac{\partial v}{\partial z}, \qquad W_{13} = -W_{31} = \frac{1}{2}\frac{\partial v}{\partial z}, \tag{5.433}$$

and so, by equation (4.126)$_1$,

$$N_1 = -\frac{\partial\theta}{\partial t}\sin\theta - \frac{1}{2}\frac{\partial v}{\partial z}\sin\theta, \tag{5.434}$$

$$N_2 = 0, \tag{5.435}$$

$$N_3 = \frac{\partial\theta}{\partial t}\cos\theta + \frac{1}{2}\frac{\partial v}{\partial z}\cos\theta. \tag{5.436}$$

Using these quantities, the dissipation function is given via (4.85), (4.123), (4.124) and (4.146)$_2$ by

$$\widehat{\mathcal{D}} = \frac{1}{2}g(\theta)\left(\frac{\partial v}{\partial z}\right)^2 + m(\theta)\frac{\partial\theta}{\partial t}\frac{\partial v}{\partial z} + \frac{1}{2}\gamma_1\left(\frac{\partial\theta}{\partial t}\right)^2, \tag{5.437}$$

where we have used the Parodi relation (4.127) to substitute for γ_2 as appropriate, and we have adopted the notation, slightly modified from Clark and Leslie [45],

$$m(\theta) = \alpha_3 \cos^2\theta - \alpha_2 \sin^2\theta, \tag{5.438}$$
$$g(\theta) = \tfrac{1}{2}\left[\alpha_4 + (\alpha_5 - \alpha_2)\sin^2\theta + (\alpha_3 + \alpha_6)\cos^2\theta\right] + \alpha_1 \sin^2\theta \cos^2\theta. \tag{5.439}$$

The expression for $\widehat{\mathcal{D}}$ is a quadratic form and can be written as

$$\widehat{\mathcal{D}} = \begin{bmatrix} X & Y \end{bmatrix} \begin{bmatrix} \tfrac{1}{2}g(\theta) & \tfrac{1}{2}m(\theta) \\ \tfrac{1}{2}m(\theta) & \tfrac{1}{2}\gamma_1 \end{bmatrix} \begin{bmatrix} X \\ Y \end{bmatrix}, \qquad \text{with} \qquad X = \frac{\partial v}{\partial z}, \quad Y = \frac{\partial \theta}{\partial t}. \tag{5.440}$$

The non-negativity of the dissipation function then requires that this quadratic form is positive semidefinite. Following the comments after (2.56), it is seen that the inequalities

$$g(\theta) \geq 0 \qquad \text{and} \qquad \gamma_1 g(\theta) - m^2(\theta) \geq 0, \tag{5.441}$$

must be fulfilled. The related magnetic potential for $\chi_a > 0$ is given analogously to (5.400) and is

$$\widehat{\Psi}_m = \tfrac{1}{2}\chi_\perp H^2 + \tfrac{1}{2}\chi_a H^2 \sin^2\theta. \tag{5.442}$$

Any gravitational potential will be neglected in this Section.

Notice that the constraints (4.118) are automatically satisfied. The remaining governing dynamic equations are derived from (4.152) and (4.153) for the case of a single orientational angle formulation using the above expressions for $\widehat{w}_F$, $\widehat{\mathcal{D}}$ and $\widehat{\Psi}_m$. The balance of angular momentum (4.152) in this example is

$$\frac{\partial}{\partial z}\left(\frac{\partial \widehat{w}_F}{\partial \theta_{,z}}\right) - \frac{\partial \widehat{w}_F}{\partial \theta} - \frac{\partial \widehat{\mathcal{D}}}{\partial \dot{\theta}} + \frac{\partial \widehat{\Psi}_m}{\partial \theta} = 0, \tag{5.443}$$

which reduces to

$$(K_1 \cos^2\theta + K_3 \sin^2\theta)\frac{\partial^2\theta}{\partial z^2} + (K_3 - K_1)\sin\theta \cos\theta \left(\frac{\partial\theta}{\partial z}\right)^2$$
$$-\gamma_1 \frac{\partial\theta}{\partial t} - m(\theta)\frac{\partial v}{\partial z} + \chi_a H^2 \sin\theta \cos\theta = 0. \tag{5.444}$$

The balance of linear momentum (4.153) is

$$\rho \dot{v}_i = \left(\frac{\partial \widehat{\mathcal{D}}}{\partial v_{i,z}}\right)_{,z} - \frac{\partial \widehat{\mathcal{D}}}{\partial \dot{\theta}}\theta_{,i} - \tilde{p}_{,i}, \qquad i = 1, 2, 3, \tag{5.445}$$

which can be written as

$$\rho\frac{\partial v}{\partial t} = \frac{\partial}{\partial z}\left[g(\theta)\frac{\partial v}{\partial z} + m(\theta)\frac{\partial\theta}{\partial t}\right] - \tilde{p}_{,x}, \tag{5.446}$$
$$0 = -\tilde{p}_{,y}, \tag{5.447}$$
$$0 = -\left[m(\theta)\frac{\partial v}{\partial z} + \gamma_1\frac{\partial\theta}{\partial t}\right]\frac{\partial\theta}{\partial z} - \tilde{p}_{,z}, \tag{5.448}$$

where $\tilde{p} = p + \hat{w}_F - \hat{\Psi}_m$. The terms $\hat{w}_F$ and $\hat{\Psi}_m$ are clearly functions of z and t and, because of the arbitrariness involved in the pressure p, we may then assume, by (5.447), that $p = p(z,t)$. Equations (5.446) to (5.448) are then consistent with setting

$$p(z,t) = \hat{\Psi}_m - \hat{w}_F - \int \left[m(\theta)\frac{\partial v}{\partial z} + \gamma_1 \frac{\partial \theta}{\partial t} \right] \frac{\partial \theta}{\partial z}\, dz. \tag{5.449}$$

It is considered that the inertia of the liquid crystal can be ignored in typical cells having small depths (cf. References [45, p.483] and [174, 220]), in which case the term on the left-hand side of equation (5.446) can be omitted. The balance of linear momentum equations then reduce to

$$\frac{\partial}{\partial z}\left[g(\theta)\frac{\partial v}{\partial z} + m(\theta)\frac{\partial \theta}{\partial t} \right] = 0. \tag{5.450}$$

Now suppose that the cell depth in Fig. 3.6 is d with $-\frac{d}{2} \le z \le \frac{d}{2}$ and assume that the boundary conditions are

$$\theta(\pm\tfrac{d}{2}, t) = 0 \quad \text{and} \quad v(\pm\tfrac{d}{2}, t) = 0, \qquad t \ge 0. \tag{5.451}$$

Conditions $(5.451)_2$ are the familiar no-slip boundary conditions on the velocity. To make the problem more tractable, Pieranski *et al.* [220] assume that θ is small in the early stages of the dynamics after the magnetic field $\mathbf{H}$ is switched on so that the main dynamic equations (5.444) and (5.450) can be linearised in θ to give

$$\xi^2\frac{\partial^2\theta}{\partial z^2} + \theta - \lambda\frac{\partial\theta}{\partial t} - \lambda_1\frac{\partial v}{\partial z} = 0, \tag{5.452}$$

$$\eta_1\frac{\partial^2 v}{\partial z^2} + \alpha_3\frac{\partial^2\theta}{\partial z\partial t} = 0, \tag{5.453}$$

where $\eta_1 = \frac{1}{2}(\alpha_3 + \alpha_4 + \alpha_6)$ is the Miesowicz viscosity given by (4.161) and the notation

$$\xi^2 = \frac{K_1}{\chi_a H^2}, \quad \lambda = \frac{\gamma_1}{\chi_a H^2}, \quad \lambda_1 = \frac{\alpha_3}{\chi_a H^2}, \tag{5.454}$$

is introduced. With the additional reasonable supposition that the dissipation function be strictly positive, the results in (5.441) show that at $\theta = 0$

$$\eta_1 > 0, \quad \text{and} \quad \gamma_1\eta_1 - \alpha_3^2 > 0. \tag{5.455}$$

Notice that these inequalities imply $\gamma_1 > \alpha_3^2/\eta_1 \ge 0$. It will be convenient to introduce the constant α defined by

$$\alpha = 1 - \frac{\alpha_3^2}{\gamma_1\eta_1}, \tag{5.456}$$

which, from the inequalities in (5.455), clearly satisfies $1 - \alpha > 0$ and

$$0 < \alpha < 1, \tag{5.457}$$

when $\alpha_3 \ne 0$.

We are now in a position to discuss solutions to the main governing dynamic equations (5.452) and (5.453). Suppose that initially there are small non-zero disturbances θ_0 and v_0 to the solutions $\theta \equiv 0$ and $v \equiv 0$ when a magnetic field of magnitude $H > H_c$ is switched on and that there are solutions of the form [220]

$$\theta(z,t) = \theta_0 \left[\cos\left(\frac{2qz}{d}\right) - \cos q \right] \exp\left(\frac{t}{\tau}\right), \tag{5.458}$$

$$v(z,t) = v_0 \left[\frac{d}{2} \sin\left(\frac{2qz}{d}\right) - z \sin q \right] \exp\left(\frac{t}{\tau}\right), \tag{5.459}$$

where θ_0 and v_0 are constants, with q and τ to be determined. These particular forms arise from considering solutions obtained by standard separation of variables techniques. The boundary conditions (5.451) are obviously satisfied for these forms of θ and v. Inserting (5.458) and (5.459) into (5.453) shows that for consistency we must have

$$v_0 = -\theta_0 \frac{\alpha_3}{q \eta_1 \tau}, \tag{5.460}$$

which allows fulfilment of (5.453). Making use of this result, (5.458) and (5.459) can now be inserted into the remaining dynamic equation (5.452) to reveal that θ and v are solutions provided the requirement

$$\left(1 - \frac{\lambda}{\tau}\right) \cos q + \frac{\lambda_1 \alpha_3}{q \eta_1 \tau} \sin q = \left[1 - 4\xi^2 \frac{q^2}{d^2} - \frac{\lambda}{\tau} + \frac{\lambda_1 \alpha_3}{\eta_1 \tau}\right] \cos\left(\frac{2qz}{d}\right), \tag{5.461}$$

is fulfilled. This condition can hold for all $-\frac{d}{2} \leq z \leq \frac{d}{2}$ only when both the left-hand side and the coefficient of the cosine term on the right-hand side are identically zero and therefore we require

$$\left(1 - \frac{\lambda}{\tau}\right) \cos q + \frac{\lambda_1 \alpha_3}{q \eta_1 \tau} \sin q = 0, \tag{5.462}$$

$$1 - 4\xi^2 \frac{q^2}{d^2} - \frac{\lambda}{\tau} + \frac{\lambda_1 \alpha_3}{\eta_1 \tau} = 0. \tag{5.463}$$

The second requirement (5.463) can be simplified by using the definition (5.428) for H_c and the parameters introduced at equations (5.454) and (5.456) to find

$$\tau = \lambda \alpha \left(1 - 4\frac{q^2}{\pi^2} \frac{H_c^2}{H^2}\right)^{-1}, \tag{5.464}$$

and putting this into the first requirement (5.462) gives a relation for q, namely,

$$\left(\frac{H}{H_c}\right)^2 = 4\frac{q^2}{\pi^2} \left[\frac{\tan q - q/(1 - \alpha)}{\tan q - q}\right], \tag{5.465}$$

noting that $1 - \alpha > 0$ by (5.457) when $\alpha_3 \neq 0$ and, again, using the definitions (5.454) and (5.456). Solving this relation for the parameter q and inserting the value into (5.464) then determines τ. Knowing τ, q and v_0 from equations (5.464), (5.465) and (5.460), respectively, yields the solutions for $\theta(z,t)$ and $v(z,t)$ via (5.458) and

(5.459) for an arbitrary θ_0. It is also worth noting that for $0 < H < \infty$ the values of q are restricted so that $q_0 < q < q_1$ where q_0 and q_1 are the roots of

$$\tan(q) - q/(1-\alpha) = 0 \quad \text{and} \quad \tan(q) - q = 0, \qquad (5.466)$$

respectively. The first positive root of $(5.466)_1$ is q_0, which is always less than $\frac{\pi}{2}$ because $0 < \alpha < 1$, and the first positive root of $(5.466)_2$ *greater* than q_0 is q_1. Notice that q takes the value $\frac{\pi}{2}$ at $H = H_c$ and that as H increases above H_c the root q approaches q_1. Clearly, q_1 is independent of the material parameters and is approximately equal to 4.493. Hence $\frac{\pi}{2} < q < q_1$ for $H_c < H < \infty$ (cf. Pieranski *et al.* [220]).

An indication of the switch-on time can now be given from the above result for τ in equation (5.464). Using the definitions (4.167), $(5.454)_2$ and (5.456) we have

$$\tau_{on} = \frac{\eta_{splay}}{\chi_a(H^2 - 4q^2 H_c^2/\pi^2)}, \qquad \eta_{splay} = \gamma_1 - \alpha_3^2/\eta_1. \qquad (5.467)$$

where H_c and q can be found from (5.428) and (5.465), respectively. However, there is an observation which sheds more light on this result. We can rewrite τ_{on} as

$$\tau_{on} = \frac{\gamma_1^*}{\chi_a(H^2 - H_c^2)}, \qquad \gamma_1^*(H) \equiv \eta_{splay}\frac{(H^2 - H_c^2)}{(H^2 - 4q^2 H_c^2/\pi^2)}, \qquad (5.468)$$

which is the analogue of the result obtained formerly at equation (5.427) when flow was ignored. Notice that $\eta_{splay} > 0$ due to the inequalities (5.455) (recalling that $\gamma_1 > 0$) and that γ_1^* is an effective viscosity that is now dependent on the field strength. It is easily seen from (5.468), and the fact that q is always restricted to be positive and below the bound q_1, that

$$\gamma_1^*(H) \to \eta_{splay} \quad \text{as} \quad H \to \infty, \qquad (5.469)$$

and therefore for $H \gg H_c$ we have the estimates

$$\tau_{on} = \frac{\gamma_1^*}{\chi_a(H^2 - H_c^2)} \sim \frac{\eta_{splay}}{\chi_a(H^2 - H_c^2)} \quad \text{and} \quad \frac{\gamma_1^*}{\gamma_1} \sim \frac{\eta_{splay}}{\gamma_1} = \alpha = 1 - \frac{\alpha_3^2}{\gamma_1\eta_1}. \qquad (5.470)$$

It is now clear that backflow reduces the effective viscosity from γ_1 to γ_1^* when $H \gg H_c$, and that this backflow reduces the switch-on time τ_{on}, as can be seen by a comparison with τ_{on} obtained in equation (5.427) when flow was neglected. For the values given in Table D.3 for MBBA, PAA and 5CB, the ratio $\gamma_1^*/\gamma_1 \sim \eta_{splay}/\gamma_1 = \alpha$ is always close to unity and so the approximation for τ_{on} given by (5.427) is going to be reasonably accurate for these materials when $H \gg H_c$. However, other materials may need to be considered on an individual basis to determine the reduction in the effective viscosity during the initial stages of switching on a field. Nevertheless, in contrast to this result, the analogue of γ_1^* for the alternative homeotropic to planar transition is considerably reduced by the effect of backflow, as we shall discuss later in Section 5.9.3. A schematic interpretation of the profile for backflow is given in Fig. 5.10 (cf. Chandrasekhar [38, p.164]).

Using the identification employed to obtain (5.420), the analogous result for τ_{on} in the case of an electric field is

$$\tau_{on} = \frac{\gamma_1^*}{\epsilon_0 \epsilon_a (E^2 - E_c^2)} = \frac{\gamma_1^* d^2}{\epsilon_0 \epsilon_a (V^2 - V_c^2)}, \quad \gamma_1^*(V) \equiv \eta_{splay} \frac{(V^2 - V_c^2)}{(V^2 - 4q^2 V_c^2/\pi^2)}, \quad (5.471)$$

where, as before, the magnitude of the electric field is E, the voltage is $V = Ed$, and $V_c = E_c d$ with $E_c = \frac{\pi}{d}\sqrt{\frac{K_1}{\epsilon_0 \epsilon_a}}$. Further, for $E \gg E_c$ we have the estimates

$$\tau_{on} = \frac{\gamma_1^* d^2}{\epsilon_0 \epsilon_a (V^2 - V_c^2)} \sim \frac{\eta_{splay} d^2}{\epsilon_0 \epsilon_a (V^2 - V_c^2)} \quad \text{and} \quad \frac{\gamma_1^*}{\gamma_1} \sim \frac{\eta_{splay}}{\gamma_1} = \alpha = 1 - \frac{\alpha_3^2}{\gamma_1 \eta_1}. \quad (5.472)$$

These results act as rough guides to the switch-on time for an electric field, given the comments mentioned previously in Section 3.5 about the usually more complex behaviour of electric fields in liquid crystals.

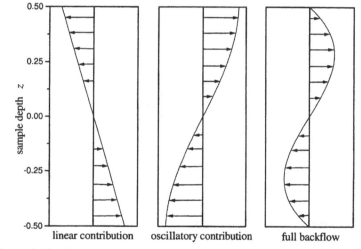

linear contribution   oscillatory contribution   full backflow

Figure 5.10: Qualitative representations of the contributions to the flow velocity $v(z, t)$ given by (5.459) at a fixed instant $t > 0$ when $v_0 > 0$ for a sample of unit depth, $-\frac{1}{2} \le z \le \frac{1}{2}$. The arrows represent the magnitude and direction of the flow. It is clear from the form of v that there is a linear contribution to the flow and an oscillatory contribution and these combine to give a backflow profile as shown which satisfies the no-slip boundary condition $(5.451)_2$. See also Fig. 5.11(a) and the comments on backflow on page 233.

Before progressing to a discussion of the homeotropic to planar transition and further comments on backflow, we reconsider the current planar to homeotropic problem when a high magnitude magnetic field is suddenly removed: this leads us to a discussion of the kickback effect.

The Kickback Effect

We now investigate what happens when a high magnitude magnetic field is suddenly switched off in the same geometry considered above, again motivated by Fig. 3.6 on page 73. The two governing dynamic equations continue to be (5.444) and (5.450) with the boundary conditions remaining the same as before, as stated in equations (5.451). The initial conditions must change of course, and are now given by

$$\theta(z,0) = \theta_0(z), \quad v(z,0) = 0, \qquad -\tfrac{d}{2} \le z \le \tfrac{d}{2}, \tag{5.473}$$

where $\theta_0(z)$ is the post-Freedericksz transition equilibrium solution for $H > H_c$ determined by equations (3.122) and (3.123), as discussed in Section 3.4.1, with H_c defined as above at equation (5.428). We shall not need to know the precise solution $\theta_0(z)$ since it will suffice for our present purposes to be aware that it is a non-zero function of z for $H > H_c$. We follow the work of Clark and Leslie [45] and Leslie [171].

For $H \gg H_c$ it can be assumed that θ is close to $\frac{\pi}{2}$ in the greater part of the sample. It therefore seems reasonable to describe the initial stages of any relaxation process that occurs upon removal of the field by the aforementioned dynamic equations when they are examined around $\theta = \frac{\pi}{2}$ with H set to zero when $t > 0$. Doing so in equations (5.444) and (5.450) leads to [171]

$$K_3\frac{\partial^2\theta}{\partial z^2} - \gamma_1\frac{\partial\theta}{\partial t} + \alpha_2\frac{\partial v}{\partial z} = 0, \tag{5.474}$$

$$\eta_2\frac{\partial^2 v}{\partial z^2} - \alpha_2\frac{\partial^2\theta}{\partial z\partial t} = 0, \tag{5.475}$$

where $\eta_2 = \frac{1}{2}(-\alpha_2+\alpha_4+\alpha_5)$ is the Miesowicz viscosity given by (4.162). As before, under the supposition that the viscous dissipation function is strictly positive, the results of (5.441) show that at $\theta = \frac{\pi}{2}$

$$\eta_2 > 0, \quad \text{and} \quad \gamma_1\eta_2 - \alpha_2^2 > 0. \tag{5.476}$$

These lead to the analogue of (5.456) with α replaced by

$$\alpha = 1 - \frac{\alpha_2^2}{\gamma_1\eta_2}, \tag{5.477}$$

so that $0 < \alpha < 1$ when $\alpha_2 \ne 0$, by the same reasoning as given earlier for establishing (5.457).

It will be supposed again that there are solutions of the forms (5.458) and (5.459) where θ_0 and v_0 are non-zero constants, except that on this occasion we expect some sort of relaxation process to occur and therefore they are modified to

$$\theta(z,t) = \theta_0\left[\cos\left(\frac{2qz}{d}\right) - \cos q\right]\exp\left(-\frac{t}{\tau}\right), \tag{5.478}$$

$$v(z,t) = v_0\left[\frac{d}{2}\sin\left(\frac{2qz}{d}\right) - z\sin q\right]\exp\left(-\frac{t}{\tau}\right), \tag{5.479}$$

with q and τ to be determined. The boundary conditions (5.451) are obviously satisfied. Inserting these forms into (5.475) shows that the condition

$$v_0 = -\theta_0 \frac{\alpha_2}{q \eta_2 \tau}, \qquad (5.480)$$

must be fulfilled for consistency in this case, which then automatically guarantees that equation (5.475) is satisfied. Further insertions into (5.474) show that θ and v are solutions provided

$$q = (1-\alpha)\tan q, \qquad (5.481)$$

$$\tau = \frac{\gamma_1 \alpha d^2}{4 K_3 q^2}. \qquad (5.482)$$

These relations follow from an argument similar in style to that used for obtaining (5.462) and (5.463), details being omitted for brevity. Solving (5.481) for q and inserting this value into (5.482) gives τ, which in turn leads to an evaluation of v_0 in terms of the original θ_0. Knowledge of q, τ and v_0 leads to solutions for $\theta(z,t)$ and $v(z,t)$ via (5.478) and (5.479) in terms of θ_0, similar to the story outlined earlier in the previous example.

The above relations for q and τ are simpler than those obtained earlier in the case when the field is switched on. Actually, as pointed out by Clark and Leslie [45] in a slightly different notation, the principle of superposition can be invoked using the relations (5.480) and (5.482) to obtain solutions that decay in time of the form

$$\theta(z,t) = \sum_{n=1}^{\infty} \theta_n \left[\cos\left(\frac{2 q_n z}{d}\right) - \cos q_n \right] \exp\left(-\frac{t}{\tau_n}\right), \qquad (5.483)$$

$$v(z,t) = -4 \frac{K_3 \alpha_2}{\gamma_1 \eta_2 \alpha d^2} \sum_{n=1}^{\infty} q_n \theta_n \left[\frac{d}{2} \sin\left(\frac{2 q_n z}{d}\right) - z \sin q_n \right] \exp\left(-\frac{t}{\tau_n}\right), \qquad (5.484)$$

where the q_n are the roots of equation (5.481), τ_n are the values derived from (5.482) via the roots q_n, and the constant coefficients θ_n are chosen to meet the initial condition (5.473)$_1$ placed upon θ, so that

$$\theta_0(z) = \sum_{n=1}^{\infty} \theta_n \left[\cos\left(\frac{2 q_n z}{d}\right) - \cos q_n \right]. \qquad (5.485)$$

The constants θ_n can be found explicitly as follows. Observe that by (5.481)

$$q_n \sin q_m \cos q_n = q_m \cos q_m \sin q_n, \qquad (5.486)$$

which leads to

$$\int_{-\frac{d}{2}}^{\frac{d}{2}} \left[\cos\left(\frac{2 q_n z}{d}\right) - \cos q_n \right] \cos\left(\frac{2 q_m z}{d}\right) dz = \begin{cases} 0, & n \neq m, \\ (q_n - \sin q_n \cos q_n)\, d/2 q_n, & n = m. \end{cases}$$
$$(5.487)$$

Therefore, multiplying both sides of (5.485) by $\cos(2q_m z/d)$ and integrating between $z = -\frac{d}{2}$ and $z = \frac{d}{2}$ yields

$$\theta_n = \frac{2q_n}{d(q_n - \sin q_n \cos q_n)} \int_{-\frac{d}{2}}^{\frac{d}{2}} \theta_0(z) \cos\left(\frac{2q_n z}{d}\right) dz, \qquad (5.488)$$

recalling that $\theta_0(z)$ is the known equilibrium solution mentioned in equation (5.473). As remarked by Leslie [45, 171, 174], for the solution presented here, it is impossible to satisfy the initial condition (5.473)$_2$ for $v(z, 0)$ because the fluid inertia term, $\rho\dot{v}$, has been ignored. As pointed out by Clark and Leslie [45], in this case the velocity must suffer a discontinuous change at $t = 0$, in accordance with Mandelstam's condition [204, Ch.26]. Similarly, since the director inertial constant σ has been set to zero, as is common practice, there must also be a discontinuity in $\partial\theta/\partial t$ at $t = 0$, by Mandelstam's condition.

For very high magnitude fields Leslie [171, 174] makes a reasonable approximation to the function $\theta_0(z)$ by simply setting

$$\theta_0(z) = \tfrac{\pi}{2}, \quad -\tfrac{d}{2} < z < \tfrac{d}{2}. \qquad (5.489)$$

It then follows from the relations (5.481) and (5.488) that

$$\theta_n = \frac{\pi \cos q_n}{\sin^2 q_n - \alpha}. \qquad (5.490)$$

Since $0 < \alpha < 1$, the positive roots of (5.481) are of the form [45]

$$q_n = \pi(n - \tfrac{1}{2}) - \delta_n, \quad \text{for some} \quad 0 < \delta_n < \tfrac{\pi}{2}, \quad n = 1, 2, 3, ..., \qquad (5.491)$$

and it is clear from the graph of (5.481) that $\delta_n \to 0$ as $n \to \infty$. From the relation (5.481) we have that

$$\tan \delta_n = (1 - \alpha)/q_n, \qquad (5.492)$$

which shows that for large n

$$\delta_n \sim \tan \delta_n \sim \frac{1 - \alpha}{\pi(n - \tfrac{1}{2})}. \qquad (5.493)$$

Therefore, using the result (5.490), we find the asymptotic relation

$$\theta_n \sim \frac{(-1)^{n-1}}{(n - \tfrac{1}{2})}, \qquad (5.494)$$

for large n, for which δ_n approaches zero. For this particular approximation we now know the complete series solutions for $\theta(z, t)$ and $v(z, t)$ via the results in equations (5.480) to (5.484) and (5.490). The asymptotic result (5.494) tells us that the rate of convergence of these series is quite slow and therefore many terms in the series are required to obtain reasonable accuracy.

As an example we have taken the appropriate physical parameter values for MBBA in Table D.3 and found the roots q_n numerically by using the software

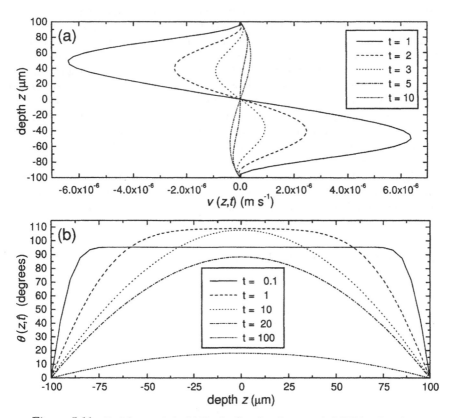

Figure 5.11: Backflow and the kickback effect for the nematic MBBA using the physical parameters listed in Table D.3. A sample depth d of 200 microns was chosen. (a) Backflow after the magnetic field is removed. The velocity v in the majority of the cell initially moves one way and then reverses direction as time t (seconds) increases. (b) The kickback effect: the director alignment angle θ initially increases above 90° after the field is removed before finally decreasing to zero. The times indicated in (a) and (b) have been selected to highlight the main characteristic features of v and θ as time progresses.

package Maple V, which easily enumerates these roots and swiftly evaluates the consequent values for the constants τ_n and θ_n. The series for $\theta(z,t)$ and $v(z,t)$ were calculated to 2000 terms for a sample of depth 200 microns $= 2 \times 10^{-4}$ m. Figure 5.11(a) shows the evolution of the velocity $v(z,t)$ for the indicated values of time t (seconds): the velocity is always zero at the boundaries and at the centre of the sample, but elsewhere the velocity initially moves one way and then reverses direction as time progresses before settling to zero. This is why such a flow effect is often called backflow. Figure 5.11(b) vividly shows the kickback effect for the angle $\theta(z,t)$: when the field is switched off, $\theta = 90°$ in the majority of the sample, then

θ initially increases by around $20°$ in the centre of the sample before eventually beginning to decrease and relax to zero. The timescales for corresponding stages of the kickback effect are qualitatively similar to those displayed by Kini, presented in Chandrasekhar [38, p.166]. As mentioned above, the solution for the velocity must have a discontinuity because the fluid inertia term has been neglected. This example is similar to that discussed by Clark and Leslie [45] and therefore it is likewise expected that the error due to this discontinuity in the flow velocity at $t = 0$ becomes negligible after a relatively short time compared to that required to attain the maximum kickback at the centre of the cell.

The kickback effect can have serious implications in twisted nematic devices (cf. Section 3.7 above). Examples of such complications in connection with experiments and displays have been reported and can be found in the brief article by Gerritsma, van Doorn and van Zanten [111]. Numerical studies showing the influence of backflow in the twisted nematic cell under the application of a strong electric field being switched on or off have been carried out by van Doorn [64].

5.9.3   Backflow in the Bend Geometry

We now turn our attention to the 'bend' homeotropic to planar Freedericksz transition geometry of Fig. 3.8 on page 78. First note that if flow is considered to be negligible then when H is near H_c an approximation for τ_{on} is given analogously to (5.427) by

$$\tau_{on} = \frac{\gamma_1}{\chi_a(H^2 - H_c^2)} = \frac{\gamma_1}{\chi_a H^2 - K_3 \pi^2/d^2}, \tag{5.495}$$

where $\chi_a > 0$ and H_c is now given by

$$H_c = \frac{\pi}{d}\sqrt{\frac{K_3}{\chi_a}}. \tag{5.496}$$

However, it is known that flow plays a significant rôle in this geometry and we shall show that the switch-on time is substantially reduced by the presence of backflow.

Motivated by the geometry and notation of Fig. 3.8, consider

$$\mathbf{n} = (\sin\theta(z,t), 0, \cos\theta(z,t)), \tag{5.497}$$
$$\mathbf{v} = (v(z,t), 0, 0), \tag{5.498}$$
$$\mathbf{H} = H(1, 0, 0). \tag{5.499}$$

By reasoning similar to that used in equations (5.429) to (5.450) to obtain the governing dynamic equations previously given in equations (5.451) to (5.457), it can be shown that there are analogous equations for this present case. We omit the details because of the identical arguments used and proceed by stating the results [220]. The dynamic equations analogous to (5.452) and (5.453) are

$$\xi^2 \frac{\partial^2 \theta}{\partial z^2} + \theta - \lambda \frac{\partial \theta}{\partial t} - \lambda_1 \frac{\partial v}{\partial z} = 0, \tag{5.500}$$

$$\eta_2 \frac{\partial^2 v}{\partial z^2} + \alpha_2 \frac{\partial^2 \theta}{\partial z \partial t} = 0, \tag{5.501}$$

where η_2 is the Miesowicz viscosity given by (4.162) and

$$\xi^2 = \frac{K_3}{\chi_a H^2}, \quad \lambda = \frac{\gamma_1}{\chi_a H^2}, \quad \lambda_1 = \frac{\alpha_2}{\chi_a H^2}. \tag{5.502}$$

The boundary conditions remain as those given in equation (5.451). Under similar suppositions as before, it is easily shown that the analogues of the inequalities in (5.455) are

$$\eta_2 > 0, \quad \text{and} \quad \gamma_1 \eta_2 - \alpha_2^2 > 0, \tag{5.503}$$

and that the constant α can be redefined as

$$\alpha = 1 - \frac{\alpha_2^2}{\gamma_1 \eta_2}, \tag{5.504}$$

where again $0 < \alpha < 1$ when $\alpha_2 \neq 0$. We assume that initially there are small non-zero disturbances θ_0 and v_0 to the solutions $\theta \equiv 0$ and $v \equiv 0$ when a field having $H > H_c$ is switched on and that they again take the forms given by (5.458) and (5.459). It is readily seen that the unknowns τ and q in the proposed solutions are given by equations (5.464) and (5.465) provided H_c and α are replaced by the quantities in equations (5.496) and (5.504), respectively, subject to the consistency requirement (cf. (5.460))

$$v_0 = -\theta_0 \frac{\alpha_2}{q \eta_2 \tau}. \tag{5.505}$$

Once q has been obtained from (5.465) for a given $H > H_c$, τ and v_0 can be evaluated for given physical material parameters and these then lead to solutions for $\theta(z,t)$ and $v(z,t)$ via (5.458) and (5.459).

The switch-on time is given by the result for τ in equation (5.464) using the relevant parameters given by the definitions (4.169), (5.502)$_2$ and (5.504). It is found that

$$\tau_{on} = \frac{\eta_{bend}}{\chi_a (H^2 - 4q^2 H_c^2/\pi^2)}, \quad \eta_{bend} = \gamma_1 - \frac{\alpha_2^2}{\eta_2}, \tag{5.506}$$

with H_c and q provided by (5.496) and the corresponding relation (5.465), respectively. The viscosity η_{bend} is positive by the positivity of γ_1 and α, when $\alpha_2 \neq 0$. This switch-on time can be rewritten as

$$\tau_{on} = \frac{\gamma_1^*}{\chi_a (H^2 - H_c^2)}, \quad \gamma_1^*(H) \equiv \eta_{bend} \frac{(H^2 - H_c^2)}{(H^2 - 4q^2 H_c^2/\pi^2)}, \tag{5.507}$$

similar to equation (5.468), with γ_1^* now being the field dependent effective viscosity. Clearly, the appropriate bounds obtained earlier from the results in (5.466) remain valid on q, in particular the upper bound q_1 is exactly the same, and therefore $H \to \infty$ as $q \to q_1$ and

$$\gamma_1^*(H) \to \eta_{bend} \quad \text{as} \quad H \to \infty. \tag{5.508}$$

For $H \gg H_c$ we then have estimates analogous to (5.470), namely,

$$\tau_{on} \sim \frac{\eta_{bend}}{\chi_a (H^2 - H_c^2)} \quad \text{and} \quad \frac{\gamma_1^*}{\gamma_1} \sim \alpha = 1 - \frac{\alpha_2^2}{\gamma_1 \eta_2}. \tag{5.509}$$

For the physical parameters for MBBA in Table D.3, the ratio $\gamma_1^*/\gamma_1 \sim \alpha$ is approximately 0.18 and therefore the effective viscosity is greatly reduced at high field strengths. This obviously enhances the switch-on time and can considerably reduce the value of τ_{on}. The effect of backflow upon the dynamics of the Freedericksz transition in the homeotropic to planar 'bend' geometry is therefore significant compared to that for the planar to homeotropic 'splay' geometry.

There are also obvious analogues of the results given by (5.471) and (5.472) for electric fields: simply replace η_{splay} by η_{bend} and K_1 by K_3 in the definition of V_c, with α given by (5.504). For example, when $V_c = E_c d$ and $E_c = \frac{\pi}{d}\sqrt{\frac{K_3}{\epsilon_0\epsilon_a}}$, we have for $E \gg E_c$

$$\tau_{on} \sim \frac{\eta_{bend}\, d^2}{\epsilon_0\epsilon_a(V^2 - V_c^2)} \quad \text{and} \quad \frac{\gamma_1^*}{\gamma_1} \sim \frac{\eta_{bend}}{\gamma_1} = \alpha = 1 - \frac{\alpha_2^2}{\gamma_1\eta_2}. \tag{5.510}$$

5.10   Light Scattering

One of the remarkable features of liquid crystals is that they scatter light strongly. It is well known that light scattering can be analysed and interpreted successfully by the dynamic continuum theory for nematic liquid crystals in terms of small amplitude orientational fluctuations. Eigenmodes and frequencies arising from the scattered light can then be investigated and determined; this knowledge allows information on the viscosities of the liquid crystal to be deduced and can lead to experiments that determine their values, as we shall see below. Extensive details on the physics of light scattering in liquid crystals can be found in the books by de Gennes and Prost [110] and Chandrasekhar [38] or the recent review by Gleeson [114].

Consider a uniformly aligned nematic liquid crystal sample where the director is initially constant and is denoted by the vector $\mathbf{n}_0$, which can be assumed to be parallel to the z-axis in a Cartesian coordinate system. Now suppose, in the usual notation, that an ingoing incident light beam of frequency ω_0, wave vector $\mathbf{k}_0$ and unit polarisation vector $\mathbf{i}$ enters the sample and that the corresponding quantities for the outgoing scattered beam are ω_1, $\mathbf{k}_1$ and $\mathbf{f}$, respectively. One such experimental set-up is as shown in Fig. 5.12; other set-ups are possible and the reader is referred to Gleeson [114] for details. This scattering process is associated with an angular frequency change ω and a wave vector change $\mathbf{q}$ defined by, respectively,

$$\omega = \omega_0 - \omega_1 \quad \text{and} \quad \mathbf{q} = \mathbf{k}_0 - \mathbf{k}_1. \tag{5.511}$$

The vector $\mathbf{q}$ is sometimes referred to as the scattering vector and the scattering angle θ is defined to be the angle between the vectors $\mathbf{k}_0$ and $\mathbf{k}_1$. The plane containing $\mathbf{k}_0$, $\mathbf{k}_1$ and $\mathbf{q}$ is called the scattering plane. In general, light scattering induces small vibrations to the orientation of the director and we suppose that these are associated with small fluctuations $\delta\mathbf{n}$ such that

$$\mathbf{n} = \mathbf{n}_0 + \delta\mathbf{n}, \tag{5.512}$$

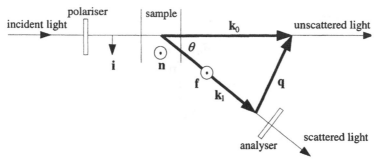

Figure 5.12: A schematic diagram of one possible light scattering experiment. The scattering angle is θ as shown where the incident beam has wave vector $\mathbf{k}_0$ and the scattered beam has wave vector $\mathbf{k}_1$. The scattering vector is defined by $\mathbf{q} = \mathbf{k}_0 - \mathbf{k}_1$ as shown. The unit polarisation vector $\mathbf{i}$ enters the sample in the direction shown and leaves the sample, in this particular case, as the vector $\mathbf{f}$ parallel to $\mathbf{n}$ (perpendicular to the page). In this geometry both $\mathbf{i}$ and $\mathbf{k}_0$ are perpendicular to $\mathbf{n}$.

where, since $\mathbf{n} \cdot \mathbf{n} = 1$, we require $\delta\mathbf{n} \cdot \mathbf{n}_0 = 0$ to a first order approximation in $\delta\mathbf{n}$. It proves convenient to resolve $\delta\mathbf{n}$ into two modes involving the displacements δn_1 and δn_2 as represented in Fig. 5.13(a): the displacement δn_1 takes place in the $\mathbf{n}_0, \mathbf{q}$-plane and consists of a superposition of bend and splay as shown in Fig. 5.13(b), while the displacement δn_2 occurs normal to the $\mathbf{n}_0, \mathbf{q}$-plane and consists of a superposition of bend and twist, shown in Fig. 5.13(c). These characteristics will become evident through the results to be developed here. For these reasons we introduce the two unit vectors

$$\mathbf{e}_2 = q_\perp^{-1}(\mathbf{n}_0 \times \mathbf{q}), \qquad \mathbf{e}_1 = \mathbf{e}_2 \times \mathbf{n}_0, \qquad (5.513)$$

where $q_\perp$ is the component of $\mathbf{q}$ perpendicular to $\mathbf{n}_0$, as depicted in Fig. 5.13(a). Relative to the Cartesian system of coordinates with basis $\{\mathbf{e}_1, \mathbf{e}_2, \hat{\mathbf{z}}\}$ we can resolve $\mathbf{q}$ into the components depicted in Fig. 5.13(a) where we see that for $q = |\mathbf{q}|$

$$q = \sqrt{q_\parallel^2 + q_\perp^2} \quad \text{with} \quad q_\parallel = q\cos\phi, \quad q_\perp = q\sin\phi, \qquad (5.514)$$

where ϕ is the angle between $\mathbf{n}_0$ and $\mathbf{q}$ as shown. It should be noted that some authors define $\mathbf{q}$ to be the vector $\mathbf{k}_1 - \mathbf{k}_0$ rather than that defined in (5.511)$_2$. The mode related to δn_1 occurs in the plane containing $\mathbf{n}_0$ and $\mathbf{q}$, and this is sometimes called the 'in-plane mode', whereas the mode related to δn_2 occurs in the plane perpendicular to $\mathbf{n}_0$ and $\mathbf{q}$, and this mode is sometimes called the 'out-of-plane mode'. The plane containing $\mathbf{n}_0$ and $\mathbf{q}$ may or may not coincide with the scattering plane (cf. the scattering geometries in [114, p.702]). The displacements can be analysed separately by a suitable choice of polarisations, one such example being pictured in Fig. 5.12 where $\mathbf{q}$ lies in the e_1e_2-plane pictured in Fig. 5.13(a), parallel to $q_\perp$, where $\phi = \frac{\pi}{2}$ and the scattering angle θ is the angle between $\mathbf{k}_0$ and $\mathbf{k}_1$.

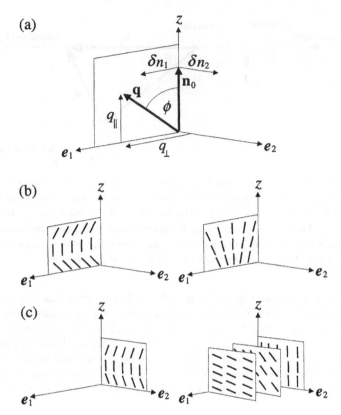

Figure 5.13: (a) δn_1 and δn_2 represent two displacements of $\mathbf{n}_0$. The displacement δn_1 takes place in the $\mathbf{n}_0$, q-plane and consists of a superposition of bend and splay as shown in (b). The displacement δn_2 occurs normal to the $\mathbf{n}_0$, q-plane and consists of a superposition of bend and twist as shown in (c).

For a given mode of vibration of angular frequency ω and scattering vector $\mathbf{q}$ we can assume

$$\delta\mathbf{n} = (\delta n_1, \delta n_2, 0) = (\delta_1, \delta_2, 0)\exp[i(\mathbf{q}\cdot\mathbf{x} - \omega t)], \qquad (5.515)$$

$$\mathbf{v} = (v_1, v_2, v_3) = (u_1, u_2, u_3)\exp[i(\mathbf{q}\cdot\mathbf{x} - \omega t)], \qquad (5.516)$$

where the quantities δ_1, δ_2, u_1, u_2 and u_3 are small. These assumed forms come from standard representations of periodic disturbances related to light scattering [38, 168, 174] when the components of $\delta\mathbf{n}$ and $\mathbf{v}$ (and the spatial vector $\mathbf{x}$) are taken relative to the basis vectors $\{\mathbf{e}_1, \mathbf{e}_2, \hat{\mathbf{z}}\}$. It follows, via Fig. 5.13, that

$$\mathbf{n} = \mathbf{n}_0 + \delta\mathbf{n} = (0, 0, 1) + \delta\mathbf{n}, \qquad (5.517)$$

with

$$\mathbf{q}\cdot\mathbf{x} = q_{\perp}x_1 + q_{\parallel}z. \qquad (5.518)$$

We now proceed to obtain the dynamic equations using the theory and notation summarised in Section 4.2.5 on page 150. First note that $\mathbf{n} \cdot \mathbf{n} = 1$ to first order in these quantities and that the constraint $\nabla \cdot \mathbf{v} = 0$ leads to the relation

$$v_1 = -q_\parallel q_\perp^{-1} v_3. \tag{5.519}$$

This relation essentially allows us to follow the results of the Orsay Group [212] and choose to write all quantities involving v_1 in terms of v_3. It should also be pointed out that the Orsay Group use the form $\exp[i(\mathbf{q} \cdot \mathbf{x} + \omega t)]$ in their perturbations (cf. [212, Eqns. (III.11)–(III.12)]) instead of the form appearing in equations (5.515) and (5.516). Of course, the final results are identical irrespective of which form is used, the results of the Orsay Group being obtained simply by replacing ω by $-\omega$.

From the above comments the constraints (4.118) are satisfied and it remains to derive the dynamic equations (4.119) and (4.120). We begin by considering the balance law for linear momentum in (4.119). The non-zero components of the symmetric rate of strain tensor A_{ij} and the skew-symmetric vorticity tensor W_{ij} can be given in terms of v_2 and v_3 by using the relation (5.519) to find

$$A_{11} = iq_\perp v_1 = -iq_\parallel v_3, \quad A_{12} = \tfrac{1}{2}iq_\perp v_2, \quad A_{13} = \tfrac{1}{2}iq_\perp^{-1}(q_\perp^2 - q_\parallel^2)v_3,$$
$$A_{23} = \tfrac{1}{2}iq_\parallel v_2, \quad A_{33} = iq_\parallel v_3, \tag{5.520}$$
$$W_{12} = -\tfrac{1}{2}iq_\perp v_2, \quad W_{13} = -\tfrac{1}{2}iq^2 q_\perp^{-1} v_3, \quad W_{23} = \tfrac{1}{2}iq_\parallel v_2.$$

To first order for the assumed forms for $\mathbf{v}$ and $\mathbf{n}$ in (5.516) and (5.517) the viscous stress (4.121) is

$$\tilde{t}_{ij} = \alpha_1 A_{33} n_i n_j + \alpha_2 N_i n_j + \alpha_3 n_i N_j + \alpha_4 A_{ij} + \alpha_5 n_j A_{i3} + \alpha_6 n_i A_{j3}, \tag{5.521}$$

where we have linearised with respect to the small quantities δn_1, δn_2, v_1, v_2 and v_3. Also

$$N_i = -i\omega \delta n_i - W_{i3}, \tag{5.522}$$

where $\delta n_3 \equiv 0$, and hence

$$\tilde{t}_{ij,j} = \alpha_1 A_{33,3} n_i + \alpha_2 N_{i,3} + \alpha_3 n_i N_{j,j} + \alpha_4 A_{ij,j} + \alpha_5 A_{i3,3} + \alpha_6 n_i A_{j3,j}. \tag{5.523}$$

Further,

$$\rho \dot{v}_i = -i\rho \omega v_i, \tag{5.524}$$

and, from (4.122), (5.520) and (5.524), the linearised expression for $\tilde{\mathbf{g}}$ is

$$\tilde{g}_i = i\omega \gamma_1 \delta n_i + \gamma_1 W_{i3} - \gamma_2 A_{i3}. \tag{5.525}$$

The arbitrary pressure appearing in equation (4.119) can be assumed to take the form $p = -w_F + p_0(x_1, z)$. With the aid of (5.523) and (5.524) the equations (4.119) then reduce to first order in $\mathbf{v}$ to

$$-i\rho \omega v_1 = -p_{0,1} + \tilde{t}_{1j,j}, \tag{5.526}$$
$$-i\rho \omega v_2 = \tilde{t}_{2j,j}, \tag{5.527}$$
$$-i\rho \omega v_3 = -p_{0,3} + \tilde{t}_{3j,j}, \tag{5.528}$$

assuming, of course, that there is no external body force $\mathbf{F}$ or generalised body force $\mathbf{G}$. The pressure contribution p_0 can be eliminated from equations (5.526) and (5.528) by differentiating (5.526) and (5.528) with respect to z and x_1, respectively, and then subtracting. This gives the equation

$$\rho\omega v_3 = -q_\perp q^{-2}(\tilde{t}_{1j,j3} - \tilde{t}_{3j,j1}),\tag{5.529}$$

once the relation (5.519) has been employed. The balance of linear momentum then reduces to the two equations (5.527) and (5.529). By (5.523), routine calculations show that equation (5.527) is

$$-i\rho\omega v_2 = \alpha_2 N_{2,3} + \alpha_4 A_{2j,j} + \alpha_5 A_{23,3},\tag{5.530}$$

where

$$
\begin{aligned}
N_{2,3} &= \omega q_\parallel \delta n_2 + \tfrac{1}{2} q_\parallel^2 v_2, &(5.531)\\
A_{2j,j} &= -\tfrac{1}{2} q^2 v_2, &(5.532)\\
A_{23,3} &= -\tfrac{1}{2} q_\parallel^2 v_2, &(5.533)
\end{aligned}
$$

while

$$
\begin{aligned}
\tilde{t}_{1j,j3} - \tilde{t}_{3j,j1} = {}& \alpha_2(N_{1,33} - N_{3,31}) - \alpha_3 N_{j,j1} + (\alpha_4 + \alpha_5)A_{13,33}\\
& + \alpha_4 A_{11,13} - (\alpha_1 + \alpha_5)A_{33,31} - (\alpha_4 + \alpha_6)A_{3j,j1},\tag{5.534}
\end{aligned}
$$

where, with the aid of (5.519) when appropriate,

$$
\begin{aligned}
N_{1,33} &= i\omega q_\parallel^2 \delta n_1 - \tfrac{1}{2} i q^2 q_\perp^{-1} q_\parallel^2 v_3, &(5.535)\\
N_{3,31} &= 0, &(5.536)\\
N_{j,j1} &= i\omega q_\perp^2 \delta n_1 - \tfrac{1}{2} i q^2 q_\perp v_3, &(5.537)\\
A_{13,33} &= -\tfrac{1}{2} i q_\perp^{-1} q_\parallel^2 (q_\perp^2 - q_\parallel^2) v_3, &(5.538)\\
A_{11,13} &= -i q_\perp^2 q_\parallel v_1 = i q_\parallel^2 q_\perp v_3, &(5.539)\\
A_{33,31} &= -i q_\parallel^2 q_\perp v_3, &(5.540)\\
A_{3j,j1} &= -\tfrac{1}{2} i q^2 q_\perp v_3. &(5.541)
\end{aligned}
$$

It then follows from equations (5.530) to (5.541) that the linear momentum balance equations (5.527) and (5.529) finally reduce to, respectively, the two simultaneous equations

$$
\begin{aligned}
[\rho\omega + iP_2(\mathbf{q})]\, v_2 - i\omega Q_2(\mathbf{q})\delta n_2 &= 0, &(5.542)\\
[\rho\omega + iP_1(\mathbf{q})]\, v_3 - i\omega Q_1(\mathbf{q})\delta n_1 &= 0, &(5.543)
\end{aligned}
$$

where

$$
\begin{aligned}
P_1(\mathbf{q}) &= q^{-2}(\eta_1 q_\perp^4 + \eta_2 q_\parallel^4 + \tfrac{1}{2}\alpha_m q_\perp^2 q_\parallel^2), &(5.544)\\
P_2(\mathbf{q}) &= \tfrac{1}{2}\alpha_4 q_\perp^2 + \eta_2 q_\parallel^2 = \eta_3 q_\perp^2 + \eta_2 q_\parallel^2, &(5.545)\\
Q_1(\mathbf{q}) &= q_\perp q^{-2}(\alpha_3 q_\perp^2 - \alpha_2 q_\parallel^2), &(5.546)\\
Q_2(\mathbf{q}) &= \alpha_2 q_\parallel, &(5.547)\\
\alpha_m &= 2(\alpha_1 + \alpha_4) + \alpha_5 + \alpha_6 + \alpha_3 - \alpha_2 = 2(\alpha_1 + \alpha_3 + \alpha_4 + \alpha_5), &(5.548)
\end{aligned}
$$

with η_1, η_2 and η_3 being the usual Miesowicz viscosities defined in Table 4.1 on page 158, and the second equality in (5.548) coming from an application of the Parodi relation (4.127).

We now turn to the balance of angular momentum equations (4.120). To first order in the approximations introduced above at (5.515) and (5.517), the nematic energy density (2.50) satisfies (cf. equation (4.131))

$$\left(\frac{\partial w_F}{\partial n_{i,j}}\right)_{,j} - \frac{\partial w_F}{\partial n_i} = (K_1 - K_2)n_{j,ji} + K_2 n_{i,jj} + (K_3 - K_2)n_{i,33}. \qquad (5.549)$$

It is now straightforward to insert the results from (5.525) and (5.549) into equation (4.120) to obtain, in the absence of any generalised body force **G**, the three equations

$$(K_1 - K_2)n_{j,j1} + K_2 n_{1,jj} + (K_3 - K_2)n_{1,33}$$
$$+ i\omega\gamma_1\delta n_1 + \gamma_1 W_{13} - \gamma_2 A_{13} = \lambda\delta n_1, \qquad (5.550)$$
$$K_2 n_{2,jj} + (K_3 - K_2)n_{2,33} + i\omega\gamma_1\delta n_2 + \gamma_1 W_{23} - \gamma_2 A_{23} = \lambda\delta n_2, \qquad (5.551)$$
$$(K_1 - K_2)n_{j,j3} - \gamma_2 A_{33} = \lambda. \qquad (5.552)$$

The Lagrange multiplier λ is easily identified from equation (5.552), and this in turn shows that we must have $\lambda\delta n_1 = \lambda\delta n_2 = 0$ to first order in the quantities δn_1, δn_2, v_2 and v_3. The two remaining governing equations arising from the balance of angular momentum are then (5.550) and (5.551) with the right-hand sides set to zero. After some tedious but simple calculations for the assumed forms of **n** and **v** in equations (5.515) and (5.516), similar in style to those used to arrive at the equations arising from the balance of linear momentum, these two equations reduce to

$$iC_1(\mathbf{q})v_3 + [-i\omega\gamma_1 + k_1(\mathbf{q})]\delta n_1 = 0, \qquad (5.553)$$
$$iC_2(\mathbf{q})v_2 + [-i\omega\gamma_1 + k_2(\mathbf{q})]\delta n_2 = 0, \qquad (5.554)$$

where, upon using the Parodi relation (4.127) appropriately in (5.554),

$$C_1(\mathbf{q}) = \tfrac{1}{2}q_\perp^{-1}\left[q_\perp^2(\gamma_1 + \gamma_2) + q_\parallel^2(\gamma_1 - \gamma_2)\right] = q_\perp^{-1}\left[\alpha_3 q_\perp^2 - \alpha_2 q_\parallel^2\right], \qquad (5.555)$$
$$C_2(\mathbf{q}) = \tfrac{1}{2}q_\parallel(\gamma_2 - \gamma_1) = \alpha_2 q_\parallel, \qquad (5.556)$$
$$k_1(\mathbf{q}) = K_1 q_\perp^2 + K_3 q_\parallel^2, \qquad (5.557)$$
$$k_2(\mathbf{q}) = K_2 q_\perp^2 + K_3 q_\parallel^2, \qquad (5.558)$$

making use of the identities (4.176) in equations (5.555) and (5.556).

Governing Dynamic Equations: Fast and Slow Modes

The governing dynamic equations are now given by the four equations (5.542), (5.543), (5.553) and (5.554). They can be collected as two sets of simultaneous equations, one representing disturbances in the in-plane mode, namely,

$$[\rho\omega + iP_1(\mathbf{q})]v_3 - i\omega Q_1(\mathbf{q})\delta n_1 = 0, \qquad (5.559)$$
$$iC_1(\mathbf{q})v_3 + [-i\omega\gamma_1 + k_1(\mathbf{q})]\delta n_1 = 0, \qquad (5.560)$$

and the other representing disturbances in the out-of-plane mode

$$[\rho\omega + iP_2(\mathbf{q})]\,v_2 - i\omega Q_2(\mathbf{q})\delta n_2 \;=\; 0, \tag{5.561}$$

$$iC_2(\mathbf{q})v_2 + [-i\omega\gamma_1 + k_2(\mathbf{q})]\,\delta n_2 \;=\; 0. \tag{5.562}$$

For non-zero solutions δn_1, δn_2, v_2 and v_3 to these sets of equations it is necessary that the related determinants of these systems vanish. Therefore it is required that

$$-i\rho\omega^2\gamma_1 + \omega\,[\rho k_\alpha(\mathbf{q}) + \gamma_1 P_\alpha(\mathbf{q}) - C_\alpha(\mathbf{q})Q_\alpha(\mathbf{q})] + ik_\alpha(\mathbf{q})P_\alpha(\mathbf{q}) = 0, \quad \alpha = 1, 2, \tag{5.563}$$

where $\alpha = 1$ for the in-plane mode and $\alpha = 2$ for the out-of-plane mode. The roots of these quadratics in ω take simplified approximate forms if some assumptions that are well motivated by physical properties are made. For example, if we set the elastic constants equal to K and replace all the viscosities, including γ_1 as appropriate, by an average viscosity η, then the functions appearing in equation (5.563) may be estimated for a general direction of $\mathbf{q}$ with $q_\perp \sim q_\parallel \sim q$ as [212]

$$P_\alpha(\mathbf{q}) \;\sim\; \eta q^2, \tag{5.564}$$

$$C_\alpha(\mathbf{q}) \;\sim\; Q_\alpha(\mathbf{q}) \sim \eta q, \tag{5.565}$$

$$k_\alpha(\mathbf{q}) \;\sim\; Kq^2, \tag{5.566}$$

for $\alpha = 1, 2$. In these circumstances, typical values of the physical parameters, in SI units, are

$$K = 10^{-11}\ \text{N}, \quad \rho = 10^3\ \text{kg m}^{-3}, \quad \eta = 0.01\ \text{Pa s}, \tag{5.567}$$

and it therefore follows that

$$\rho k_\alpha(\mathbf{q}) \ll \gamma_1 P_\alpha(\mathbf{q}) - C_\alpha(\mathbf{q})Q_\alpha(\mathbf{q}). \tag{5.568}$$

Equation (5.568) encourages us to neglect the term involving $\rho k_\alpha(\mathbf{q})$ in (5.563), leading to the approximate roots ω given by

$$\omega \doteq -i\frac{\gamma_1 P_\alpha(\mathbf{q}) - C_\alpha(\mathbf{q})Q_\alpha(\mathbf{q})}{2\gamma_1\rho}\left[1 \mp \left\{1 - \frac{2\gamma_1\rho P_\alpha(\mathbf{q})k_\alpha(\mathbf{q})}{[\gamma_1 P_\alpha(\mathbf{q}) - C_\alpha(\mathbf{q})Q_\alpha(\mathbf{q})]^2}\right\}\right], \tag{5.569}$$

where (5.568) has also allowed us to make a first order approximation to the square root of the usual quadratic discriminant. The two types of eigenfrequencies become, for $\alpha = 1, 2$,

$$\omega_{S\alpha} \;\sim\; -i\frac{k_\alpha(\mathbf{q})P_\alpha(\mathbf{q})}{\gamma_1 P_\alpha(\mathbf{q}) - C_\alpha(\mathbf{q})Q_\alpha(\mathbf{q})}, \tag{5.570}$$

$$\omega_{F\alpha} \;\sim\; -i\frac{\gamma_1 P_\alpha(\mathbf{q}) - C_\alpha(\mathbf{q})Q_\alpha(\mathbf{q})}{\gamma_1\rho}, \tag{5.571}$$

after making further use of the approximation (5.568) to arrive at (5.571). In view of (5.568), solutions corresponding to frequencies with the subscripts S are called 'slow' modes, and those with subscripts F are called 'fast' modes, the subscript α

referring of course to whether the in-plane ($\alpha = 1$) or out-of-plane ($\alpha = 2$) modes are being considered. It should also be observed that the elastic constants only appear in $\omega_{S\alpha}$.

In both slow and fast modes it is observed that any small disturbances of the forms proposed in equations (5.515) and (5.516) will be overdamped and decay exponentially with time. This can be seen from the following observations. From the eigenfrequencies (5.570) and (5.571) we obtain

$$-i\omega_{S\alpha} = \frac{-k_\alpha(\mathbf{q})P_\alpha(\mathbf{q})}{\gamma_1 P_\alpha(\mathbf{q}) - C_\alpha(\mathbf{q})Q_\alpha(\mathbf{q})}, \qquad -i\omega_{F\alpha} = \frac{-[\gamma_1 P_\alpha(\mathbf{q}) - C_\alpha(\mathbf{q})Q_\alpha(\mathbf{q})]}{\gamma_1 \rho}.$$
$$(5.572)$$

It is clear that

$$k_\alpha(\mathbf{q}) \geq 0, \qquad \alpha = 1, 2, \tag{5.573}$$

because K_1, K_2 and K_3 are non-negative. Further, it can be shown, using (5.514), that

$$
\begin{aligned}
P_1(\mathbf{q}) &= q^2 g(\phi - \tfrac{\pi}{2}), & (5.574)\\
P_2(\mathbf{q}) &= q^2 h(\phi - \tfrac{\pi}{2}), & (5.575)\\
\gamma_1 P_1(\mathbf{q}) - C_1(\mathbf{q})Q_1(\mathbf{q}) &= q^2 \left[\gamma_1 g(\phi - \tfrac{\pi}{2}) - m^2(\phi - \tfrac{\pi}{2})\right], & (5.576)\\
\gamma_1 P_2(\mathbf{q}) - C_2(\mathbf{q})Q_2(\mathbf{q}) &= q^2 \left[\gamma_1 h(\phi - \tfrac{\pi}{2}) - \alpha_2^2 \sin^2(\phi - \tfrac{\pi}{2})\right], & (5.577)
\end{aligned}
$$

where the functions g, h and m are precisely those defined earlier at equations (5.201) to (5.203) on page 193. If it is assumed that the elastic constants and the dissipation function are all strictly positive then the inequalities in equations (5.205), (5.206) and (5.573) all become strictly positive inequalities which hold for any angle θ (or ϕ). It therefore follows from equations (5.573) to (5.577) that

$$k_\alpha(\mathbf{q}) > 0, \quad P_\alpha(\mathbf{q}) > 0, \quad \gamma_1 P_\alpha(\mathbf{q}) - C_\alpha(\mathbf{q})Q_\alpha(\mathbf{q}) > 0, \qquad \alpha = 1, 2, \tag{5.578}$$

and hence we obtain from (5.572)

$$-i\omega_{S\alpha} < 0, \quad \text{and} \quad -i\omega_{F\alpha} < 0, \qquad \alpha = 1, 2. \tag{5.579}$$

These results in (5.579) show that the disturbances proposed in (5.515) and (5.516) will decay exponentially with time.

It proves convenient to introduce, in the standard notation, the positive quantities

$$u_{S\alpha} = i\omega_{S\alpha}, \quad u_{F\alpha} = i\omega_{F\alpha}. \qquad \alpha = 1, 2. \tag{5.580}$$

For the approximations introduced above at equations (5.564) to (5.567), we have

$$u_{S\alpha} \sim K q^2 \eta^{-1}, \quad u_{F\alpha} \sim \eta q^2 \rho^{-1}, \quad \text{and} \quad \frac{u_{S\alpha}}{u_{F\alpha}} \sim \frac{K\rho}{\eta^2} \sim 10^{-4}. \tag{5.581}$$

The slow modes tend to be dominant in light scattering experiments from which information on the nematic viscosities can be obtained, as we now proceed to demonstrate.

Response Functions, Power Spectrum and Intensities

As reviewed by Chandrasekhar [38], we suppose that the uncoupled displacements δn_1 and δn_2 arise from forces G_1 and G_2 which can be represented through the relations

$$\delta n_\alpha = \chi_\alpha(\mathbf{q}, \omega)G_\alpha, \qquad \alpha = 1, 2. \tag{5.582}$$

The terms $\chi_\alpha(\mathbf{q}, \omega)$ are sometimes called the response functions [212]. It is well known in the literature that the fluctuation-dissipation theorem results in the power spectrum $I_\alpha(\mathbf{q}, \omega)$ being given by the relation [38, 212]

$$I_\alpha(\mathbf{q}, \omega) = \frac{k_B T}{\pi \omega} \text{Im}(\chi_\alpha(\mathbf{q}, \omega)), \qquad \alpha = 1, 2, \tag{5.583}$$

where k_B is the Boltzmann constant, T is the thermodynamic temperature (in kelvin) and Im denotes the imaginary part. We shall see later that the integral of the power spectrum will turn out to be of use in the experimental determination of physical parameters. For the generalised body force $\mathbf{G} = (G_1, G_2, 0)$, we can return to equation (4.120) and modify equations (5.550) to (5.552) accordingly to find that the governing equations (5.560) and (5.562) may be replaced by

$$iC_1(\mathbf{q})v_3 + [-i\omega\gamma_1 + k_1(\mathbf{q})]\,\delta n_1 - G_1 \;=\; 0, \tag{5.584}$$
$$iC_2(\mathbf{q})v_2 + [-i\omega\gamma_1 + k_2(\mathbf{q})]\,\delta n_2 - G_2 \;=\; 0, \tag{5.585}$$

with equations (5.559) and (5.561) remaining as stated above. Using the relations (5.582), there are non-zero solutions to the analogues of the systems of equations (5.559) to (5.562) provided the related determinants are zero. This requires fulfilment of the conditions

$$\chi_\alpha(\mathbf{q}, \omega) = \frac{\rho\omega + iP_\alpha(\mathbf{q})}{[\rho\omega + iP_\alpha(\mathbf{q})]\,[-i\omega\gamma_1 + k_\alpha(\mathbf{q})] - \omega C_\alpha(\mathbf{q})Q_\alpha(\mathbf{q})}, \qquad \alpha = 1, 2. \tag{5.586}$$

Neglecting the contributions involving the product $\rho k_\alpha(\mathbf{q})$, because of the inequality (5.568) above, it is observed that

$$[\rho\omega + iP_\alpha(\mathbf{q})]\,[-i\omega\gamma_1 + k_\alpha(\mathbf{q})] - \omega C_\alpha(\mathbf{q})Q_\alpha(\mathbf{q}) \;\sim\; -i\gamma_1\rho(\omega + iu_{S\alpha})(\omega + iu_{F\alpha}), \tag{5.587}$$

and hence the response functions can be approximated by

$$\chi_\alpha(\mathbf{q}, \omega) \;=\; \frac{-[\rho\omega + iP_\alpha(\mathbf{q})]}{i\gamma_1\rho(\omega + iu_{S\alpha})(\omega + iu_{F\alpha})}$$
$$=\; \frac{\rho\omega + iP_\alpha(\mathbf{q})}{\gamma_1\rho(u_{F\alpha} - u_{S\alpha})} \left[(\omega + iu_{S\alpha})^{-1} - (\omega + iu_{F\alpha})^{-1}\right]. \tag{5.588}$$

It is expected that $u_{S\alpha} \ll u_{F\alpha}$, as discussed above in terms of $\omega_{S\alpha}$ and $\omega_{F\alpha}$ at equation (5.581), and therefore using the relation (5.583) with (5.588), and a further application of (5.581)$_3$, we obtain [212]

$$I_\alpha(\mathbf{q}, \omega) \approx \frac{k_B T}{\pi \gamma_1 \rho u_{F\alpha}} \left[\frac{P_\alpha(\mathbf{q})}{\omega^2 + u_{S\alpha}^2} - \frac{C_\alpha(\mathbf{q})Q_\alpha(\mathbf{q})}{\gamma_1(\omega^2 + u_{F\alpha}^2)}\right]. \tag{5.589}$$

For fixed $\mathbf{q}$, $I_\alpha(\mathbf{q}, \omega)$ is therefore a superposition of two Lorentzian lines of widths $u_{S\alpha}$ and $u_{F\alpha}$. However, since $u_{S\alpha} \ll u_{F\alpha}$ by (5.581)$_3$, the second term in the brackets of (5.589) can be neglected and therefore the power spectrum can be approximated by

$$I_\alpha(\mathbf{q}, \omega) = \frac{k_B T P_\alpha(\mathbf{q})}{\pi \gamma_1 \rho u_{F\alpha}(\omega^2 + u_{S\alpha}^2)} \tag{5.590}$$

By equations (5.570), (5.571) and (5.580),

$$\gamma_1 \rho u_{F\alpha} = \frac{k_\alpha(\mathbf{q}) P_\alpha(\mathbf{q})}{u_{S\alpha}}, \tag{5.591}$$

which allows (5.590) to be rewritten as

$$I_\alpha(\mathbf{q}, \omega) = \frac{k_B T u_{S\alpha}}{\pi k_\alpha(\mathbf{q})(\omega^2 + u_{S\alpha}^2)}. \tag{5.592}$$

Light scattering is therefore controlled by the slow modes $u_{S\alpha}$, $\alpha = 1, 2$.

The integrated intensities are given by

$$\int_{-\infty}^{\infty} I_\alpha(\mathbf{q}, \omega) \, d\omega = \frac{k_B T}{k_\alpha(\mathbf{q})}, \qquad \alpha = 1, 2. \tag{5.593}$$

Further, from equations (5.570), (5.571), (5.580) and the definitions of P_α, Q_α and C_α we have the results

$$\frac{k_1(\mathbf{q})}{u_{S1}} = \gamma_1 - \frac{(\alpha_3 q_\perp^2 - \alpha_2 q_\parallel^2)^2}{\eta_1 q_\perp^4 + \eta_2 q_\parallel^4 + \frac{1}{2}\alpha_m q_\perp^2 q_\parallel^2} \equiv \eta_1^{eff}(\mathbf{q}), \tag{5.594}$$

$$\frac{k_2(\mathbf{q})}{u_{S2}} = \gamma_1 - \frac{\alpha_2^2 q_\parallel^2}{\eta_3 q_\perp^2 + \eta_2 q_\parallel^2} \equiv \eta_2^{eff}(\mathbf{q}), \tag{5.595}$$

where we have introduced definitions for the effective viscosities $\eta_1^{eff}(\mathbf{q})$ and $\eta_2^{eff}(\mathbf{q})$. Notice also that the bend and splay elastic constants are involved in (5.594) and that the bend and twist constants arise in (5.595), corresponding, respectively, to the displacements pictured in Fig. 5.13(b) and (c). The integrated intensities (5.593) lead to evaluations of the elastic constants K_1, K_2 and K_3. The observed experimental line widths $u_{S\alpha}$ allow measurements of viscosities via the results (5.594) and (5.595), once the elastic constants (or $k_\alpha(\mathbf{q})$) are determined; alternatively, if some viscosities are known, then these same results allow the measurement of elastic constants. Of course, if the elastic constants are known from other data, for example Freedericksz transitions, then the effective viscosities can be measured by observation of the corresponding line widths using (5.594) and (5.595).

Two special cases are worth mentioning: when $q_\perp = 0$ or when $q_\parallel = 0$. We have

$$\eta_1^{eff} = \eta_2^{eff} = \eta_{bend}, \qquad \text{when} \quad q_\perp = 0, \tag{5.596}$$

$$\eta_1^{eff} = \eta_{splay} \quad \text{and} \quad \eta_2^{eff} = \gamma_1, \qquad \text{when} \quad q_\parallel = 0, \tag{5.597}$$

where η_{bend} and η_{splay} are the usual bend and splay viscosities appearing in Table 4.1 on page 158. The measurement of the line widths u_{S1} and u_{S2} then lead, via the relations (5.557), (5.558), (5.594) and (5.595), to the quantities

$$u_{S1} = u_{S2} = \frac{K_3 q_{\parallel}^2}{\eta_{bend}}, \qquad \text{when} \quad q_{\perp} = 0, \qquad (5.598)$$

$$u_{S1} = \frac{K_1 q_{\perp}^2}{\eta_{splay}}, \quad u_{S2} = \frac{K_2 q_{\perp}^2}{\gamma_1}, \quad \text{when} \quad q_{\parallel} = 0. \qquad (5.599)$$

The scattering geometry in Fig. 5.12 corresponds to the case when $q_{\parallel} = 0$, as can be seen from Fig. 5.13. In this particular geometry, η_{splay} can be measured if the splay elastic constant K_1 is known, and vice-versa, by $(5.599)_1$; this is similarly the case for the usual rotational viscosity γ_1 and the twist elastic constant K_2, by $(5.599)_2$. As mentioned above, other geometries are feasible and have been reviewed by Gleeson [114].

The results of this Section on light scattering can be summarised as follows. Light scattering is dominated by the behaviour of the slow modes in which a distorted director configuration relaxes exponentially to an equilibrium director profile. Measurements of the integrated intensities of the light scattering determine information about the elastic constants via equation (5.593). Experimental measurements of the line widths of the scattered light provide information on the ratios of elastic constants and certain viscosity coefficients, as given by equations (5.594) and (5.595).

Leslie [168] has reviewed further aspects and developments of the theory outlined in this Section, including the application of magnetic fields, and has indicated some of the similarities of the above approach with that used for shear wave reflectance experiments, such as those carried out by Martinoty and Candau [193]. The mathematical analysis for light scattering is very similar to that for shear wave reflectance at a nematic–solid crystal interface which is relevant to ultrasound experiments; de Jeu [137] briefly outlines the experimental set-up and states the results for measuring three combinations of nematic viscosities, while a mathematical modification of the equations introduced above in this Section have been given in the review by Leslie [168], which the interested reader should consult for further details and associated references. A broader view on the above topics and results has been collected by Stephen and Straley [258]. For other brief comments on the physics and results of light scattering techniques the reader is referred to the shorter reviews by de Jeu [137], Moscicki [205] and Dunmur [68, §5.2], as well as Gleeson [114], who also states the results for the fluctuation modes in the presence of an electric field.

Chapter 6

Theory of Smectic C Liquid Crystals

6.1 Introduction

In this Chapter we shall present a brief account of static and dynamic theory for smectic C liquid crystals based upon the nonlinear continuum theory proposed in 1991 by Leslie, Stewart and Nakagawa [173]. This continuum theory originates from concepts similar to those discussed in Chapters 2 and 4 for nematic liquid crystals and many of the results and ideas for smectic C liquid crystal theory are analogous extensions of those employed in nematic theory. It is for these reasons that only a brief outline of the derivation of the results for smectics will be necessary, especially at this introductory level: attention will be drawn to relevant parallels with nematics at appropriate stages where the reader will be referred to earlier Chapters for more details. However, it should be pointed out that some of these results are by no means trivial extensions of nematic theory and that the interested reader should consult the relevant cited literature for full details: the objective here is to give an informative introduction to smectic C liquid crystal theory and some elementary applications. The applications and implications of this smectic liquid crystal theory are not as fully resolved as they are for the theory of nematic liquid crystals and therefore we restrict our discussion of applications to a relatively few cases which demonstrate how the theory can be used. Smectic liquid crystals are currently being investigated intensively by experimentalists and theoreticians and consequently the literature in this area is presently under development and expanding rapidly, which means that we can hope to present here only some basic applications and introductory results. Sections 6.2 and 6.3 concentrate on non-chiral smectic C (SmC) liquid crystals and we defer the discussion on chiral smectic C (SmC*) liquid crystals until Section 6.4; SmC* liquid crystals are also called ferroelectric smectic C liquid crystals.

Section 6.2 introduces the basic mathematical description of SmC liquid crystals and proceeds to discuss the associated elastic energy, magnetic and electric energies, equilibrium equations and an elementary Freedericksz transition. The dynamic theory will be summarised in Section 6.3. It is worth emphasising now that in this theory the viscous stress is identical for both SmC and SmC* liquid crystals and that the main distinction between chiral and non-chiral smectics is essentially given by the difference between the associated elastic energies and, as is often the case, electric and ferroelectric energies: SmC and SmC* liquid crystals share the same set of viscosity coefficients in the model to be introduced below. A short discussion on the smectic C viscosities is contained in Section 6.3.2 which is followed by examples of flow alignment for SmC liquid crystals in Section 6.3.3. Section 6.4 introduces a basic description of SmC* liquid crystals and the simplest forms of energies that may be associated with them. Comments on a simple model for a director reorientation are discussed in Section 6.4.3. The Chapter closes with Section 6.5, which mentions various topics concerned with SmC and SmC* liquid crystals.

6.2   Static Theory of Smectic C

SmC liquid crystals are layered structures in which the director **n** makes an angle θ with respect to the layer normal, this angle usually being temperature dependent. We shall consider only incompressible SmC liquid crystals in the isothermal state in which case it can be assumed that θ is a fixed angle; θ is sometimes called the *smectic tilt angle* or *smectic cone angle*. It will also be assumed that the smectic layers are spaced equidistantly in these materials with interlayer distance a, perhaps in the range $20 \sim 80$ Å. For many situations these assumptions on the smectic cone angle and interlayer distance appear reasonable, but it should be mentioned that they may be considered too restrictive in some circumstances. Nevertheless, the particular nonlinear theory presented here has proved useful in practical problems. The basic mathematical description of SmC follows that described by de Gennes and Prost [110, p.347] and employs two vectors, namely, a unit vector **a** defining the layer normal and a unit vector **c**, sometimes called the *c-director*, which is the unit orthogonal projection of **n** onto the smectic planes as shown in Fig. 6.1. It is easily seen that for **n** we have the relation

$$\mathbf{n} = \mathbf{a}\cos\theta + \mathbf{c}\sin\theta, \tag{6.1}$$

which will be useful in many instances. It proves mathematically convenient to introduce the unit vector **b** defined by

$$\mathbf{b} = \mathbf{a} \times \mathbf{c}, \tag{6.2}$$

which is clearly orthogonal to **a** and **c**. (The vector **b** is parallel to the direction of the spontaneous polarisation in SmC* liquid crystals, discussed in Section 6.4.) The vectors **b** and **c** are obviously tangential to the smectic layers. The two vectors **a** and **c** are clearly subject to the constraints

$$\mathbf{a} \cdot \mathbf{a} = 1, \qquad \mathbf{c} \cdot \mathbf{c} = 1, \qquad \mathbf{a} \cdot \mathbf{c} = 0. \tag{6.3}$$

Also, in the absence of any singularities or defects in the smectic layers the unit layer normal **a** is additionally subject to the constraint

$$\nabla \times \mathbf{a} = \mathbf{0}, \tag{6.4}$$

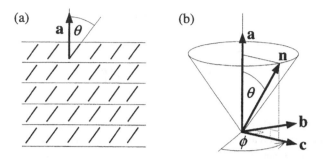

Figure 6.1: The vectors used to describe the layer structure and director alignment in smectic C liquid crystals. (a) The usual director, represented by a short bold line, makes a fixed angle θ to the local unit layer normal $\mathbf{a}$. This angle θ is usually called the smectic tilt angle or smectic cone angle. (b) The vector $\mathbf{c}$ is the unit orthogonal projection of the director $\mathbf{n}$ onto the smectic planes as shown and $\mathbf{b} = \mathbf{a} \times \mathbf{c}$ is a unit vector introduced for convenience. The director $\mathbf{n}$ may take any orientation on the surface of a 'fictitious' cone of semi-vertical angle θ as indicated; the orientation of $\mathbf{n}$ around this cone can be deduced from knowledge of the vectors $\mathbf{a}$ and $\mathbf{c}$ using the relation (6.1). In simple alignments, this orientation can be described by the 'phase' angle ϕ of the c-director measured relative to some fixed axis within the smectic planes.

first identified by Oseen [215] for smectic A (SmA) liquid crystals (where $\theta \equiv 0$) and later exploited by the Orsay Group [213] for general planar layers of SmC. This constraint, which is also valid for SmC* liquid crystals, can be understood as follows. If the smectic layers are incompressible and equidistant with interlayer distance a, then the integral

$$\frac{1}{a} \int_A^B \mathbf{a} \cdot d\mathbf{x},\tag{6.5}$$

represents a measure of the number of layers crossed by an observer travelling along the path from point A to point B. In well aligned layers free from defects the layer normal $\mathbf{a}$ can be considered differentiable. In particular, if we consider the local scenario pictured in Fig. 6.2 for planar layers seen in a two-dimensional cross-section, then the path from A to B can form a closed loop Γ so that

$$\oint_\Gamma \mathbf{a} \cdot d\mathbf{x} = 0,\tag{6.6}$$

because the number of layers crossed 'going up' equals the number of layers crossed 'coming down.' Since $\mathbf{a}$ is differentiable in the absence of defects we can apply Stokes' Theorem to obtain

$$\int_S (\nabla \times \mathbf{a}) \cdot \boldsymbol{\nu} \, dS = \oint_\Gamma \mathbf{a} \cdot d\mathbf{x} = 0,\tag{6.7}$$

where $\boldsymbol{\nu}$ is the unit outward normal to the area S enclosed by the loop Γ. Since Γ, and therefore the area S, is arbitrary, the result in equation (6.7) shows that the

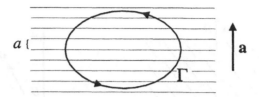

Figure 6.2: The lines represent layers of SmC liquid crystal which have the common unit layer normal **a** and interlayer distance a. When the layers are well aligned with no defects the line integral of **a** over an arbitrary loop Γ given by equation (6.6) must vanish, and so by Stokes' Theorem the constraint $\nabla \times \mathbf{a} = \mathbf{0}$ must hold.

constraint $\nabla \times \mathbf{a} = \mathbf{0}$ in (6.4) must hold at all points in the material. Notice that this result relies on the layers being locally parallel and that it is possible, by a suitably modified argument, to obtain the same result when the smectic layers are parallel but not necessarily spaced equidistantly.

The static (and dynamic) theory of SmC will be developed using the vectors **a** and **c** which are subject to the constraints contained in equations (6.3) and (6.4). These constraints will lead to four Lagrange multipliers in the theory: three scalar function multipliers arising from the three constraints in (6.3) and one vector function multiplier arising from the vector constraint in (6.4). Knowledge of the behaviour of **a** and **c** is sufficient to derive the orientation of the usual director **n** through the relation (6.1).

Observe that the constraints (6.3) and (6.4) lead to the relations (cf. equation (2.10))

$$a_i a_{i,j} = 0, \qquad c_i c_{i,j} = 0, \qquad a_i c_{i,j} + c_i a_{i,j} = 0, \qquad \epsilon_{ijk} a_{k,j} = 0. \qquad (6.8)$$

6.2.1 The Elastic Energy for Smectic C

As in the nematic theory discussed in Chapter 2, it will be assumed that there is a free energy density associated with distortions of the director **n**, which can be related to distortions of the vectors **a** and **c** in SmC liquid crystals. A detailed derivation of this free energy density will not be given here. Nevertheless, after stating some results for the energy, a physical interpretation of the elastic terms will be given which leads to an illustrative construction of the energy. The energy density is of the form

$$w = w(\mathbf{a}, \mathbf{c}, \nabla \mathbf{a}, \nabla \mathbf{c}), \qquad (6.9)$$

with the total free energy being given by

$$W = \int_V w(\mathbf{a}, \mathbf{c}, \nabla \mathbf{a}, \nabla \mathbf{c}) dV, \qquad (6.10)$$

where V is the sample volume. It will be supposed that the energy density w is quadratic in the gradients of **a** and **c**. Similar to the construction of the nematic energy, the energy density must be invariant to arbitrary superposed rigid body rotations and we consequently require

$$w(\mathbf{a}, \mathbf{c}, \nabla\mathbf{a}, \nabla\mathbf{c}) = w(Q\mathbf{a}, Q\mathbf{c}, Q\nabla\mathbf{a}Q^T, Q\nabla\mathbf{c}Q^T), \qquad (6.11)$$

where Q is any proper ($\det Q = 1$) orthogonal matrix, Q^T being its transpose; this requirement must necessarily hold for SmC* also. Moreover, for non-chiral SmC the requirement in (6.11) must additionally hold for any orthogonal matrix Q ($\det Q = \pm 1$). The energy has to be invariant to the simultaneous changes in sign

$$\mathbf{a} \rightarrow -\mathbf{a} \quad \text{and} \quad \mathbf{c} \rightarrow -\mathbf{c}, \qquad (6.12)$$

this invariance arising from a consideration of the symmetry invariance required when $\mathbf{n} \rightarrow -\mathbf{n}$. The resulting energy density w for non-chiral SmC then takes the form stated by Leslie, Stewart, Carlsson and Nakagawa [172]

$$
\begin{aligned}
w = {} & \tfrac{1}{2}K_1(\nabla\cdot\mathbf{a})^2 + \tfrac{1}{2}K_2(\nabla\cdot\mathbf{c})^2 + \tfrac{1}{2}K_3(\mathbf{a}\cdot\nabla\times\mathbf{c})^2 + \tfrac{1}{2}K_4(\mathbf{c}\cdot\nabla\times\mathbf{c})^2 \\
& + \tfrac{1}{2}K_5(\mathbf{b}\cdot\nabla\times\mathbf{c})^2 + K_6(\nabla\cdot\mathbf{a})(\mathbf{b}\cdot\nabla\times\mathbf{c}) + K_7(\mathbf{a}\cdot\nabla\times\mathbf{c})(\mathbf{c}\cdot\nabla\times\mathbf{c}) \\
& + K_8(\nabla\cdot\mathbf{c})(\mathbf{b}\cdot\nabla\times\mathbf{c}) + K_9(\nabla\cdot\mathbf{a})(\nabla\cdot\mathbf{c}),
\end{aligned} \qquad (6.13)
$$

where the K_i, $i = 1, 2, ...9$, are elastic constants, surface terms being omitted; recall that $\mathbf{b} = \mathbf{a} \times \mathbf{c}$. As in nematic theory, we shall always suppose that

$$w(\mathbf{a}, \mathbf{c}, \nabla\mathbf{a}, \nabla\mathbf{c}) \geq 0. \qquad (6.14)$$

This energy density can also be written in an equivalent Cartesian component form when surface terms which can be written as a divergence are neglected (cf. the comments after equation (2.50) on page 21). This form is often useful in calculations and can be expressed as [172]

$$
\begin{aligned}
w = {} & \tfrac{1}{2}K_1(a_{i,i})^2 + \tfrac{1}{2}(K_2 - K_4)(c_{i,i})^2 + \tfrac{1}{2}(K_3 - K_4)c_{i,j}c_jc_{i,k}c_k + \tfrac{1}{2}K_4 c_{i,j}c_{i,j} \\
& + \tfrac{1}{2}(K_5 - K_3)(c_ja_{i,j}c_j)^2 + K_6 a_{i,i}(c_ja_{j,k}c_k) - K_7 c_{i,j}c_jc_{i,k}a_k \\
& + (K_8 - K_7)c_{i,i}(c_ja_{j,k}c_k) + K_9 a_{i,i}c_{j,j}.
\end{aligned} \qquad (6.15)
$$

Notice that the constraint (6.4) is equivalent to the condition $a_{i,j} = a_{j,i}$ in Cartesian component form. Other equivalent forms for the energy in terms of any two of the vectors $\mathbf{a}$, $\mathbf{b}$ and $\mathbf{c}$ are available [172] and these allow comparisons with earlier results obtained by other workers, particularly the Orsay Group [213], Rapini [228], Dahl and Lagerwall [62] and Nakagawa [209]. It is worth remarking here that three surface terms have been identified for the SmC phase, namely [172],

$$
\begin{aligned}
S_1 &\equiv \nabla\cdot[\mathbf{c}(\nabla\cdot\mathbf{c}) - (\mathbf{c}\cdot\nabla)\mathbf{c}] = (c_ic_{j,j} - c_jc_{i,j})_{,i}, & (6.16) \\
S_2 &\equiv \nabla\cdot[\mathbf{a}(\nabla\cdot\mathbf{c}) - (\mathbf{a}\cdot\nabla)\mathbf{c}] = (a_ic_{j,j} - a_jc_{i,j})_{,i}, & (6.17) \\
S_3 &\equiv \nabla\cdot[(\nabla\cdot\mathbf{a})\mathbf{a}] = (a_{i,i})^2 - a_{i,j}a_{i,j} & (6.18) \\
&= -2(\mathbf{b}\cdot\nabla\times\mathbf{c})(\mathbf{c}\cdot\nabla\times\mathbf{b}) - 2\left[\tfrac{1}{2}(\mathbf{c}\cdot\nabla\times\mathbf{c} - \mathbf{b}\cdot\nabla\times\mathbf{b})\right]^2. & (6.19)
\end{aligned}
$$

The equality in (6.19) arises in Section 6.4.1 below and its derivation requires detailed manipulations of the identities contained in [35, 172].

For small perturbations to planar aligned layers of SmC the nonlinear energy coincides with the approximate energy introduced by the Orsay Group [213] and it is worthwhile here to pursue this version since it gives insight into the physical interpretation of the elastic constants. The Orsay Group, in their formulation of the elastic energy, describe small deformations to planar aligned smectic layers which initially have the Cartesian z-axis parallel to the layer normal $\mathbf{a}$ with the c-director parallel to the x-axis: see Fig. 6.3(a). The elastic deformations are described by setting

$$\mathbf{a} = \widehat{\mathbf{a}} + \boldsymbol{\Omega} \times \widehat{\mathbf{a}}, \qquad \mathbf{b} = \widehat{\mathbf{b}} + \boldsymbol{\Omega} \times \widehat{\mathbf{b}}, \qquad \mathbf{c} = \widehat{\mathbf{c}} + \boldsymbol{\Omega} \times \widehat{\mathbf{c}}, \qquad (6.20)$$

where

$$\widehat{\mathbf{a}} = (0,0,1), \quad \widehat{\mathbf{b}} = (0,1,0), \quad \widehat{\mathbf{c}} = (1,0,0) \quad \text{and} \quad \boldsymbol{\Omega} = (\Omega_x, \Omega_y, \Omega_z), \qquad (6.21)$$

with $\boldsymbol{\Omega}$ being an arbitrary small rotation of the smectic layers. We note for reference that

$$\mathbf{a} = (\Omega_y, -\Omega_x, 1), \qquad \mathbf{b} = (-\Omega_z, 1, \Omega_x), \qquad \mathbf{c} = (1, \Omega_z, -\Omega_y), \qquad (6.22)$$

and that the constraint (6.4) forces the requirements

$$\Omega_{x,z} = \Omega_{y,z} = 0, \qquad \Omega_{x,x} + \Omega_{y,y} = 0. \qquad (6.23)$$

Inserting these quantities into the energy given by equation (6.13), using the constraints (6.3) and (6.4) (now expressed via the results in equation (6.23)) and ignoring quantities that enter through the above surface terms S_1, S_2, and S_3, gives the version of the energy derived by the Orsay Group (see Reference [172] for details)

$$\begin{aligned} w \;=\; & \tfrac{1}{2}A_{12}(\Omega_{y,x})^2 + \tfrac{1}{2}A_{21}(\Omega_{x,y})^2 - A_{11}(\Omega_{x,x})^2 + \tfrac{1}{2}B_1(\Omega_{z,x})^2 + \tfrac{1}{2}B_2(\Omega_{z,y})^2 \\ & + \tfrac{1}{2}B_3(\Omega_{z,z})^2 + B_{13}\Omega_{z,x}\Omega_{z,z} - C_1\Omega_{x,x}\Omega_{z,x} + C_2\Omega_{x,y}\Omega_{z,y}, \end{aligned} \qquad (6.24)$$

provided we set

$$\begin{aligned} &K_1 = A_{21}, && K_2 = B_2, && K_3 = B_1, \\ &K_4 = B_3, && K_5 = 2A_{11} + A_{12} + A_{21} + B_3, && K_6 = -(A_{11} + A_{21} + \tfrac{1}{2}B_3), \qquad (6.25) \\ &K_7 = -B_{13}, && K_8 = C_1 + C_2 - B_{13}, && K_9 = -C_2. \end{aligned}$$

The above elastic constants are those used by the Orsay Group, except that for later notational convenience we have set $A_{11} = -\tfrac{1}{2}A_{11}^{Orsay}$ and $C_1 = -C_1^{Orsay}$. As we shall see below, the constants A_{12}, A_{21} and A_{11} are related to bending of the smectic layers (cf. de Gennes and Prost [110, p.346]) while the constants B_1, B_2, B_3 and B_{13}, originally introduced by Saupe [239] in an earlier description of smectics, are related to the reorientation of the c-director within or across layers. The constants C_1 and C_2 are related to various couplings of these deformations. The formulation of the energy in (6.24) is not entirely convenient for calculations and is relevant only for small distortions, which is why alternative nonlinear vector formulations have been sought.

It was identified by Carlsson, Stewart and Leslie [32] (see also [172]) that the SmC energy given by equation (6.13) has a particularly convenient equivalent form when constructed in terms involving only the vectors **b** and **c**. It is given by

$$
\begin{aligned}
w \; = \; & \tfrac{1}{2}A_{12}(\mathbf{b}\cdot\nabla\times\mathbf{c})^2 + \tfrac{1}{2}A_{21}(\mathbf{c}\cdot\nabla\times\mathbf{b})^2 + A_{11}(\mathbf{b}\cdot\nabla\times\mathbf{c})(\mathbf{c}\cdot\nabla\times\mathbf{b}) \\
& + \tfrac{1}{2}B_1(\nabla\cdot\mathbf{b})^2 + \tfrac{1}{2}B_2(\nabla\cdot\mathbf{c})^2 + \tfrac{1}{2}B_3\left[\tfrac{1}{2}(\mathbf{b}\cdot\nabla\times\mathbf{b}+\mathbf{c}\cdot\nabla\times\mathbf{c})\right]^2 \\
& + B_{13}(\nabla\cdot\mathbf{b})\left[\tfrac{1}{2}(\mathbf{b}\cdot\nabla\times\mathbf{b}+\mathbf{c}\cdot\nabla\times\mathbf{c})\right] \\
& + C_1(\nabla\cdot\mathbf{c})(\mathbf{b}\cdot\nabla\times\mathbf{c}) + C_2(\nabla\cdot\mathbf{c})(\mathbf{c}\cdot\nabla\times\mathbf{b}).
\end{aligned} \tag{6.26}
$$

This formulation, which shares some of the features of that first introduced by Rapini [228], has two illustrative advantages: firstly, it can be related directly to the Orsay Group formulation and, secondly, a physical interpretation of the basic elastic constants and their related deformations can be visualised easily. This version of the energy can be constructed by simple combinations of the basic deformations, as we shall now show with the aid of Fig. 6.3.

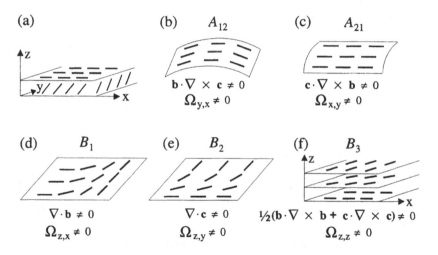

Figure 6.3: The five basic distortions for SmC liquid crystals. The bold lines parallel to the smectic planes represent the *c*-director. Each distortion is related to an elastic constant as indicated. There are also four 'coupling' terms involving combinations of these distortions, finally leading to a total energy having nine terms: see the text for details.

We follow the physical interpretation contained in [32]. In Fig. 6.3 the bold lines parallel to the smectic planes represent the orientation of the *c*-director. Figure 6.3(a) shows the undistorted state. In Fig. 6.3(b) the smectic plane is allowed to bend in such a way that the layer normal **a** changes as an observer travels along the direction of the *c*-director. From Fig. 6.3(b) it is clear that to achieve this distortion the z-component of *c*-director must change with respect to x; since $\mathbf{c} = (1, \Omega_z, -\Omega_y)$ (by equations (6.20) and (6.21)) it follows that the component Ω_y of the rotation Ω

must change with respect to x. This particular distortion therefore corresponds to $\Omega_{y,x} \neq 0$, or, equivalently, $\mathbf{b} \cdot \nabla \times \mathbf{c} \neq 0$, as can be seen by comparing the coefficients of the A_{12} terms in equations (6.24) and (6.26). A physical interpretation of the elastic constant A_{12} is therefore provided by Fig. 6.3(b). If the smectic plane is allowed to bend in such a way that $\mathbf{a}$ changes along the direction perpendicular to the c-director as pictured in Fig. 6.3(c), then the vector $\mathbf{b} = (-\Omega_z, 1, \Omega_x)$, which is perpendicular to $\mathbf{c}$, must change its z component with respect to y. This distortion therefore corresponds to $\Omega_{x,y} \neq 0$, which, by comparing the A_{21} terms in equations (6.24) and (6.26), is equivalent to saying $\mathbf{c} \cdot \nabla \times \mathbf{b} \neq 0$. The elastic constant A_{21} therefore has the physical interpretation depicted by Fig. 6.3(c). Two other possible distortions include those where the smectic layers may remain planar while the c-director rotates as an observer travels either parallel or perpendicular to the original local alignment of the c-director, as shown in Fig. 6.3(d) and Fig. 6.3(e), respectively. In Fig. 6.3(d) the y-component of $\mathbf{c}$ changes with respect to x, leading to the conclusion that $\Omega_{z,x} \neq 0$, equivalent to requiring $\nabla \cdot \mathbf{b} \neq 0$, as seen by comparing the B_1 terms in (6.24) and (6.26). The elastic constant B_1 is therefore related to the distortion of the c-director shown in Fig. 6.3(d). By similar reasoning, the y-component of $\mathbf{c}$ must change with respect to y to obtain the distortion shown in Fig. 6.3(e), leading to the requirement that $\Omega_{z,y} \neq 0$, equivalent to $\nabla \cdot \mathbf{c} \neq 0$. Therefore the distortion represented in Fig. 6.3(e) is related to the elastic constant B_2. Finally, we can consider the case when the c-director remains constant within each smectic layer but rotates as an observer travels along the direction of the layer normal, as shown in Fig. 6.3(f). This requires the y-component of $\mathbf{c}$ to change with respect to z, which demands that $\Omega_{z,z} \neq 0$. By comparing the B_3 terms in equations (6.24) and (6.26), this is equivalent to requiring $\frac{1}{2}(\mathbf{b} \cdot \nabla \times \mathbf{b} + \mathbf{c} \cdot \nabla \times \mathbf{c}) \neq 0$. The distortion pictured in Fig. 6.3(f) is therefore related to the elastic constant B_3. Five basic distortions have now been identified and depicted in Fig. 6.3.

We are now in a position to construct and justify the elastic energy given by equation (6.26). There are two groups of distortion terms which transform differently under the transformation (6.12). The first consists of the three terms $\nabla \cdot \mathbf{c}$, $\mathbf{b} \cdot \nabla \times \mathbf{c}$ and $\mathbf{c} \cdot \nabla \times \mathbf{b}$ and the second consists of the two terms $\nabla \cdot \mathbf{b}$ and $\frac{1}{2}(\mathbf{b} \cdot \nabla \times \mathbf{b} + \mathbf{c} \cdot \nabla \times \mathbf{c})$, recalling that $\mathbf{b} = \mathbf{a} \times \mathbf{c}$. As the elastic energy must be unaltered under the symmetry invariance requirement (6.12) we can only accept contributions to the energy which are squares of these basic five deformation terms or are formed by the product of two terms which come from the same group of terms listed above. It is obvious that there can only be four such products and when these are combined with the five squares of the basic distortion terms we arrive at the nine term energy stated in equation (6.26). The A_{11} term represents a coupling of the two distortions related to the A_{12} and A_{21} terms and the B_{13} term arises from a coupling of the two distortions related to the B_1 and B_3 terms. The C_1 term emerges as a coupling between the two distortions related to the B_2 and A_{12} terms while the C_2 term results from a coupling of the two distortions related to the B_2 and A_{21} terms. The above account briefly justifies and establishes the energy in terms of the vectors $\mathbf{b}$ and $\mathbf{c}$. The other equivalent versions of the energy mentioned above can be obtained via mathematical identities and the reader is referred to Reference [172] for details.

To fulfil the requirement (6.14) the energy w must be non-negative, as in nematic theory. This leads to restrictions on the values of the elastic constants. Notice that w given by equation (6.26) is actually a quadratic form in the five basic deformation operators. As an example, we follow Carlsson *et al.* [32] and imagine that a deformation of the smectic can be made where only the distortions related to layer bending occur. If we set $X_1 = (\mathbf{b} \cdot \nabla \times \mathbf{c})$ and $X_2 = (\mathbf{c} \cdot \nabla \times \mathbf{b})$ then the energy terms related to the A_{ij} terms can be written as the quadratic form

$$\tfrac{1}{2}A_{12}X_1^2 + \tfrac{1}{2}A_{21}X_2^2 + A_{11}X_1X_2 = \tfrac{1}{2}\begin{bmatrix} X_1 & X_2 \end{bmatrix}\begin{bmatrix} A_{12} & A_{11} \\ A_{11} & A_{21} \end{bmatrix}\begin{bmatrix} X_1 \\ X_2 \end{bmatrix}. \tag{6.27}$$

As already mentioned after equation (2.56), this quantity is non-negative if and only if the determinants of all the principal submatrices of the above symmetric matrix are non-negative. In this case the principal submatrices are the first entry in the matrix and the matrix itself, which leads to the conclusion that $A_{12} \geq 0$ and $A_{12}A_{21} - A_{11}^2 \geq 0$. The right-hand side of the expression for the quadratic form in (6.27) is identical if A_{12} is interchanged with A_{21} simultaneously with an interchange in the rôles of X_1 and X_2, so that we also have $A_{21} \geq 0$. By making analogous arguments for the terms involving the three other coupling terms in w in (6.26) we finally arrive at the inequalities (cf. [32, 213, 228])

$$A_{12}, A_{21}, B_1, B_2, B_3 \; \geq \; 0, \tag{6.28}$$
$$A_{12}A_{21} - A_{11}^2 \; \geq \; 0, \tag{6.29}$$
$$B_1 B_3 - B_{13}^2 \; \geq \; 0, \tag{6.30}$$
$$B_2 A_{12} - C_1^2 \; \geq \; 0, \tag{6.31}$$
$$B_2 A_{21} - C_2^2 \; \geq \; 0. \tag{6.32}$$

Notice that if $A_{12} = 0$ or $A_{21} = 0$, then $A_{11} = 0$ also, by (6.29). Similarly, from the above inequalities, we can establish: if $B_1 = 0$ or $B_3 = 0$ then $B_{13} = 0$; if $B_2 = 0$ or $A_{12} = 0$ then $C_1 = 0$; if $B_2 = 0$ or $A_{21} = 0$ then $C_2 = 0$.

It is possible to derive many other inequalities for these elastic constants, but we concentrate on some useful general inequalities which were first observed by Atkin and Stewart [9]. Clearly,

$$(A_{12} \pm A_{11})^2 = A_{12}^2 + A_{11}^2 \pm 2A_{12}A_{11} \geq 0. \tag{6.33}$$

Adding the inequality (6.29) to these inequalities gives

$$A_{12}(A_{12} + A_{21} \pm 2A_{11}) \geq 0. \tag{6.34}$$

By (6.28), $A_{12} \geq 0$ and therefore from (6.34) we can deduce that

$$A_{12} + A_{21} + 2A_{11} \geq 0 \quad \text{and} \quad A_{12} + A_{21} - 2A_{11} \geq 0, \tag{6.35}$$

which also implies that

$$|A_{11}| \leq \tfrac{1}{2}(A_{12} + A_{21}). \tag{6.36}$$

The inequality $(6.35)_1$ was first derived by Carlsson *et al.* [32] for the special case when the smectic tilt angle θ is assumed small (see the small angle approximations in equations (6.41) below), whereas the inequalities in (6.35) and (6.36) are valid for any tilt angle θ. Inequality (6.35)$_1$ has proved particularly useful in certain Freedericksz transitions and stability arguments for SmC liquid crystals confined to a wedge or cylindrical geometries [9, 32]. Similarly, by considering the quantities $(B_1 \pm B_{13})^2$, $(B_2 \pm C_1)^2$, $(B_2 \pm C_2)^2$ and adding the inequalities (6.30), (6.31) and (6.32) to each quantity, respectively, and using the results in (6.28), we can derive the additional inequalities

$$B_1 + B_3 \pm 2B_{13} \geq 0 \quad \text{and} \quad |B_{13}| \leq \tfrac{1}{2}(B_1 + B_3), \tag{6.37}$$

$$B_2 + A_{12} \pm 2C_1 \geq 0 \quad \text{and} \quad |C_1| \leq \tfrac{1}{2}(B_2 + A_{12}), \tag{6.38}$$

$$B_2 + A_{21} \pm 2C_2 \geq 0 \quad \text{and} \quad |C_2| \leq \tfrac{1}{2}(B_2 + A_{21}). \tag{6.39}$$

It is very common for the quantities $A_{12}+A_{11}$ and $A_{21}+A_{11}$ to appear in calculations and it is important to realise that they cannot be negative simultaneously because of inequality (6.35)$_1$.

One final set of properties for the elastic constants is worth noting. As the smectic tilt angle θ tends to zero the elastic energy ought to converge to that for the SmA phase, given by [110, 258]

$$w_{A\,elast} = \tfrac{1}{2}K_1(\nabla \cdot \mathbf{a})^2, \tag{6.40}$$

where $K_1 \geq 0$ is the splay elastic constant of nematic theory. By employing the symmetry of the SmC phase, Carlsson et al. [32] were able use the techniques introduced by Dahl and Lagerwall [62] in order to postulate the tilt angle dependence of the SmC elastic constants for small θ, based upon arguments similar to those used in Section 6.3.2 below when the SmC viscosities are discussed. The results showed that

$$A_{12} = K_1 + \overline{A}_{12}\theta^2, \quad A_{21} = K_1 + \overline{A}_{21}\theta^2, \quad A_{11} = -K_1 + \overline{A}_{11}\theta^2,$$

$$B_1 = \overline{B}_1\theta^2, \qquad\qquad B_2 = \overline{B}_2\theta^2, \qquad\qquad B_3 = \overline{B}_3\theta^2, \tag{6.41}$$

$$B_{13} = \overline{B}_{13}\theta^3, \qquad\qquad C_1 = \overline{C}_1\theta, \qquad\qquad C_2 = \overline{C}_2\theta,$$

where the elastic constants K_1, $\overline{A}_i$, $\overline{B}_i$ and $\overline{C}_i$ are assumed to be only weakly temperature dependent (recall that θ is temperature dependent). These approximations may be useful when trying to establish what ought to be the dominant elastic constants in the SmC phase for particular problems.

One valuable identity is worth recording. It has been proved in References [228] and [172, Eqns. (A17), (A21)] that

$$(\nabla \cdot \mathbf{a})^2 = (\mathbf{b} \cdot \nabla \times \mathbf{c})^2 + (\mathbf{c} \cdot \nabla \times \mathbf{b})^2 - 2(\mathbf{b} \cdot \nabla \times \mathbf{c})(\mathbf{c} \cdot \nabla \times \mathbf{b}). \tag{6.42}$$

This identity allows us to prove that the inequalities stated above at equations (6.28), (6.29), (6.35) and (6.36) for the SmC elastic constants will also be valid for the 'barred' constants which appear in the expressions stated in (6.41). This is accomplished by inserting the expressions (6.41) into the energy density w in (6.26) and repeating the arguments used above for the resulting quadratic forms in K_1 and the barred constants.

Estimates and Evaluations for the SmC Elastic Constants

Some of the elastic constants of the SmC phase are expected to be roughly of the same order of magnitude as those in the nematic phase, while others are more open to speculation at the present time due to a scarcity of available experimental data. Estimates for the smectic elastic constants have been obtained by using a basic comparison with the elastic constants for nematics as suggested, for example, by Lagerwall and Dahl [157]. In this case, a typical SmC material having a smectic cone angle $\theta \approx 22.5°$ would be expected to have the B_i elastic constants lying in the ranges, in piconewtons (pN) (1 pN $= 10^{-12}$ N) [28],

$$1 \leq B_1 \leq 10, \qquad 5 \leq B_2 \leq 100, \qquad 1 \leq B_3 \leq 10, \qquad 0.1 \leq B_{13} \leq 1. \qquad (6.43)$$

Schiller and Pelzl [244] were able to use experimental data to estimate an effective elastic constant for the SmC liquid crystal 5-n-heptyl-2-[4-n-nonyloxyphenyl]-pyrimidine which, at 47°C, has the dielectric anisotropy $\epsilon_a = 0.47$ and smectic cone angle $\theta = 20°$. They chose to set $B_1 = B_2$ and defined the effective elastic constant $B = B_1 \cos^2\theta + B_3 \sin^2\theta$ to find that their results predicted a value of $B = 6.4 \times 10^{-12}$ N.

Measurements of the layer bending constants A_{12}, A_{21} and A_{11}, and the coupling constants C_1 and C_2, are rare. Nevertheless, preliminary results for a combination of the A_i constants for a SmC liquid crystal have been reported by Findon [87] and Findon and Gleeson [88]. In the notation introduced earlier they found that

$$\overline{A}_{21} + \overline{A}_{11} \sim (-1.3 \pm 0.2) \times 10^{-4} \text{ N}, \qquad (6.44)$$

perhaps larger than expected, for the particular material M3 (see [87] for the chemical composition of this liquid crystal), although it should be remembered that these results are valid where $A_{21} + A_{11} \sim [\overline{A}_{21} + \overline{A}_{11}]\theta^2$ when θ is small. (Note that there is a minus sign missing in Table 1 in [88].) This result was obtained via a Freedericksz transition by using a 'wedge' geometry, with wedge angle $\beta = 2 \times 10^{-3}$ radians, in the experiment suggested by Carlsson, Stewart and Leslie [32]. By the inequalities (6.28) and (6.35)$_1$ (in terms of the 'barred' quantities), the result (6.44) implies that for this particular material we must have

$$\overline{A}_{12} + \overline{A}_{11} > 0, \qquad \overline{A}_{11} < 0, \qquad \overline{A}_{12} > -\overline{A}_{11} > \overline{A}_{21} > 0, \qquad (6.45)$$

with, of course, the other analogous inequalities in equations (6.29), (6.35)$_2$ and (6.36) also being valid. In particular, since $\overline{A}_{11} < 0$, we must have $\overline{A}_{21} \neq 0$ by (6.29), which ensures that $\overline{A}_{21} > 0$.

Notice that the elastic constants A_i, B_i and C_i all have the dimensions of energy per unit length (J m^{-1} = N).

The Elastic Energy for SmC$_M$

We close this Section by commenting on a special case of the SmC energy. If the invariance condition (6.12) is changed to the requirement that the energy remains invariant to the transformation $\mathbf{a} \to -\mathbf{a}$ and $\mathbf{c} \to -\mathbf{c}$ *separately*, then the nine-term energy density (6.13) reduces to six terms, namely,

$$w = \tfrac{1}{2}K_1(\nabla \cdot \mathbf{a})^2 + \tfrac{1}{2}K_2(\nabla \cdot \mathbf{c})^2 + \tfrac{1}{2}K_3(\mathbf{a} \cdot \nabla \times \mathbf{c})^2 + \tfrac{1}{2}K_4(\mathbf{c} \cdot \nabla \times \mathbf{c})^2$$
$$+ \tfrac{1}{2}K_5(\mathbf{b} \cdot \nabla \times \mathbf{c})^2 + K_6(\nabla \cdot \mathbf{a})(\mathbf{b} \cdot \nabla \times \mathbf{c}). \tag{6.46}$$

This corresponds to the energy for what has been termed the smectic C_M (SmC$_M$) phase of liquid crystals, the subscript M denoting the name of W.L. McMillan, who was first to propose such a phase. This phase has been described by Brand and Pleiner [26], who also looked at some hydrodynamic properties.

6.2.2 The Magnetic and Electric Energies

It is common practice in elementary problems to assume that the magnetic and electric energy densities for SmC liquid crystals are the same as those defined for nematics in Chapter 2. The discussion and comments in Section 2.3 on the influence of magnetic and electric fields remain valid, except that smectic C materials may have different magnitudes of dielectric anisotropy or magnetic anisotropy. However, these values for smectic C materials may be approximated by known results from nematic theory. The director $\mathbf{n}$ appearing in the magnetic and electric energies is simply replaced by using the relation $\mathbf{n} = \mathbf{a}\cos\theta + \mathbf{c}\sin\theta$ stated in equation (6.1). When the contribution independent of the orientation of $\mathbf{a}$ and $\mathbf{c}$ is omitted from equation (2.98), which is often the case for the same reasons mentioned on page 30, the magnetic energy density in SI units is obtained from equation (2.99) to find

$$w_{mag} = -\tfrac{1}{2}\mu_0\Delta\chi \left(\mathbf{a} \cdot \mathbf{H}\cos\theta + \mathbf{c} \cdot \mathbf{H}\sin\theta\right)^2, \tag{6.47}$$

where $\mathbf{H}$ is the magnetic field, μ_0 is the permeability of free space and $\Delta\chi$ is the unitless magnetic anisotropy (see (2.97)$_3$). It is also possible to give a magnetic energy in terms of the magnetic induction where it can be seen from equation (2.102) that we have

$$w_{mag} = -\tfrac{1}{2}\mu_0^{-1}\Delta\chi \left(\mathbf{a} \cdot \mathbf{B}\cos\theta + \mathbf{c} \cdot \mathbf{B}\sin\theta\right)^2, \tag{6.48}$$

where $\mathbf{B}$ is the magnetic induction.

If, as above, the contribution independent of the orientation of $\mathbf{a}$ and $\mathbf{c}$ is neglected then the electric energy density follows from analogous reasoning using equation (2.87) to give

$$w_{elec} = -\tfrac{1}{2}\epsilon_0\epsilon_a \left(\mathbf{a} \cdot \mathbf{E}\cos\theta + \mathbf{c} \cdot \mathbf{E}\sin\theta\right)^2, \tag{6.49}$$

where $\mathbf{E}$ is the electric field, ϵ_0 is the permittivity of free space and ϵ_a is the unitless dielectric anisotropy, which may be negative or positive. Typical values for ϵ_a may be estimated to lie in the range $|\epsilon_a| \lesssim 10$ for SmC materials [110, p.389] (cf. page 257). The expression (6.49) may be adopted when the dielectric biaxiality of the SmC phase is considered small: see Lagerwall [158, pp.212–213] and the brief discussion on page 311 below.

It is worth remarking that the forms for the electric and magnetic energies in equations (2.86), (2.98) and (2.100) may nevertheless prove useful in some instances, as mentioned in Section 2.3.

The comments on fields and units in Section 2.3.3 should also be consulted if the magnetic or electric energies are to be set in Gaussian units. For a description in Gaussian units we simply replace $\mu_0 \Delta \chi$ appearing in the above energy (6.47) by χ_a; recall, however, that in numerical examples of data we must set $\mu_0 = 1$ in Gaussian units, and, from equation (2.105), remember that numerically $\Delta \chi = 4\pi \chi_a$ where $\Delta \chi$ is measured relative to SI units and χ_a is in cgs units.

Similarly, we may simply replace $\frac{1}{2} \epsilon_0 \epsilon_a$ in (6.49) by $\epsilon_a / 8\pi$ to obtain the electric energy in Gaussian units (cf. (2.111)). The conversion formulae in Table D.2 may also be used if results in the literature are to be compared.

6.2.3 Equilibrium Equations

The derivation of the equilibrium equations for SmC liquid crystals parallels that outlined in Section 2.4 for nematic and cholesteric liquid crystals, this approach being based on work by Ericksen [73, 74]. The energy density will be described in terms of the vectors **a** and **c**, and the equilibrium equations and static theory will be phrased in this formulation; these vectors turn out to be particularly suitable for the mathematical description of statics and dynamics. We assume that the variation of the total energy at equilibrium satisfies a principle of virtual work for a given volume V of SmC liquid crystal of the form postulated by Leslie, Stewart and Nakagawa [173]

$$\delta \int_V w \, dV = \int_V (\mathbf{F} \cdot \delta \mathbf{x} + \mathbf{G}^a \cdot \Delta \mathbf{a} + \mathbf{G}^c \cdot \Delta \mathbf{c}) \, dV$$
$$+ \int_S (\mathbf{t} \cdot \delta \mathbf{x} + \mathbf{s}^a \cdot \Delta \mathbf{a} + \mathbf{s}^c \cdot \Delta \mathbf{c}) \, dS, \qquad (6.50)$$

where

$$\Delta a_i = \delta a_i + a_{i,j} \delta x_j, \qquad \Delta c_i = \delta c_i + c_{i,j} \delta x_j, \qquad (6.51)$$

with $\delta \mathbf{x}$ being the virtual displacements (cf. equations (2.126) and (2.127) in Section 2.4.2). In the above, S is the boundary surface, w is the SmC energy density, $\mathbf{F}$ is the external body force per unit volume, $\mathbf{G}^a$ and $\mathbf{G}^c$ are generalised external body forces per unit volume related to **a** and **c**, respectively, $\mathbf{t}$ is the surface traction per unit area, and $\mathbf{s}^a$ and $\mathbf{s}^c$ are generalised surface tractions per unit area related to **a** and **c**, respectively. The virtual displacements $\delta \mathbf{x}$ and the variations in the vectors **a** and **c** are also subject to the constraints (cf. equation (2.128))

$$(\delta x_i)_{,i} = 0, \qquad (6.52)$$
$$a_i \delta a_i = 0, \qquad a_i \Delta a_i = 0, \qquad (6.53)$$
$$c_i \delta c_i = 0, \qquad c_i \Delta c_i = 0, \qquad (6.54)$$
$$a_i \delta c_i + c_i \delta a_i = 0, \qquad a_i \Delta c_i + c_i \Delta a_i = 0. \qquad (6.55)$$

The first of these conditions is due to the assumption of incompressibility and the others follow from the constraints (6.3). By carrying out calculations which parallel those required in equations (2.130) to (2.134) we arrive at the relation

$$\delta \int_V w\, dV = \int_V \left\{ \left[\frac{\partial w}{\partial a_i} - \left(\frac{\partial w}{\partial a_{i,j}} \right)_{,j} \right] \Delta a_i + \left[\frac{\partial w}{\partial c_i} - \left(\frac{\partial w}{\partial c_{i,j}} \right)_{,j} \right] \Delta c_i \right.$$
$$\left. + \left[\left(\frac{\partial w}{\partial a_{k,j}} a_{k,i} \right)_{,j} + \left(\frac{\partial w}{\partial c_{k,j}} c_{k,i} \right)_{,j} \right] \delta x_i \right\} dV$$
$$+ \int_S \left[\frac{\partial w}{\partial a_{i,j}} \Delta a_i + \frac{\partial w}{\partial c_{i,j}} \Delta c_i - \left(\frac{\partial w}{\partial a_{k,j}} a_{k,i} + \frac{\partial w}{\partial c_{k,j}} c_{k,i} \right) \delta x_i \right] dS_j.$$
$$(6.56)$$

To establish the equilibrium equations it only remains to take into account the four constraints in equations (6.3) and (6.4) and the assumption that the mass density is constant, that is $\rho(\mathbf{x}) = \rho_0$, ρ_0 a constant. These constraints may be written as

$$\psi_1 \ = \ \rho - \rho_0 = 0, \tag{6.57}$$
$$\psi_2 \ = \ \tfrac{1}{2}(a_i a_i - 1) = 0, \tag{6.58}$$
$$\psi_3 \ = \ \tfrac{1}{2}(c_i c_i - 1) = 0, \tag{6.59}$$
$$\psi_4 \ = \ a_i c_i = 0, \tag{6.60}$$
$$\psi_{5p} \ = \ \epsilon_{pjk} a_{k,j} = 0, \quad p = 1, 2, 3. \tag{6.61}$$

First observe that for ψ_4 we have, using the general relation (2.123),

$$0 \equiv \lambda_4 \delta \psi_4 \ = \ \lambda_4 \left(\frac{\partial \psi_4}{\partial a_i} \delta a_i + \frac{\partial \psi_4}{\partial c_i} \delta c_i + \psi_{4,i} \delta x_i \right)$$
$$= \ \lambda_4 \left(\frac{\partial \psi_4}{\partial a_i} \delta a_i + \frac{\partial \psi_4}{\partial a_i} a_{i,j} \delta x_j + \frac{\partial \psi_4}{\partial c_i} \delta c_i + \frac{\partial \psi_4}{\partial c_i} c_{i,j} \delta x_j \right)$$
$$= \ \lambda_4 (c_i \Delta a_i + a_i \Delta c_i), \tag{6.62}$$

which is an extension of the results in equation (2.137) and the integrand in equation (A.7). In conjunction with this observation, and because the constraint ψ_1 is the same as that in (2.135), the constraints ψ_1 to ψ_4 lead to a result similar in form to equation (2.138) on page 37, namely,

$$0 \ = \ \int_V \left[(\lambda_1 \rho)_{,i} \delta x_i + \lambda_2 a_i \Delta a_i + \lambda_3 c_i \Delta c_i + \lambda_4 (c_i \Delta a_i + a_i \Delta c_i) \right] dV$$
$$- \int_S \lambda_1 \rho \delta_{ij} \delta x_i dS_j, \tag{6.63}$$

which can be obtained by analogy with the expressions manipulated in Appendix A at equations (A.6) and (A.7). Each λ_i is a scalar function Lagrange multiplier corresponding to each constraint ψ_i, respectively, $i = 1, 2, 3, 4$. Regarding ψ_{5p},

$p = 1, 2, 3$, first notice that the analogue of (2.137), or (6.62) above, in this case is given by

$$0 \equiv \lambda_{5p} \delta \psi_{5p} = \lambda_{5p} \left(\frac{\partial \psi_{5p}}{\partial a_{r,s}} \delta(a_{r,s}) + \frac{\partial \psi_{5p}}{\partial a_{r,s}} a_{r,si} \delta x_i \right), \tag{6.64}$$

with λ_{5p}, $p = 1, 2, 3$, being scalar function Lagrange multipliers. By using the definition for the operator Δ given in $(6.51)_1$ and the corresponding result for $\mathbf{u} = \mathbf{a}$ in equation (2.121) we therefore obtain

$$
\begin{aligned}
0 &= \lambda_{5p} \epsilon_{pjk} \left[\delta(a_{k,j}) + a_{k,ji} \delta x_i \right] \\
&= \lambda_{5p} \epsilon_{pjk} \left[\Delta(a_{k,j}) \right] \\
&= \lambda_{5p} \epsilon_{pjk} \left[(\delta a_k + a_{k,i} \delta x_i)_{,j} - a_{k,i} (\delta x_i)_{,j} \right] \\
&= \lambda_{5p} \epsilon_{pjk} \left[(\Delta a_k)_{,j} - a_{k,i} (\delta x_i)_{,j} \right]. \tag{6.65}
\end{aligned}
$$

Integrating by parts then shows

$$
\begin{aligned}
0 &= \int_V \lambda_{5p} \epsilon_{pjk} \left[(\Delta a_k)_{,j} - a_{k,i} (\delta x_i)_{,j} \right] dV \\
&= \int_V \left[(\lambda_{5p} \epsilon_{pjk} a_{k,i})_{,j} \delta x_i - (\lambda_{5p})_{,j} \epsilon_{pji} \Delta a_i \right] dV \\
&\quad + \int_S \left[-\lambda_{5p} \epsilon_{pjk} a_{k,i} \delta x_i + \lambda_{5p} \epsilon_{pji} \Delta a_i \right] dS_j, \tag{6.66}
\end{aligned}
$$

relabelling indices as appropriate. For ease of notation later, we redefine the Lagrange multipliers by

$$p = \lambda_1 \rho, \quad \gamma = -\lambda_2, \quad \tau = -\lambda_3, \quad \mu = -\lambda_4, \quad \beta_p = -\lambda_{5p}, \tag{6.67}$$

where p can be considered as the arbitrary pressure arising from the assumed incompressibility. Making the necessary replacements in equations (6.63) and (6.66) and adding these two equations produces the result

$$
\begin{aligned}
0 &= \int_V \left[\{ p_{,i} - (\beta_p \epsilon_{pjk} a_{k,i})_{,j} \} \delta x_i - \{ \gamma a_i + \epsilon_{ijp} \beta_{p,j} \} \Delta a_i \right. \\
&\quad \left. -\tau c_i \Delta c_i - \mu (c_i \Delta a_i + a_i \Delta c_i) \right] dV \\
&\quad + \int_S \left[\{ -p \delta_{ij} + \beta_p \epsilon_{pjk} a_{k,i} \} \delta x_i + \epsilon_{ijp} \beta_p \Delta a_i \right] dS_j, \tag{6.68}
\end{aligned}
$$

observing that $\epsilon_{pji} = -\epsilon_{ijp}$. We are now in a position to add equation (6.68) to the right-hand side of equation (6.56), since (6.68) constitutes a zero contribution. This then allows a reformulation of equation (6.56) in the full incompressible case, giving

$$
\begin{aligned}
\delta \int_V w \, dV &= -\int_V \left[t_{ij,j} \delta x_i + (\Pi_i^a + \gamma a_i + \mu c_i + \epsilon_{ijp} \beta_{p,j}) \Delta a_i + (\Pi_i^c + \tau c_i + \mu a_i) \Delta c_i \right] dV \\
&\quad + \int_S \left[t_{ij} \delta x_i + s_{ij}^a \Delta a_i + s_{ij}^c \Delta c_i \right] dS_j, \tag{6.69}
\end{aligned}
$$

where

$$t_{ij} = -p\delta_{ij} + \beta_p \epsilon_{pjk} a_{k,i} - \frac{\partial w}{\partial a_{k,j}} a_{k,i} - \frac{\partial w}{\partial c_{k,j}} c_{k,i}, \tag{6.70}$$

$$s_{ij}^a = \epsilon_{ijp}\beta_p + \frac{\partial w}{\partial a_{i,j}}, \tag{6.71}$$

$$s_{ij}^c = \frac{\partial w}{\partial c_{i,j}}, \tag{6.72}$$

$$\Pi_i^a = \left(\frac{\partial w}{\partial a_{i,j}}\right)_{,j} - \frac{\partial w}{\partial a_i}, \tag{6.73}$$

$$\Pi_i^c = \left(\frac{\partial w}{\partial c_{i,j}}\right)_{,j} - \frac{\partial w}{\partial c_i}. \tag{6.74}$$

The components t_{ij} form the stress tensor and the components s_{ij}^a and s_{ij}^c belong to torque stresses. Recalling that dS_j can be replaced by $\nu_j dS$ where ν is the outward unit normal to the surface S, we can use the constraints $(6.53)_2$, $(6.54)_2$ and $(6.55)_2$ and equate the corresponding surface terms appearing as coefficients of δx_i, Δa_i and Δc_i in the surface integrals in equations (6.50) and (6.69). Doing so allows us to express the surface forces in terms of the corresponding stress tensors and the outward unit normal ν as

$$t_i = t_{ij}\nu_j, \tag{6.75}$$

$$s_i^a = \alpha_1 a_i + \alpha_3 c_i + s_{ij}^a \nu_j, \tag{6.76}$$

$$s_i^c = \alpha_2 c_i + \alpha_3 a_i + s_{ij}^c \nu_j, \tag{6.77}$$

where the scalar functions α_1, α_2 and α_3 arise from the constraints $(6.53)_2$, $(6.54)_2$ and $(6.55)_2$, respectively.

The Equilibrium Equations

Following the method adopted for nematics, we are now in a position to equate the coefficients of δx_i, Δa_i and Δc_i appearing on the right-hand sides of the volume integrals in the principle of virtual work (6.50) and equation (6.69). This yields three sets of differential equations for equilibrium in the bulk of the sample, namely,

$$t_{ij,j} + F_i = 0, \tag{6.78}$$

$$\left(\frac{\partial w}{\partial a_{i,j}}\right)_{,j} - \frac{\partial w}{\partial a_i} + G_i^a + \gamma a_i + \mu c_i + \epsilon_{ijk}\beta_{k,j} = 0, \tag{6.79}$$

$$\left(\frac{\partial w}{\partial c_{i,j}}\right)_{,j} - \frac{\partial w}{\partial c_i} + G_i^c + \tau c_i + \mu a_i = 0. \tag{6.80}$$

Equation (6.78) represents a balance of forces, while, similar to the static theory of nematics, equations (6.79) and (6.80) are equivalent to a balance of moments; see also Remark (i) below. Notice that the 'a-equations' in (6.79) are coupled to the 'c-equations' in (6.80) via the multiplier μ. Recall that the Lagrange multipliers γ, μ and τ are scalar valued functions while β is a vector function.

A Simplification

A major simplification to these equations, which allows us to neglect equation (6.78), is possible if we consider the body forces $\mathbf{F}$, $\mathbf{G}^a$ and $\mathbf{G}^c$ to be specified functions of $\mathbf{x}$, $\mathbf{a}$ and $\mathbf{c}$ and follow a similar procedure to that outlined for nematics at equations (2.149) to (2.153). Suppose that

$$F_i = \frac{\partial \Psi}{\partial x_i}, \quad G_i^a = \frac{\partial \Psi}{\partial a_i}, \quad G_i^c = \frac{\partial \Psi}{\partial c_i}, \tag{6.81}$$

where $\Psi(\mathbf{a}, \mathbf{c}, \mathbf{x})$ is a scalar energy density function, an obvious special case being $\Psi \equiv 0$ when external body forces and generalised external body forces are absent. Substituting $(6.81)_1$ and (6.70) into the balance of forces (6.78) gives

$$\frac{\partial \Psi}{\partial x_i} - p_{,i} + \epsilon_{kpj}\beta_{p,j}a_{k,i} - \left(\frac{\partial w}{\partial a_{k,j}}\right)_{,j} a_{k,i} - \frac{\partial w}{\partial a_{k,j}} a_{k,ij}$$
$$- \left(\frac{\partial w}{\partial c_{k,j}}\right)_{,j} c_{k,i} - \frac{\partial w}{\partial c_{k,j}} c_{k,ij} = 0, \tag{6.82}$$

using the constraint (6.4) in the term involving the multiplier β. We can then use equations (6.79) and (6.80), with $\mathbf{G}^a$ and $\mathbf{G}^c$ given by $(6.81)_{2,3}$, to substitute for $(\partial w/\partial a_{k,j})_{,j}$ and $(\partial w/\partial c_{k,j})_{,j}$ in (6.82) to find that, when partial derivatives commute and the constraints (6.3) and (6.4) are taken into account via the relations (6.8),

$$\frac{\partial \Psi}{\partial a_j}a_{j,i} + \frac{\partial \Psi}{\partial c_j}c_{j,i} + \frac{\partial \Psi}{\partial x_i} - p_{,i} - \frac{\partial w}{\partial a_j}a_{j,i} - \frac{\partial w}{\partial a_{k,j}}a_{k,ji} - \frac{\partial w}{\partial c_j}c_{j,i} - \frac{\partial w}{\partial c_{k,j}}c_{k,ji} = 0. \tag{6.83}$$

This expression can, as in the nematic case, be expressed more conveniently by appropriately applying the general rule (2.123) to find

$$(\Psi - p - w)_{,i} = 0, \tag{6.84}$$

recalling that $w = w(\mathbf{a}, \mathbf{c}, \nabla\mathbf{a}, \nabla\mathbf{c})$. The balance of forces (6.78) therefore provides an expression for the pressure p, namely,

$$p + w - \Psi = p_0, \tag{6.85}$$

where p_0 is an arbitrary constant. Similar to the situation in nematic theory (see equation (2.153)), this has the important consequence that if $\mathbf{F}$, $\mathbf{G}^a$ and $\mathbf{G}^c$ are given by the expressions in (6.81), then we have to solve only equations (6.79) and (6.80). In this case, equation (6.78) can be ignored and need not concern us, unless there is a desire to compute forces. Further, it is seen that equations (6.3), (6.4), (6.79) and (6.80) give twelve equations in the twelve unknowns a_i, c_i, β_i, γ, μ and τ, the pressure p being given via equation (6.85).

Remark (i)

As in the nematic liquid crystal case, it is clear that (6.78) represents a balance of forces at equilibrium. This can be seen by applying arguments that parallel those in Remark (i) on page 40. Similarly, it can be shown that the equilibrium equations (6.79) and (6.80) actually arise from a balance of moments, as has been demonstrated by Stewart and McKay [267], by means of an appropriate extension to the Ericksen identity (B.6) (cf. [181]). For the present, however, we restrict our attention to the derivation of the couple stress tensor l_{ij} and its associated couple stress vector l_i. Analogous to the discussion for the nematic case in Remark (i) on page 40 when the balance of moments was discussed, consider an arbitrary, infinitesimal, rigid body rotation ω for which

$$\delta \mathbf{x} = \omega \times \mathbf{x}, \qquad \Delta \mathbf{a} = \omega \times \mathbf{a}, \qquad \Delta \mathbf{c} = \omega \times \mathbf{c}. \tag{6.86}$$

It can be shown that putting these expressions into the principle of virtual work (6.50) leads to an extension of the result in equation (2.163) given by

$$\int_V (\mathbf{x} \times \mathbf{F} + \mathbf{a} \times \mathbf{G}^a + \mathbf{c} \times \mathbf{G}^c)\, dV + \int_S (\mathbf{x} \times \mathbf{t} + \mathbf{a} \times \mathbf{s}^a + \mathbf{c} \times \mathbf{s}^c)\, dS = \mathbf{0}. \tag{6.87}$$

The generalised body and surface forces are related to the body moment $\mathbf{K}$ and couple stress vector $\mathbf{l}$ by, respectively,

$$\mathbf{K} = \mathbf{a} \times \mathbf{G}^a + \mathbf{c} \times \mathbf{G}^c, \qquad \mathbf{l} = \mathbf{a} \times \mathbf{s}^a + \mathbf{c} \times \mathbf{s}^c. \tag{6.88}$$

It then follows from (6.71), (6.72), (6.76) and (6.77) that, contracting the alternators when appropriate,

$$l_i = l_{ij}\nu_j = \epsilon_{ipq}\left(a_p s_q^a + c_p s_q^c\right), \tag{6.89}$$

$$l_{ij} = \epsilon_{ipq}\left(a_p s_{qj}^a + c_p s_{qj}^c\right) = \beta_p a_p \delta_{ij} - \beta_i a_j + \epsilon_{ipq}\left(a_p \frac{\partial w}{\partial a_{q,j}} + c_p \frac{\partial w}{\partial c_{q,j}}\right). \tag{6.90}$$

Remark (ii)

The expressions for $\mathbf{\Pi}^a$ and $\mathbf{\Pi}^c$ in (6.73) and (6.74), which also appear in the equilibrium equations (6.79) and (6.80), can be reformulated in general vector form following the methodology of Nakagawa [210]. It is a straightforward exercise to find the general form when

$$w = w_{elas} + w_{mag}, \tag{6.91}$$

where w_{elas} is the SmC elastic free energy density given by (6.13) and w_{mag} is the magnetic energy density given by (6.47). The results are

$$
\begin{aligned}
\mathbf{\Pi}^a \;=\; & K_1\nabla(\nabla \cdot \mathbf{a}) - K_3(\mathbf{a} \cdot \nabla \times \mathbf{c})(\nabla \times \mathbf{c}) - K_5(\mathbf{b} \cdot \nabla \times \mathbf{c})(\mathbf{c} \times \nabla \times \mathbf{c}) \\
& +K_6\{\nabla(\mathbf{b} \cdot \nabla \times \mathbf{c}) - (\nabla \cdot \mathbf{a})(\mathbf{c} \times \nabla \times \mathbf{c})\} \\
& -K_7(\mathbf{c} \cdot \nabla \times \mathbf{c})(\nabla \times \mathbf{c}) - K_8(\nabla \cdot \mathbf{c})(\mathbf{c} \times \nabla \times \mathbf{c}) + K_9\nabla(\nabla \cdot \mathbf{c}) \\
& +\mu_0\Delta\chi \cos\theta \{(\mathbf{a} \cdot \mathbf{H})\cos\theta + (\mathbf{c} \cdot \mathbf{H})\sin\theta\}\,\mathbf{H},
\end{aligned} \tag{6.92}
$$

$$
\begin{aligned}
\mathbf{\Pi}^c \;=\; & K_2\nabla(\nabla \cdot \mathbf{c}) - K_3\nabla \times \{(\mathbf{a} \cdot \nabla \times \mathbf{c})\mathbf{a}\} \\
& -K_4\left[\nabla \times \{(\mathbf{c} \cdot \nabla \times \mathbf{c})\mathbf{c}\} + (\mathbf{c} \cdot \nabla \times \mathbf{c})(\nabla \times \mathbf{c})\right] \\
& +K_5\left[(\mathbf{b} \cdot \nabla \times \mathbf{c})(\mathbf{a} \times \nabla \times \mathbf{c}) - \nabla \times \{(\mathbf{b} \cdot \nabla \times \mathbf{c})\mathbf{b}\}\right] \\
& +K_6\left[(\nabla \cdot \mathbf{a})(\mathbf{a} \times \nabla \times \mathbf{c}) - \nabla \times \{(\nabla \cdot \mathbf{a})\mathbf{b}\}\right] \\
& -K_7\left[\nabla \times \{(\mathbf{a} \cdot \nabla \times \mathbf{c})\mathbf{c} + (\mathbf{c} \cdot \nabla \times \mathbf{c})\mathbf{a}\} + (\mathbf{a} \cdot \nabla \times \mathbf{c})(\nabla \times \mathbf{c})\right] \\
& +K_8\left[\nabla(\mathbf{b} \cdot \nabla \times \mathbf{c}) - \nabla \times \{(\nabla \cdot \mathbf{c})\mathbf{b}\} + (\nabla \cdot \mathbf{c})(\mathbf{a} \times \nabla \times \mathbf{c})\right] \\
& +K_9\nabla(\nabla \cdot \mathbf{a}) + \mu_0\Delta\chi \sin\theta \{(\mathbf{a} \cdot \mathbf{H})\cos\theta + (\mathbf{c} \cdot \mathbf{H})\sin\theta\}\,\mathbf{H}.
\end{aligned} \tag{6.93}
$$

Recall that θ is the constant smectic tilt angle. These results were stated by Kedney and Stewart [140] in terms of the electric energy: simply replace $\mu_0\Delta\chi$ by $\epsilon_0\epsilon_a$ and $\mathbf{H}$ by $\mathbf{E}$ in the above expressions. The elastic constants K_i can of course be replaced by the equivalent Orsay constants if desired using the relations (6.25).

Remark (iii)

It is possible to obtain a reformulation of the equilibrium equations which, to some extent, parallels the reformulation for nematics carried out in Section 2.7. We do not pursue this aspect here, but refer the reader to the article by Leslie [181] which contains full details of the results, including the situation for generalised curvilinear coordinates. When $\mathbf{a}$ is a *fixed* constant vector in Cartesian coordinates we can set

$$
w = w(a_i, a_{i,j}, c_i, c_{i,j}) = \widehat{w}(\phi, \phi_{,i}), \qquad \text{and} \qquad \Psi(a_i, c_i, x_i) = \widehat{\Psi}(\phi, x_i), \tag{6.94}
$$

where ϕ is the orientation angle that the c-director makes relative to some fixed axis within the smectic planes, as shown in Fig. 6.1(b). In the special case mentioned above when $\mathbf{F}$, $\mathbf{G}^a$ and $\mathbf{G}^c$ satisfy (6.81), with $c_i(\mathbf{x}) = \widehat{c}_i(\phi(\mathbf{x}))$ (cf. Section 2.7), the equilibrium equations (6.79) and (6.80) reduce to

$$
\left(\frac{\partial\widehat{w}}{\partial\phi_{,i}}\right)_{,i} - \frac{\partial\widehat{w}}{\partial\phi} + \frac{\partial\widehat{\Psi}}{\partial\phi} = 0, \tag{6.95}
$$

where $\widehat{w}$ is the reformulated bulk energy density and $\widehat{\Psi}$ is the reformulated version of any appropriate relevant scalar density function such as an electric or magnetic potential: for example, $\widehat{\Psi} = -\widehat{w}_{mag}$ is one possibility, where $\widehat{w}_{mag}$ is the reformulated magnetic energy density in terms of ϕ. The repeated index is summed as appropriate (cf. equation (2.220) in the case for nematic liquid crystals).

Comments on the Lagrange Multipliers

The vector Lagrange multiplier $\boldsymbol{\beta}$ has a physical interpretation, as pointed out by Leslie [177]. If we consider planar layers subject to an external body moment, for example via an external field, or, as we shall see below in Section 6.3, perhaps subject to flow, then equation (6.89) shows that the smectic layers are subject to the moments $l_i = l_{ij}\nu_j$ where $\boldsymbol{\nu}$ is the unit outward surface normal. If the smectic layers are constrained to be planar, for example by boundary plates parallel to the xy-plane with the smectic layer normal $\mathbf{a} = (0,0,1)$ coincident with the z-axis, then, by (6.90),

$$l_i = l_{ij}a_j = a_p\beta_p a_i - \beta_i + B_i, \qquad B_i = \epsilon_{ipq}\left(a_p\frac{\partial w}{\partial a_{q,j}} + c_p\frac{\partial w}{\partial c_{q,j}}\right)a_j. \qquad (6.96)$$

In this example we then have, in an obvious notation,

$$l_x = -\beta_x + B_x, \qquad l_y = -\beta_y + B_y, \qquad l_z = 0 + B_z. \qquad (6.97)$$

From these expressions it is seen that the vector $\boldsymbol{\beta}$ provides a mechanism through which the couple stress transmits the required torques to maintain the parallel planar alignment of the sample. In other words, $\boldsymbol{\beta}$ can be interpreted as contributing to the torques which resist destabilising forces on the planar layers that occur when external fields or forces are applied, and therefore $\boldsymbol{\beta}$ allows the initial static planar geometry to be preserved under disturbances induced by such external agencies. More generally, if a particular type of smectic layering is assumed for geometrical or other reasons, not necessarily planar layering, then the vector $\boldsymbol{\beta}$ provides a way for the couple stress to transmit any required necessary torques needed to maintain the assumed initial layer configuration in equilibrium.

The Lagrange multipliers γ, τ and μ reflect the importance of the constraints $(6.3)_1$, $(6.3)_2$ and $(6.3)_3$, respectively. To obtain the final differential equations these multipliers must be eliminated or evaluated explicitly in terms of the solution. For example, taking the scalar product of the set of c-equations in (6.80) with the vector $\mathbf{c}$ yields τ while taking the scalar product with $\mathbf{a}$ delivers μ, because $\mathbf{c}$ and $\mathbf{a}$ are unit vectors and $\mathbf{a} \cdot \mathbf{c} = 0$. The expression for μ can then be inserted into the a-equations in (6.79), leaving only γ and $\boldsymbol{\beta}$ to be determined.

The multipliers γ and $\boldsymbol{\beta}$ may be found by inspection in elementary examples. However, in more complicated cases the following technique has proved worthwhile (cf. the comments in Section 6.2.4 and the explicit examples worked out in Sections 6.2.5 and 6.3.3 below). It is well known that for any vector $\mathcal{F} \in C^1(\mathbb{R}^3)$,

$$\nabla \cdot \mathcal{F} = 0 \qquad \text{if and only if} \qquad \mathcal{F} = \nabla \times \mathcal{G}, \qquad (6.98)$$

for some vector field $\mathcal{G} \in C^2(\mathbb{R}^3)$. The vector field $\mathcal{G}$, called the vector potential, is unique apart from the addition of the gradient of an arbitrary scalar field. For example $\nabla\Omega$, where Ω is a scalar field, can be added to $\mathcal{G}$ without changing the result in (6.98) because $\nabla \times (\nabla\Omega) = \mathbf{0}$. Since $\nabla \cdot (\nabla \times \boldsymbol{\beta}) = 0$, the divergence of the a-equations in (6.79) can be taken to eliminate $\boldsymbol{\beta}$, leaving a differential equation for the scalar function multiplier γ. If we set the left-hand side of the a-equations to be in the form $\mathcal{F} + \nabla \times \boldsymbol{\beta}$, then taking the divergence of both sides of equation (6.79) gives $\nabla \cdot \mathcal{F} = 0$. This gives the differential equation to solve for γ. Once a solution for γ has been identified, the result in equation (6.98) can be applied to conclude that there exists a unique (apart from the gradient of a scalar field) differentiable vector field $\mathcal{G}$ which satisfies $\mathcal{F} = \nabla \times \mathcal{G}$. We can then set $\boldsymbol{\beta} = -\mathcal{G}$. It then follows that $\mathcal{F} + \nabla \times \boldsymbol{\beta} = \mathbf{0}$, so that the a-equations (6.79) are satisfied: the solution for γ guarantees the existence of $\boldsymbol{\beta}$ and the consequent fulfilment of the a-equations. In many instances it remains only to determine the solution to the c-equations. (Solutions for γ do not necessarily need to be found explicitly in some simple cases since knowledge that the differential equation has a solution is sufficient to guarantee the existence of the vector multiplier $\boldsymbol{\beta}$, which sometimes does not need to be calculated explicitly.) In cases where γ has been found, the construction of the vector $\mathcal{G} = -\boldsymbol{\beta}$ can be carried out as follows (cf. Jeffreys and Jeffreys [133, p.224] or Rutherford [235, p.96]). In Cartesian coordinates, set

$$\mathcal{F} = (\mathcal{F}_1(x,y,z), \mathcal{F}_2(x,y,z), \mathcal{F}_3(x,y,z)), \tag{6.99}$$

and choose an arbitrary fixed point (x_0, y_0, z_0) in the region under discussion. Then $\mathcal{G}$ is given by

$$\mathcal{G} = \left(\int_{z_0}^{z} \mathcal{F}_2(x,y,\widetilde{z})\, d\widetilde{z} - \int_{y_0}^{y} \mathcal{F}_3(x,\widetilde{y},z_0)\, d\widetilde{y}, \; -\int_{z_0}^{z} \mathcal{F}_1(x,y,\widetilde{z})\, d\widetilde{z}, \; 0 \right). \tag{6.100}$$

It can be verified directly that $\mathcal{F} = \nabla \times \mathcal{G}$ using the fact that $\nabla \cdot \mathcal{F} = 0$. This solution for $\mathcal{G}$ is unique apart from terms involving $\nabla\Omega$ for an arbitrary scalar field Ω. Applying boundary conditions provides a means for determining Ω, if required, as in potential theory problems. This procedure, with suitable modifications, can also be applied in other coordinate systems.

6.2.4   Focal Conic Defects: Dupin and Parabolic Cyclides

There are well known static defect structures in SmA and SmC liquid crystals called focal conics which have been observed experimentally. A key test for the above theory of SmC liquid crystals is that it is able to provide solutions that predict smectic layers exhibiting these focal conics. The constraint $\nabla \times \mathbf{a} = \mathbf{0}$, discussed at equation (6.4), is known to restrict the possible arrangements of locally parallel and equidistant smectic layers (note, as observed earlier, that this constraint in the general case does not necessarily require equidistant layers). There are six known types of well behaved surfaces which possess a *unit* layer normal $\mathbf{a}$ satisfying $\nabla \times \mathbf{a} = \mathbf{0}$ *and* can be stacked equidistantly. They are: parallel planes, concentric cylinders, concentric spheres, circular tori of revolution, Dupin cyclides and parabolic cyclides. Equilibrium solutions for planes and cylinders, together with the necessary Lagrange multipliers, can be derived in a straightforward manner. However, equilibrium solutions for concentric spheres, circular tori, Dupin cyclides and parabolic cyclides are technically involved and intricate. Moreover, these solutions depend crucially upon correctly obtaining all the necessary scalar and vector Lagrange multipliers. We shall not attempt a full account of these equilibrium solutions here, but rather present a brief summary of some solutions to the equilibrium equations for Dupin and parabolic cyclides found by, respectively, Nakagawa [210] and Stewart, Leslie and Nakagawa [260]. These solutions require restrictions on the nine elastic constants of the SmC phase (for example, restricting the energy to be that for the SmC$_M$ phase given by (6.46)). It should also be mentioned that equilibrium solutions for cylinders, spheres and circular tori have been obtained by McKay [187] by a similar approach. The book by Kléman [151] contains a basic review on cyclides and other defects. A more extensive review on Dupin and parabolic cyclides, including stereoviews, has been compiled by Bouligand [23]. A general summary of the more general aspects of both Dupin and parabolic cyclides, including the topic of blending them with surfaces, has been given by Hirst [127].

The Dupin Cyclides

Dupin cyclides first originated in the geometrical investigations of Dupin [69]. Early discussions on Dupin cyclides in the context of liquid crystals go back to the work of Friedel [94], while experimental observations of Dupin cyclides were discussed by Bragg [25], who also mentioned planes, cylinders and circular tori of revolution. The reader is referred to the books by Gray and Goodby [119] and de Gennes and Prost [110] for some other details on observations; some relevant and clear experimental images in the context of liposomes may also be found in the article by Zasadzinski, Scriven and Davis [285]. Theoretical investigations of Dupin cyclides have also been made by, among others, Bouligand [21, 22], Kléman [150], Kléman and Lavrentovich [152] and Nakagawa [210]. We concentrate here on the theoretical results presented by Leslie, Stewart and Nakagawa [173], based upon the exposition offered by Nakagawa [210].

The geometrical construction of the Dupin cyclides is built upon an ellipse and a branch of a hyperbola, as considered by Maxwell [195], who examined them in a different context before the discovery of liquid crystals. These two curves correspond to line defects in SmC (or SmA). The branch of the hyperbola is in a plane perpendicular to the ellipse and has its vertex placed at one of the foci of the ellipse in such a way that the hyperbola is also in the plane containing the two foci of the ellipse. Just as the hyperbola passes through a focus of the ellipse, the ellipse passes through a focus of the hyperbola. The geometrical set-up of the ellipse and hyperbola can be seen in Fig. 6.4(a) where a cross-section of some typical equidistantly arranged Dupin cyclide surfaces is also shown. In the special case where the ellipse becomes a circle, the hyperbola becomes a straight line and the smectic layers then form circular tori of revolution as shown in Fig. 6.4(b). It is for these reasons that the ellipse and hyperbola, which are central to the construction of the Dupin cyclides, are sometimes referred to as a pair of focal conics. For a more detailed description and graphical illustration of the geometrical construction of Dupin cyclides the reader is recommended to consult the book by Hartshorne and Stuart [122] and the aforementioned references to Dupin cyclides.

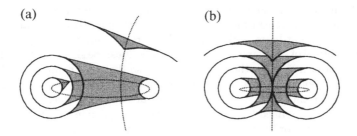

Figure 6.4: (a) Cross-section through equidistantly arranged Dupin cyclide surfaces. A branch of a hyperbola passes through a focus of an ellipse as discussed in the text; the ellipse also passes through a focus of the hyperbola. (b) A special limiting case where the ellipse becomes a circle and the hyperbola becomes a straight line which passes through the centre of the circle. The Dupin cyclide layers become concentric tori of revolution, or parts thereof.

The focal conics consisting of an ellipse and hyperbola which enable the construction of a Dupin cyclide may be described in Cartesian coordinates by

$$\text{ellipse:} \qquad \frac{x^2}{a^2} + \frac{y^2}{b^2} = 1, \qquad z = 0, \tag{6.101}$$

$$\text{hyperbola:} \qquad \frac{x^2}{c^2} - \frac{z^2}{b^2} = 1, \qquad y = 0, \tag{6.102}$$

where $c^2 = a^2 - b^2$ and the eccentricity of the ellipse is defined by $e = c/a$. The general equation for a Dupin cyclide surface based on these conics is [90, 173, 210]

$$(x^2 + y^2 + z^2 - r^2)^2 - 2a^2(1 + e^2)(x^2 + r^2)$$
$$-2a^2(1 - e^2)(y^2 - z^2) + 8a^2 erx + a^4(1 - e^2)^2 = 0, \tag{6.103}$$

where r is a real parameter. The 'inner' parts (to be displayed later in Fig. 6.5) of the Dupin cyclides obtained from equation (6.103) can be parametrised as [210]

$$x = \frac{a\cos v(a\cosh u - r) + c\cosh u(r - c\cos v)}{a\cosh u - c\cos v}, \tag{6.104}$$

$$y = \frac{b\sin v(a\cosh u - r)}{a\cosh u - c\cos v}, \tag{6.105}$$

$$z = \frac{b\sinh u(r - c\cos v)}{a\cosh u - c\cos v}, \tag{6.106}$$

where the parameter ranges for r, u and v are given by

$$c\cos v \leq r \leq a\cosh u, \tag{6.107}$$

$$-\infty < u < \infty, \tag{6.108}$$

$$0 \leq v \leq 2\pi. \tag{6.109}$$

A transformation from the Cartesian coordinate system to the local (r, u, v) frame can then be made. It can be shown that this transformed coordinate system is orthogonal with unit basis vectors $\{\hat{\mathbf{r}}, \hat{\mathbf{u}}, \hat{\mathbf{v}}\}$ where the r-direction is parallel to the local layer normal of the Dupin cyclide. Moreover, it can be shown [210] that $\nabla r = \hat{\mathbf{r}}$ is the unit layer normal to the cyclide surface, so that $\mathbf{a} = \hat{\mathbf{r}}$ automatically fulfils the necessary requirements $\mathbf{a} \cdot \mathbf{a} = 1$ and $\nabla \times \mathbf{a} = \mathbf{0}$, stated at equations $(6.3)_1$ and (6.4). It consequently follows that for a fixed value of r, the 'inner' part of a Dupin cyclide surface is obtained by varying the values of u and v in the above expressions for the x, y and z coordinates of the surface. An alternative parametrisation for the full Dupin cyclide surfaces is available in the book by Forsyth [90, p.326], but care needs to be exercised in the context of parallel layers for liquid crystals because complicated restrictions on the range of the parameters must be introduced.

It can be shown that for the six term elastic energy for the SmC_M phase given by equation (6.46), the vectors

$$\mathbf{a} = \hat{\mathbf{r}} \qquad \text{and} \qquad \mathbf{c} = \hat{\mathbf{u}}, \tag{6.110}$$

satisfy the transformed version of the equilibrium equations (6.79) and (6.80), with suitable scalar and vector Lagrange multipliers being derived explicitly by Nakagawa [210]. Clearly, $\mathbf{c} \cdot \mathbf{c} = 1$ and $\mathbf{a} \cdot \mathbf{c} = 0$ in addition to the above properties for $\mathbf{a}$, so that all the constraints in (6.3) and (6.4) are satisfied. Nakagawa has also shown that these vectors provide solutions for the nine term energy given by equation (6.13) for the SmC phase provided constraints are imposed on the additional elastic constants, namely, $B_{13} = 0$ and $C_1 + C_2 = 0$ (or, equivalently, $K_7 = 0$ and $K_8 = 0$, using the expressions (6.25)). The solutions (6.110) provide static solutions representing parts of the Dupin cyclide surfaces. It is anticipated that a suitable coordinate transformation will also provide solutions for the remaining 'outer' parts of these surfaces.

The parametrisation introduced at equations (6.104), (6.105) and (6.106) can be used to obtain surface plots of the inner Dupin cyclide surfaces. For ease of illustration, the graphs in Fig. 6.5 have been produced by fixing $a = 5$, $b = 4$ and $c = 3$ and plotting the surfaces and cross-sections for various integer values of r, for example, $r = -2, -1, 0, 1, \ldots$. These layers are representational of the local layers in the material and smaller increments in r will simply lead to more layers being displayed. The top part of Fig. 6.5 shows the ellipse (6.101) in the xy-plane with a branch of the hyperbola (6.102) passing through one of the foci of the ellipse, this hyperbola lying in the xz-plane. The equally spaced smectic layers can be built around these defect lines, as has been partially shown in the top part of Fig. 6.5. This stacking of layers is only partly indicated so that some of the complicated arrangement of the layers can be visualised. Each individual layer is a Dupin cyclide surface, or part of one. The actual arrangement of the smectic layers can be deduced from cross-sections of the full three-dimensional sample. The middle part of Fig. 6.5 is a cross-section taken at the plane $z = 0$. This shows the elliptical defect and how the equally spaced layers of smectic match up near the ellipse; the hyperbola is perpendicular to the page and passes through the focus of the ellipse as indicated by a dot at the point $(3, 0, 0)$ (in Cartesian coordinates).

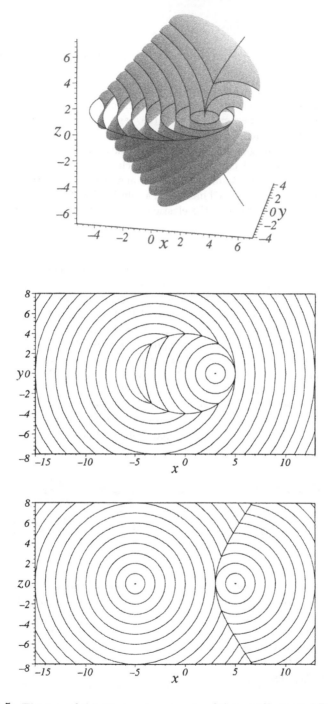

Figure 6.5: The top of this Figure shows parts of the equally spaced Dupin cyclide surfaces using the parametrisation given by equations (6.104) to (6.109). The defects are located along the hyperbola and the ellipse. The middle part of the Figure is a cross-section in the plane $z = 0$ showing the arrangement of the smectic layers around the elliptical defect, while the lower part shows a cross-section in the plane $y = 0$ around the hyperbola; details are contained in the text.

The bottom part of Fig. 6.5 depicts a cross-section taken in the plane $y = 0$ showing the hyperbola, and displays how the smectic layers match up near the hyperbola; the ellipse is perpendicular to the page and passes through the two points indicated by dots at $(-5, 0, 0)$ and $(5, 0, 0)$.

As mentioned by Nakagawa [210], solutions should still be available in the special case $e = 0$ when the focal conics reduce to a circle and a straight line that is orthogonal to the plane containing the circle and passes through its centre, as shown in Fig. 6.4(b). McKay [187] has found a solution for the relevant full circular tori surfaces in this case provided the elastic constant $K_9 = -C_2$ is set to zero.

The Parabolic Cyclides

The parabolic cyclides and their associated focal conic defects in smectic liquid crystals have been discussed by Kléman [150]. Micrographs of experiments displaying parabolic focal conic defects in SmA, together with a further discussion on parabolic cyclide surfaces, have been reported by Rosenblatt, Pindak, Clark and Meyer [234]. Rather surprisingly, Bragg [25] did not consider parabolic cyclides as possible structures for liquid crystals, despite considering Dupin cyclides, as mentioned above. Here, we present a summary of the parabolic cyclide equilibrium solutions found by Stewart, Leslie and Nakagawa [260] for the SmC$_M$ phase.

The description of parabolic cyclide surfaces is based upon two confocal parabolas in mutually perpendicular planes, with the vertex of one parabola passing through the focus of the other. These parabolas represent line defects in smectic liquid crystals. Parts of some typical parabolic cyclide surfaces are pictured in

Figure 6.6: The two types of parabolic cyclide surface: only parts of these surfaces are depicted. The left plot shows the 'tunnel' type and the middle plot shows the 'bridge' type. The bridge can be omitted in the context of smectic liquid crystals, leaving that part of the parabolic cyclide surface of relevance shown in the right plot. These plots were obtained using the parametrisation in equations (6.114), (6.115) and (6.116) with $\ell = -4$ and, respectively, $\mu = 1$ for the tunnel type and $\mu = -1$ for the bridge type.

Fig. 6.6. The left plot in this Figure shows part of a 'tunnel' type of parabolic surface. The middle plot shows part of a 'bridge' type of parabolic surface and the right plot shows the surface when the bridge is removed from the middle plot. Parallel layers of these surfaces can be stacked together when they consist of both forms of parabolic surface, provided any bridges are removed. This stacking of layers of

typical parabolic cyclide surfaces is pictured in the upper part of Fig. 6.7, where
some of the internal structure of the equally spaced layers of parabolic cyclides, to
be discussed in more detail below, is evident: the bold curves represent parts of the
confocal parabolic defect lines.

We can describe the parabolas in Cartesian coordinates by

$$y^2 = 4\ell(x+\ell), \quad z = 0, \tag{6.111}$$
$$z^2 = -4\ell x, \quad\quad y = 0, \tag{6.112}$$

where ℓ is assumed to be a fixed non-zero real constant. These parabolas have
latera recta equal to -4ℓ; the focus of parabola (6.111) is at the origin and the
focus of parabola (6.112) is at the point $(-\ell, 0, 0)$. The general Cartesian equation
of a parabolic cyclide may then be written as [127, 260]

$$x(x^2 + y^2 + z^2) + (x^2 + y^2)(\ell - \mu) - z^2(\ell + \mu) - (x - \mu + \ell)(\ell + \mu)^2 = 0, \tag{6.113}$$

where μ is a real parameter. As shown in [260], varying μ produces parallel families
of parabolic cyclide surfaces. In contrast to the parametrisation introduced above
for the Dupin cyclides, for each fixed value of μ a complete parabolic cyclide surface
can be parametrised. Specifically, we have [260]

$$x = \frac{\mu(\theta^2 + t^2 - 1) + \ell(t^2 - \theta^2 - 1)}{1 + \theta^2 + t^2}, \tag{6.114}$$
$$y = \frac{2t[\ell(\theta^2 + 1) + \mu]}{1 + \theta^2 + t^2}, \tag{6.115}$$
$$z = \frac{2\theta(\ell t^2 - \mu)}{1 + \theta^2 + t^2}, \tag{6.116}$$

where the parameter ranges for μ, θ and t are

$$-\infty < \mu < \infty, \quad -\infty < \theta < \infty, \quad -\infty < t < \infty. \tag{6.117}$$

It is now clear that fixing a value for μ and varying θ and t will generate a complete
parabolic cyclide surface. For some values of μ only parts of the surface will be
relevant for liquid crystals, as pointed out by Rosenblatt *et al.* [234], but we shall
not go into such details here: briefly, 'bridges' appear as shown in the middle part of
Fig. 6.6 which are thought to have no physical relevance to smectic liquid crystals.
(The above parametrisation is based on that of Forsyth [90], but note that the
surface equation and parametrisation introduced in [90] contain some misprints: a
full derivation is given in Reference [260].)

Similar to the Dupin cyclides discussed above, a transformation from Cartesians
to the local (μ, θ, t) coordinate system can be made. This new coordinate system
is orthogonal with unit basis vectors $\{\widehat{\mu}, \widehat{\theta}, \widehat{t}\}$, where the μ-direction is parallel to
the local layer normal of the parabolic cyclide surface. It can be shown [260], using
the surface equation (6.113), that $\nabla\mu = \widehat{\mu}$ coincides precisely with the unit layer
normal, and therefore $\mathbf{a} = \widehat{\mu}$ fulfils the requirements $\mathbf{a} \cdot \mathbf{a} = 1$ and $\nabla \times \mathbf{a} = \mathbf{0}$. By
setting the vectors

$$\mathbf{a} = \widehat{\mu} \quad \text{and} \quad \mathbf{c} = \widehat{\theta}, \tag{6.118}$$

it is seen that the further relations $\mathbf{c} \cdot \mathbf{c} = 1$ and $\mathbf{a} \cdot \mathbf{c} = 0$ hold, so that all the constraints (6.3) and (6.4) are satisfied. Stewart *et al.* [260] were then able to show that the vectors in (6.118) are solutions to the transformed version of the equilibrium equations (6.79) and (6.80) for the SmC$_M$ phase having elastic energy given by (6.46), with all the necessary scalar and vector Lagrange multipliers being found explicitly. A full discussion of these multipliers and the relevant parabolic cyclide surfaces can be found in Reference [260].

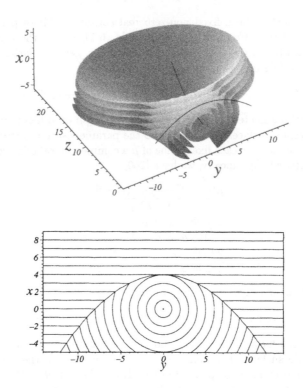

Figure 6.7: Parts of equally spaced parabolic cyclide surfaces are shown in the upper illustration. These were obtained from the parametrisation introduced at equations (6.114) to (6.117) with the mutually orthogonal parabolic defects (6.111) and (6.112) indicated by the two bold curves. The lower illustration shows a cross-section of the upper figure in the plane $z = 0$: the parabolic defect is evident and the corresponding smectic layers are arranged as shown with the other parabola being perpendicular to the page and passing through the point $(0,0,0)$. By symmetry, there is a similar cross-section in the plane $y = 0$.

The parallel layers that make up the configuration displayed in Fig. 6.7 require parabolic cyclides that do not possess 'bridges' (or cusp points) and ones that possess them but have them removed, as shown in the left and right plots pictured in Fig. 6.6. Such parts of the parabolic cyclides can be stacked equidistantly and parallel to each other as shown in the upper part of Fig. 6.7. Here, the two parabolas representing the defect lines can be seen in this partial view of the layering. In this Figure we have set $\ell = -4$, and plotted the relevant surfaces for $\mu = -2, -1, 0, 1, 2$. As in the Dupin cyclides, smaller increments in μ will lead to more layers being represented. The lower part of Fig. 6.7 shows a cross-section of the parabolic cyclide layers taken in the $z = 0$ plane: the parabolic defect (6.111) in the xy-plane is clear while the other mutually orthogonal parabolic defect (6.112) passes perpendicular to the page through the point $(0, 0, 0)$, as indicated by a dot. It is clear from the symmetry of the upper part of the Figure that there is a similar cross-section in the plane $y = 0$. The relation between the lower and upper parts of this Figure should enable the reader to have a reasonable picture of the three-dimensional structure of the parabolic focal conic defect in SmC or SmA. Full details on obtaining such plots for the parametrisation given by equations (6.114) to (6.116), including the restrictions on θ and t near the defects, have been given by Stewart [259].

6.2.5   A Freedericksz Transition in Bookshelf Smectic C

The concept of a Freedericksz transition for SmC liquid crystals is a natural extension to that discussed earlier for nematics in Section 3.4. A comprehensive discussion on Freedericksz transitions in planar aligned layers of SmC liquid crystals has been given by Rapini [228], who identified twelve possible geometries involving cells having different combinations of smectic layer alignments and applied magnetic fields. It should also be mentioned that instabilities in SmC liquid crystals induced by magnetic fields were also examined by Meirovitch, Luz and Alexander [196]. We shall concentrate on one particular geometry considered by Kedney and Stewart [140] to illustrate an application of the equilibrium equations (6.79) and (6.80). The general procedure outlined here deals explicitly with the Lagrange multipliers appearing in these equilibrium equations and it is similar to the procedure that secured solutions for the cyclides in References [210, 260], as mentioned briefly in Section 6.2.4. As remarked below, a simpler approach is available for the particular example of a Freedericksz transition to be discussed in this Section, but the procedure here has particular merits and it is for this reason that a more detailed exposition is developed below: see the Remark on page 280.

Consider the case when a sample of SmC liquid crystal is confined between two parallel plates in the bookshelf geometry shown in Fig. 6.8 where the sample is assumed to be of unit depth in the x- and y-directions but of depth d in the z-direction. Suppose also that the director $\mathbf{n}$ is strongly anchored parallel to the boundaries at both plates and further suppose that on these boundaries the projection of the director $\mathbf{n}$ onto the layer boundary makes an angle β to the x-direction, called the in-plane surface twist angle [139]. For simplicity, we shall assume that β is equivalent to the fixed smectic tilt angle θ, as shown in the Figure; other values for β are feasible. In general, there will also be an out-of-plane surface tilt angle ζ which the director $\mathbf{n}$ makes relative to the boundary surface [139], but in this example $\zeta = 0$; see the comments on surface tilt and surface twist on page 279. It is assumed that a uniform magnetic field $\mathbf{H}$ is applied perpendicular to the plates

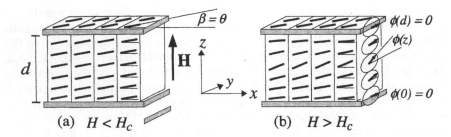

Figure 6.8: A magnetic field **H** is applied across a sample of depth d of SmC liquid crystal in the 'bookshelf' geometry, as shown. For simplicity, the director is assumed to be strongly anchored at the surfaces in such a way that $\phi(0) = \phi(d) = 0$ where $\phi(z)$ is the orientation angle of the c-director defined by $(6.120)_2$: this indicates that the director makes an in-plane surface twist angle $\beta = \theta$ to the x-axis at the boundaries, equivalent to the constant smectic tilt angle. Other values for β are feasible: see page 279. (a) When $H < H_c$, where H_c is the Freedericksz threshold defined by (6.143), only $\phi \equiv 0$ is available and the director alignment is uniformly constant across the sample. (b) There is a Freedericksz transition at $H = H_c$ and for $H > H_c$ the c-director reorients within the yz-plane, as shown schematically.

in the z-direction so that

$$\mathbf{H} = H(0, 0, 1), \qquad H = |\mathbf{H}|, \tag{6.119}$$

where H is the magnitude of the field and it is supposed that the magnetic anisotropy $\Delta\chi$ is positive. Set

$$\mathbf{a} = (1, 0, 0), \qquad \mathbf{c} = (0, \cos\phi(z), \sin\phi(z)), \tag{6.120}$$

where $\phi(z)$ is the angle that the c-director makes relative to the y-axis within the yz-plane. This orientation angle ϕ (also called the phase angle of the c-director) satisfies the boundary conditions

$$\phi(0) = \phi(d) = 0, \tag{6.121}$$

which match the boundary alignment depicted in Fig. 6.8.

The total energy integrand is

$$w = w_{elas} + w_{mag}, \tag{6.122}$$

where w_{elas} is the elastic free energy density for the SmC phase given by (6.13) (or, equivalently, by (6.15)) and w_{mag} is the magnetic energy density given by (6.47). At this stage it is possible, in this particularly simple geometry, to derive the governing equilibrium equation (6.138) directly by using the result at equation (6.95) with $i = 3$ and the energy density w in (6.122) expressed as $\widehat{w}(\phi, \phi')$, $\widehat{\Psi}$ being set to zero in (6.95) because the magnetic potential has been incorporated via the magnetic energy w_{mag}. However, we shall derive equation (6.138) below by illustrating the rôles of the Lagrange multipliers. Using the expressions (6.73) and (6.74) with the definition of w given by (6.122), the equilibrium equations (6.79) and (6.80) are

$$\Pi_i^a + \gamma a_i + \mu c_i + \epsilon_{ijk}\beta_{k,j} = 0, \qquad (6.123)$$

$$\Pi_i^c + \tau c_i + \mu a_i = 0, \qquad (6.124)$$

when the general external forces $\mathbf{G}^a$ and $\mathbf{G}^c$ are absent. Notice that $\mathbf{\Pi}^a$ and $\mathbf{\Pi}^c$ are functions of z only because $\mathbf{a}$ and $\mathbf{c}$ have the forms stated in equation (6.120). Taking the scalar product of the c-equations in (6.124) with $\mathbf{a}$ given by (6.120)$_1$ shows that

$$\mu(z) = -a_i\Pi_i^c = -\Pi_1^c, \qquad (6.125)$$

from which it follows that the first of the c-equations ($i = 1$) in (6.124) is trivially satisfied. Taking the scalar product of (6.124) with $\mathbf{c}$ given by (6.120)$_2$ yields

$$\tau(z) = -\Pi_2^c \cos\phi - \Pi_3^c \sin\phi. \qquad (6.126)$$

Inserting this into (6.124) then reduces the c-equations to

$$\Pi_2^c - \Pi_2^c \cos^2\phi - \Pi_3^c \sin\phi\,\cos\phi = 0, \qquad (6.127)$$

$$\Pi_3^c - \Pi_2^c \sin\phi\,\cos\phi - \Pi_3^c \sin^2\phi = 0. \qquad (6.128)$$

Multiplying (6.127) by $\sin\phi$ and (6.128) by $\cos\phi$ and subtracting provides the governing equilibrium equation for $\phi(z)$ given by

$$\Pi_2^c \sin\phi - \Pi_3^c \cos\phi = 0, \qquad (6.129)$$

provided the Lagrange multipliers for γ and β can be found so that the a-equations (6.123) can be satisfied. To see that this is possible in this instance, the result for μ in (6.125) can be inserted into the a-equations (6.123) and then the procedure outlined on pages 266 to 267 for obtaining the Lagrange multipliers γ and β can be followed. Taking the divergence of the left-hand side of (6.123) with $\mathbf{a}$ and $\mathbf{c}$ given by (6.120) gives

$$\frac{\partial\gamma}{\partial x} + \frac{\partial}{\partial z}\left(\Pi_3^a - \Pi_1^c \sin\phi\right) = 0. \qquad (6.130)$$

Integrating this expression with respect to x provides the multiplier γ in terms of the solution $\phi(z)$, namely,

$$\gamma(x, y, z) = -x\frac{d}{dz}\left[\Pi_3^a(z) - \Pi_1^c(z) \sin\phi(z)\right] + f(y, z), \qquad (6.131)$$

where $f(y, z)$ is an arbitrary function of y and z. Hence the existence of a vector $\boldsymbol{\beta}$ is guaranteed by the result in equation (6.98), with $\boldsymbol{\beta} = -\mathbf{\mathcal{G}}$ where $\mathbf{\mathcal{G}}$ is given by equation (6.100) with

$$\mathbf{\mathcal{F}} = \mathbf{\Pi}^a + \gamma\mathbf{a} + \mu\mathbf{c}, \qquad (6.132)$$

$\gamma(x, y, z)$ and $\mu(z)$ given by (6.131) and (6.125), respectively. The vector $\boldsymbol{\beta}$ is unique apart from the addition of the gradient of an arbitrary scalar field. Its components can be calculated explicitly using the formula (6.100) with $x_0 = 0$, $y_0 = 0$ and $z_0 = 0$, to reveal

$$\beta_1 = -\int_0^z [\Pi_2^a(\tilde{z}) + \mu(\tilde{z}) \cos \phi(\tilde{z})] \, d\tilde{z} + y[\Pi_3^a(0) + \mu(0) \sin \phi(0)], \quad (6.133)$$

$$\beta_2 = \int_0^z [\Pi_1^a(\tilde{z}) + \gamma(x, y, \tilde{z})] \, d\tilde{z}, \quad (6.134)$$

$$\beta_3 = 0. \quad (6.135)$$

The components of $\boldsymbol{\Pi}^a$ can be obtained explicitly from the general vector formula (6.92) if desired, with, of course, $\mathbf{a}$ and $\mathbf{c}$ given by (6.120). It is a simple exercise to use (6.125), (6.130) and (6.131) to verify that $\boldsymbol{\mathcal{F}} + \nabla \times \boldsymbol{\beta} = \mathbf{0}$ where $\boldsymbol{\mathcal{F}}$ is defined by (6.132), indicating that the a-equations (6.123) are fulfilled. It now follows that solving the equilibrium equation (6.129) for $\phi(z)$ will result in a full solution to the equilibrium equations for $\mathbf{a}$ and $\mathbf{c}$ given by (6.120). Although we have calculated all the Lagrange multipliers explicitly in this example, it is not often required in simple static problems to know $\boldsymbol{\beta}$ explicitly since it is frequently sufficient to know that γ and $\boldsymbol{\beta}$ can be found if desired once a solution to the c-equations has been determined, as mentioned previously.

The general expression for $\boldsymbol{\Pi}^c$ given by equation (6.93) can be used for $\mathbf{a}$ and $\mathbf{c}$ given by (6.120) (with $\mathbf{b} = \mathbf{a} \times \mathbf{c}$) to find that

$$\Pi_2^c = -K_3 \left[\phi'' \sin \phi + (\phi')^2 \cos \phi \right], \quad (6.136)$$

$$\Pi_3^c = K_2 \left[\phi'' \cos \phi - (\phi')^2 \sin \phi \right] + \mu_0 \Delta \chi H^2 \sin^2\theta \sin\phi, \quad (6.137)$$

where a prime denotes differentiation with respect to z. The component Π_1^c need not be calculated explicitly because equation (6.124) is identically satisfied when $i = 1$ by the result in equation (6.125), as remarked upon earlier. For the purposes of comparison with other results in the literature it is convenient here to set $K_2 = B_2$ and $K_3 = B_1$, using the relations (6.25) for the elastic constants in terms of the Orsay constants. Doing so, and employing the results (6.136) and (6.137), finally reduces the equilibrium equation (6.129) to

$$\left[B_2 \cos^2\phi + B_1 \sin^2\phi \right] \phi'' + (B_1 - B_2)(\phi')^2 \sin\phi \cos\phi + \mu_0 \Delta \chi H^2 \sin^2\theta \sin\phi \cos\phi = 0, \quad (6.138)$$

with its accompanying boundary conditions (6.121). This equation is identical in form to equation (3.111) on page 74 with its boundary conditions (3.105). If, in equation (3.111), we make the substitutions

$$K_1 \to B_2, \quad K_3 \to B_1, \quad \chi_a \to \mu_0 \Delta \chi \sin^2\theta, \quad \theta_m \to \phi_m, \quad \theta(z) \to \phi(z), \quad (6.139)$$

then the problem reduces to that for the first classical Freedericksz transition in nematics discussed in Section 3.4.1 for the nematic 'splay' geometry (note that K_1 and K_3 in (6.139) refer to the nematic elastic constants in (3.111)). Consequently, the full nonlinear analysis of solutions to the differential equation (6.138) parallels that for the classical Freedericksz transition in nematics given from equations (3.111) to (3.133). Briefly, there are two solutions: $\phi \equiv 0$ and the symmetric nonlinear distorted solution given by equation (3.118) with the substitutions (6.139) being made when, without loss of generality, it is assumed that $\phi'(\frac{d}{2}) = 0$, $\phi(\frac{d}{2}) = \phi_m > 0$. The solution to the nonlinear problem obeying $\phi(z) = \phi(d - z)$ for $0 \leq z \leq d$ is then conveniently given in a reformulated form using equations (3.122) and (3.123) by

$$\sqrt{\mu_0 \Delta \chi} \sin\theta \, Hz = \int_0^\psi G(\phi_m, \lambda) d\lambda, \qquad \psi = \sin^{-1}\left(\frac{\sin\phi}{\sin\phi_m}\right), \tag{6.140}$$

with the condition for determining ϕ_m at $z = \frac{d}{2}$ given by

$$\sqrt{\mu_0 \Delta \chi} \sin\theta H \frac{d}{2} = \int_0^{\frac{\pi}{2}} G(\phi_m, \lambda) \, d\lambda, \tag{6.141}$$

where $G(\phi_m, \lambda)$ is now the integrand defined by

$$G(\phi_m, \lambda) = \left[\frac{B_2 + (B_1 - B_2)\sin^2\phi_m \sin^2\lambda}{1 - \sin^2\phi_m \sin^2\lambda}\right]^{\frac{1}{2}}. \tag{6.142}$$

The solution (6.140) is valid for $0 \leq z \leq \frac{d}{2}$: as in the nematic case, the solution for $\frac{d}{2} \leq z \leq d$ is obtained by simply replacing z by $d - z$ in (6.140). The solution (6.140) first becomes available at the critical field strength H_c given by

$$H_c = \frac{\pi}{d \sin\theta} \sqrt{\frac{B_2}{\mu_0 \Delta \chi}}, \tag{6.143}$$

obtained by identical reasoning to that used to find the threshold value at equation (3.126) for nematics. As in the nematic case, for a fixed depth d and a given field magnitude $H > H_c$, the corresponding value for $\phi_m > 0$ can be obtained from the requirement (6.141). This value of ϕ_m is then inserted into (6.140) to provide the full nonlinear distorted solution $\phi(z)$. In general, this procedure has to be carried out numerically.

The distorted solution (6.140) is energetically favoured for $H > H_c$, following from the result (3.133) for nematics, when W and ΔW are given by (3.109) and (3.129) with the appropriate replacements for the parameters given by (6.139): cf. the energy densities given by inserting the expressions (6.120) into (6.13) and (6.47). Hence there is a Freedericksz transition and the critical threshold for its onset is given by H_c in (6.143). This particular value of H_c has also been discussed by Rapini [228, p.243] and may provide a means of measuring the elastic constant B_2.

It is possible to make the one-constant approximation $B_1 = B_2 \equiv B$ and suitably amend the work contained on pages 79 to 82 using (6.139) to obtain qualitative plots for the behaviour of $\phi(z)$ for $H > H_c$, analogous to those presented in Figs. 3.9 and 3.10 for nematics.

Surface Tilt and Surface Twist

The in-plane surface twist angle β indicated in Fig. 6.8 at the boundary surfaces has been set equal to θ in order to simplify the above analysis. However, in the bookshelf geometry, any angle having absolute value less than or equal to θ may be feasible as a value for β on the surfaces, in which case, by considering the geometry of the bookshelf set-up, the boundary conditions on ϕ may be replaced by, for example, $\phi(0) = \phi(d) = \phi_0 \equiv \cos^{-1}(\tan\beta/\tan\theta)$, when $-\theta \leq \beta \leq \theta$ [139]. Also, for example, when $-\theta \leq \beta \leq \theta$, there will be an out-of-plane surface tilt angle $\zeta = \sin^{-1}(\sin\theta\sin\phi_0)$ which the director $\mathbf{n}$ makes relative to the boundary surface [139]. The analysis then proceeds in a similar way to that carried out for a nematic with pretilt at the boundaries, as discussed in Section 3.4.2. The details are left to the reader. The angles β and ζ, and their restrictions, will of course be modified if the smectic layers themselves, instead of being in the bookshelf geometry, are tilted by a layer tilt angle δ relative to the boundary surface; details can be found in [139].

Remarks on the Electric Field Case

The threshold in equation (6.143) above has an analogous formulation in terms of an electric field. Using the substitution $\mu_0\Delta\chi H^2\sin^2\theta \mapsto \epsilon_0\epsilon_a E^2\sin^2\theta$, replacing the magnetic field by an electric field for a material with positive dielectric anisotropy gives the critical threshold

$$E_c = \frac{\pi}{d\sin\theta}\sqrt{\frac{B_2}{\epsilon_0\epsilon_a}}. \tag{6.144}$$

This form of threshold has been obtained and discussed by Pelzl, Schiller and Demus [219], who also go on to discuss their theoretical and experimental results for some SmC materials. Although the dynamic theory for SmC liquid crystals has yet to be reached in Section 6.3, it seems appropriate to record here that these authors additionally considered switch-on and switch-off times (rise and decay times, respectively) and arrived at results analogous to those for a nematic liquid crystal given by equations (5.420) and (5.424) which were encountered when the dynamics of the Freedericksz transition for nematic liquid crystals was discussed in Section 5.9.1. These results for SmC are

$$\tau_{on} = \frac{2\lambda_5}{\epsilon_0\epsilon_a\sin^2\theta(E^2 - E_c^2)} = \frac{2\lambda_5 d^2}{\epsilon_0\epsilon_a\sin^2\theta(V^2 - V_c^2)} = \frac{2\lambda_5 d^2}{B_2\pi^2(V^2/V_c^2 - 1)}, \tag{6.145}$$

$$\tau_{off} = 2\frac{\lambda_5}{B_2}\left(\frac{d}{\pi}\right)^2, \tag{6.146}$$

where $V_c = E_c d$, and $\lambda_5 > 0$ is a rotational viscosity which will be introduced below in Section 6.3. Pelzl *et al.* [219] have also used the switch-off time given by (6.146) to evaluate the viscosity λ_5 (equivalent to $\frac{1}{2}\lambda$ in their notation) for two different SmC liquid crystals.

Remark

In simple planar alignments of SmC in straightforward geometries, Freedericksz transitions may be discussed more simply by expressing the total energy in terms of a single orientation angle, similar to that used for nematic theory: see Remark (iii) on page 265 and equation (6.95). However, the exercise above expounds the methodology for determining the Lagrange multipliers and such techniques have proved beneficial in more complex geometries such as the Dupin and parabolic cyclides discussed briefly above in Section 6.2.4.

Freedericksz Transitions in Other Geometries

The above example in Section 6.2.5 serves as a prototype for more complicated geometrical arrangements of the smectic layers in the context of Freedericksz transitions. We conclude this Section by pointing out that there are results for other experimental set-ups which may allow for the possible measurement of the elastic constants B_1, B_2 and B_3, as well as combinations of the layer bending constants A_{12}, A_{21} and A_{11}. For example, Freedericksz transitions involving concentric cylindrical layers of SmC liquid crystal have been discussed by Carlssón et al. [32] for samples confined in a wedge, with the axis of the wedge coinciding with the common axis of the layers; experiments making use of this geometry to measure the combination $\overline{A}_{21} + \overline{A}_{11}$ have been carried out by Findon [87] and Findon and Gleeson [88], as mentioned on page 257. Freedericksz transitions for samples between two coaxial concentric circular cylinders have also been investigated theoretically by Atkin and Stewart [9], who also examined a cone and plate geometry involving a special arrangement of concentric spherical layers of SmC liquid crystal [8]. Some of these results have been summarised and presented in a review by Atkin and Stewart [10].

6.2.6   Smectic Layer Compression

The smectic layers may themselves compress or dilate in a full general theory of deformations in smectic liquid crystals. These effects have not been considered in the theory presented above because, in many instances, it may be assumed that layer compression or dilation may be neglected in basic planar layered geometries of SmC liquid crystals. Nevertheless, for the sake of completeness, we record for the interested reader some details from the literature. The additional term for smectic layer compression in SmA liquid crystals, which also serves as a first approximation to the compression term in SmC for the planar geometry pictured in Fig. 6.3, is of the form mentioned by de Gennes [106] and de Gennes and Prost [110, pp.345–346]

$$w_{comp} = \frac{1}{2}\overline{B}\left(\frac{\partial u}{\partial z}\right)^2, \tag{6.147}$$

where $u = u(x, y, z)$ is the vertical displacement of the layers relative to their original state and $\overline{B}$ is the associated layer compression constant. This supplementary energy density term is simply added to any of the forms w for the elastic energy density for SmC given by the expressions stated in Section 6.2.1. Although (6.147) is often considered adequate for determining the onset of critical phenomena, it is, nevertheless, usually amended to

$$w_{comp} = \frac{1}{2}\overline{B}\left\{\frac{\partial u}{\partial z} - \frac{1}{2}\left[\left(\frac{\partial u}{\partial x}\right)^2 + \left(\frac{\partial u}{\partial y}\right)^2\right]\right\}^2, \tag{6.148}$$

as mentioned by Johnson and Saupe [138], in order to describe post-transitional effects. An early review of the basic model for SmA, including pictures of experimental studies, has been made by Ribotta [231]. Recent applications of the energy (6.148) in the description of layer instabilities and undulations in SmA can be found in the articles by, for example, Fukuda and Onuki [96], Geer *et al.* [103] and Singer [251]. Observations of layer undulations in SmA under dilative or compressive stresses have been reported by Ribotta and Durand [232].

We record here that to obtain a common energy density describing layer undulations for SmA in terms of $u(x, y, z)$, we may make the approximation $\mathbf{a} = (-\partial u/\partial x, -\partial u/\partial y, 1)$ [149] and construct the energy as the sum of $w_{A\,elast}$ and w_{comp} to give, from (6.40) and (6.148), (for example, see [103])

$$ w_A = \frac{1}{2}K_1\left(\frac{\partial^2 u}{\partial x^2} + \frac{\partial^2 u}{\partial y^2}\right)^2 + \frac{1}{2}\overline{B}\left\{\frac{\partial u}{\partial z} - \frac{1}{2}\left[\left(\frac{\partial u}{\partial x}\right)^2 + \left(\frac{\partial u}{\partial y}\right)^2\right]\right\}^2, \qquad (6.149) $$

where the term in $\overline{B}$ may be simplified to that given by (6.147) in elementary studies [106], [110, p.343 & pp.361–364], especially when considering layer undulations depending only on x and z, as has been considered by Clark and Meyer [47]. A discussion of the related magnetic energy in terms of the derivatives of u can be found in [110]. For example, if the magnetic field is of the form $\mathbf{H} = (H, 0, 0)$ where $H = |\mathbf{H}|$ then the magnetic energy, in the usual notation, is given by (2.99) with $\mathbf{n}$ replaced by the above approximation for $\mathbf{a}$ (cf. the expression (6.47) above when $\theta = 0$), to yield

$$ w_{mag} = -\frac{1}{2}\mu_0\Delta\chi H^2\left(\frac{\partial u}{\partial x}\right)^2, \qquad (6.150) $$

exactly the form used by de Gennes and Prost [110, p.363] when considering the geometry in Fig. 6.3 for SmA with $\Delta\chi > 0$.

A comprehensive account of the general energy for smectic phases in terms of the smectic layer displacement has been given by Stallinga and Vertogen [257] (who employ the notation u_z for what is represented here by u). A simplified version of an energy density for planar layers of SmC in terms of the derivatives of u when layer undulations are assumed to vanish at the boundaries in z has been identified by Stewart [264] in the geometry of Fig. 6.3: the result has been based on the above elastic energy descriptions for SmC and is consistent with the more general theory presented in [257]. For example, if $u(x, y, z) = f(x)g(y)h(z)$ and $h(0) = h(d) = 0$ where the boundaries are at $z = 0, d$, then one possibility, in this geometry of planar layers, is to construct an energy density for SmC in the form [264]

$$ w_C = \frac{1}{2}A_{12}\left(\frac{\partial^2 u}{\partial x^2}\right)^2 + \frac{1}{2}(B_2 + A_{21} + 2C_2)\left(\frac{\partial^2 u}{\partial y^2}\right)^2 + \frac{1}{2}[B_1 - 2(A_{11} + C_1)]\frac{\partial^2 u}{\partial x^2}\frac{\partial^2 u}{\partial y^2} $$
$$ + \frac{1}{2}\overline{B}\left\{\frac{\partial u}{\partial z} - \frac{1}{2}\left[\left(\frac{\partial u}{\partial x}\right)^2 + \left(\frac{\partial u}{\partial y}\right)^2\right]\right\}^2. \qquad (6.151) $$

This energy was obtained by taking the approximations $\mathbf{a} = (-\partial u/\partial x, -\partial u/\partial y, 1)$, $\mathbf{b} = (-\partial u/\partial y, 1, \partial u/\partial y)$ and $\mathbf{c} = (1, \partial u/\partial y, \partial u/\partial x)$ in the expression (6.26), adding the layer compression energy (6.148), and applying some integral identities. It can also be derived from the expression (6.24) by setting $\Omega_x = \Omega_z = \partial u/\partial y$ and $\Omega_y = -\partial u/\partial x$ and neglecting terms involving $\partial^2 u/\partial z^2$ since they are considered to be dominated by the term $\overline{B}(\partial u/\partial z)^2$ [110, p.343]; this assumes a variables separable form for u which allows the integration of the term involving B_{13} so that it is effectively an evaluation at a boundary. There is no term involving B_3 in this elementary energy (and therefore the coupling term B_{13} cannot be expected to appear in the bulk energy) which means that a rotation of the c-director as an observer travels across the layers in the z-direction may not materialise to this order of approximation: see Fig. 6.3(f) and the comments that accompany it. Notice that the constraints (6.3) hold to the first order of magnitude in derivatives of u. The constraint $\nabla \times \mathbf{a} = \mathbf{0}$ stated at equation (6.4) no longer holds identically, although it can be shown that for small disturbances it is zero to first order because the derivatives of u with respect to z are close to the squares of derivatives in x and y for small layer distortions [264]: this is to be expected if the smectic layers are themselves able to become distorted in a general manner and is in accordance with the results for SmA [149]. The inequalities in (6.28) and (6.39)$_1$ guarantee that the coefficients in the first two expressions on the right-hand side of (6.151) are non-negative while the approximations in (6.41) show that the coefficient of the third expression tends to the non-negative splay elastic constant K_1 as the smectic cone angle θ tends to zero. Actually, the expressions in (6.41) demonstrate that the layer undulation energy for SmC in (6.151) reduces to that in (6.149) for SmA as θ tends to zero, that is, $w_C \to w_A$ as $\theta \to 0$. This shows that (6.151) is one possible consistent extension from the energy for SmA to an energy for SmC in a planar layer geometry. Other versions of a distortion energy for SmC in terms of u and the phase angle of the c-director are available: see Johnson and Saupe [138], who also include experimental micrographs of SmC textures displaying undulations.

The incorporation of a magnetic energy in terms of u to the above energy for SmC has also been investigated [264]. For example, in the geometry of Fig. 6.3, if we set $\mathbf{H} = (H, 0, 0)$ then one possibility for the additional magnetic energy density may be written as

$$w_{mag} = -\frac{1}{2}\mu_0 \Delta\chi H^2 \left\{ \cos(2\theta)\left(\frac{\partial u}{\partial x}\right)^2 + \left[1 - \left(\frac{\partial u}{\partial y}\right)^2\right]\sin^2\theta \right\}. \tag{6.152}$$

This result requires second order approximations to be made on $\mathbf{a}$ and $\mathbf{c}$ rather than the first order ones introduced above. It is seen that w_{mag} reduces to the magnetic energy density for SmA given by (6.150) as the smectic cone angle $\theta \to 0$. Application of this magnetic energy that are relevant to layer undulations in SmC have been presented in reference [264], where it was shown that the distortions of the layers at a critical threshold are anticipated to be independent of y, although post-threshold behaviour may require dependence on y; a special case will be given below. The consistency of the form in (6.152) with other results that involve a magnetic energy for SmA which is dependent on x and y has also been investigated (for example, in the degenerate case when $\theta = \frac{\pi}{2}$ we recover the magnetic energy density relevant to SmA discussed by de Gennes and Prost [110, p.344]), but the details are beyond the scope of this book.

Recall that the elastic constants A_i, B_i and C_i have the dimensions of energy per unit length ($\text{J m}^{-1} = \text{N}$), whereas the layer compression constant $\overline{B}$ introduced here must have dimensions of energy per unit volume ($\text{J m}^{-3} = \text{N m}^{-2}$). The compression constant $\overline{B}$ has been estimated at 10^7 dyn cm^{-2} (equivalent to 10^6 N m^{-2} in SI units) for SmA liquid crystals by Kléman and Parodi [149]. Measurements by Collin, Gallani and Martinoty [51] for the liquid crystal TBBA indicate that $\overline{B} \sim 8.95 \times 10^7$ N m^{-2} in its SmA phase, while in its SmC phase $\overline{B} \sim 8.47 \times 10^6$ N m^{-2}. The layer compression constant $\overline{B}$ has been measured experimentally by Shibahara *et al.* [249, 250] for various materials exhibiting SmC* phases (these phases are introduced below in Section 6.4); in some of these materials (for example, DOBAMBC [249]) the value of $\overline{B}$ can be lower in the SmC* phase than in the SmA phase, away from the transition temperature from SmA to SmC*, while in others (for example 10BIMF9 [250]) $\overline{B}$ can be higher in the SmC* phase than in the SmA phase.

Penetration Length

We shall now investigate a property of smectic liquid crystals which is quite extraordinary compared to the properties of nematic liquid crystals: any small distortion of the smectic layers will penetrate across the layers over a distance which is very much greater than the wavelength or magnitude of the initial distortion along the smectic layers. This is in contrast to the situation for nematics in which it is well known that similar effects have a characteristic length comparable in magnitude to the boundary distortion [110, pp.115–116]. In the simplest case, consider a semi-infinite sample of SmA liquid crystal having a boundary condition that corresponds to a sinusoidal undulation of the smectic layers in the geometry shown in Fig. 6.9, as described by de Gennes and Prost [110, pp.354–356]. It is assumed that u and its derivatives are small and that the distortion, in the geometry of the Figure, is independent of y; this enables us to consider a sample that is of unit depth in the y-direction. Consequently, the relevant energy density to be minimised is provided by (6.149) when the contribution involving the square of $\partial u / \partial x$ in the energy is neglected. Given that u depends only upon x and z, the energy density is then

$$w_A = \frac{1}{2} K_1 \left(\frac{\partial^2 u}{\partial x^2} \right)^2 + \frac{1}{2} \overline{B} \left(\frac{\partial u}{\partial z} \right)^2. \tag{6.153}$$

The wavelength of the undulation $u(x, z)$ along the x-axis is set to be $2\pi/k$ for some wave number k, in which case the boundary condition at $z = 0$ may be taken as

$$u(x, 0) = u_0 \sin(kx), \tag{6.154}$$

for some fixed k and constant $|u_0| \ll 1$. The amplitude of the undulation is considered small, in the sense that $|u_0|$ is less than the smectic interlayer distance a (perhaps around $20 \sim 80$ Å: cf. page 6), and $k|u_0| \ll 1$ to ensure small derivatives: if $|u_0| > d$ then nonlinear effects may have to be included in the model [110]. The general Euler–Lagrange equation for equilibrium solutions to integrals having integrands of the form

$$f(u, u_x, u_z, u_{xx}, u_{xz}, u_{zz}), \tag{6.155}$$

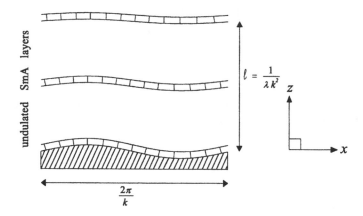

Figure 6.9: A small sinusoidal undulation $u_0 \sin(kx)$, with wavelength $2\pi/k$, at a boundary surface of a semi-infinite sample of SmA liquid crystal has an extensive effect which penetrates far inside the sample in the region $z > 0$. The penetration length of this effect is defined by $\ell = 1/\lambda k^2$, where $\lambda = \sqrt{K_1/B}$, and is generally much greater than the wavelength of the undulation at the boundary.

where the subscripts denote partial differentiation with respect to the indicated variable or variables, is given by [254, p.202]

$$\frac{\partial f}{\partial u} - \frac{\partial}{\partial x}\frac{\partial f}{\partial u_x} - \frac{\partial}{\partial z}\frac{\partial f}{\partial u_z} + \frac{\partial^2}{\partial x^2}\frac{\partial f}{\partial u_{xx}} + \frac{\partial^2}{\partial x \partial z}\frac{\partial f}{\partial u_{xz}} + \frac{\partial^2}{\partial z^2}\frac{\partial f}{\partial u_{zz}} = 0. \quad (6.156)$$

This result can be applied to w_A to find that the equilibrium equation is

$$\frac{\partial^2 u}{\partial z^2} - \lambda^2 \frac{\partial^4 u}{\partial x^4} = 0, \qquad \text{where} \qquad \lambda = \sqrt{\frac{K_1}{B}}. \quad (6.157)$$

The parameter λ is a length scale and also appears below when the Helfrich–Hurault transition is investigated. We can use the separation of variables technique and search for solutions of the form

$$u(x, z) = f(x)g(z). \quad (6.158)$$

Inserting this expression into (6.157) and dividing throughout by $f(x)g(z)$ shows that

$$\frac{1}{g}\frac{d^2 g}{dz^2} = \lambda^2 \frac{1}{f}\frac{d^4 f}{dx^4} \equiv C, \quad (6.159)$$

where C is a constant; this constant arises because, in the first equality appearing in (6.159), the left-hand side is independent of x and the right-hand side is independent of z, which shows that both sides must equal the same constant. The differential equation for f is

$$\frac{d^4 f}{dx^4} - C\lambda^{-2} f = 0. \quad (6.160)$$

It has the eigenvalues $\pm\lambda^{-\frac{1}{2}}C^{\frac{1}{4}}$ and $\pm i\lambda^{-\frac{1}{2}}C^{\frac{1}{4}}$. However, by the imposed boundary condition (6.154), it is clear that $C = \lambda^2 k^4$ since f must then be given by

$$f(x) = f_0 \sin(kx). \qquad (6.161)$$

The equation for $g(z)$ is then

$$\frac{d^2 g}{dz^2} - \lambda^2 k^4 g = 0. \qquad (6.162)$$

The eigenvalues for this differential equation are $\pm\lambda k^2$. However, since the solution must decay as z tends to infinity, we must have

$$g(z) = g_0 \exp(-\lambda k^2 z), \qquad g(0) = g_0. \qquad (6.163)$$

We can now set the product $f_0 g_0$ to be equal to u_0 to arrive at the solution

$$u(x, z) = u_0 \sin(kx) \exp\left(-\frac{z}{\ell}\right), \qquad (6.164)$$

where the *penetration length* ℓ is defined by

$$\ell = \frac{1}{\lambda k^2}. \qquad (6.165)$$

The sinusoidal undulation at the boundary is seen to diminish exponentially across the smectic layers in the z-direction over a distance characterised by the penetration length ℓ.

The penetration length ℓ is generally much larger than the wavelength of the undulation at the boundary. For example, if we consider the case when

$$2\pi/k = 10\mu m, \qquad \lambda = 20 \text{ Å}, \qquad (6.166)$$

then $\ell \sim 1.3$ mm, which is considerably greater than the wavelength of 10 μm at the boundary. This penetration length is also substantially larger than that encountered in nematic samples (cf. [110, p.116]) where the analogue of ℓ is of the same order as the boundary distortion.

The foregoing result for SmA has its theoretical analogue for SmC. We can adopt the energy density (6.151), under the same assumption that u is small (so that the layer compression contribution involving the square of $\partial u/\partial x$ is, as above, neglected), to find that the penetration length ℓ is again given by equation (6.165), except that λ is now replaced by $\lambda = \left(A_{12}/\overline{B}\right)^{1/2}$.

In general, boundary conditions, including natural boundary conditions and the effects of weak anchoring, may be derived to supplement the general Euler–Lagrange equation (6.156): this is accomplished by standard methods from the calculus of variations by considering the vanishing of the first variation of the total energy, similar in style to the methodology used in Sections 2.6 and 2.7.2. See also the comments at the top of page 291.

The Helfrich–Hurault Transition

Consider the experimental set-up displayed in Fig. 6.10 where a magnetic field **H** is applied in the x-direction parallel to a system of planar aligned layers of SmC liquid crystal as shown. Suppose that $\Delta\chi > 0$, which indicates that the directorprefers to align parallel to **H**, and assume that initially **c** is parallel to **H**.

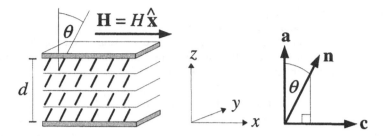

Figure 6.10: The arrangement of a planar aligned sample of SmC liquid crystal described in cartesian coordinates. The director **n** is tilted at the smectic cone angle θ to the layer normal **a**; **c** is the unit orthogonal projection of **n** onto the smectic planes. A magnetic field **H** is applied in the x-direction. When $\Delta\chi > 0$, the smectic layers will begin to distort, as shown in Fig. 6.11 below, when $H \gtrsim H_c$, where H_c is defined by equation (6.177).

One can imagine that as the magnitude $H = |\mathbf{H}|$ increases, there will be a critical value $H = H_c$ at which the smectic layers will desire to undulate as **n** attempts to align as much as possible with the magnetic field. This anticipated transition from planar layers to undulated or distorted layers is called the Helfrich–Hurault transition and its onset is expected to occur at a critical field magnitude $H_c > 0$. A schematic diagram showing the anticipated behaviour of the smectic layers is shown in Fig. 6.11. Effects such as this were first examined by Helfrich [126] and Hurault [130] when they investigated infinite samples of cholesteric liquid crystals under the influence of magnetic fields. Chandrasekhar [38] and de Gennes and Prost [110] have reviewed the Helfrich–Hurault transition in SmA. We now pursue a special case described in [264] in order to determine H_c for this example which involves SmC.

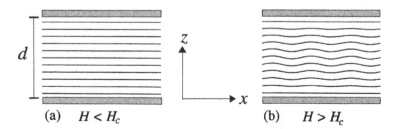

Figure 6.11: A schematic diagram of the anticipated undulations of the smectic layers under the influence of a magnetic field for the geometry introduced in Fig. 6.10. As H increases through the value H_c, defined by equation (6.177), the smectic layers will begin to distort. For $H < H_c$ no layer undulations occur, while for $H > H_c$ the layers will undulate or distort.

For simplicity, it is supposed that the smectic layers are bounded by plates at $z = 0$ and $z = d$, as shown in Fig. 6.10, and that $u = u(x, z)$ since it is expected that, close to the onset of any critical effect, layer distortions need not depend upon y in a first approximation [264]. It is also assumed that the smectic layers will have negligible distortions near the boundary surfaces. We can then set

$$\mathbf{a} = \left(-\frac{\partial u}{\partial x}, 0, 1 \right), \qquad \mathbf{c} = \left(1, 0, \frac{\partial u}{\partial x} \right). \tag{6.167}$$

The relevant energy for this problem is given by $w = w_C + w_{mag}$ via equations (6.151) and (6.152), except that we choose to adopt the form (6.147) for the compression term because we are only seeking a critical threshold for displacements u which are small. The energy is then given by

$$w = \frac{1}{2} A_{12} \left(\frac{\partial^2 u}{\partial x^2} \right)^2 + \frac{1}{2} \overline{B} \left(\frac{\partial u}{\partial z} \right)^2 - \frac{1}{2} \mu_0 \Delta \chi H^2 \left[\cos(2\theta) \left(\frac{\partial u}{\partial x} \right)^2 + \sin^2 \theta \right]. \tag{6.168}$$

Now consider possible displacements which vanish at the boundaries at $z = 0$ and $z = d$ of the form

$$u = u_0 \sin(kx) \sin(\pi z/d), \tag{6.169}$$

where $|u_0| \ll 1$. For a periodic function f having period P, introduce the average $\langle f \rangle$ defined by

$$\langle f \rangle = \frac{1}{P} \int_0^P f(m) \, dm. \tag{6.170}$$

For example, any functions of the form $\sin^2(\psi)$, $\cos^2(\psi)$ or $G = constant$ result in

$$\langle \sin^2 \rangle = \langle \cos^2 \rangle = \tfrac{1}{2}, \qquad \langle G \rangle = G. \tag{6.171}$$

We can now take the average of the energy density in (6.168) over a sample of unit depth in y to find that

$$\langle w \rangle = \tfrac{1}{8} u_0^2 \left[\overline{B} \pi^2/d^2 + A_{12} k^4 - \mu_0 \Delta \chi H^2 k^2 \cos(2\theta) \right] - \tfrac{1}{2} \mu_0 \Delta \chi H^2 \sin^2(\theta). \tag{6.172}$$

Also, for the undistorted state $u \equiv 0$,

$$\langle w(u \equiv 0) \rangle = -\tfrac{1}{2} \mu_0 \Delta \chi H^2 \sin^2(\theta), \tag{6.173}$$

and therefore a comparison of energies between the distorted and undistorted states gives

$$\begin{aligned} \Delta \langle w \rangle &= \langle w(u) \rangle - \langle w(u \equiv 0) \rangle \\ &= \frac{1}{8} u_0^2 \left[\overline{B} \pi^2/d^2 + A_{12} k^4 - \mu_0 \Delta \chi H^2 k^2 \cos(2\theta) \right]. \end{aligned} \tag{6.174}$$

The critical field H_c is found by minimising $\Delta \langle w \rangle$ over non-zero values of k and then determining the least value of H above which $\Delta \langle w \rangle$ will become negative, which indicates the system's preference for adopting the distorted variable solution u rather than the undistorted zero solution. The right-hand side of (6.174) can be minimised with respect to non-zero values of k to find that its minimum occurs at $k = k_x$ given by

$$k_x^2 = \frac{\mu_0 \Delta \chi H^2}{2 A_{12}} \cos(2\theta), \qquad (6.175)$$

assuming that $A_{12} > 0$ (recall that $A_{12} \geq 0$ by (6.28)). For $k = k_x$ we can write

$$\Delta \langle w \rangle = \frac{1}{8} u_0^2 \left[\overline{B} \frac{\pi^2}{d^2} - \frac{(\mu_0 \Delta \chi)^2 H^4}{4 A_{12}} \cos^2(2\theta) \right]. \qquad (6.176)$$

As H increases from zero, it is seen that the averaged energy difference $\Delta \langle w \rangle$ will decrease through zero and that the critical magnetic field strength H_c will be given by

$$
\begin{aligned}
\mu_0 \Delta \chi H_c^2 \cos(2\theta) &= 2\sqrt{\overline{B} A_{12}} \frac{\pi}{d} \\
&\equiv 2\pi \frac{A_{12}}{\lambda d},
\end{aligned}
\qquad (6.177)
$$

where

$$\lambda = \sqrt{\frac{A_{12}}{\overline{B}}}, \qquad (6.178)$$

can be introduced as a length scale. The product $\overline{B} A_{12}$ can be determined at H_c via equation (6.177). Notice also that H_c is proportional to $d^{-\frac{1}{2}}$, whereas the critical Freedericksz threshold in each of the classical Freedericksz transitions is proportional to d^{-1} (see Section 3.4.1 and equation (6.143) in Section 6.2.5). Additionally, the wave number k_x defined in (6.175), becomes k_c at $H = H_c$, where

$$k_c^2 = \frac{\pi}{\lambda d}. \qquad (6.179)$$

Similar to the observation by de Gennes and Prost [110, p.363], it follows that the optimal wavelength q of the distortion is given by

$$q \equiv \frac{2\pi}{k_c} = 2\sqrt{\pi}\sqrt{\lambda d}, \qquad (6.180)$$

which indicates that q is proportional to the geometric mean of the sample depth d and the length scale λ.

The results from equations (6.41) can be inserted into (6.177) and (6.178) to find that for the SmA phase, which occurs at $\theta = 0$, we have

$$\mu_0 \Delta \chi H_c^2 = 2\pi \frac{K_1}{\lambda d}, \qquad \lambda = \sqrt{\frac{K_1}{\overline{B}}}, \qquad (6.181)$$

which is precisely the result mentioned by de Gennes and Prost [110, p.363] for the critical field strength at the onset of the Helfrich–Hurault transition in SmA, K_1 being the usual splay constant arising in (6.40) for the SmA elastic energy. This shows that the threshold derived in equation (6.177) is one possible extension of the result from the SmA case to SmC.

Following de Gennes and Prost [110], we can estimate λ to be of the order of a microscopic length scale and, as an example, set $\lambda = 20$ Å. This allows the smectic layer bending constant A_{12} to be estimated from equation (6.178) if $\overline{B}$ is known. To obtain an estimate for a typical SmC material we can take the value of $\overline{B}$ stated above on page 284 for the liquid crystal TBBA in its SmC phase [51], namely $\overline{B} \sim 8.47 \times 10^6$ N m^{-2}, to find from (6.178) that

$$A_{12} \approx 3.39 \times 10^{-11} \text{ N}, \tag{6.182}$$

larger than, but comparable to, the Frank elastic splay constant K_1 of nematic theory which is often used in the description of SmA. We can adopt the values used by de Gennes and Prost [110] for SmA and estimate a typical critical threshold H_c for SmC by using the value for μ_0 in SI units from Table D.1 and setting

$$\lambda = 20 \text{ Å}, \qquad \theta = 22°, \qquad d = 1 \text{ mm}, \qquad \Delta\chi = 1.3 \times 10^{-6}, \tag{6.183}$$

and use the above estimate in (6.182) for A_{12}, to find that the critical value for the field strength is given via (6.177) by $H_c \approx 95 \times 10^5$ A m^{-1}, which is about three times the magnitude of that for a typical SmA [110]. For example, when the relevant parameters are as stated in (6.183), we can estimate $K_1 = 5 \times 10^{-12}$ N to find that from equation (6.181)$_1$ we have $H_c \approx 31 \times 10^5$ A m^{-1} for a typical SmA.

The critical magnetic field magnitudes derived here are relatively large, as commented upon by de Gennes and Prost, and are really given only as examples. However, there are analogous results for electric fields using the identification $\mu_0 \Delta\chi H^2 \longleftrightarrow \epsilon_0 \epsilon_a E^2$ in equations (6.177) and (6.181). In this context, the interested reader may consult the previously mentioned article by Geer *et al.* [103], which addresses electric field effects in a chiral SmA liquid crystal and also investigates the post-threshold behaviour of the layers as they 'buckle' for $E > E_c$. As an example for a critical field E_c, consider the values

$$\lambda = 20 \text{ Å}, \qquad \theta = 22°, \qquad d = 10 \text{ μm}, \qquad \epsilon_a = 10, \tag{6.184}$$

with A_{12} given by (6.182) and ϵ_0 as stated in Table D.1. Using the aforementioned identification, the result in equation (6.177) then delivers the critical electric field magnitude $E_c^{HH} \approx 12 \times 10^6$ V m^{-1} for the onset of the Helfrich–Hurault transition in SmC. This is known to be well within the range of experimentally feasible values (cf. [103], where the fields reported in the analysis of observed post-threshold effects in a Helfrich–Hurault transition in a particular SmA* material were similar in magnitude to this value). This is considerably higher than the electric field required for a classical Freedericksz transition where the smectic layers themselves remain intact; for example, for the values in (6.184), and taking $B_2 = 5 \times 10^{-12}$ N as a typical smectic constant, the critical threshold for a classical Freedericksz transition is given by equation (6.144) as $E_c^F = 0.2 \times 10^6$ V m^{-1}, considerably lower than the Helfrich–Hurault critical threshold. In this example, it would appear that the Helfrich–Hurault transition may well have a qualitatively higher magnitude threshold than the usual Freedericksz threshold for SmC discussed in the previous Section.

The above analysis assumes that the undulations of the layers vanish at the boundaries. Ishikawa and Lavrentovich [131] have modelled a lamellar system of cholesteric liquid crystal which allows layer undulations near the boundaries. This was achieved by adding a finite surface anchoring energy to a bulk energy that is essentially of the same form as that stated above for SmA. This idea could perhaps also be modified to model general SmA or SmC liquid crystals. The results in [131] seem to indicate that the incorporation of finite surface anchoring leads to a decrease in the theoretical threshold for the onset of the Helfrich–Hurault transition.

Some comments on the dynamics of undulations, not considered here, are mentioned on page 319. It should also be emphasised that the numerical thresholds given above are speculative and depend on the values of physical parameters, many of which are currently unknown for most materials. Nevertheless, these results demonstrate the methodology which is used by numerous authors and are in the spirit and general flavour of many of the articles in the literature.

6.3 Dynamic Theory of Smectic C

The dynamic theory for incompressible SmC liquid crystals introduced by Leslie, Stewart and Nakagawa [173] will now be summarised. The technical details are beyond the scope of this text and we only give a review of the basic features to show that this theory is a natural development of previous concepts from the dynamic theory of nematic liquid crystals, which was fully derived in detail in Chapter 4. The dynamic theory rests upon the assumptions made in modelling the SmC phase outlined in Section 6.2 above. It has been highlighted by Leslie [183], in his brief review of this theory, that the assumption of fixed layer spacing can be interpreted as supposing that variations in layer thickness require high energy. Consequently, it seems reasonable to accept the assumption of fixed smectic layer thickness when the smectic is not under extremes of strain. Similarly, the assumption that the smectic tilt angle θ is constant excludes the possibility of some thermal or pre-transitional effects. However, aspects of the theory given here have proved useful in some practical contexts, some of which are mentioned below, and this encourages further developments to the theory. A brief outline of the derivation of this theory will be given next in Section 6.3.1. Section 6.3.2 will make some comments on the viscosities for SmC and SmC$_M$. Some simple flow alignments will be discussed in Section 6.3.3.

6.3.1 Dynamic Equations for SmC Liquid Crystals

The dynamic theory for SmC liquid crystals is based upon the same three conservation laws for mass, linear momentum and angular momentum that were employed in the derivation of the dynamic theory of nematics developed in Section 4.2.2. They are given by equations (4.29), (4.30) and (4.31) and, as before, the incompressibility assumption applied to the mass conservation law leads to the familiar constraint that the velocity vector $\mathbf{v}$ satisfies

$$v_{i,i} = 0, \qquad (6.185)$$

with the density ρ being constant. The balances of linear momentum and angular momentum are again given by equations (4.41) and (4.45) and can be written as, respectively,

$$\rho \dot{v}_i \;=\; \rho F_i + t_{ij,j}, \tag{6.186}$$

$$\rho K_i + \epsilon_{ijk} t_{kj} + l_{ij,j} \;=\; 0, \tag{6.187}$$

where $\mathbf{F}$ is the external body force per unit mass, t_{ij} is the stress tensor, $\mathbf{K}$ is the external body moment per unit mass, l_{ij} is the couple stress tensor and a superposed dot represents the material time derivative defined in equation (4.5). As in the derivation of the dynamic theory for nematics, the director inertial term has been omitted because it is considered to be negligible in many circumstances. Also, as in nematic theory, there seems to be a convention in the literature that F_i appears in static theory while ρF_i appears in dynamic theory: this means that in statics F_i represents the external body force per unit volume while it represents the external body force per unit mass in dynamics. A similar statement applies to the external body moment K_i. The derivation of the SmC dynamic theory relies on the assumption that the rate of viscous dissipation $\mathcal{D}$ per unit volume is positive, as was supposed in a similar way for nematics at equations (4.46) to (4.58), with w_F now replaced by the smectic elastic energy density w given by equation (6.13) (or (6.15)). This leads to the analogue of equation (4.51) obtained by considering the rate of work done on an arbitrary volume, giving

$$t_{ij} v_{i,j} + l_{ij} w_{i,j} - w_i \epsilon_{ijk} t_{kj} = \dot{w} + \mathcal{D}, \tag{6.188}$$

where the vector $\mathbf{w}$ is the local angular velocity of a liquid crystal material element, satisfying

$$\dot{\mathbf{a}} = \mathbf{w} \times \mathbf{a} \qquad \text{and} \qquad \dot{\mathbf{c}} = \mathbf{w} \times \mathbf{c}, \tag{6.189}$$

these being the smectic equivalents corresponding to $\dot{\mathbf{n}} = \mathbf{w} \times \mathbf{n}$ given in equation (4.4). (The symbol $\boldsymbol{\omega}$ was used in [173] to denote the local angular velocity of the liquid crystal material element, but in this text we have adopted the notation defined in Section 4.2.1 on page 134.)

Motivated by the static theory for SmC, it seems reasonable, from equations (6.70) and (6.90), to set

$$t_{ij} \;=\; -p\delta_{ij} + \beta_p \epsilon_{pjk} a_{k,i} - \frac{\partial w}{\partial a_{k,j}} a_{k,i} - \frac{\partial w}{\partial c_{k,j}} c_{k,i} + \tilde{t}_{ij}, \tag{6.190}$$

$$l_{ij} \;=\; \beta_p a_p \delta_{ij} - \beta_i a_j + \epsilon_{ipq}\left(a_p \frac{\partial w}{\partial a_{q,j}} + c_p \frac{\partial w}{\partial c_{q,j}} \right) + \tilde{l}_{ij}, \tag{6.191}$$

where $\tilde{t}_{ij}$ and $\tilde{l}_{ij}$ denote the dynamic contributions to the stress and couple stress, respectively. Following a similar procedure to that for nematics, it is first supposed that

$$\tilde{t}_{ij} \quad \text{and} \quad \tilde{l}_{ij} \qquad \text{are functions of} \qquad a_i,\, c_i,\, w_i \text{ and } v_{i,j}. \tag{6.192}$$

Material frame-indifference requires that the dependence in equation (6.192) be replaced by (cf. the nematic case in Section 4.2.3 at equation (4.67))

$$\tilde{t}_{ij} \quad \text{and} \quad \tilde{l}_{ij} \qquad \text{are functions of} \qquad a_i,\, c_i,\, A_i,\, C_i \text{ and } D_{ij}, \tag{6.193}$$

where

$$D_{ij} = \tfrac{1}{2}(v_{i,j} + v_{j,i}), \qquad W_{ij} = \tfrac{1}{2}(v_{i,j} - v_{j,i}), \qquad (6.194)$$
$$A_i = \dot{a}_i - W_{ik}a_k, \qquad C_i = \dot{c}_i - W_{ik}c_k, \qquad (6.195)$$

with the symmetry condition (6.12) imposing additional restrictions on the functional dependence stated at equation (6.193). Here, D_{ij} and W_{ij} are the usual rate of strain tensor and vorticity tensor, respectively: we have changed the notation of the rate of strain tensor from A_{ij}, used in the theory of nematics, to D_{ij} in smectic theory to avoid possible notational confusion with the vector A_i. The quantities **A** and **C** are the co-rotational time fluxes of the vectors **a** and **c** and are the smectic analogues of the nematic counterpart **N** given by equations (4.10). It can also be shown by arguments analogous to those used to obtain the results in equations (4.8) to (4.10) that

$$\mathbf{A} = \boldsymbol{\omega} \times \mathbf{a}, \qquad \mathbf{C} = \boldsymbol{\omega} \times \mathbf{c}, \qquad (6.196)$$

where $\boldsymbol{\omega}$ is the relative angular velocity introduced at equation (4.2).

The above constitutive assumptions (6.190) and (6.191) enable the identity (6.188) to be reduced to the inequality

$$\mathcal{D} = \tilde{t}_{ij}v_{i,j} + \tilde{l}_{ij}w_{i,j} - w_i\epsilon_{ijk}\tilde{t}_{kj} \geq 0, \qquad (6.197)$$

upon supposing that the viscous dissipation $\mathcal{D}$ is always positive. This particular reduction requires some identities involving the elastic energy density and the quantities a_i, w_i, $a_{i,j}$, $w_{i,j}$ and $v_{i,j}$ (see Reference [173] for details). The inequality (6.197) and the constitutive assumptions (6.192) immediately yield the result

$$\tilde{l}_{ij} = 0, \qquad (6.198)$$

since the positivity of $\mathcal{D}$ must be preserved under changes in sign of $w_{i,j}$, as is the case in nematics (cf. equations (4.59) and (4.60)). This permits a further reduction of (6.197) to

$$\mathcal{D} = \tilde{t}_{ij}v_{i,j} - w_i\epsilon_{ijk}\tilde{t}_{kj} \geq 0. \qquad (6.199)$$

Under the assumption that $\tilde{t}_{ij}$ is a linear function of A_i, C_i and the rate of strain D_{ij}, and that the symmetry requirement (6.11) must also hold, it can be shown that the dependence in (6.193) forces $\tilde{t}_{ij}$ to be an isotropic function of the variables listed there, analogous to the case for nematic liquid crystals (cf. Section 4.2.3). It then follows that $\tilde{t}_{ij}$ consists of forty-one terms [266]: they can be obtained by arguments analogous to those used to obtain the nematic viscous stress at equations (4.71) to (4.74). Four of these terms equate to zero by a simple application of the inequality (6.199), analogous to showing $\mu_2 = 0$ in nematic theory at equation (4.84). A further five can be shown to be linear combinations of the other terms by means of rather involved vector identities, which leaves thirty-two terms. However, this number can be reduced further to twenty terms if we apply Onsager relations [266]. Employing the notation

$$D_i^a = D_{ij}a_j, \quad \text{and} \quad D_i^c = D_{ij}c_j, \qquad (6.200)$$

the final expression for the viscous stress $\tilde{t}_{ij}$ can be written as

$$\tilde{t}_{ij} = \tilde{t}_{ij}^s + \tilde{t}_{ij}^{ss}, \tag{6.201}$$

where $\tilde{t}_{ij}^s$ and $\tilde{t}_{ij}^{ss}$ are the symmetric and skew-symmetric parts of the viscous stress given by [173, 265]

$$
\begin{aligned}
\tilde{t}_{ij}^s ={} & \mu_0 D_{ij} + \mu_1 a_p D_p^a a_i a_j + \mu_2 (D_i^a a_j + D_j^a a_i) + \mu_3 c_p D_p^c c_i c_j \\
& + \mu_4 (D_i^c c_j + D_j^c c_i) + \mu_5 c_p D_p^a (a_i c_j + a_j c_i) \\
& + \lambda_1 (A_i a_j + A_j a_i) + \lambda_2 (C_i c_j + C_j c_i) + \lambda_3 c_p A_p (a_i c_j + a_j c_i) \\
& + \kappa_1 (D_i^a c_j + D_j^a c_i + D_i^c a_j + D_j^c a_i) \\
& + \kappa_2 \left[a_p D_p^a (a_i c_j + a_j c_i) + 2 a_p D_p^a a_i a_j \right] \\
& + \kappa_3 \left[c_p D_p^c (a_i c_j + a_j c_i) + 2 a_p D_p^c c_i c_j \right] \\
& + \tau_1 (C_i a_j + C_j a_i) + \tau_2 (A_i c_j + A_j c_i) \\
& + 2 \tau_3 c_p A_p a_i a_j + 2 \tau_4 c_p A_p c_i c_j,
\end{aligned}
\tag{6.202}
$$

$$
\begin{aligned}
\tilde{t}_{ij}^{ss} ={} & \lambda_1 (D_j^a a_i - D_i^a a_j) + \lambda_2 (D_j^c c_i - D_i^c c_j) + \lambda_3 c_p D_p^a (a_i c_j - a_j c_i) \\
& + \lambda_4 (A_j a_i - A_i a_j) + \lambda_5 (C_j c_i - C_i c_j) + \lambda_6 c_p A_p (a_i c_j - a_j c_i) \\
& + \tau_1 (D_j^a c_i - D_i^a c_j) + \tau_2 (D_j^c a_i - D_i^c a_j) + \tau_3 a_p D_p^a (a_i c_j - a_j c_i) \\
& + \tau_4 c_p D_p^c (a_i c_j - a_j c_i) + \tau_5 (A_j c_i - A_i c_j + C_j a_i - C_i a_j).
\end{aligned}
\tag{6.203}
$$

There are twenty viscosity coefficients: the twelve viscosities μ_0 to μ_5 and λ_1 to λ_6 are associated with contributions to the dynamic stress which are even in the vector $\mathbf{c}$ or do not contain $\mathbf{c}$, while the remaining eight viscosities κ_1 to κ_3 and τ_1 to τ_5 are linked to the terms which are odd in the vector $\mathbf{c}$.

It is possible to rearrange some of the above equations to prove more convenient ways of expressing the final dynamic equations. It is observed that the intrinsic torque arising from the expression in equation (6.203) may be expressed as

$$\epsilon_{ijk} \tilde{t}_{kj}^{ss} = \epsilon_{ijk} (a_j \tilde{g}_k^a + c_j \tilde{g}_k^c), \tag{6.204}$$

where it proves convenient to define the quantities

$$
\begin{aligned}
\tilde{g}_i^a ={} & -2 \left(\lambda_1 D_i^a + \lambda_3 c_i c_p D_p^a + \lambda_4 A_i + \lambda_6 c_i c_p A_p + \tau_2 D_i^c \right. \\
& \left. + \tau_3 c_i a_p D_p^a + \tau_4 c_i c_p D_p^c + \tau_5 C_i \right), \\
\tilde{g}_i^c ={} & -2 \left(\lambda_2 D_i^c + \lambda_5 C_i + \tau_1 D_i^a + \tau_5 A_i \right).
\end{aligned}
\tag{6.205}
\tag{6.206}
$$

These quantities then allow the viscous dissipation inequality (6.199) to be more conveniently formulated as

$$\mathcal{D} = \tilde{t}_{ij}^s D_{ij} - \tilde{g}_i^a A_i - \tilde{g}_i^c C_i \geq 0, \tag{6.207}$$

using the results in equations (6.189) and (6.196) combined with some rearrangements of terms by arguments analogous to those used to obtain the equalities in equation (4.81) which exploited the result (4.2). This viscous dissipation inequality leads to restrictions upon the various smectic viscosity coefficients and is the smectic analogue of the inequality (4.82) for nematics.

Following the form of equation (4.100), the external body moment may be written as

$$\rho K_i = \epsilon_{ijk}(a_j G_k^a + c_j G_k^c), \tag{6.208}$$

where $\mathbf{G}^a$ and $\mathbf{G}^c$ are the external body forces introduced in a similar way to those mentioned in the static theory of SmC at equation (6.50), and appearing throughout Section 6.2.3. Employing the constitutive relations (6.190) and (6.191) and the results (6.198) and (6.208) to the balance of angular momentum equation (6.187), it is possible to use the methodology detailed for nematic theory in Section 4.2.4 to obtain the smectic analogues of the balance of angular momentum for nematics given by (4.102). These equations in turn allow a convenient formulation for the balance of linear momentum (6.186), similar to that for nematics obtained at equation (4.106). Rather than give the technical details here, we shall, for brevity and convenience, now state these final results for the governing dynamic equations for SmC liquid crystals and list them together with the relevant constraints. Further details are contained in References [173, 265].

Dynamic Equations for SmC

The results from Reference [173] can be summarised as follows. The vectors **a** and **c** are subject to the constraints

$$\mathbf{a} \cdot \mathbf{a} = 1, \quad \mathbf{c} \cdot \mathbf{c} = 1, \quad \mathbf{a} \cdot \mathbf{c} = 0, \quad \nabla \times \mathbf{a} = 0, \tag{6.209}$$

and the velocity vector **v** must satisfy

$$v_{i,i} = 0. \tag{6.210}$$

The governing dynamic equations consist of the balance of linear momentum

$$\rho \dot{v}_i = \rho F_i - \tilde{p}_{,i} + G_k^a a_{k,i} + G_k^c c_{k,i} + \tilde{g}_k^a a_{k,i} + \tilde{g}_k^c c_{k,i} + \tilde{t}_{ij,j}, \tag{6.211}$$

with

$$\tilde{p} = p + w, \tag{6.212}$$

and the balance of angular momentum, which reduces to the two coupled sets of equations

$$\left(\frac{\partial w}{\partial a_{i,j}}\right)_{,j} - \frac{\partial w}{\partial a_i} + G_i^a + \tilde{g}_i^a + \gamma a_i + \mu c_i + \epsilon_{ijk}\beta_{k,j} \;=\; 0, \tag{6.213}$$

$$\left(\frac{\partial w}{\partial c_{i,j}}\right)_{,j} - \frac{\partial w}{\partial c_i} + G_i^c + \tilde{g}_i^c + \tau c_i + \mu a_i \;=\; 0, \tag{6.214}$$

where the scalar functions γ, μ and τ and the vector function $\boldsymbol{\beta}$ are Lagrange multipliers, as introduced and discussed in Section 6.2.3. In the above equations F_i is the external body force per unit mass, G_i^a and G_i^c are generalised external body forces per unit volume related to **a** and **c**, respectively, p is the arbitrary pressure, w is the elastic energy density for SmC given by (6.13) or (6.15) (using the identities for the elastic constants in (6.25) if required), $\tilde{g}_i^a$ and $\tilde{g}_i^c$ are the dynamic contributions given by (6.205) and (6.206), and $\tilde{t}_{ij}$ is the viscous stress given via equations (6.201), (6.202) and (6.203). A superposed dot represents the usual material time derivative (4.5). The comments on the physical significance of the Lagrange multipliers made on page 266 are equally valid for dynamic problems in SmC liquid crystals. The notation introduced at equations (6.73) and (6.74), namely Π_i^a and Π_i^c, often proves convenient when using equations (6.213) and (6.214).

It is observed, in the above dynamic theory, that equations (6.209), (6.210), (6.211), (6.213) and (6.214) provide sixteen equations in the sixteen unknowns a_i, c_i, v_i, p, β_i, γ, μ and τ.

Dynamic Equations for SmC$_{\mathrm{M}}$

The dynamic equations for SmC$_{\mathrm{M}}$ are as given above by equations (6.209) to (6.214) for SmC, except that all the relevant terms must additionally obey the transformation **a** $\rightarrow$ $-$**a** and **c** $\rightarrow$ $-$**c** separately, as indicated in the discussion on page 258. In this case, the elastic energy w is given by equation (6.46) (or, equivalently, by the first six terms in equation (6.15)) while, under this symmetry, the terms $\tilde{t}_{ij}^s$, $\tilde{t}_{ij}^{ss}$, $\tilde{g}_i^a$ and $\tilde{g}_i^c$ must be replaced by

$$
\begin{aligned}
\tilde{t}_{ij}^s = {} & \mu_0 D_{ij} + \mu_1 a_p D_p^a a_i a_j + \mu_2 (D_i^a a_j + D_j^a a_i) + \mu_3 c_p D_p^c c_i c_j \\
& + \mu_4 (D_i^c c_j + D_j^c c_i) + \mu_5 c_p D_p^a (a_i c_j + a_j c_i) \\
& + \lambda_1 (A_i a_j + A_j a_i) + \lambda_2 (C_i c_j + C_j c_i) + \lambda_3 c_p A_p (a_i c_j + a_j c_i),
\end{aligned} \tag{6.215}
$$

$$
\begin{aligned}
\tilde{t}_{ij}^{ss} = {} & \lambda_1 (D_j^a a_i - D_i^a a_j) + \lambda_2 (D_j^c c_i - D_i^c c_j) + \lambda_3 c_p D_p^a (a_i c_j - a_j c_i) \\
& + \lambda_4 (A_j a_i - A_i a_j) + \lambda_5 (C_j c_i - C_i c_j) + \lambda_6 c_p A_p (a_i c_j - a_j c_i),
\end{aligned} \tag{6.216}
$$

$$
\tilde{g}_i^a = -2 \left(\lambda_1 D_i^a + \lambda_3 c_i c_p D_p^a + \lambda_4 A_i + \lambda_6 c_i c_p A_p \right), \tag{6.217}
$$

$$
\tilde{g}_i^c = -2 \left(\lambda_2 D_i^c + \lambda_5 C_i \right). \tag{6.218}
$$

These results are obtained by simply omitting all the terms connected to the τ_i and κ_i viscosities in equations (6.202), (6.203), (6.205) and (6.206), since they are odd in **a** and odd in **c** when they are present in $\tilde{t}_{ij}$. This leaves twelve viscosities: μ_0, μ_1, μ_2, μ_3, μ_4, μ_5, λ_1, λ_2 λ_3, λ_4, λ_5 and λ_6.

6.3.2   The Smectic C Viscosities

Measurements for smectic viscosities are presently rather scarce in the literature although some early measurements, interpreted via dynamic equations similar to those for nematics, have been made by some workers such as Galerne, Martinand, Durand and Veyssie [100], among others: see pages 300 and 301 below. Nevertheless, theoretical information about the smectic viscosities can be gained by further analysis of the viscous stress $\tilde{t}_{ij}$, allowing a physical interpretation to be made which may be of use to both theoreticians and experimentalists. To this end, we follow the classification introduced by Carlsson, Leslie and Clark [36] for the smectic viscosity coefficients.

Each of the twenty smectic viscosity coefficients introduced by Leslie, Stewart and Nakagawa [173] can be classified into one of four groups identified from the form of $\tilde{t}_{ij}$ provided via equations (6.201), (6.202) and (6.203). The first group consists solely of μ_0, which is associated with a term in the viscous stress which is independent of the vectors **a** and **c** and therefore corresponds to the usual isotropic contribution to the viscous stress. The second group consists of those viscosities which are connected to the terms which are independent of the vector **c** and only depend on **a**. This group consists of μ_1, μ_2, λ_1 and λ_4 whose terms, being independent of **c**, suggest that they ought to be present in the SmA phase when the smectic tilt angle θ is zero, and for this reason they are called smectic A-like viscosities. The five viscosities μ_0, μ_1, μ_2, λ_1 and λ_4 are therefore anticipated to be connected to the dynamical properties of SmA liquid crystals, this number of viscosities agreeing with the number proposed in an earlier dynamic theory for SmA by Martin, Parodi and Pershan [192]. The third group of viscosities is connected to the terms which only depend upon the vector **c** and consists of μ_3, μ_4, λ_2 and λ_5. They have been designated as nematic-like viscosities since their associated contributions resemble those for nematics. However, as pointed out in [36], the similarity with nematics should not be taken for direct comparisons because the viscous stress in nematics is described in terms of **n**, while in SmC these terms are expressed using the vector **c**. The fourth group consists of the remaining eleven viscosities μ_5, λ_3, λ_6, κ_1, κ_2, κ_3, τ_1, τ_2, τ_3, τ_4, and τ_5. Their associated terms in the viscous stress may be called coupling terms, since they depend upon both **a** and **c**. These last terms have no counterparts in the theory of nematic liquid crystals. In summary, we have the four groups

$$\text{isotropic}: \quad \mu_0, \tag{6.219}$$

$$\text{smectic A-like}: \quad \mu_1, \mu_2, \lambda_1, \lambda_4, \tag{6.220}$$

$$\text{nematic-like}: \quad \mu_3, \mu_4, \lambda_2, \lambda_5, \tag{6.221}$$

$$ac\text{-coupling}: \quad \mu_5, \lambda_3, \lambda_6, \kappa_1, \kappa_2, \kappa_3, \tau_1, \tau_2, \tau_3, \tau_4, \tau_5. \tag{6.222}$$

Carlsson *et al.* [36] go on to investigate the smectic tilt angle dependence of the smectic viscosity coefficients using the ideas introduced by Dahl and Lagerwall [62]. The resultant approximations for the viscosities may prove useful in theoretical investigations for SmC (and SmC*) materials. For temperatures near the SmA to SmC phase transition temperature T_{AC}, the smectic tilt angle θ may be considered to be relatively small. If the smectic layer normal **a** is unchanged while the changes θ to $-\theta$ and **c** to $-\mathbf{c}$ are carried out simultaneously, then the description of the SmC material ought to remain intact; therefore the viscous stress must also be invariant to such changes (cf. equation (6.1)). It can then be concluded that terms in $\tilde{t}_{ij}$ which are odd in **c** must have corresponding viscosity coefficients which are odd in θ, in order to preserve the correct symmetry invariance for SmC under the above simultaneous changes in signs of θ and **c**. Similarly, the terms which are even in **c** must have corresponding viscosity coefficients which are even in θ. In particular, the terms associated with the even viscosities μ_0, μ_1, μ_2, λ_1 and λ_4 do not contain $\mathbf{c}$ and should remain in the SmA phase where $\theta \equiv 0$. This forces these five viscosities to be independent of θ (to their highest order in any expansion involving small θ). There are seven other terms which are even in $\mathbf{c}$ and their viscosity coefficients are μ_3, μ_4, μ_5, λ_2, λ_3, λ_5 and λ_6. The μ_3 term depends on the fourth power of $\mathbf{c}$ while the other six depend only on the second power of $\mathbf{c}$. It is therefore expected that the μ_3 term will disappear more rapidly than the other six terms as θ tends to zero. This allows the smectic tilt angle dependence of these seven viscosities to be approximated by

$$\mu_3 = \overline{\mu}_3 \theta^4, \quad \mu_4 = \overline{\mu}_4 \theta^2, \quad \mu_5 = \overline{\mu}_5 \theta^2, \tag{6.223}$$

$$\lambda_2 = \overline{\lambda}_2 \theta^2, \quad \lambda_3 = \overline{\lambda}_3 \theta^2, \quad \lambda_5 = \overline{\lambda}_5 \theta^2, \quad \lambda_6 = \overline{\lambda}_6 \theta^2, \tag{6.224}$$

where the constant coefficients $\overline{\mu}_i$ and $\overline{\lambda}_i$ are assumed to be only weakly temperature dependent, similar to the description employed for the smectic elastic constants given by the expressions at equation (6.41). The remaining eight viscosities κ_1, κ_2, κ_3, τ_1, τ_2 τ_3, τ_4 and τ_5 are associated with terms which are odd in $\mathbf{c}$. The terms linked with κ_3 and τ_4 depend on the third power of $\mathbf{c}$ while the other six depend linearly on $\mathbf{c}$. These observations lead to the approximations

$$\kappa_1 = \overline{\kappa}_1 \theta, \quad \kappa_2 = \overline{\kappa}_2 \theta, \quad \kappa_3 = \overline{\kappa}_3 \theta^3, \tag{6.225}$$

$$\tau_1 = \overline{\tau}_1 \theta, \quad \tau_2 = \overline{\tau}_2 \theta, \quad \tau_3 = \overline{\tau}_3 \theta, \quad \tau_4 = \overline{\tau}_4 \theta^3, \quad \tau_5 = \overline{\tau}_5 \theta, \tag{6.226}$$

where, as above, the constant coefficients $\overline{\kappa}_i$ and $\overline{\tau}_i$ are only considered to be weakly temperature dependent. The classifications and approximations introduced above may help in identifying the dominant viscosities in many theoretical investigations of SmC liquid crystals.

It should also be remarked that Osipov, Sluckin and Terentjev [216] have applied a statistical approach for deriving elementary properties of the smectic viscosities. Their results coincide with the above expressions in equations (6.223) to (6.226) for small θ. Additionally, they give physical reasons for further approximations in elementary calculations, concluding that the viscosities λ_3, λ_6, τ_3 and τ_4, although not identically zero, may be set to zero in very basic and simple problems for the SmC phase because they expect these four viscosities to be much smaller than the remaining sixteen. These authors also note that there are four viscosity coefficients which may be considered as rotational viscosities. These particular viscosities are the ones which appear exclusively in the skew-symmetric contribution $\tilde{t}_{ij}^{ss}$ to the viscous stress $\tilde{t}_{ij}$, namely,

$$\text{rotational viscosities}: \quad \lambda_4, \lambda_5, \lambda_6, \tau_5. \tag{6.227}$$

The viscosity coefficient λ_4 is related to the rotation of the local smectic layer normal $\mathbf{a}$ and is the only rotational viscosity to appear in the SmA classification on page 297. The key rotational viscosity in SmC is λ_5, with the coefficients λ_6 and τ_5 being 'ac-coupling' rotational viscosities.

We now define two viscosity combinations which repeatedly appear in calculations [17, 113, 177, 183] and often prove convenient. They are

$$\eta_1 = \tfrac{1}{2}\left(\mu_0 + \mu_2 - 2\lambda_1 + \lambda_4\right), \tag{6.228}$$
$$\eta_2 = \tfrac{1}{2}\left(\mu_4 + \mu_5 + 2\lambda_2 - 2\lambda_3 + \lambda_5 + \lambda_6\right). \tag{6.229}$$

Notice, from the lists in equations (6.219) to (6.222), that η_1 only contains isotropic and smectic A-like viscosities, while η_2 only consists of viscosities which are nematic-like or are involved with the coupling of the **a** and **c** vectors. Further, for a sample of SmC where the smectic tilt angle θ can be considered small, we observe from the expressions in equations (6.223) to (6.226) that η_1 is independent of θ, whereas η_2 may be approximated by

$$\eta_2 = \overline{\eta}_2\theta^2, \qquad \text{where} \qquad \overline{\eta}_2 = \tfrac{1}{2}\left(\overline{\mu}_4 + \overline{\mu}_5 + 2\overline{\lambda}_2 - 2\overline{\lambda}_3 + \overline{\lambda}_5 + \overline{\lambda}_6\right). \tag{6.230}$$

For other relevant combinations of the smectic viscosities which arise naturally, the reader should consult the article by Schneider and Kneppe [247]. These authors have identified and discussed, in relation to a general viscosity function, possibly relevant experimental combinations and have also suggested possible experiments for their measurement.

Many inequalities involving the smectic viscosities can be derived using the viscous dissipation inequality (6.207). For example, elementary consideration of some basic flow alignments has shown that

$$\lambda_5 \geq 0, \tag{6.231}$$
$$\eta_1 \geq 0, \tag{6.232}$$
$$\eta_1 + \eta_2 \geq 0. \tag{6.233}$$

It will be shown on page 304 in Section 6.3.3 that λ_5 can be identified as a rotational viscosity that is *a priori* non-negative: this viscosity bears some similarities to the rotational viscosity γ_1 from the theory of nematics ($2\lambda_5$ plays the rôle of γ_1 in simple geometrical set-ups). From the physical point of view, it appears that λ_5 is the most important viscosity. It is related to the azimuthal rotation of the usual director **n** as it moves around the smectic cone, as depicted in Fig. 6.1. The other two inequalities in (6.232) and (6.233) above have been derived by Gill and Leslie [113, p.1910].

Rather than give an exhaustive account of inequalities for the viscosities, we quote a selection from Gill [112] which may prove useful:

$$\lambda_4 \geq 0, \tag{6.234}$$

$$\lambda_4 + \lambda_5 + \lambda_6 \geq 0, \tag{6.235}$$

$$\mu_0 + \mu_2 \geq 0, \tag{6.236}$$

$$\mu_0 + \mu_4 \geq 0, \tag{6.237}$$

$$\mu_0 + \mu_2 + \mu_4 + \mu_5 \geq 0, \tag{6.238}$$

$$2\mu_0 + \mu_1 + 2\mu_2 \geq 0, \tag{6.239}$$

$$2\mu_0 + \mu_3 + 2\mu_4 \geq 0, \tag{6.240}$$

$$(\mu_0 + \mu_2)(\mu_0 + \mu_4) \geq \kappa_1^2, \tag{6.241}$$

$$(\mu_0 + \mu_2)\lambda_4 \geq \lambda_1^2, \tag{6.242}$$

$$(\mu_0 + \mu_2)\lambda_5 \geq \tau_1^2, \tag{6.243}$$

$$(\mu_0 + \mu_4)\lambda_4 \geq \tau_2^2, \tag{6.244}$$

$$(\mu_0 + \mu_4)\lambda_5 \geq \lambda_2^2, \tag{6.245}$$

$$\lambda_4\lambda_5 \geq \tau_5^2. \tag{6.246}$$

These results were obtained by using the viscous dissipation inequality (6.207) to obtain quadratic forms analogous to those used to obtain the inequalities for the nematic viscosities at equations (4.91) to (4.95) in Section 4.2.3. Expressions (6.241) to (6.246) are of the form $xy \geq z^2$ with $x \geq 0$ and $y \geq 0$. Since $(x - y)^2 = x^2 + y^2 - 2xy \geq 0$, we can then write $(x + y)^2 \geq 4xy \geq 4z^2$, which implies that $x + y \pm 2z \geq 0$. Hence the following linear restrictions on the SmC viscosities must hold:

$$2\mu_0 + \mu_2 + \mu_4 \pm 2\kappa_1 \geq 0, \tag{6.247}$$

$$\mu_0 + \mu_2 + \lambda_4 \pm 2\lambda_1 \geq 0, \tag{6.248}$$

$$\mu_0 + \mu_2 + \lambda_5 \pm 2\tau_1 \geq 0, \tag{6.249}$$

$$\mu_0 + \mu_4 + \lambda_4 \pm 2\tau_2 \geq 0, \tag{6.250}$$

$$\mu_0 + \mu_4 + \lambda_5 \pm 2\lambda_2 \geq 0, \tag{6.251}$$

$$\lambda_4 + \lambda_5 \pm 2\tau_5 \geq 0. \tag{6.252}$$

Notice that these linear restrictions are consequences of the previous inequalities and are not to be taken as equivalences: counterexamples can be easily constructed to show this.

Galerne, Martinand, Durand and Veyssie [100] have measured some of the smectic viscosities for the SmC liquid crystal DOBCP at 103°C. From the experimental data, Leslie and Gill [178] deduced that, in the context of light scattering, in terms of the theoretical description given here we have

$$\lambda_5 = 0.0300 \quad \text{Pa s}, \tag{6.253}$$

$$|\lambda_5 - \lambda_2| = 0.0325 \quad \text{Pa s}, \tag{6.254}$$

$$|\tau_5 - \tau_1| = 0.0273 \quad \text{Pa s}, \tag{6.255}$$

$$2\eta_1 = \mu_0 + \mu_2 - 2\lambda_1 + \lambda_4 = 0.0377 \quad \text{Pa s}, \tag{6.256}$$

$$\mu_0 + \mu_4 - 2\lambda_2 + \lambda_5 = 0.0533 \quad \text{Pa s}, \tag{6.257}$$

$$\kappa_1 - \tau_1 - \tau_2 + \tau_5 = 0.0366 \quad \text{Pa s}. \tag{6.258}$$

(There are some minor miscalculations in [178] which have been corrected above.) The viscosity λ_5 has also been measured in SmC* materials (see Section 6.4 below) by various techniques. For example, experiments by Escher, Geelhaar and Böhm [83] indicate that at 30°C, for the SmC* mixtures produced by Merck stated below, $\overline{\lambda}_5$ takes the values

$$\text{ZLI-3234}: \quad \overline{\lambda}_5 \approx 0.100 \quad \text{Pa s}, \tag{6.259}$$

$$\text{ZLI-3488}: \quad \overline{\lambda}_5 \approx 0.105 \quad \text{Pa s}, \tag{6.260}$$

$$\text{ZLI-3489}: \quad \overline{\lambda}_5 \approx 0.195 \quad \text{Pa s}. \tag{6.261}$$

(The viscosity γ defined in [83] is equivalent to $2\overline{\lambda}_5$. Recall that $\lambda_5 = \overline{\lambda}_5 \theta^2$ for small cone angles θ, by equation $(6.224)_3$.)

The viscosities τ_1 and τ_5 often appear together in problems. Preliminary mathematical results indicate that it may be reasonable to assume $\tau_5 > \tau_1$ for SmC materials, in which case the modulus sign may be omitted in equation (6.255), although acceptance of this inequality should perhaps await experimental confirmation: see the elementary argument used to justify this inequality at equation (6.286) in Section 6.3.3. The reader is referred to Carlsson *et al.* [36] for some speculative theoretical suggestions for various SmC viscosity values and restrictions, including some preliminary estimates based upon a comparison with nematic viscosities. It has also been suggested from physical considerations [36] that $\tau_5 > 0$.

All the classifications and approximations introduced in this Section apply equally well to the viscous stress for SmC_M, where only twelve viscosities are present, as discussed on page 296. In this case all the τ_i and κ_i viscosities appearing above are to be set to zero.

6.3.3   Simple Flow Alignment in Smectic C

Flow effects in smectic C liquid crystals are often quite complex. Also, in many instances, the highly nonlinear differential equations resulting from the above dynamic theory require numerical methods for their solution. Nevertheless, some key properties concerning the response of the c-director to flow can be derived analytically in straightforward flow problems when some basic approximations and assumptions are made. For example, following the assumptions made for a simple flow alignment in Section 5.2 or the Zwetkoff experiment in Section 5.4, the gradients $a_{i,j}$ and $c_{i,j}$ of the vectors $\mathbf{a}$ and $\mathbf{c}$, together with the elastic energy w, may be ignored in the most basic situation, analogous to omitting spatial derivatives of $\mathbf{n}$ in the theory for nematics when considering simple anisotropic fluids. We consider two relatively elementary configurations considered by Leslie, Stewart and Naka-gawa [173] in the context of the above theory for SmC liquid crystals and derive the governing dynamic equations in detail in order to demonstrate the applicability of the dynamic theory outlined above. Although these equations will be readily solved, their derivation requires several terms in the viscous stress tensor $\tilde{t}_{ij}$ for SmC liquid crystals.

Planar Homeotropic Aligned Sample of SmC

Consider a sample of SmC liquid crystal between parallel boundary plates with the smectic layers arranged everywhere parallel to the plates. This geometrical arrangement is often called the planar homeotropic alignment for SmC [17, 113]. A practical method for orienting such samples has been discussed by Beresnev, Blinov, Osipov and Pikin [15]. Suppose that the upper plate is set in motion to create a simple shear flow, as shown in Fig. 6.12. In a preliminary investigation such as this we can ignore the influence of the boundary plates upon the director alignment and therefore consider setting, in the geometry of Fig. 6.12,

$$\mathbf{a} = (0,0,1), \qquad \mathbf{c} = (\cos\phi(t), \sin\phi(t), 0), \qquad (6.262)$$

where the velocity is given by

$$\mathbf{v} = (kz, 0, 0), \qquad k \text{ a positive constant.} \qquad (6.263)$$

The constraints (6.209) and (6.210) are clearly fulfilled.

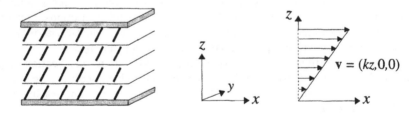

Figure 6.12: A simple shear flow in a planar homeotropic aligned sample of SmC liquid crystal.

It follows from (6.262) and (6.263) that the non-zero components of D_{ij}, W_{ij}, A_i, C_i, D_i^a and D_i^c are

$$D_{13} = D_{31} = \tfrac{1}{2}k, \quad W_{13} = -W_{31} = \tfrac{1}{2}k, \quad D_1^a = \tfrac{1}{2}k, \quad D_3^c = \tfrac{1}{2}k\cos\phi, \qquad (6.264)$$

$$A_1 = -\tfrac{1}{2}k, \quad C_1 = -\frac{d\phi}{dt}\sin\phi, \quad C_2 = \frac{d\phi}{dt}\cos\phi, \quad C_3 = \tfrac{1}{2}k\cos\phi, \qquad (6.265)$$

leading to

$$c_p D_p^a = -c_p A_p = \tfrac{1}{2} k \cos \phi, \tag{6.266}$$
$$a_p D_p^a = c_p D_p^c = 0. \tag{6.267}$$

From these quantities it is seen that $\tilde{t}_{ij,j} = 0$, because $\mathbf{a}$ and $\mathbf{c}$ have no spatial dependence. Further, the material time derivative of the velocity satisfies $\dot{v}_i = 0$. In the absence of any external body force F_i and generalised body forces G_i^a and G_i^c, the balance of linear momentum (6.211) consequently reduces to

$$0 = -\tilde{p}_{,i}, \tag{6.268}$$

from which it can be deduced that

$$\tilde{p} = p(t), \tag{6.269}$$

where $p(t)$ is an arbitrary function of time t, the elastic free energy w being zero because spatial gradients of $\mathbf{a}$ and $\mathbf{c}$ are zero. The balance of linear momentum equations are therefore completely satisfied by setting the arbitrary pressure p to be a function of time only.

We now turn to the angular momentum balance equations (6.213) and (6.214). From the definitions in (6.73) and (6.74) and the expressions (6.92) and (6.93) we have $\mathbf{\Pi}^a = \mathbf{\Pi}^c = \mathbf{0}$. Taking the scalar product of the c-equations (6.214) with $\mathbf{a}$ and $\mathbf{c}$ separately therefore gives the two Lagrange multipliers

$$\mu(t) = -\tilde{g}_3^c = k(\lambda_2 + \lambda_5) \cos \phi, \tag{6.270}$$
$$\tau(t) = -c_i \tilde{g}_i^c = -k(\tau_5 - \tau_1) \cos \phi, \tag{6.271}$$

the right-hand side equalities being obtained from the expression (6.206) and the relevant quantities in (6.264) and (6.265). For these multipliers, the third of the c-equations (6.214) is identically satisfied ($i = 3$). The remaining two equations are

$$\tilde{g}_1^c + \tau \cos \phi = 0 \quad \text{and} \quad \tilde{g}_2^c + \tau \sin \phi = 0. \tag{6.272}$$

Multiplying $(6.272)_1$ by $\sin \phi$ and $(6.272)_2$ by $\cos \phi$ and subtracting these results delivers the governing dynamic equation for $\phi(t)$, namely,

$$\tilde{g}_1^c \sin \phi - \tilde{g}_2^c \cos \phi = 0. \tag{6.273}$$

This will turn out to be the key dynamic equation for this problem, provided the Lagrange multipliers γ and β can be found so that the a-equations (6.213) are satisfied. Before going on to discuss the governing equation (6.273), we therefore proceed to examine the a-equations (6.213) in order to identify these remaining multipliers.

The a-equations (6.213) reduce to

$$\tilde{g}_i^a + \gamma a_i + \mu c_i + \epsilon_{ijk} \beta_{k,j} = 0, \tag{6.274}$$

with $\mu(t)$ provided by equation (6.270). The procedure outlined at equations (6.98) to (6.100) can be applied to find that we may set $\mathcal{F} = \tilde{g}^a + \gamma a + \mu c$ in formula (6.100), which leads to finding $\beta = -\mathcal{G}$ where $\mathcal{G}$ is defined by (6.100). Applying this procedure by taking the divergence of equations (6.274), we see that to fulfil the condition $\nabla \cdot \mathcal{F} = 0$ we can choose the multiplier γ to be an arbitrary function of x and y (notice that $\tilde{g}^a$ and μc are functions of time t only). It then follows from (6.100) that we can set

$$\gamma = \gamma(x, y), \tag{6.275}$$

$$\beta_1 = -(z - z_0)[\tilde{g}_2^a + \mu \sin \phi] + (y - y_0)\tilde{g}_3^a + \int_{y_0}^y \gamma(x, \tilde{y}) \, d\tilde{y}, \tag{6.276}$$

$$\beta_2 = (z - z_0)[\tilde{g}_1^a + \mu \cos \phi], \tag{6.277}$$

$$\beta_3 = 0, \tag{6.278}$$

where (x_0, y_0, z_0) is any fixed point within the sample. This allows a direct verification that these Lagrange multipliers lead to the fulfilment of the a-equations (6.274).

It now only remains to solve the dynamic equation (6.273) for $\phi(t)$: once such a solution has been found, all of the a- and c-equations will be satisfied, following from the above discussion. Actually, all the constraints and dynamic equations (6.209) to (6.214) will then be fulfilled. From (6.206) and the relevant quantities in (6.264) and (6.265) it is seen that

$$\tilde{g}_1^c = 2\lambda_5 \frac{d\phi}{dt} \sin \phi + k(\tau_5 - \tau_1), \tag{6.279}$$

$$\tilde{g}_2^c = -2\lambda_5 \frac{d\phi}{dt} \cos \phi. \tag{6.280}$$

Inserting these results into equation (6.273) finally reveals the governing dynamic equation to be [173]

$$2\lambda_5 \frac{d\phi}{dt} + k(\tau_5 - \tau_1) \sin \phi = 0. \tag{6.281}$$

At this point it is worth remarking on the dissipation inequality (6.207) for this problem in the special case when $k = 0$, that is, in the absence of any velocity gradient. In this case only the terms involving C_i remain in $\mathcal{D}$, leading to the result

$$\mathcal{D} = -\tilde{g}_i^c C_i = 2\lambda_5 \left(\frac{d\phi}{dt}\right)^2 \geq 0, \tag{6.282}$$

which shows that $\lambda_5 \geq 0$. If $\lambda_5 = 0$ then equation (6.281) is reduced to a static problem: henceforth it is will be assumed that $\lambda_5 > 0$.

The solution to equation (6.281) is obtained by separation of variables and is given explicitly by

$$\tan\left(\frac{\phi}{2}\right) = A \exp\left\{-\frac{(\tau_5 - \tau_1)}{2\lambda_5} kt\right\}, \tag{6.283}$$

for some constant A which is determined from a given initial condition. Hence, given that $\lambda_5 > 0$, we see that as $t \to \infty$

$$\phi(t) \to 0 \qquad \text{if } \tau_5 > \tau_1, \tag{6.284}$$

while

$$\phi(t) \rightarrow \pi \qquad \text{if } \tau_5 < \tau_1, \tag{6.285}$$

if $A > 0$; $\phi(t) \rightarrow -\pi$ if $\tau_5 < \tau_1$ and $A < 0$. This result shows that if there is a perturbation to the original alignment at $\phi \equiv 0$ where $\mathbf{c} = (1,0,0)$, then this director alignment is stable if $\tau_5 > \tau_1$, but is unstable if $\tau_5 < \tau_1$. Intuitively, it seems natural that the c-director ought to align with the flow in the x-direction where originally $\phi = 0$, rather than begin to reorient against the flow in an attempt to realign at $\phi = \pm \pi$. It therefore seems reasonable to suppose that

$$\tau_5 > \tau_1, \tag{6.286}$$

as adopted by Gill and Leslie [113] for these very reasons.

Bookshelf Aligned Sample of SmC

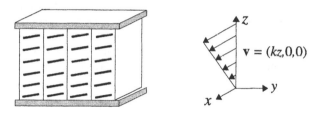

Figure 6.13: A simple shear flow of SmC in the bookshelf geometry.

Now consider a sample of SmC liquid crystal in the bookshelf geometry pictured in Fig. 6.13 where the smectic layers are everywhere perpendicular to parallel boundary plates. Suppose that the upper plate is set in motion to drive a simple shear as shown in the Figure. As in the planar homoeotropic example described above, it is assumed that the influence of the boundary plates upon the director can be ignored in such a preliminary situation and that for this configuration we can set

$$\mathbf{a} = (0,1,0), \qquad \mathbf{c} = (\cos\phi(t), 0, \sin\phi(t)), \tag{6.287}$$

where the velocity vector $\mathbf{v}$ is again given by equation (6.263). Calculations similar to those given in the previous example ultimately yield the dynamic equation [173]

$$2\lambda_5 \frac{d\phi}{dt} + k[\lambda_5 + \lambda_2 \cos(2\phi)] = 0, \tag{6.288}$$

which may or may not produce a flow alignment, depending upon the relative magnitudes of λ_2 and λ_5.

   Firstly, consider the case when $|\lambda_2| \geq \lambda_5 > 0$. We can then define the acute angle ϕ_0 given by the relation

$$\cos(2\phi_0) = -\frac{\lambda_5}{\lambda_2}, \tag{6.289}$$

(cf. equation (5.137) for an analogous definition of the Leslie angle for shear flow in a nematic). In this case, equation (6.288) can be integrated to find that the solution satisfies

$$\tan\phi = \tan\phi_0 \tanh\left\{\tfrac{1}{2}\tan(2\phi_0)kt + A\right\}, \qquad (6.290)$$

where A is a constant of integration, determined by an initial condition. Therefore, if $\lambda_2 > 0$, so that $\cos(2\phi_0) < 0$, then we see that $\tan(2\phi_0) < 0$ (because ϕ_0 is an acute angle) and

$$\phi(t) \to -\phi_0 \quad \text{or} \quad \pi - \phi_0, \qquad \text{as } t \to \infty, \qquad (6.291)$$

while if $\lambda_2 < 0$, with $\cos(2\phi_0) > 0$, then

$$\phi(t) \to \phi_0 \quad \text{or} \quad \phi_0 - \pi, \qquad \text{as } t \to \infty. \qquad (6.292)$$

In both possibilities there is therefore a flow alignment where the c-director makes an angle ϕ_0 to the direction of flow, as indicated by these results. This is somewhat similar to the flow alignment that arises in nematics: see Section 5.2.

Now consider the case when $|\lambda_2| < \lambda_5$. No flow alignment can occur when this inequality holds and therefore a non-stationary flow must develop. As remarked by Gill and Leslie [113], it is expected that this effect may be similar in nature to the 'tumbling' effect exhibited by some nematic materials as described by Carlsson [30].

Remarks

We close this Section by drawing attention to some work on flow in smectic liquid crystals which employs the above theory and tries to extend the problems discussed in earlier Chapters for nematics, such as backflow effects and light scattering. Shear flows of SmC, SmC_M, SmC* and SmC_M^* incorporating boundary conditions have been investigated by Gill and Leslie [113], who considered both planar homeotropic and bookshelf aligned samples. These authors also examined periodic fluctuations in the SmC_M phase [180] in the context of light scattering. Detailed physical interpretations of flow behaviour in systems where the smectic layers are planar have been given by Carlsson, Leslie and Clark [36]. Flow and backflow effects in SmC (cf. Sections 5.9.2 and 5.9.3 for details in nematics) have been investigated by Blake and Leslie [17, 179] and Barratt and Duffy [12]. Further general comments on the above theory may be found in the brief review articles by Leslie [177, 183], while a full and detailed derivation of the dynamic theory has been carried out by Stewart and Atkin [265].

6.4   Theory of Smectic C* Liquid Crystals

Chiral smectic C liquid crystals (SmC*) have a twist axis perpendicular to the usual smectic C layers and are known to be ferroelectric. SmC* liquid crystals were first shown to be ferroelectric in experiments carried out by Meyer, Liébert, Strzelecki and Keller [200] in 1975, after speculation on the possibility of such ferroelectric liquid crystals by McMillan [190] in 1973. In general, these ferroelectric liquid crystals possess a spontaneous polarisation **P** which, in terms of the model for SmC introduced above, can be written as a vector parallel to the vector $\mathbf{b} = \mathbf{a} \times \mathbf{c}$ as

$$\mathbf{P} = P_0\mathbf{b} \quad \text{or} \quad \mathbf{P} = -P_0\mathbf{b}, \quad (6.293)$$

where $P_0 = |\mathbf{P}|$, following on from the convention of Dahl and Lagerwall [62]. We say that the spontaneous polarisation **P** is positive if it has the same direction as **b** and negative if it has the opposite direction to **b**: this is the generally accepted sign convention [110, 158]: see Fig. 6.14. In locally planar samples of SmC* the

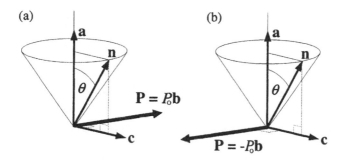

Figure 6.14: Definition of the sign convention for the spontaneous polarisation **P** in SmC* liquid crystals. $P_0 > 0$ is always equal to $|\mathbf{P}|$. Recall that $\mathbf{b} = \mathbf{a} \times \mathbf{c}$. (a) positive spontaneous polarisation, and (b) negative spontaneous polarisation.

director **n** exhibits a helical structure with helical pitch p as shown in Fig. 6.15 when the spontaneous polarisation is positive: the director **n** rotates around the smectic cone as an observer travels along the direction of the layer normal **a**. The pitch p is typically estimated to be of order $1 \sim 10\ \mu$m and the number of smectic layers over one pitch length is often of the order 10^3. The spontaneous polarisation **P** is always perpendicular to the plane containing **a** and **n** and rotates relative to the planes of the smectic layers in a helical manner as shown in the Figure. A cross indicates that **P** is directed into the page when **P** is a positive spontaneous polarisation and, as an observer travels 'upwards' in the Figure, **P** rotates in a clockwise sense when 'looking upwards' along the axis of the helix. In this Figure the helical structure shown is right-handed, but there is no *a priori* reason why a positive polarisation should not produce a left-handed helix: there is at present no known necessary relation between the sense of **P** and the sense of the helix, as mentioned by Lagerwall [158, p.75].

Typical values for P_0 may be in the range $10 \sim 10^3\ \mu$C m^{-2}, although some SmC* materials are known to have P_0 values as high as 2200 μC m^{-2} [158, p.76]. From the data presented by Dumrongrattana and Huang [67, Fig.1(a)], estimates can be made for the following material parameters for the SmC* liquid crystal DOBAMBC (originally investigated in this context by Meyer *et al.* [200]) at 88.7°C:

$$P_0 \sim 42\ \mu\text{C m}^{-2}, \qquad \theta \sim 21°, \qquad p \sim 2\ \mu\text{m}, \quad (6.294)$$

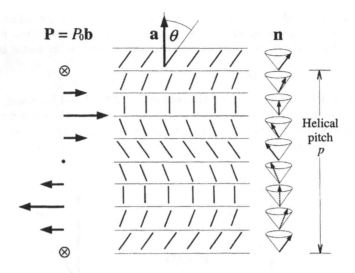

Figure 6.15: A schematic representation of a SmC* liquid crystal in the case when the spontaneous polarisation **P** is positive, using the sign convention mentioned in the text and in Fig. 6.14. The director **n** rotates around the smectic cone as an observer travels upwards in the direction of the layer normal **a**: the corresponding spontaneous polarisation **P** = P_0**b** may rotate clockwise within the smectic planes as indicated, as discussed in the text. The helical pitch is denoted by p.

where θ is the usual smectic cone angle. DOBAMBC has a negative polarisation [158, pp.75–76], according to the sign convention introduced above, and therefore the spontaneous polarisation for this particular liquid crystal is **P** = $-P_0$**b**, where P_0 is given by $(6.294)_1$. When an electric field is applied across a sample of SmC* liquid crystal it is known that the spontaneous polarisation **P** has a preference to align parallel to, and in the same direction as, the applied electric field.

There is a vast and expansive literature on SmC* liquid crystals and it is impossible to cover many of their interesting properties here. This is especially the situation since the possibility of using them for fast switching devices was first raised by Clark and Lagerwall [49] in 1980. By using a sufficiently thin sample of bookshelf aligned SmC*, with depth less than the pitch p, it is possible to suppress the helical structure through appropriate boundary conditions and then exploit the polarisation properties of SmC* via electric fields. This gives rise to the possibility of producing what is called a surface stabilised ferroelectric liquid crystal (SSFLC) device which exhibits bistability and has the capacity to switch very much faster than twisted nematic devices (cf. Section 3.7). However, it is well known that SSFLC devices are difficult to fabricate and that the switching process involves complex behaviour. For example, 'chevrons' and the formation of defects may occur [233], phenomena beyond the scope of our discussion here. In Section 6.4.1 we shall state the bulk elastic energy for SmC* and introduce the energy contributions relevant to applied electric fields. Section 6.4.2 will comment on the static and dynamic theory and in Section 6.4.3 we shall examine a dynamic equation of some relevance to both SSFLC devices and certain samples of non-chiral SmC liquid crystals. A convenient source of reference on ferroelectric liquid crystals is the text by Lagerwall [158] which introduces many of the complex phenomena of SmC* and other liquid crystals in an elegant way.

6.4.1 Energies for Smectic C*

The main energy terms for a simple model of ferroelectric smectic C liquid crystals will be introduced here. We mention the elastic energy and the energies arising from electric fields.

The Elastic Energy for Smectic C*

The expression w for the elastic energy density of the SmC phase can be extended to that required for the SmC* phase by introducing two additional terms, as discussed by Carlsson, Stewart and Leslie [35]. These extra chiral terms may be written as

$$w_1^* = A_{11}\delta\left(\mathbf{c}\cdot\nabla\times\mathbf{c}-\mathbf{b}\cdot\nabla\times\mathbf{b}\right) = 2A_{11}\delta b_i c_{j,i} a_j = 2A_{11}\delta\epsilon_{ipk}a_p c_k c_{j,i} a_j\,,\quad(6.295)$$

$$w_2^* = \tfrac{1}{2}B_3 q\left(\mathbf{c}\cdot\nabla\times\mathbf{c}+\mathbf{b}\cdot\nabla\times\mathbf{b}\right) = -B_3 q b_i c_{i,j} a_j = -B_3 q\epsilon_{ipk}a_p c_k c_{i,j} a_j\,,\quad(6.296)$$

where the second equalities in (6.295) and (6.296) are consequences of identities found in references [35, 172], the wave vector q satisfies

$$q = 2\pi/p\,,\quad(6.297)$$

and p is the helical pitch discussed above. These terms correspond to the terms D_2 and $-D_3$, respectively, discussed by de Gennes and Prost [110, p.378]. Two additional terms are also mentioned by de Gennes and Prost: $D_1(\mathbf{c}\cdot\nabla\mathbf{c}\cdot\mathbf{b}) = -D_1(\nabla\cdot\mathbf{b})$ (also discussed by Carlsson *et al.* [35]) and $D_4(\mathbf{b}\cdot\nabla\gamma)$, where γ is related to the dilation of the smectic layers. The term D_1 is a divergence term which can be converted to a surface term via the divergence theorem (1.19), while D_4 is a term incompatible with constant smectic interlayer distance: D_4 can be omitted when the layers are considered incompressible, as is the case here. The term $-w_2^*$ corresponds to the c-director rotating in a positive sense as an observer moves along the direction of the layer normal $\mathbf{a}$ and is therefore responsible for the helical ordering of the c-director that appears in the SmC* phase: see also Fig. 6.3(f) and note that $\Omega_{z,z} = -\tfrac{1}{2}(\mathbf{c}\cdot\nabla\times\mathbf{c}+\mathbf{b}\cdot\nabla\times\mathbf{b})$, using the expressions contained in equations (6.22) and (6.23). The elastic energy density we adopt for SmC* liquid crystals therefore takes the form $w_f = w + w_1^* + w_2^*$ where w is any one of the equivalent SmC energy expressions given by (6.13), (6.15) or (6.26). For example, in Cartesian component form we have from (6.15), the right-hand sides of

(6.295) and (6.296) and the relations (6.25)

$$
\begin{aligned}
w_f = {} & \tfrac{1}{2}A_{21}(a_{i,i})^2 + \tfrac{1}{2}(B_2 - B_3)(c_{i,i})^2 + \tfrac{1}{2}(B_1 - B_3)c_{i,j}c_jc_{i,k}c_k + \tfrac{1}{2}B_3c_{i,j}c_{i,j} \\
& + \tfrac{1}{2}(A_{12} + A_{21} + 2A_{11} - B_1 + B_3)(c_ia_{i,j}c_j)^2 \\
& - (A_{11} + A_{21} + \tfrac{1}{2}B_3)a_{i,i}(c_ja_{j,k}c_k) + B_{13}c_{i,j}c_jc_{i,k}a_k \\
& + (C_1 + C_2)c_{i,i}(c_ja_{j,k}c_k) - C_2a_{i,i}c_{j,j} \\
& + 2A_{11}\delta\epsilon_{ipk}a_pc_kc_{j,i}a_j - B_3q\epsilon_{ipk}a_pc_kc_{i,j}a_j.
\end{aligned} \tag{6.298}
$$

Ignoring surface contributions entering via divergence terms, there is also an equivalent bulk elastic energy formulation in terms of the vectors **b** and **c** which can be written via the expressions (6.19), (6.26), (6.295) and (6.296) as [35]

$$
\begin{aligned}
w_f = {} & \tfrac{1}{2}A_{12}(\mathbf{b} \cdot \nabla \times \mathbf{c})^2 + \tfrac{1}{2}A_{21}(\mathbf{c} \cdot \nabla \times \mathbf{b})^2 \\
& - A_{11}\left[\tfrac{1}{2}(\mathbf{c} \cdot \nabla \times \mathbf{c} - \mathbf{b} \cdot \nabla \times \mathbf{b}) - \delta\right]^2 \\
& + \tfrac{1}{2}B_1(\nabla \cdot \mathbf{b})^2 + \tfrac{1}{2}B_2(\nabla \cdot \mathbf{c})^2 + \tfrac{1}{2}B_3\left[\tfrac{1}{2}(\mathbf{b} \cdot \nabla \times \mathbf{b} + \mathbf{c} \cdot \nabla \times \mathbf{c}) + q\right]^2 \\
& + B_{13}(\nabla \cdot \mathbf{b})\left[\tfrac{1}{2}(\mathbf{b} \cdot \nabla \times \mathbf{b} + \mathbf{c} \cdot \nabla \times \mathbf{c})\right] \\
& + C_1(\nabla \cdot \mathbf{c})(\mathbf{b} \cdot \nabla \times \mathbf{c}) + C_2(\nabla \cdot \mathbf{c})(\mathbf{c} \cdot \nabla \times \mathbf{b}),
\end{aligned} \tag{6.299}
$$

where an additional irrelevant constant contribution $-A_{11}\delta^2 + \tfrac{1}{2}B_3q^2$ has been added in order to simplify the result, recalling that the wave vector q is defined by (6.297). Other general forms for the energy are possible [172] and although any one will suffice, the most convenient to adopt is determined by the particular problem under consideration.

There are grounds for setting $\delta = 0$ in w_1^* when considering planar or bookshelf aligned samples, because allowing δ to be non-zero may result in distortions of the smectic layers which could lead to either a change to the supposed constant layer thickness or a series of dislocations [35, 110]. Consequently, in the simple cases discussed here for the model introduced throughout this Chapter, it may be considered reasonable to set $\delta = 0$ in a first approximation. Nevertheless, setting $\delta \neq 0$ may be acceptable and appropriate when tackling more complex examples, despite some concerns about its interpretation at present. It is worth remarking that Gill and Leslie [113] found the term w_1^* to be of no importance in the simple shear flow problems that they investigated.

Energies for SmC* Arising from an Electric Field

To a good approximation, the dielectric energy density defined by w_{elec} in equation (6.49) may be used when discussing both SmC and SmC* liquid crystals. However, there is an additional important electric field contribution due to the spontaneous polarisation present in the SmC* phase given by

$$
w_{pol} = -\mathbf{P} \cdot \mathbf{E}. \tag{6.300}
$$

The energy density w_{pol} is clearly minimised when **P** and **E** are parallel to each other and have the same sense of direction, indicating that **P** has a preference to align in the same direction parallel to the electric field. The total electric energy density for SmC* due to an applied electric field $\mathbf{E}$ is therefore

$$w_{elec} + w_{pol} = -\tfrac{1}{2}\epsilon_0\epsilon_a \left(\mathbf{a} \cdot \mathbf{E}\cos\theta + \mathbf{c} \cdot \mathbf{E}\sin\theta\right)^2 - \mathbf{P} \cdot \mathbf{E}. \qquad (6.301)$$

The unitless dielectric anisotropy ϵ_a is expected to be of a similar order to that for SmC materials and may be negative or positive, depending upon the particular SmC* liquid crystal. Many SmC* materials are known to have negative dielectric anisotropy (see Table 6.1).

The above formulation assumes that the dielectric biaxiality $\partial\epsilon$ of the SmC* liquid crystal may be neglected. However, it should be pointed out that although in some instances it can be acceptably negligible compared to the uniaxial approximation ϵ_a (especially, in some materials, near the transition temperature to SmA* [158, p.212]), it may nevertheless have a dominant rôle to play and has been incorporated into many models of cell switching, for example by Maltese, Piccolo and Ferrara [191] and Brown, Dunn and Jones [28]. Measurements of dielectric biaxiality for several SmC materials have been reported by Jones and Raynes [139]. Convenient references on the definition of these parameters for SmC* may be found in the text by Lagerwall [158, p.187] or the article by Jones and Raynes [139].

Total Energy for SmC*

The total energy density for the SmC* liquid crystal phase in the presence of an applied electric field $\mathbf{E}$ is given by

$$w^* = w_f + w_{elec} + w_{pol}, \qquad (6.302)$$

where w_f is defined by either of the expressions (6.298) or (6.299) (or an equivalent form) and $w_{elec} + w_{pol}$ is given by (6.301).

The addition of a layer compression energy w_{comp}, such as those of the forms introduced at equations (6.147) and (6.148), can also be added to w^*. Some preliminary theoretical results involving the onset of layer undulations in a Helfrich–Hurault transition in SmC* using such an energy obtained, for example, via (6.147), (6.299), (6.301) and (6.302), has been reported by Stewart [263]. Also, the smectic layer compression constant $\overline{B}$ has been measured for various materials that exhibit SmA and SmC* phases: see the comments and references on page 284.

Comments

Notice that w_{pol} couples linearly with $\mathbf{E}$ while w_{elec} couples quadratically with $\mathbf{E}$. In some instances, $\tfrac{1}{2}\epsilon_0|\epsilon_a||\mathbf{E}|^2$ may be considerably smaller than $P_0|\mathbf{E}|$, for example when the magnitude of the electric field is comparatively small. In such cases it is common to neglect the dielectric contribution w_{elec}. Similarly, when high magnitude fields are applied, the contribution w_{pol} may be neglected. Nevertheless, many examples necessarily employ both w_{elec} and w_{pol}.

Typical values for some of the elastic constants and other parameters for SmC* liquid crystals may be gleaned from the work of MacGregor [184] who has collected

and obtained some useful information on the SmC* liquid crystal SCE3, produced
by Merck. Other results have been reported, especially concerning the elastic constant B_3, sometimes referred to as the smectic 'twist' constant. For example, Yang,
Bradberry and Sambles [282] have measured B_3 and some of the other material
parameters for the SmC* liquid crystal SCE13, also from Merck. Some of these
results for SCE3 and SCE13 are tabulated in Table 6.1.

	SmC* liquid crystal	
	SCE3 at 15°C	SCE13 at 52.5°C
helical pitch p		9.16 μm
smectic cone angle θ	25.5°	9.65°
spontaneous polarisation P_0	88 μC m^{-2}	135 μC m^{-2}
dielectric constants		
$\epsilon_\parallel$	3.10	4.02
$\epsilon_\perp$	5.04	4.74
$\epsilon_a = \epsilon_\parallel - \epsilon_\perp$	-1.94	-0.72 [139]
elastic constants		
A	4×10^{-12} N	
B_1	1×10^{-12} N	
B_2	4×10^{-12} N	
B_3	1×10^{-12} N	1.41×10^{-12} N

Table 6.1: Some material parameters for the SmC* liquid crystals SCE3, as discussed by Mac-
Gregor [184], and SCE13, discussed by Yang, Bradberry and Sambles [282]. The smectic layer
bending constants A_{11}, A_{12} and A_{21} were all set equal to A in the model used by MacGregor to
fit the experimental data. The value of ϵ_a for SCE13 has been estimated from the data presented
by Jones and Raynes [139]. From the descriptions contained in [184, 282], it would appear that
both these materials have positive spontaneous polarisation, indicating that $\mathbf{P} = P_0 \mathbf{b}$.

6.4.2   Static and Dynamic Theory for Smectic C*

The static equilibrium equations for SmC* are precisely those given by equations
(6.78), (6.79) and (6.80) in Section 6.2.3, provided w is replaced by w^* defined in
equation (6.302). Of course, the constraints (6.3) and (6.4) must also be fulfilled. If
the body forces $\mathbf{F}$, $\mathbf{G}^a$ and $\mathbf{G}^c$ are as specified in equation (6.81), then, as before, the
balance of forces (6.78) provides an expression for the pressure p, given by equation
(6.85), again with w replaced by w^*. As in the case for SmC liquid crystals, when
this simplification is possible it means that only equations (6.79) and (6.80) (the
'a-equations' and 'c-equations') need to be considered when seeking equilibrium
solutions.

Similarly, the dynamic equations for SmC* are exactly those given by equations
(6.209) to (6.214), with w^* replacing w. In particular, the viscous stress $\tilde{t}_{ij}$ given
by equations (6.201) to (6.203) is *identical* for both SmC and SmC* liquid crystals.
Consequently, all the results, comments and discussion on the smectic C viscosities

in Section 6.3.2 are equally valid for the SmC* phase described by the above model. In particular, the results from equations (6.253) to (6.261) for smectic viscosity values may well serve as useful estimates in theoretical studies of SmC* liquid crystals.

6.4.3 Director Reorientation in Smectic C*

For a simple example of dynamics induced by an electric field we consider a sample of bookshelf aligned SmC* liquid crystal in the arrangement shown in Fig. 6.13. We neglect flow by setting $\mathbf{v} = \mathbf{0}$ and, for the moment, suppose that the sample depth d between the two boundary plates is sufficiently small so that the helical structure of the SmC* alignment is largely suppressed by appropriate boundary conditions. Additionally, suppose that an electric field $\mathbf{E}$ is applied across the boundary plates in the direction of z, employing the geometry of Fig. 6.13. This is related to the experimental situation initially considered by Clark and Lagerwall [49], known as the SSFLC device, which was mentioned earlier. A general review of the switching of SSFLC devices and their electro-optics can be found in Lagerwall [158, Ch.6].

We begin by setting

$$\mathbf{E} = E(0,0,1), \tag{6.303}$$
$$\mathbf{a} = (0,1,0), \tag{6.304}$$
$$\mathbf{c} = (\cos\phi(z,t), 0, \sin\phi(z,t)), \tag{6.305}$$
$$\mathbf{b} = (\sin\phi(z,t), 0, -\cos\phi(z,t)), \tag{6.306}$$
$$\mathbf{P} = P_0\mathbf{b}, \tag{6.307}$$

where $P_0 = |\mathbf{P}| > 0$ and it is supposed that we are modelling a liquid crystal with positive polarisation, as defined above: if the material has negative polarisation then simply replace P_0 by $-P_0$ in the results that follow. Note that $\mathbf{v} = \mathbf{0}$ implies, in the notation introduced earlier for dynamic theory in Section 6.3.1,

$$D_i^a = D_i^c = A_i = 0, \quad \text{and} \quad C_i = \frac{\partial c_i}{\partial t}. \tag{6.308}$$

From these observations and equations (6.205) and (6.206) it follows that

$$\tilde{g}_i^a = -2\tau_5\frac{\partial c_i}{\partial t}, \quad \text{and} \quad \tilde{g}_i^c = -2\lambda_5\frac{\partial c_i}{\partial t}. \tag{6.309}$$

We now introduce the quantities

$$\Pi_i^a = \left(\frac{\partial w^*}{\partial a_{i,j}}\right)_{,j} - \frac{\partial w^*}{\partial a_i}, \tag{6.310}$$

$$\Pi_i^c = \left(\frac{\partial w^*}{\partial c_{i,j}}\right)_{,j} - \frac{\partial w^*}{\partial c_i}, \tag{6.311}$$

and deal firstly with the c-equations (6.214), having now replaced w by w^*. In the absence of any external body forces these equations are

$$\Pi_i^c + \tilde{g}_i^c + \tau c_i + \mu a_i = 0. \tag{6.312}$$

Taking the scalar product of (6.312) with $\mathbf{a}$ and $\mathbf{c}$ gives, respectively,

$$\mu = -\Pi_2^c, \tag{6.313}$$
$$\tau = -\Pi_i^c c_i = -\Pi_1^c \cos\phi - \Pi_3^c \sin\phi, \tag{6.314}$$

upon observing that $\tilde{g}_i^c c_i$ and $\tilde{g}_i^c a_i$ are both identically zero. The second of the c-equations ($i = 2$) in (6.312) is automatically satisfied for these Lagrange multipliers, and therefore the remaining c-equations may be written as

$$\Pi_1^c + \tilde{g}_1^c + \tau\cos\phi = 0, \tag{6.315}$$
$$\Pi_3^c + \tilde{g}_3^c + \tau\sin\phi = 0. \tag{6.316}$$

Multiplying (6.315) by $\sin\phi$ and (6.316) by $\cos\phi$ and subtracting the resulting equations then eliminates τ and yields the governing dynamic equation

$$\Pi_1^c \sin\phi - \Pi_3^c \cos\phi + \tilde{g}_1^c \sin\phi - \tilde{g}_3^c \cos\phi = 0. \tag{6.317}$$

The quantities $\tilde{g}_1^c$, $\tilde{g}_3^c$, Π_1^c and Π_3^c can be calculated explicitly from equations (6.298), (6.301), (6.302), (6.309) and (6.311) to find

$$\tilde{g}_1^c = 2\lambda_5 \sin\phi \frac{\partial\phi}{\partial t}, \tag{6.318}$$

$$\tilde{g}_3^c = -2\lambda_5 \cos\phi \frac{\partial\phi}{\partial t}, \tag{6.319}$$

$$\Pi_1^c = -B_1\left[\cos\phi\left(\frac{\partial\phi}{\partial z}\right)^2 + \sin\phi\frac{\partial^2\phi}{\partial z^2}\right] - P_0 E, \tag{6.320}$$

$$\Pi_3^c = -B_2\left[\sin\phi\left(\frac{\partial\phi}{\partial z}\right)^2 - \cos\phi\frac{\partial^2\phi}{\partial z^2}\right] + \epsilon_a\epsilon_0 E^2 \sin^2\theta \sin\phi, \tag{6.321}$$

where θ is the usual smectic cone angle, noting that we do not need to calculate Π_2^c explicitly for our present purposes. (The quantities Π_1^c and Π_3^c may also be derived in a way similar to that given at equations (6.136) and (6.137) by using a straightforward modification of the expression (6.93) to cover the case for SmC* liquid crystals.) Inserting these results into (6.317) finally reveals the governing dynamic equation to be

$$2\lambda_5\frac{\partial\phi}{\partial t} = [B_1 \sin^2\phi + B_2 \cos^2\phi]\frac{\partial^2\phi}{\partial z^2} + (B_1 - B_2)\left(\frac{\partial\phi}{\partial z}\right)^2 \sin\phi\cos\phi$$
$$+ P_0 E \sin\phi + \epsilon_0\epsilon_a E^2 \sin^2\theta \sin\phi\cos\phi. \tag{6.322}$$

In the special one-constant approximation $B_1 = B_2 \equiv B$ this equation reduces to

$$2\lambda_5\frac{\partial\phi}{\partial t} = B\frac{\partial^2\phi}{\partial z^2} + P_0 E \sin\phi + \epsilon_0\epsilon_a E^2 \sin^2\theta \sin\phi\cos\phi. \tag{6.323}$$

This dynamic equation and simplified versions of it, arising from alternative derivations, occur frequently in the literature on SmC* liquid crystals, as will be discussed below in a little more detail. However, before doing so we make two important remarks.

The first remark is to notice that suitable Lagrange multipliers γ and β may be found so that the relevant a-equations in (6.213) with $w = w^*$ are fulfilled. The details are omitted since they are similar to those given for the identification of γ and β given at equations (6.130) to (6.135) when a Freedericksz transition for SmC in the bookshelf geometry was examined, and at equations (6.274) to (6.278), when a simple flow in a planar homeotropic alignment of SmC was considered. The only remaining equations to be satisfied, apart from (6.322) or (6.323), are therefore the balance of linear momentum equations (6.211). This leads to the second remark which touches upon a more delicate mathematical issue arising from the particular problem modelled here. These particular balance of linear momentum equations, under the assumptions that external body forces are neglected, $\mathbf{v} = \mathbf{0}$ and $w^* = w^*(z, t)$, are

$$p_{,x} = \tilde{t}_{13,3}, \tag{6.324}$$

$$p_{,y} = \tilde{t}_{23,3}, \tag{6.325}$$

$$p_{,z} = -(w^*)_{,z} + \tilde{g}_k^c c_{k,z} + \tilde{t}_{33,3}. \tag{6.326}$$

It is generally impossible to find p such that these equations are satisfied for a non-constant spatially dependent solution $\phi(z, t)$. For example, it is necessary that $\nabla \times (\nabla p) = \mathbf{0}$, which leads to the requirements

$$\tilde{t}_{13,33} = \tilde{t}_{23,33} = 0. \tag{6.327}$$

These requirements are not fulfilled for many of the known spatially dependent solutions to equation (6.323): for example, the solution (6.337) given below does not satisfy the requirements contained in (6.327). Of course, this inconsistency is removed in the special case when $\phi = \phi(t)$. The incompatibility appears to arise from the assumption that $\mathbf{v} = \mathbf{0}$, and it is known that the inconsistency is resolved by introducing a non-zero velocity $\mathbf{v}$, which leads to backflow. In other words, to be fully *mathematically* consistent with the theory for smectics introduced above, this particular dynamical problem outlined here should incorporate flow if spatial dependence of the director is considered influential due to elastic effects. This will lead to a slight modification of the dynamical contributions $\tilde{g}_i^c$ and $\tilde{g}_i^a$ to the angular momentum equations and will lead to consistent equations from the balance of linear momentum after elimination of the pressure p. This is precisely the methodology that was employed by, among others, Blake and Leslie [17] and Barratt and Duffy [12] when considering SmC liquid crystals. Nevertheless, despite its shortcomings, equation (6.323) has appeared extremely frequently in the literature and has been extensively exploited by many because it is considered to be a good physical approximation to the basic dynamics of SmC* when the effects of flow are assumed to be negligible. For details on the full bookshelf problem including backflow, the interested reader is referred to Blake and Leslie [17, p.325]. Woods, Mottram and Stewart [280] have also adapted the work in [17] to model director relaxation and backflow in the electric field case for SmC* arranged in a planar homeotropic geometry.

We now discuss the two cases $\phi = \phi(t)$ and $\phi = \phi(z, t)$.

Solution for $\phi = \phi(t)$

When $\phi = \phi(t)$ we can set $p = p(t)$ for any arbitrary function of time t so that the linear momentum equations (6.324) to (6.326) are satisfied. All the smectic dynamic equations are then consistent and reduce via (6.323) to the single dynamic equation

$$2\lambda_5 \frac{d\phi}{dt} = P_0 E \sin\phi + \epsilon_0 \epsilon_a E^2 \sin^2\theta \sin\phi \cos\phi. \tag{6.328}$$

This simplified model equation was used by Xue, Handschy and Clark [281] for modelling the electro-optic response in an SSFLC device. It has the solution [281]

$$\frac{t}{\tau} = \frac{1}{1-\alpha^2} \left\{ \ln\left[\frac{\tan(\phi/2)}{\tan(\phi_0/2)}\right] + \alpha \ln\left[\frac{(1+\alpha\cos\phi)\sin\phi_0}{(1+\alpha\cos\phi_0)\sin\phi}\right] \right\}, \tag{6.329}$$

where $\phi_0 = \phi(0)$ and, for convenience, the notation

$$\tau = \frac{2\lambda_5}{P_0 E}, \tag{6.330}$$

$$\alpha = \epsilon_0 \epsilon_a \frac{E}{P_0} \sin^2\theta, \tag{6.331}$$

is introduced. This solution can be obtained by separating the variables and using a standard integration result from Reference [116, 2.561.2]. The constant α provides a measure of the balance between the ferroelectric and the dielectric influences, while $|\tau|$ is one possible measure of the response time. The ferroelectric term dominates the behaviour when α is small. Remarks on the physical interpretation and validity of this solution can be found in Reference [281].

In the special case when the dielectric term involving ϵ_a is neglected in the dynamic equation (6.328), α may be set to zero in the solution (6.329). The dynamic problem then has the solution

$$\phi(t) = 2\tan^{-1}\left[\tan(\phi_0/2)e^{t/\tau}\right]. \tag{6.332}$$

The response time $|\tau|$ is inversely proportional to the magnitude of the electric field. This is in contrast to the purely dielectric case where the analogous response time is inversely proportional to the square of the field magnitude. For example, if we set $P_0 = 0$ in equation (6.328) then the solution is given by

$$\phi(t) = \tan^{-1}\left[\tan(\phi_0)e^{t/\tau}\right], \qquad \tau = \frac{2\lambda_5}{\epsilon_a \epsilon_0 E^2 \sin^2\theta}. \tag{6.333}$$

(This can be compared with the results in Section 5.9 where the dielectric 'switch-on' times for nematics were shown to be proportional to the inverse of the square of the field magnitude). As highlighted by Lagerwall [158, p.179], the ferroelectric liquid crystal electro-optic effect obtained via the polarisation will be faster than the usual dielectric effect at low voltages: compare the corresponding response times given by equations (6.330) and (6.333)$_2$ for low magnitude fields and the forms of the electric field contributions to (6.328). Further, the dielectric effect will eventually become faster when very high magnitude fields are applied. It is well known that SmC* liquid crystal cells can switch very quickly at moderate field magnitudes, but it is the linear coupling to the field that makes switching in SmC* devices unique: they can switch equally fast in both directions as the field is reversed, a feature that would be impossible if the coupling to the electric field were purely quadratic.

Typical values for SmC* liquid crystals may be inserted into (6.330) to estimate the response time $|\tau|$. Taking, for example,

$$2\lambda_5 = 0.05 \text{ Pa s}, \qquad P_0 = 100 \ \mu\text{C m}^{-2}, \qquad E = 10 \text{ V } \mu\text{m}^{-1}, \qquad (6.334)$$

leads to a response time of around 50 μs. This is known to be up to around one thousand times faster than conventional switch-on times for twisted nematic types of displays.

Solution for $\phi = \phi(z,t)$

In the absence of flow we may approximate the dynamics by equation (6.323) when the one-constant approximation $B_1 = B_2 \equiv B$ is invoked, bearing in mind the comments after equation (6.327). Restricting our attention to the special case when $\epsilon_a < 0$ (which is known to be the situation in many commercially available SmC* materials), we can reformulate equation (6.323) as

$$2\lambda_5 \frac{\partial \phi}{\partial t} = B \frac{\partial^2 \phi}{\partial z^2} + a \sin \phi - b \sin \phi \cos \phi, \qquad (6.335)$$

where

$$a = P_0 E, \qquad \text{and} \qquad b = \epsilon_0 |\epsilon_a| E^2 \sin^2 \theta. \qquad (6.336)$$

Equation (6.335) is known to have an exact soliton-like solution, or solitary wave (also called a moving domain wall), when $|a/b| < 1$, and this solution was reported by Schiller, Pelzl and Demus [246] and Cladis and van Saarloos [43]. It is given by

$$\phi(z,t) = 2 \tan^{-1} \left[\exp \left\{ \sqrt{\frac{b}{B}} (z \pm vt) \right\} \right], \qquad (6.337)$$

with wave speed v defined by

$$v = \frac{|a|}{2\lambda_5} \sqrt{\frac{B}{b}} = \frac{P_0}{2\lambda_5 \sin \theta} \sqrt{\frac{B}{\epsilon_0 |\epsilon_a|}}. \qquad (6.338)$$

In this solution (6.337), the plus sign is taken when $a > 0$ and the negative sign is taken when $a < 0$. Changing the sign of a is, by (6.336)$_1$, equivalent to reversing the sign of the electric field. One observes that reversing the sign of the electric field then reverses the direction of the velocity of the solitary wave. For example, when $a > 0$

$$\phi \to 0 \quad \text{as} \quad t \to -\infty \qquad \text{and} \qquad \phi \to \pi \quad \text{as} \quad t \to +\infty, \qquad (6.339)$$

while for $a < 0$

$$\phi \to \pi \quad \text{as} \quad t \to -\infty \qquad \text{and} \qquad \phi \to 0 \quad \text{as} \quad t \to +\infty. \qquad (6.340)$$

Notice also that the equation (6.335) is invariant to a change in the sign of z and therefore z can be replaced by $-z$ in (6.337) for an alternative solution.

It is also possible to introduce a characteristic length to this problem when $E > 0$ ($a > 0$) defined by [158, p.176]

$$\xi = \sqrt{\frac{B}{P_0 E}}. \qquad (6.341)$$

This allows the governing dynamic equation (6.335) to be written in dimensionless form as

$$\frac{\partial \phi}{\partial T} = \frac{\partial^2 \phi}{\partial Z^2} + \sin \phi - \frac{b}{a} \sin \phi \cos \phi, \qquad (6.342)$$

where

$$T = t/\tau, \quad Z = z/\xi, \qquad (6.343)$$

and τ is defined by (6.330). The solution to (6.342) is easily found via (6.337) to be

$$\phi(Z,T) = 2 \tan^{-1} \left[\exp \left\{ \sqrt{\frac{b}{a}} Z + T \right\} \right]. \qquad (6.344)$$

(Suitably modified definitions of τ and ξ can be introduced to produce a similar type of soliton-like solution when $a < 0$.) The characteristic length ξ expresses a balance between elastic and electric effects. The applied electric field will align $\mathbf{P}$ along the field direction except in a narrow layer of order ξ in depth where boundary conditions will dictate most of the alignment of $\mathbf{P}$. From (6.341) it is seen that this layer thickness decreases proportionally to $E^{-\frac{1}{2}}$, and this will obviously become smaller than any sample depth d for sufficiently high magnitude fields. As pointed out by Lagerwall [158, p.177], ξ has to be much smaller than the wavelength of light in order for the sample to be optically homogeneous, in which case the response time $|\tau|$ given via (6.330) is applicable. That this is indeed the case can be seen by choosing, as typical values,

$$B = 5 \times 10^{-12} \text{ N}, \qquad P_0 = 100 \ \mu\text{C m}^{-2}, \qquad E = 10 \text{ V } \mu\text{m}^{-1}. \qquad (6.345)$$

These show that $\xi \sim 7 \times 10^{-8}$ m, which is below the wavelength of light (approximately in the range 0.39×10^{-6}m $\sim 0.74 \times 10^{-6}$m).

Cladis, Brand and Finn [42] were the first to discuss solitary waves in connection with SmC* liquid crystals. The particular form of solution given by (6.337) has been employed in the context of SmC and SmC* liquid crystals by Schiller, Pelzl and Demus [246] (where B may be the elastic constant B_3, depending upon the particular problem and geometry being examined) and in the reviews by Cladis and van Saarloos [43] and Maclennan, Clark and Handschy [188]. Schiller et al. [246] also discuss options when modelling the case $|a/b| > 1$. A full Painlevé analysis for equation (6.335) has been carried out by Stewart [261], who also tabulated the solutions used by the aforementioned authors in the framework of a more general mathematical setting. Extensions to equation (6.335) may include another spatial coordinate or other additional terms to the right-hand side and have been investigated numerically by Maclennan, Qi Jiang and Clark [189].

The forms of the equation and solution given by (6.335) and (6.337) also arise in the theory of non-chiral SmC liquid crystals when an applied electric field is tilted with respect to the planes of the smectic layers, as discussed by Schiller *et al.* [246]. They are also relevant when both electric and magnetic fields are applied parallel to the smectic layers across a sample of SmC*, as envisaged by van Saarloos, van Hecke and Holyst [236]. In both these circumstances the coefficients a and b must obviously be different due to the tilt of the field or the inclusion of an additional magnetic field: details may be found in References [236, 246]. Some further elementary stability analysis has been performed by van Saarloos *et al.* [236] and Stewart [262].

6.5   Comments on Theories of Smectics

With the exception of Section 6.2.6, the static and dynamic theory presented in this Chapter has assumed that the smectic layers themselves do not compress or dilate. This assumption is reasonable in simple cases of planar layers where the interlayer distance may not change. However, a full theory that is able to describe more complex phenomena than those treated here must take into account the additional possibility of smectic layer compression, as discussed in Section 6.2.6 for the static case of the well known results for layer undulations and distortions of smectic layers subjected to an applied magnetic field, leading to the well-known Helfrich–Hurault transition or deformation. There are also some very preliminary results on the Helfrich–Hurault transition in SmC [264] and SmC* [263] based upon the static theory presented in this Chapter and set in the context of the smectic layer displacement u. McKay and Leslie [186] have also extended the static theory presented in this Chapter and developed a simplified continuum theory for SmC liquid crystals allowing layer dilation and compression. A general mathematical theory involving a more comprehensive account of layer dilation, compression and distortions in SmC liquid crystals has been proposed by Blake and Virga [16]. A linearised theory for the dynamics of SmA can be found in de Gennes and Prost [110].

The rôle of permeation has not been mentioned in this Chapter. This effect occurs when there is a mass transport through the structure [110, p.413]. At this stage, it would appear that an additional equation or term is perhaps needed as a supplement to the theory presented here in order to describe this phenomenon. Such a term for smectics was first discussed by Helfrich [123] and later by de Gennes [108], and some details can be found in de Gennes and Prost [110, pp.435–445] for the case of SmA liquid crystals. The modelling of dynamics of layer undulations has also been carried out by some authors. Ben-Abraham and Oswald [14] and Chen and Jasnow [39] have examined dynamic aspects of SmA undulations using models based on the static theory described in Section 6.2.6 which incorporate flow and the influence of permeation. Experimental observations of a boundary layer in permeative flow of SmA around an obstacle have been reported by Clark [48]. Some more recent experimental and theoretical results involving permeation with compression and dilation of the smectic layers in a flow problem around a solid obstacle where there is a transition from SmA to SmC have been presented by Walton, Stewart and Towler [277] and Towler *et al.* [269].

A general elasticity theory for various smectic mesophases, including SmC and SmC*, has been suggested by Stallinga and Vertogen [257]. This work discusses both bulk and surface energy terms in a more general setting for smectics, including smectic C-type phases which are not mentioned here: their results for the bulk energy integrand for incompressible SmC and SmC* liquid crystals coincide with those presented in this Chapter.

A splay or bending distortion may create a polarisation in liquid crystals. This phenomenon is called the flexoelectric effect and was first studied theoretically by Meyer [197] in the context of nematic liquid crystals; convenient summaries can be found in Chandrasekhar [38, pp.205–212] or de Gennes and Prost [110, pp.135–139]. The flexoelectric effect for SmC liquid crystals has been investigated by Dahl and Lagerwall [62] and a brief development can be found in [110, pp.347–349]. Flexoelectric effects in smectics have also been discussed by Lagerwall [158].

It should also be mentioned that an earlier attempt at SmC dynamic theory was made by Schiller [245], whose approach differs from that outlined here, and results in fewer viscosity coefficients. Among much more recent theoretical descriptions which include nonlinear hydrodynamics and smectic layer compression effects in SmC and SmC* liquid crystals is the work by Pleiner and Brand [224]; the reader is referred to their article and its references for further details.

Appendix A

Results Employing Variational Methods

In this Appendix we establish some relations for constraints which are required in Section 2.4.2. We can employ the result in (2.125) with f given by $f\,(\mathbf{x}, \rho(\mathbf{x}))$ and u_i replaced by $\rho(\mathbf{x})$ to find that

$$\delta \int_V f(\mathbf{x}, \rho(\mathbf{x})) dV = \int_V \left[f_{,\rho} \delta\rho + f_{,i} \delta x_i + f(\delta x_i)_{,i} \right] dV. \tag{A.1}$$

This result can then be used to find that the conservation of mass forces the mass density $f = \rho(\mathbf{x})$ to satisfy

$$0 = \delta \int_V \rho\, dV = \int_V \left[\delta\rho + (\rho \delta x_i)_{,i} \right] dV, \tag{A.2}$$

and so, since V may be any arbitrary volume, admissible variations must satisfy

$$\delta\rho + (\rho \delta x_i)_{,i} = 0. \tag{A.3}$$

Clearly, if the material is incompressible then ρ is constant and the above condition reduces to

$$(\delta x_i)_{,i} = 0, \tag{A.4}$$

as stated in equation $(2.128)_1$.

To obtain equation (2.138) we first observe that equation (2.137) gives, for incompressible materials,

$$0 = \int_V \lambda_m \left(\frac{\partial \psi_m}{\partial n_i} \delta n_i + \frac{\partial \psi_m}{\partial x_i} \delta x_i \right) dV, \quad m = 1, 2, \tag{A.5}$$

for scalar function Lagrange multipliers λ_m. For the constraint $\psi_1 = 0$ defined in equation (2.135) we can use the mass conservation requirement (A.3) (which must hold in all cases), (A.4) and integration by parts to see that equation (A.5) leads to

$$0 = -\int_V \lambda_1 \rho\, (\delta x_i)_{,i} dV = -\int_S \lambda_1 \rho \delta_{ij} \delta x_i dS_j + \int_V (\lambda_1 \rho)_{,i} \delta x_i dV. \tag{A.6}$$

For the constraint $\psi_2 = 0$ in equation (2.136) we can write $\psi_{2,i}$ in terms of its dependent variables n_i in a similar fashion to that used to obtain equation (2.131) to find that (A.5) becomes

$$0 = \int_V \lambda_2 \left(\frac{\partial \psi_2}{\partial n_i} \delta n_i + \frac{\partial \psi_2}{\partial n_j} n_{j,i} \delta x_i \right) dV = \int_V \lambda_2 n_i \Delta n_i dV. \qquad (A.7)$$

Adding equations (A.6) and (A.7) gives

$$0 = \int_V [\lambda_2 n_i \Delta n_i + (\lambda_1 \rho)_{,i} \delta x_i] dV - \int_S \lambda_1 \rho \delta_{ij} \delta x_i \, dS_j, \qquad (A.8)$$

which is precisely the result in equation (2.138). The above technique has been generalised by Ericksen [73, 74] for arbitrary constraints.

Appendix B

Identities

A Tensor Identity

In the first part of this Appendix we derive the identity obtained by Ericksen [73] stated in equation (2.160). We shall make use of the invariance of the energy to arbitrary superposed rigid body rotations as prescribed by equation (2.6) for any proper orthogonal tensor Q. If we set

$$Q_{ij} = \delta_{ij} + R_{ij}, \qquad R_{ij} + R_{ji} = 0, \tag{B.1}$$

with $R = (R_{ij})$ an arbitrary infinitesimal skew-symmetric tensor, then it is straightforward to verify that Q_{ij} is proper orthogonal to first order in R_{ij}. With this choice, the relationship (2.6) becomes, to first order in R_{ij},

$$w(n_i, n_{i,j}) = w(n_i + R_{ip}n_p, n_{i,j} + R_{ip}n_{p,j} + R_{jp}n_{i,p}). \tag{B.2}$$

Expanding the right-hand side of this expression in a Taylor series to first order in R_{ij} (in a similar style to that used for equation (2.12)) then reveals that

$$\frac{\partial w}{\partial n_i} R_{ij} n_j + \frac{\partial w}{\partial n_{i,j}} (R_{ip} n_{p,j} + R_{jp} n_{i,p}) = 0, \tag{B.3}$$

or, after some rearrangement and relabelling of the indices,

$$R_{kj} \left(n_j \frac{\partial w}{\partial n_k} + n_{j,p} \frac{\partial w}{\partial n_{k,p}} + n_{p,j} \frac{\partial w}{\partial n_{p,k}} \right) = 0. \tag{B.4}$$

However, since R is skew-symmetric it can be written as [161, p.43]

$$R_{kj} = -\epsilon_{kji} r_i = \epsilon_{ijk} r_i, \tag{B.5}$$

where $\mathbf{r}$ is the axial vector associated with R. The expression in (B.4) therefore becomes the identity in equation (2.160), namely,

$$\epsilon_{ijk} \left(n_j \frac{\partial w}{\partial n_k} + n_{j,p} \frac{\partial w}{\partial n_{k,p}} + n_{p,j} \frac{\partial w}{\partial n_{p,k}} \right) = 0, \tag{B.6}$$

given that $\mathbf{r}$ arising from R must be arbitrary.

Reynolds' Transport Theorem

Let V be any closed volume moving with the fluid being considered and let $\mathfrak{F}(\mathbf{x}, t)$ be any function that depends on $\mathbf{x}$ and time t. Then Reynolds' transport theorem states that [4, p.85]

$$\frac{D}{Dt}\int_V \mathfrak{F}(\mathbf{x}, t)\, dV = \int_V \left[\dot{\mathfrak{F}} + \mathfrak{F}\,(\nabla\cdot\mathbf{v})\right] dV, \tag{B.7}$$

where both D/Dt and the superposed dot represent the material time derivative (4.5) and $\mathbf{v}$ is the velocity.

If we put $\mathfrak{F} = \rho\mathcal{F}$ into the above result then we see that

$$
\begin{aligned}
\frac{D}{Dt}\int_V \rho\mathcal{F}(\mathbf{x}, t)\, dV &= \int_V \left[\overline{\dot{\rho\mathcal{F}}} + \rho\mathcal{F}\,(\nabla\cdot\mathbf{v})\right] dV \\
&= \int_V \left[\rho\dot{\mathcal{F}} + \mathcal{F}\,(\dot{\rho} + \rho(\nabla\cdot\mathbf{v}))\right] dV \\
&= \int_V \rho\dot{\mathcal{F}}\, dV,
\end{aligned}
\tag{B.8}
$$

where we have used the equation of continuity in the form of (4.33) to arrive at the last equality.

It is also seen from the result in (B.7) that if we are dealing with an incompressible material where, by necessity, $\nabla\cdot\mathbf{v} = 0$, then

$$\frac{D}{Dt}\int_V \mathfrak{F}(\mathbf{x}, t)\, dV = \int_V \dot{\mathfrak{F}}\, dV. \tag{B.9}$$

Appendix C

Physical Components in Cylindrical Polar Coordinates

Let e_i, $i = 1, 2, 3$, represent the basis vectors in the usual cylindrical polar coordinate system $(u_1, u_2, u_3) = (r, \theta, z)$, with $e_1 = e_r$, $e_2 = e_\theta$ and $e_3 = e_z$. The Cartesian components of the position vector $\mathbf{r}$ are

$$\mathbf{r} = (r \cos \theta, r \sin \theta, z), \tag{C.1}$$

and the scale factors h_i, $i = 1, 2, 3$, in the usual notation, are given by

$$h_1 = 1, \quad h_2 = r, \quad h_3 = 1. \tag{C.2}$$

The physical components in cylindrical polar coordinates of the gradient of any scalar function p are then given by [115, p.135]

$$\nabla p = \sum_{i=1}^{3} \frac{1}{h_i} \frac{\partial p}{\partial u_i} e_i = \frac{\partial p}{\partial r} e_r + \frac{1}{r} \frac{\partial p}{\partial \theta} e_\theta + \frac{\partial p}{\partial z} e_z, \tag{C.3}$$

and the physical components of the term $(\mathbf{v} \cdot \nabla)\mathbf{n}$ (cf. page 135) are [115, p.136]

$$
\begin{aligned}
(\mathbf{v} \cdot \nabla)\mathbf{n} &= \sum_{i=1}^{3} \sum_{j=1}^{3} \left[\frac{v_j}{h_j} \frac{\partial n_i}{\partial u_j} + \frac{n_j}{h_i h_j} \left(v_i \frac{\partial h_i}{\partial u_j} - v_j \frac{\partial h_j}{\partial u_i} \right) \right] e_i, \\
&= \left[v_1 \frac{\partial n_1}{\partial r} + \frac{v_2}{r} \left(\frac{\partial n_1}{\partial \theta} - n_2 \right) + v_3 \frac{\partial n_1}{\partial z} \right] e_r \\
&\quad + \left[v_1 \frac{\partial n_2}{\partial r} + \frac{v_2}{r} \left(\frac{\partial n_2}{\partial \theta} + n_1 \right) + v_3 \frac{\partial n_2}{\partial z} \right] e_\theta \\
&\quad + \left[v_1 \frac{\partial n_3}{\partial r} + \frac{v_2}{r} \frac{\partial n_3}{\partial \theta} + v_3 \frac{\partial n_3}{\partial z} \right] e_z,
\end{aligned}
\tag{C.4}
$$

where v_i and n_i are the physical components of $\mathbf{v}$ and $\mathbf{n}$. Notice that the physical components of the material time derivative (4.5) of the velocity vector $\mathbf{v}$ in cylindrical coordinates are given by $\dot{\mathbf{v}} = \partial \mathbf{v}/\partial t + (\mathbf{v} \cdot \nabla)\mathbf{v}$ via an application of the formula in (C.4) with $\mathbf{n} = \mathbf{v}$.

For convenience, we record here that the divergence of **n** is given by [115, p.140]

$$\nabla \cdot \mathbf{n} = \frac{1}{r}\frac{\partial}{\partial r}(rn_1) + \frac{1}{r}\frac{\partial n_2}{\partial \theta} + \frac{\partial n_3}{\partial z}, \tag{C.5}$$

the physical components of the curl of **n** may be written as [115, p.141]

$$\nabla \times \mathbf{n} = \frac{1}{r}\begin{vmatrix} \mathbf{e}_r & r\mathbf{e}_\theta & \mathbf{e}_z \\ \frac{\partial}{\partial r} & \frac{\partial}{\partial \theta} & \frac{\partial}{\partial z} \\ n_1 & rn_2 & n_3 \end{vmatrix}, \tag{C.6}$$

and the Laplacian of the scalar function f is given by [115, p.141]

$$\nabla^2 f = \frac{1}{r}\frac{\partial}{\partial r}\left(r\frac{\partial f}{\partial r}\right) + \frac{1}{r^2}\frac{\partial^2 f}{\partial \theta^2} + \frac{\partial^2 f}{\partial z^2}. \tag{C.7}$$

In cylindrical polar coordinates, for any vector **v** the physical components of $[\nabla \mathbf{v}]$ are given by (cf. the definition (1.17) and [115, pp.109,140])

$$[\nabla \mathbf{v}]_{11} = v_{1,r}, \quad [\nabla \mathbf{v}]_{12} = \frac{v_{1,\theta} - v_2}{r}, \quad [\nabla \mathbf{v}]_{13} = v_{1,z}, \tag{C.8}$$

$$[\nabla \mathbf{v}]_{21} = v_{2,r}, \quad [\nabla \mathbf{v}]_{22} = \frac{v_{2,\theta} + v_1}{r}, \quad [\nabla \mathbf{v}]_{23} = v_{2,z}, \tag{C.9}$$

$$[\nabla \mathbf{v}]_{31} = v_{3,r}, \quad [\nabla \mathbf{v}]_{32} = \frac{v_{3,\theta}}{r}, \quad [\nabla \mathbf{v}]_{33} = v_{3,z}. \tag{C.10}$$

The physical components of the rate of strain tensor $\mathsf{A} = \frac{1}{2}\left[\nabla \mathbf{v} + (\nabla \mathbf{v})^T\right]$ are [115, p.140],

$$A_{11} = \frac{\partial v_1}{\partial r}, \quad A_{22} = \frac{v_1}{r} + \frac{1}{r}\frac{\partial v_2}{\partial \theta}, \quad A_{33} = \frac{\partial v_3}{\partial z}, \tag{C.11}$$

$$A_{12} = A_{21} = \frac{1}{2}\left[\frac{1}{r}\frac{\partial v_1}{\partial \theta} + r\frac{\partial}{\partial r}\left(\frac{v_2}{r}\right)\right], \tag{C.12}$$

$$A_{13} = A_{31} = \frac{1}{2}\left(\frac{\partial v_1}{\partial z} + \frac{\partial v_3}{\partial r}\right), \tag{C.13}$$

$$A_{23} = A_{32} = \frac{1}{2}\left(\frac{\partial v_2}{\partial z} + \frac{1}{r}\frac{\partial v_3}{\partial \theta}\right). \tag{C.14}$$

The non-zero physical components of the vorticity tensor $\mathsf{W} = \frac{1}{2}\left[\nabla \mathbf{v} - (\nabla \mathbf{v})^T\right]$ are [115, p.140]

$$W_{12} = -W_{21} = \frac{1}{2r}\left[\frac{\partial v_1}{\partial \theta} - \frac{\partial}{\partial r}(rv_2)\right], \tag{C.15}$$

$$W_{13} = -W_{31} = \frac{1}{2}\left(\frac{\partial v_1}{\partial z} - \frac{\partial v_3}{\partial r}\right), \tag{C.16}$$

$$W_{23} = -W_{32} = \frac{1}{2}\left(\frac{\partial v_2}{\partial z} - \frac{1}{r}\frac{\partial v_3}{\partial \theta}\right). \tag{C.17}$$

The divergence of the stress tensor T with Cartesian components T_{ij} is defined in this text by $[\nabla \cdot \mathsf{T}]_i = T_{ij,j}$. The physical components of this divergence in cylindrical polar coordinates are [115, p.144]

$$[\nabla \cdot \mathsf{T}]_1 = \frac{\partial T_{11}}{\partial r} + \frac{1}{r}\frac{\partial T_{12}}{\partial \theta} + \frac{\partial T_{13}}{\partial z} + \frac{T_{11}}{r} - \frac{T_{22}}{r}, \tag{C.18}$$

$$[\nabla \cdot \mathsf{T}]_2 = \frac{\partial T_{21}}{\partial r} + \frac{1}{r}\frac{\partial T_{22}}{\partial \theta} + \frac{\partial T_{23}}{\partial z} + \frac{T_{21}}{r} + \frac{T_{12}}{r}, \tag{C.19}$$

$$[\nabla \cdot \mathsf{T}]_3 = \frac{\partial T_{31}}{\partial r} + \frac{1}{r}\frac{\partial T_{32}}{\partial \theta} + \frac{\partial T_{33}}{\partial z} + \frac{T_{31}}{r}, \tag{C.20}$$

where we note that Goodbody [115] has chosen to define the divergence $[\nabla \cdot \mathsf{T}]_i$ by $T_{ji,j}$ where T_{ji} is the transpose of T_{ij}.

Appendix D

Tables

Fundamental constant	Symbol	Gaussian units	SI units
speed of light	c	2.998×10^{10} cm s^{-1}	2.998×10^{8} m s^{-1}
permeability of free space	μ_0	1	$4\pi \times 10^{-7} \doteq 12.566 \times 10^{-7}$ H m^{-1}
permittivity of free space	ϵ_0	1	$\mu_0^{-1} c^{-2} \doteq 8.854 \times 10^{-12}$ F m^{-1}
acceleration of free fall	g	980.665 cm s^{-2}	9.80665 m s^{-2}
Boltzmann constant	k_B	1.380658×10^{-16} erg K^{-1}	1.380658×10^{-23} J K^{-1}

Table D.1: Important constants. The abbreviations are: henry (H), farad (F), kelvin (K) and joule (J).

Quantity	Symbol	Gaussian units	Equivalent in SI units
density	ρ	1 g cm^{-3}	10^3 kg m^{-3}
relative density*	d	1	1
specific volume	$v = 1/\rho$	1 cm^3 g^{-1}	10^{-3} m^3 kg^{-1}
magnetic anisotropy	$\Delta\chi$	1	4π
dielectric anisotropy	ϵ_a	1	1
force	F	1 dyn	10^{-5} N
surface tension	γ	1 dyn cm^{-1}	10^{-3} N m^{-1}
pressure	p	1 dyn cm^{-2}	10^{-1} Pa
dynamic viscosity	$\alpha_i, \eta_i, \gamma_i$	1 P	10^{-1} Pa s
magnetic induction	B	1 G	10^{-4} T
magnetic field	H	1 Oe	$\frac{1}{4\pi} 10^3$ A m^{-1}
electric potential	V	1 statvolt	$10^{-6} c$ V
electric field	E	1 statvolt cm^{-1}	$10^{-4} c$ V m^{-1}
electric polarisation	P	1 statcoul cm^{-2}	$10^3 c^{-1}$ C m^{-2}
energy	W	1 erg	10^{-7} J
energy density	w	1 erg cm^{-3}	10^{-1} J m^{-3}

Table D.2: Conversion factors for common quantities [132, 206]. These allow simple conversions from Gaussian cgs units into standard SI units and vice-versa. $c \doteq 2.998 \times 10^8$ represents the numerical value of the speed of light in SI units. The abbreviations used are: poise (P), gauss (G), oersted (Oe), newton (N), pascal second (Pa s), tesla (T), ampere (A), volt (V), coulomb (C), joule (J). *This quantity was formerly known as specific gravity; it is the density of the material divided by the (maximum) density of water (given by 1 g cm^{-3}, equivalent to 1000 kg m^{-3}).

Quantity	MBBA near 25°C	PAA near 122°C	5CB near 26°C
viscosities (Pa s)			
η_1	0.0240 [153]	0.0024 [202]	0.0204 [154]
η_2	0.1361 [153]	0.0092 [202]	0.1052 [154]
η_3	0.0413 [153]	0.0034 [202]	0.0326 [154]
η_{12}	-0.0181 [153]	0.0043 [258]	-0.0060 [154]
γ_1	0.1093 [155]	0.0067 [258]	0.0777 [155]
$\gamma_2 = \eta_1 - \eta_2$	-0.1121	-0.0068	-0.0848
α_1	-0.0181 [155]	0.0043 [258]	-0.0060
α_2	-0.1104 [155]	-0.0069 [258]	-0.0812
α_3	-0.001104 [155]	-0.0002 [258]	-0.0036
α_4	0.0826 [155]	0.0068 [258]	0.0652
α_5	0.0779 [155]	0.0047 [258]	0.0640
α_6	-0.0336 [155]	-0.0023 [258]	-0.0208
elastic constants (N)			
K_1	6×10^{-12} [121]	6.9×10^{-12} [137]	6.2×10^{-12} [68, p.220]
K_2	3.8×10^{-12} [121]	3.8×10^{-12} [137]	3.9×10^{-12} [68, p.220]
K_3	7.5×10^{-12} [121]	11.9×10^{-12} [137]	8.2×10^{-12} [68, p.220]
surface tension ($N\,m^{-1}$)			
γ	35.5×10^{-3} [68, p.487]	38×10^{-3} [258]	32.6×10^{-3} [68, p.488]
dielectric constants			
$\epsilon_\parallel$	4.7 [258]	5.538 [258]	18.5 [68, p.525]
$\epsilon_\perp$	5.4 [258]	5.705 [258]	7 [68, p.525]
$\epsilon_a = \epsilon_\parallel - \epsilon_\perp$ (unitless)	-0.7	-0.167	11.5 [68, p.525]
magnetic anisotropy			
$\Delta\chi$ (unitless) SI units	1.219×10^{-6}	1.48×10^{-6}	1.43×10^{-6}
χ_a cgs units	0.97×10^{-7} [258]	1.18×10^{-7} [258]	1.14×10^{-7} [29]
density ($kg\,m^{-3}$)			
ρ	1088 [258]	1168 [258]	1020
specific volume ($m^3\,kg^{-1}$)			
$v = 1/\rho$	9.19×10^{-4}	8.56×10^{-4}	9.8×10^{-4} [68, p.154]
T_{NI}	45.1°C [155]	135.5°C [137]	35.4°C [155]

Table D.3: Various physical parameters for the nematic phases of MBBA, PAA and 5CB given in SI units; χ_a has also been given in cgs units for convenience. The nematic to isotropic transition temperature is denoted by T_{NI}. Uncited data have been calculated using Table D.2 and the relations in equations (4.170) to (4.175). Some authors interchange the definitions of η_1 and η_2: see Section 4.4 for details. As highlighted in Section 2.3.3, caution should be exercised when analysing data for the magnetic anisotropy $\Delta\chi$, especially when converting from cgs units to SI units or vice-versa. References to [68] and [258] refer to the EMIS review of data and the collected data reported in the review by Stephen and Straley, respectively, which the reader should consult should more detailed information and references be required.

Bibliography

[1] M. Abramowitz and I.A. Stegun, Handbook of Mathematical Functions, Dover, New York, 1970.

[2] H. Anton, Elementary Linear Algebra, 7th Edition, Wiley, New York, 1994.

[3] A. Anzelius, Über die Bewegung der anisotropen Flüssigkeiten, *Uppsala Univ. Arsskr., Mat. och Naturvet.*, 1, 1–84 (1931).

[4] R. Aris, Vectors, Tensors and the Basic Equations of Fluid Mechanics, Dover, New York, 1989.

[5] R.J. Atkin, Poiseuille Flow of Liquid Crystals of the Nematic Type, *Arch. Rat. Mech. Anal.*, 38, 224–240 (1970).

[6] R.J. Atkin and F.M. Leslie, Couette Flow of Nematic Liquid Crystals, *Q. Jl. Mech. Appl. Math.*, 23, S3–S24 (1970).

[7] R.J. Atkin and P.J. Barratt, Some solutions in the magnetohydrostatic theory for nematic liquid crystals, *Q. Jl. Mech. Appl. Math.*, 26, 109–128 (1973).

[8] R.J. Atkin and I.W. Stewart, Freedericksz transitions in spherical droplets of smectic C liquid crystals, *Q. Jl. Mech. Appl. Math.*, 47, 231–245 (1994).

[9] R.J. Atkin and I.W. Stewart, Non-linear solutions for smectic C liquid crystals in wedge and cylinder geometries, *Liq. Cryst.*, 22, 585–594 (1997).

[10] R.J. Atkin and I.W. Stewart, Theoretical studies of Freedericksz transitions in SmC liquid crystals, *Euro. Jnl. of Applied Mathematics*, 8, 253–262 (1997).

[11] G. Barbero and L.R. Evangelista, An Elementary Course on the Continuum Theory for Nematic Liquid Crystals, World Scientific, Singapore, 2001.

[12] P.J. Barratt and B.R. Duffy, A backflow effect in smectic C liquid crystals in a bookshelf geometry, *Liq. Cryst.*, 21, 865–869 (1996).

[13] V.V. Belyaev, Relationship between nematic viscosities and molecular structure, Section 8.4 in Reference [68].

[14] S.I. Ben-Abraham and P. Oswald, Dynamic Aspects of the Undulation Instability in Smectic A Liquid Crystals, *Mol. Cryst. Liq. Cryst.*, 94, 383–399 (1983).

[15] L.A. Beresnev, L.M. Blinov, M.A. Osipov and S.A. Pikin, Ferroelectric Liquid-Crystals, *Mol. Cryst. Liq. Cryst.*, **158**, 3–150 (1988).

[16] G.I. Blake and E.G. Virga, On the equilibrium of smectic C liquid crystals, *Continuum Mech. Thermodyn.*, **8**, 323–339 (1996).

[17] G.I. Blake and F.M. Leslie, A backflow effect in smectic C liquid crystals, *Liq. Cryst.*, **25**, 319–327 (1998).

[18] L.M. Blinov, Electro-optical and Magneto-optical Properties of Liquid Crystals, Wiley, Chichester, 1983.

[19] L.M. Blinov, A.Y. Kabayenkov and A.A. Sonin, Experimental studies of the anchoring energy of nematic liquid crystals, *Liq. Cryst.*, **5**, 654–661 (1989).

[20] T. Borzsonyi, A. Buka, A.P. Krekhov and L. Kramer, Response of a homeotropic nematic liquid crystal to rectilinear oscillatory shear, *Phys. Rev. E*, **58**, 7419–7427 (1998).

[21] Y. Bouligand, Recherches sur les textures des états mésomorphes. 1. Les arrangements focaux dans les smectiques: rappels et considérations théoriques, *J. de Physique*, **33**, 525–547 (1972).

[22] Y. Bouligand, Defects and Textures in Liquid Crystals, *Dislocations in Solids*, **5**, 300–347 (1980).

[23] Y. Bouligand, Defects and Textures, in Handbook of Liquid Crystals, Volume 1, D. Demus, J. Goodby, G.W. Gray, H.-W. Spiess and V. Vill (Eds.), 406–453, Wiley-VCH, Weinheim, Germany, 1998.

[24] D.E. Bourne and P.C. Kendall, Vector Analysis and Cartesian Tensors, 3rd Edition, Chapman and Hall, London, 1992.

[25] W.H. Bragg, Liquid Crystals, *Nature* (Supplement), **133**, 445–456 (1934).

[26] H.R. Brand and H. Pleiner, Hydrodynamic and electrohydrodynamic properties of the smectic C_M phase in liquid crystals, *J. Phys. II France*, **1**, 1455–1464 (1991).

[27] F. Brochard, L. Léger and R.B. Meyer, Freedericksz Transition of a Homeotropic Nematic Liquid Crystal in Rotating Magnetic Fields, *J. de Physique Colloq.*, **36 C1**, 209–213 (1975).

[28] C.V. Brown, P.E. Dunn and J.C. Jones, The effect of the elastic constants on the alignment and electro-optic behaviour of smectic C liquid crystals, *Euro. Jnl. of Applied Mathematics*, **8**, 281–291 (1997).

[29] J.D. Bunning, D.A. Crellin and T.E. Faber, The effect of molecular biaxiality on the bulk properties of some nematic liquid crystals, *Liq. Cryst.*, **1**, 37–51 (1986).

[30] T. Carlsson, Theoretical Investigation of the Shear Flow of Nematic Liquid Crystals with the Leslie Viscosity $\alpha_3 > 0$: Hydrodynamic Analogue of First Order Phase Transitions, *Mol. Cryst. Liq. Cryst.*, **104**, 307–334 (1984).

[31] T. Carlsson and F.M. Leslie, Behaviour of biaxial nematics in the presence of electric and magnetic fields: Evidence of bistability, *Liq. Cryst.*, **10**, 325–340 (1991).

[32] T. Carlsson, I.W. Stewart and F.M. Leslie, Theoretical studies of smectic C liquid crystals confined in a wedge: Stability considerations and Frederiks transitions, *Liq. Cryst.*, **9**, 661–678 (1991).

[33] T. Carlsson, F.M. Leslie and J.S. Laverty, Flow Properties of Biaxial Nematic Liquid Crystals, *Mol. Cryst. Liq. Cryst.*, **210**, 95–127 (1992).

[34] T. Carlsson, F.M. Leslie and J.S. Laverty, Biaxial Nematic Liquid Crystals – Flow Properties and Evidence of Bistability in the Presence of Electric and Magnetic Fields, *Mol. Cryst. Liq. Cryst.*, **212**, 189–196 (1992).

[35] T. Carlsson, I.W. Stewart and F.M. Leslie, An elastic energy for the ferroelectric chiral smectic C* phase, *J. Phys. A: Math. Gen.*, **25**, 2371–2374 (1992).

[36] T. Carlsson, F.M. Leslie and N.A. Clark, Macroscopic theory for the flow behavior of smectic-C and smectic-C* liquid crystals, *Phys. Rev. E*, **51**, 4509–4525 (1995).

[37] T. Carlsson and F.M. Leslie, The development of theory for flow and dynamic effects for nematic liquid crystals, *Liq. Cryst.*, **26**, 1267–1280 (1999).

[38] S. Chandrasekhar, Liquid Crystals, 2nd Edition, Cambridge University Press, Cambridge, 1992.

[39] Hsuan-Yi Chen and D. Jasnow, Layer dynamics of freely standing smectic-*A* films, *Phys. Rev. E*, **61**, 493–503 (2000).

[40] P.E. Cladis and M. Kléman, Non-singular disclinations of strength $S = +1$ in nematics, *J. de Physique*, **33**, 591–598 (1972).

[41] P.E. Cladis and S. Torza, Stability of Nematic Liquid Crystals in Couette Flow, *Phys. Rev. Lett.*, **35**, 1283–1286 (1975).

[42] P.E. Cladis, H.R. Brand and P.L. Finn, "Soliton switch" in chiral smectic liquid crystals, *Phys. Rev. A*, **28**, 512–514 (1983).

[43] P.E. Cladis and W. van Saarloos, Some Nonlinear Problems in Anisotropic Systems, in Solitons in Liquid Crystals, L. Lam and J. Prost (Eds.), 110–150, Springer-Verlag, New York, 1992.

[44] M.G. Clark, Algebraic derivation of the free-energy of a distorted nematic liquid crystal, *Molecular Physics*, **31**, 1287–1289 (1976).

[45] M.G. Clark and F.M. Leslie, A calculation of orientational relaxation in nematic liquid crystals, *Proc. R. Soc. Lond.* A, **361**, 463–485 (1978).

[46] M.G. Clark, F.C. Saunders, I.A. Shanks and F.M. Leslie, A Study of Flow Alignment Instability During Rectilinear Oscillatory Shear of Nematics, *Mol. Cryst. Liq. Cryst.*, **70**, 195–222 (1981).

[47] N.A. Clark and R.B. Meyer, Strain-induced instability of monodomain smectic A and cholesteric liquid crystals, *Appl. Phys. Lett.*, **22**, 493–494 (1973).

[48] N.A. Clark, Observation of Extended Boundary Layers in the Permeative Flow of a Smectic-A Liquid Crystal around an Obstacle, *Phys. Rev. Lett.*, **40**, 1663–1666 (1978).

[49] N.A. Clark and S.T. Lagerwall, Submicrosecond bistable electro-optic switching in liquid crystals, *Appl. Phys. Lett.*, **36**, 899–901 (1980).

[50] B.D. Coleman and W. Noll, On Certain Steady Flows of General Fluids, *Arch. Rat. Mech. Anal.*, **3**, 289–303 (1959).

[51] D. Collin, J.L. Gallani and P. Martinoty, Abnormal Sound Damping in the Smectic-C Phase of Terephthal-*bis-p-p'*-Butylaniline (TBBA): Evidence for Anharmonic Effects, *Phys. Rev. Lett.*, **58**, 254–257 (1987).

[52] P.J. Collings, Liquid Crystals: Nature's Delicate Phase of Matter, Adam Hilger, Bristol, 1990.

[53] P.J. Collings and M. Hird, Introduction to Liquid Crystals, Taylor and Francis, London, 1997.

[54] C.A. Coulson, Electricity, 5th Edition, Oliver and Boyd, Edinburgh, 1958.

[55] R. Courant and D. Hilbert, Methods of Mathematical Physics, Volume 1, Interscience Publishers, New York, 1953.

[56] S.C. Cowin, The Theory of Polar Fluids, *Advances in Applied Mechanics*, **14**, 279–347 (1974).

[57] P.K. Currie, Couette Flow of a Nematic Liquid Crystal in the Presence of a Magnetic Field, *Arch. Rat. Mech. Anal.*, **37**, 222–242 (1970).

[58] P.K. Currie, Parodi's Relation as a Stability Condition for Nematics, *Mol. Cryst. Liq. Cryst.*, **28**, 335–338 (1974).

[59] P.K. Currie, Viscometric flows of anisotropic fluids, *Rheol. Acta*, **16**, 205–212 (1977).

[60] P.K. Currie, Apparent viscosity during viscometric flow of nematic liquid crystals, *J. de Physique*, **40**, 501–505 (1979).

[61] P.K. Currie and G.P. MacSithigh, The Stability and Dissipation of Solutions for Shearing Flow of Nematic Liquid Crystals, *Q. Jl. Mech. Appl. Math.*, **32**, 499–511 (1979).

[62] I. Dahl and S.T. Lagerwall, Elastic and flexoelectric properties of chiral smectic-C phase and symmetry considerations on ferroelectric liquid-crystal cells, *Ferroelectrics*, **58**, 215–243 (1984).

[63] H.J. Deuling, Deformation of Nematic Liquid Crystals in an Electric Field, *Mol. Cryst. Liq. Cryst.*, **19**, 123–131 (1972).

[64] C.Z. van Doorn, Dynamic behvior of twisted nematic liquid-crystal layers in switched fields, *J. Appl. Phys.*, **46**, 3738–3745 (1975).

[65] E. Dubois-Violette, E. Guyon, I. Janossy, P. Pieranski and P. Manneville, Theory and experiments on plane shear flow instabilities in nematics, *J. Mécanique*, **16**, 733–767 (1977).

[66] E. Dubois-Violette, G. Durand, E. Guyon, P. Manneville and P. Pieranski, Instabilities in nematic liquid crystals, *Solid State Physics* (Supplement), **14**, 147–208 (1978).

[67] S. Dumrongrattana and C.C. Huang, Polarization and Tilt-Angle Measurements near the Smectic-A-Chiral-Smectic-C Transition of p-(n-decycloxybenzylidene)-p-amino-(2-methyl-butyl)cinnamate (DOBAMBC), *Phys. Rev. Lett.*, **56**, 464–467 (1986).

[68] D.A. Dunmur, A. Fukuda and G.R. Luckhurst (Eds.), Physical Properties of Liquid Crystals: Nematics, EMIS Datareviews Series No. 25, The Institution of Electrical Engineers (INSPEC), London, 2001.

[69] C. Dupin, Applications de géométrie et de méchanique, à la marine, aux ponts et chaussées, etc., Bachelier, Paris, 1822.

[70] I.E. Dzyaloshinskii, Theory of disclinations in liquid crystals, *Sov. Phys. JETP*, **31**, 773–777 (1970).

[71] J.L. Ericksen, Anisotropic Fluids, *Arch. Rat. Mech. Anal.*, **4**, 231–237 (1960).

[72] J.L. Ericksen, Transversely Isotropic Fluids, *Kolloid-Zeit.*, **173**, 117–122 (1960).

[73] J.L. Ericksen, Conservation Laws for Liquid Crystals, *Trans. Soc. Rheol.*, **5**, 23–34 (1961).

[74] J.L. Ericksen, Hydrostatic Theory of Liquid Crystals, *Arch. Rat. Mech. Anal.*, **9**, 371–378 (1962).

[75] J.L. Ericksen, Nilpotent Energies in Liquid Crystals, *Arch. Rat. Mech. Anal.*, **10**, 189–196 (1962).

[76] J.L. Ericksen, Inequalities in Liquid Crystal Theory, *The Physics of Fluids*, **9**, 1205–1207 (1966).

[77] J.L. Ericksen, General Solutions in the Hydrostatic Theory of Liquid Crystals, *Trans. Soc. Rheol.*, **11**, 5–14 (1967).

[78] J.L. Ericksen, Twisting of liquid crystals, *J. Fluid Mech.*, **27**, 59–64 (1967).

[79] J.L. Ericksen, A Boundary-Layer Effect in Viscometry of Liquid Crystals, *Trans. Soc. Rheol.*, **13**, 9–15 (1969).

[80] J.L. Ericksen, On Equations of Motion for Liquid Crystals, *Q. Jl. Mech. Appl. Math.*, **29**, 203–208 (1976).

[81] J.L. Ericksen, Equilibrium Theory of Liquid Crystals, *Advances in Liquid Crystals*, **2**, 233–298 (1976).

[82] J.L. Ericksen, Liquid Crystals with Variable Degree of Orientation, *Arch. Rat. Mech. Anal.*, **113**, 97–120 (1991).

[83] C. Escher, T. Geelhaar and E. Böhm, Measurement of the rotational viscosity of ferroelectric liquid crystals based on a simple dynamical model, *Liq. Cryst.*, **3**, 469–484 (1988).

[84] J.L. Fergason, U.S. Patent Application Serial No. 113,948, filed on 9th February 1971: see U.S. Patent No. 3,918,796, 11th November 1975, and the related Patent in Reference [85].

[85] J.L. Fergason, U.S. Patent No. 3,731,986, 8th May 1973.

[86] J. Ferguson and Z. Kembłowski, Applied Fluid Rheology, Elsevier, London, 1991.

[87] A. Findon, Freedericksz Transition Studies of the Smectic A and Smectic C Liquid Crystalline Phases, *Ph.D. Thesis*, Department of Physics and Astronomy, University of Manchester, 1995.

[88] A. Findon and H.F. Gleeson, Elastic constants of an achiral smectic-C material, *Ferroelectrics*, **277**, 35–45 (2002).

[89] J. Fisher and A.G. Fredrickson, Interfacial Effects on the Viscosity of a Nematic Mesophase, *Mol. Cryst. Liq. Cryst.*, **8**, 267–284 (1969).

[90] A.R. Forsyth, Lectures on the Differential Geometry of Curves and Surfaces, Cambridge University Press, Cambridge, 1912.

[91] F.C. Frank, On The Theory of Liquid Crystals, *Discuss. Faraday Soc.*, **25**, 19–28 (1958).

[92] C. Fraser, Theoretical investigation of Fréedericksz transitions in twisted nematics with surface tilt, *J. Phys. A: Math. Gen.*, **11**, 1439–1448 (1978).

[93] V. Freedericksz and V. Zolina, Forces causing the orientation of an anisotropic liquid, *Trans. Faraday Soc.*, **29**, 919–930 (1933).

[94] G. Friedel, Les états mésomorphes de la matière, *Ann. Phys. (Paris)*, **18**, 273–474 (1922).

[95] J. Friedel and P.G. de Gennes, Boucles de disclination dans les cristaux liquides, *C.R. Acad. Sc. Paris B*, **268**, 257–259 (1969).

[96] J. Fukuda and A. Onuki, Dynamics of Undulation Instability in Lamellar Systems, *J. Phys. II, France*, **5**, 1107–1113 (1995).

[97] C. Gähwiller, The viscosity coefficients of a room-temperature liquid crystal (MBBA), *Physics Letters*, **36**A, 311–312 (1971).

[98] C. Gähwiller, Temperature Dependence of Flow Alignment in Nematic Liquid Crystals, *Phys. Rev. Lett.*, **28**, 1554–1556 (1972).

[99] C. Gähwiller, Direct Determination of the Five Independent Viscosity Coefficients of Nematic Liquid Crystals, *Mol. Cryst. Liq. Cryst.*, **20**, 301–318 (1973).

[100] Y. Galerne, J.L. Martinand, G. Durand and M. Veyssie, Quasielectric Rayleigh Scattering in a Smectic-C Liquid Crystal, *Phys. Rev. Lett.*, **29**, 562–564 (1972).

[101] H. Gasparoux and J. Prost, Détermination Directe de l'Anisotropie Magnétique de Cristaux Liquides Nématiques, *J. de Physique*, **32**, 953–962 (1971).

[102] L. Gattermann and A. Ritschke, Über Azoxyphenoläther, *Ber. Deutsch. Chem. Ges.*, **23**, 1738–1750 (1890).

[103] R.E. Geer, S.J. Singer, J.V. Selinger, B.R. Ratna and R. Shashidhar, Electric-field-induced layer buckling in chiral smectic-A liquid crystals, *Phys. Rev. E*, **57**, 3059–3062 (1998).

[104] I.M. Gelfand and S.V. Fomin, Calculus of Variations, Revised English Edition, Prentice-Hall, Englewood Cliffs, New Jersey, 1963.

[105] P.G. de Gennes, Structure des cloisons de Grandjean-Cano, *C.R. Acad. Sc. Paris B*, **266**, 571–573 (1968).

[106] P.G. de Gennes, Conjectures sur l'état smectique, *J. de Physique Colloq.*, **30** C4, 65–71 (1969).

[107] P.G. de Gennes, Structures en domaines dans un nématique sous champ magnétique, *Solid State Commun.*, **8**, 213–216 (1970).

[108] P.G. de Gennes, Viscous flow in smectic A liquid crystals, *The Physics of Fluids*, **17**, 1645–1654 (1974).

[109] P.G. de Gennes, Nematodynamics, in Molecular Fluids, R. Balian and G. Weill (Eds.), 373–400, Gordon and Breach, London, 1976.

[110] P.G. de Gennes and J. Prost, The Physics of Liquid Crystals, 2nd Edition, Clarendon Press, Oxford, 1993.

[111] C.J. Gerritsma, C.Z. van Doorn and P. van Zanten, Transient effects in the electrically controlled light transmission of a twisted nematic layer, *Phys. Lett. A*, **48**, 263–264 (1974).

[112] S.P.A. Gill, Theoretical studies of certain phenomena in smectic liquid crystals induced by shear flow, infinitesimal progressive waves and magnetic fields, *Ph.D. Thesis*, Department of Mathematics, University of Strathclyde, Glasgow, 1993.

[113] S.P.A. Gill and F.M. Leslie, Shear flow and magnetic field effects on smectic C, C^*, C_M and C_M^* liquid crystals, *Liq. Cryst.*, **14**, 1905–1923 (1993).

[114] H.F. Gleeson, Light Scattering from Liquid Crystals, in Handbook of Liquid Crystals, Volume 1, D. Demus, J. Goodby, G.W. Gray, H.-W. Spiess and V. Vill (Eds.), 699–718, Wiley-VCH, Weinheim, Germany, 1998.

[115] A.M. Goodbody, Cartesian Tensors, Ellis Horwood, Chichester, 1982.

[116] I.S. Gradshteyn and I.M. Ryzhik, Table of Integrals, Series and Products, 4th Edition, Academic Press, San Diego, 1980.

[117] G.W. Gray, K.J. Harrison and J.A. Nash, New Family of Nematic Liquid Crystals for Displays, *Electronics Letters*, **9**, 130–131 (1973).

[118] G.W. Gray, K.J. Harrison, J.A. Nash, J. Constant, D.S. Hulme, J. Kirton and E.P. Raynes, Stable, Low Melting Nematogens of Positive Dielectric Anisotropy for Display Devices, in Liquid Crystals and Ordered Fluids, J.F. Johnson and R.S. Porter (Eds.), **2**, 617–643, Plenum Press, New York, 1974.

[119] G.W. Gray and J.W. Goodby, Smectic Liquid Crystals: Textures and Structures, Leonard Hill, Glasgow, 1984.

[120] E.A. Guggenheim, Thermodynamics, 7th Edition, North-Holland, Amsterdam, 1985.

[121] I. Haller, Elastic Constants of the Nematic Liquid Crystalline Phase of *p*-Methoxybenzylidene-*p-n*-Butylaniline (MBBA), *J. Chem. Phys.*, **57**, 1400–1405 (1972).

[122] N.H. Hartshorne and A. Stuart, Crystals and the Polarising Microscope, 4th Edition, Edward Arnold, London, 1970.

[123] W. Helfrich, Capillary Flow of Cholesteric and Smectic Liquid Crystals, *Phys. Rev. Lett.*, **23**, 372–374 (1969).

[124] W. Helfrich, Conduction-Induced Alignment of Nematic Liquid Crystals: Basic Model and Stability Considerations, *J. Chem. Phys.*, **51**, 4092–4105 (1969).

[125] W. Helfrich and M. Schadt, Swiss Patent No. 532,261, 4th December 1970.

[126] W. Helfrich, Electrohydrodynamic and dielectric instabilities of cholesteric liquid crystals, *J. Chem. Phys.*, **55**, 839–842 (1971).

[127] A.E. Hirst, Blending Cones and Planes by Using Cyclides of Dupin, *Bulletin of the Institute of Mathematics and its Applications*, **26**, 41–46 (1990).

[128] S.J. Hogan, T. Mullin and P. Woodford, Rectilinear low-frequency shear of homogeneously aligned nematic liquid crystals, *Proc. Roy..Soc. Lond. A*, **441**, 559–573 (1993).

[129] S.C. Hunter, Mechanics of Continuous Media, Ellis Horwood, Chichester, 1976.

[130] J.P. Hurault, Static distortions of a cholesteric planar structure induced by magnetic or ac electric fields, *J. Chem. Phys.*, **59**, 2068–2075 (1973).

[131] T. Ishikawa and O.D. Lavrentovich, Undulations in a confined lamellar system with surface anchoring, *Phys. Rev. E*, **63**, 030501-1–030501-4 (2001).

[132] J.D. Jackson, Classical Electrodynamics, 3rd Edition, Wiley, New York, 1998.

[133] H. Jeffreys and B. Jeffreys, Methods of Mathematical Physics, 3rd Edition, Cambridge University Press, Cambridge, 1972.

[134] J.T. Jenkins, Cholesteric Energies, *J. Fluid Mech.*, **45**, 465–475 (1971).

[135] J.T. Jenkins and P.J. Barratt, Interfacial effects in the static theory of nematic liquid crystals, *Q. Jl. Mech. Appl. Math.*, **27**, 111–127 (1974).

[136] B. Jerome, P. Pieranski and M. Boix, Bistable anchoring of nematics on SiO films, *Europhys. Lett.*, **5**, 693–696 (1988).

[137] W.H. de Jeu, Physical Properties of Liquid Crystalline Materials, Gordon and Breach, New York, 1980.

[138] D. Johnson and A. Saupe, Undulation instabilities in smectic C phases, *Phys. Rev. A*, **15**, 2079–2085 (1977).

[139] J.C. Jones and E.P. Raynes, Measurement of the biaxial permittivities for several smectic C host materials used in ferroelectric liquid crystal devices, *Liq. Cryst.*, **11**, 199–217 (1992).

[140] P.J. Kedney and I.W. Stewart, The onset of layer deformations in non-chiral smectic C liquid crystals, *Z. Angew. Math. Phys.*, **45**, 882–898 (1994).

[141] P.J. Kedney and F.M. Leslie, Switching in a simple bistable nematic cell, *Liq. Cryst.*, **24**, 613–618 (1998).

[142] H. Kelker and B. Scheurle, A Liquid-crystalline (Nematic) Phase with a Particularly Low Solidification Point, *Angew. Chem. Internat. Edit.*, **8**, 884–885 (1969).

[143] H. Kelker, History of Liquid Crystals, *Mol. Cryst. Liq. Cryst.*, **21**, 1–48 (1973).

[144] H. Kelker and P.M. Knoll, Some pictures of the history of liquid crystals, *Liq. Cryst.*, **5**, 19–42 (1989).

[145] P.K. Khabibullaev, E.V. Gevorkyan and A.S. Lagunov, Rheology of Liquid Crystals, Allerton Press, New York, 1994.

[146] I.C. Khoo, Liquid Crystals: Physical Properties and Nonlinear Optical Phenomena, Wiley, New York, 1995.

[147] U.D. Kini, Magnetic and Electric Field Induced Periodic Deformations in Nematics, *J. de Physique II*, **5**, 1841–1861 (1995).

[148] U.D. Kini, Crossed fields induced periodic deformations in nematics: effect of weak anchoring, *Liq. Cryst.*, **21**, 713–726 (1996).

[149] M. Kléman and O. Parodi, Covariant Elasticity for Smectics A, *J. de Physique*, **36**, 671–681 (1975).

[150] M. Kléman, Energetics of the focal conics of smectic phases, *J. de Physique*, **38**, 1511–1518 (1977).

[151] M. Kléman, Points, Lines and Walls in Liquid Crystals, Magnetic Systems and Various Ordered Media, Wiley, Chichester, 1983.

[152] M. Kléman and O.D. Lavrentovich, Curvature energy of a focal conic domain with arbitrary eccentricity, *Phys. Rev. E*, **61**, 1574–1578 (2000).

[153] H. Kneppe and F. Schneider, Determination of the Viscosity Coefficients of the Liquid Crystal MBBA, *Mol. Cryst. Liq. Cryst.*, **65**, 23–38 (1981).

[154] H. Kneppe, F. Schneider and N.K. Sharma, A Comparative Study of the Viscosity Coefficients of Some Nematic Liquid Crystals, *Ber. Bunsenges. Phys. Chem.*, **85**, 784–789 (1981).

[155] H. Kneppe, F. Schneider and N.K. Sharma, Rotational viscosity γ_1 of nematic liquid crystals, *J. Chem. Phys.*, **77**, 3203–3208 (1982).

[156] A.P. Krekhov and L. Kramer, Flow-alignment instability and slow director oscillations in nematic liquid crystals under oscillatory flow, *Phys. Rev. E*, **53**, 4925–4932 (1996).

[157] S.T. Lagerwall and I. Dahl, Ferroelectric Liquid Crystals, *Mol. Cryst. Liq. Cryst.*, **114**, 151–187 (1984).

[158] S.T. Lagerwall, Ferroelectric and Antiferroelectric Liquid Crystals, Wiley-VCH, Weinheim, Germany, 1999.

[159] L.D. Landau and E.M. Lifshitz, Fluid Mechanics, Course of Theoretical Physics Volume 6, Pergamon, London, 1959.

[160] O. Lehmann, Über fliessende Krystalle, *Zeitschrift für Physikalische Chemie*, **4**, 462–472 (1889).

[161] D.C. Leigh, Nonlinear Continuum Mechanics, McGraw-Hill, New York, 1968.

[162] F.M. Leslie, Some Constitutive Equations for Anisotropic Fluids, *Q. Jl. Mech. Appl. Math.*, **19**, 357–370 (1966).

[163] F.M. Leslie, Some Constitutive Equations for Liquid Crystals, *Arch. Rat. Mech. Anal.*, **28**, 265–283 (1968).

[164] F.M. Leslie, Some thermal effects in cholesteric liquid crystals, *Proc. Roy. Soc. A*, **307**, 359–372 (1968).

[165] F.M. Leslie, Continuum Theory of Cholesteric Liquid Crystals, *Mol. Cryst. Liq. Cryst.*, **7**, 407–420 (1969).

[166] F.M. Leslie, Distortion of Twisted Orientation Patterns in Liquid Crystals by Magnetic Fields, *Mol. Cryst. Liq. Cryst.*, **12**, 57–72 (1970).

[167] F.M. Leslie, Distorted twisted orientation patterns in nematic liquid crystals, *Pramana, Suppl. No.* **1**, 41–55 (1975).

[168] F.M. Leslie, Theory of Flow Phenomena in Liquid Crystals, *Advances in Liquid Crystals*, **4**, 1–81 (1979).

[169] F.M. Leslie, Viscometry of Nematic Liquid Crystals, *Mol. Cryst. Liq. Cryst.*, **63**, 111–127 (1981).

[170] F.M. Leslie, Some Topics In Equilibrium Theory of Liquid Crystals, in Theory and Applications of Liquid Crystals, J.L. Ericksen and D. Kinderlehrer (Eds.), 211–234, Springer-Verlag, New York, 1987.

[171] F.M. Leslie, Theory of Flow Phenomena in Nematic Liquid Crystals, in Theory and Applications of Liquid Crystals, J.L. Ericksen and D. Kinderlehrer (Eds.), 235–254, Springer-Verlag, New York, 1987.

[172] F.M. Leslie, I.W. Stewart, T. Carlsson and M. Nakagawa, Equivalent smectic C liquid crystal energies, *Continuum Mech. Thermodyn.*, **3**, 237–250 (1991).

[173] F.M. Leslie, I.W. Stewart and M. Nakagawa, A Continuum Theory for Smectic C Liquid Crystals, *Mol. Cryst. Liq. Cryst.*, **198**, 443–454 (1991).

[174] F.M. Leslie, Liquid Crystal Devices, Instituut Wiskundige Dienstverlening, Technische Universiteit Eindhoven (1992).

[175] F.M. Leslie, Continuum theory for nematic liquid crystals, *Continuum Mech. Thermodyn.*, **4**, 167–175 (1992).

[176] F.M. Leslie, J.S. Laverty and T. Carlsson, Continuum Theory for Biaxial Nematic Liquid Crystals, *Q. Jl. Mech. Appl. Math.*, **45**, 595–606 (1992).

[177] F.M. Leslie, Some flow effects in continuum theory for smectic liquid crystals, *Liq. Cryst.*, **14**, 121–130 (1993).

[178] F.M. Leslie and S.P.A. Gill, Some topics from continuum theory for smectic liquid crystals, *Ferroelectrics*, **148**, 11–24 (1993).

[179] F.M. Leslie and G.I. Blake, Flow and backflow effects in a continuum theory for smectic C liquid crystals, *Mol. Cryst. Liq. Cryst.*, **262**, 403–415 (1995).

[180] F.M. Leslie and S.P.A. Gill, Damped periodic fluctuations in certain smectic liquid crystals, in IUTAM Symposium on Anisotropy, Inhomogeneity and Nonlinearity in Solid Mechanics, D.F. Parker and A.H. England (Eds.), 133–138, Kluwer Academic Publishers, The Netherlands, 1995.

[181] F.M. Leslie, Elasticity of smectic C liquid crystals, in Contemporary Research in the Mechanics and Mathematics of Materials, R.C. Batra and M.F. Beatty (Eds.), 226–235, CIMNE, Barcelona, 1996.

[182] F.M. Leslie, Continuum Theory for Liquid Crystals, in Handbook of Liquid Crystals, Volume 1, D. Demus, J. Goodby, G.W. Gray, H.-W. Spiess and V. Vill (Eds.), 25–39, Wiley-VCH, Weinheim, Germany, 1998.

[183] F.M. Leslie, A Theory of Flow in Smectic Liquid Crystals, in Advances in the Flow and Rheology of Non-Newtonian Fluids, Part A, D.A. Siginer, D. De Kee and R.P. Chhabra (Eds.), 591–611, Elsevier, Amsterdam, 1999.

[184] A.R. MacGregor, A Method for Computing the Optical Properties of a Smectic C* Liquid Crystal Cell with a Chevron Layer Structure, *J. Mod. Optics*, **37**, 919–935 (1990).

[185] J.G. McIntosh, F.M. Leslie and D.M. Sloan, Stability for shearing flow of nematic liquid crystals, *Cont. Mech. Thermodyn.*, **9**, 293–308 (1997).

[186] G. McKay and F.M. Leslie, A continuum theory for smectic liquid crystals allowing layer dilation and compression, *Euro. Jnl. of Applied Mathematics*, **8**, 273–280 (1997).

[187] G. McKay, Cylindrical, Spherical and Toroidal Layering of Smectic C Liquid Crystals, *Mol. Cryst. Liq. Cryst.*, **366**, 403–412 (2001).

[188] J.E. Maclennan, N.A. Clark and M.A. Handschy, Solitary Waves in Ferroelectric Liquid Crystals, in Solitons in Liquid Crystals, L. Lam and J. Prost (Eds.), 151–190, Springer-Verlag, New York, 1992.

[189] J.E. Maclennan, Qi Jiang and N.A. Clark, Computer simulation of domain growth in ferroelectric liquid crystals, *Phys. Rev. E*, **52**, 3904–3914 (1995).

[190] W.L. McMillan, Simple Molecular Theory of the Smectic C Phase, *Phys. Rev. A*, **8**, 1921–1929 (1973).

[191] P. Maltese, R. Piccolo and V. Ferrara, An addressing effective computer model for surface stabilized ferroelectric liquid crystal cells, *Liq. Cryst.*, **15**, 819–834 (1993).

[192] P.C. Martin, O. Parodi and P.S. Pershan, Unified Hydrodynamic Theory for Crystals, Liquid Crystals and Normal Fluids, *Phys. Rev. A*, **6**, 2401–2420 (1972).

[193] P. Martinoty and S. Candau, Determination of Viscosity Coefficients of a Nematic Liquid Crystal Using a Shear Waves Reflectance Technique, *Mol. Cryst. Liq. Cryst.*, **14**, 243–271 (1971).

[194] A.F. Martins, Measurement of viscoelastic coefficients for nematic mesophases using magnetic resonance, Section 8.3 in Reference [68].

[195] J.C. Maxwell, On the Cyclide, *Q. Jl. Pure and Appl. Math.*, **9**, 111–126 (1868).

[196] E. Meirovitch, Z. Luz and S. Alexander, Magnetic instabilities of smectic-C liquid crystals, *Phys. Rev. A*, **15**, 408–416 (1977).

[197] R.B. Meyer, Piezoelectric effects in liquid crystals, *Phys. Rev. Lett.*, **22**, 918–921 (1969).

[198] R.B. Meyer, Point Disclinations at a Nematic–Isotropic Liquid Interface, *Mol. Cryst. Liq. Cryst.*, **16**, 355–369 (1972).

[199] R.B. Meyer, On the Existence of Even Indexed Disclinations in Nematic Liquid Crystals, *Phil. Mag.*, **27**, 405–424 (1973).

[200] R.B. Meyer, L. Liébert, L. Strzelecki and P. Keller, Ferroelectric Liquid Crystals, *J. de Physique Lett.*, **36**, L69–L71 (1975).

[201] M. Miesowicz, Der Einfluß des magnetischen Feldes auf die Viskosität der Flüssigkeiten in der nematischen Phase, *Bull. Acad. Pol. A (Poland)*, 228–247 (1936).

[202] M. Miesowicz, The Three Coefficients of Viscosity of Anisotropic Liquids, *Nature*, **158**, 27 (1946).

[203] M. Miesowicz, Liquid Crystals in my Memories and Now – the Role of Anisotropic Viscosity in Liquid Crystal Research, *Mol. Cryst. Liq. Cryst.*, **97**, 1–11 (1983).

[204] N. Minorsky, Nonlinear Oscillations, Van Nostrand, Princeton, 1962.

[205] J.K. Moscicki, Measurements of viscosities in nematics, Section 8.2 in Reference [68].

[206] B.M. Moskowitz, Fundamental Physical Constants and Conversion Factors, in Global Earth Physics: A Handbook of Physical Constants, T.J. Ahrens (Ed.), American Geophysical Union, Washington DC, 346–355 (1995).

[207] I. Müller, Thermodynamics, Pitman, London, 1985.

[208] T. Mullin and T. Peacock, Hydrodynamic instabilities in nematic liquid crystals under oscillatory shear, *Proc. Roy. Soc. Lond. A*, **455**, 2635–2653 (1999).

[209] M. Nakagawa, On the Elastic Theory of Ferroelectric SmC* Liquid Crystals, *J. Phys. Soc. Japan*, **58**, 2346–2354 (1989).

[210] M. Nakagawa, A Theoretical Study of Smectic Focal Domains, *J. Phys. Soc. Japan*, **59**, 81–89 (1990).

[211] J. Nehring, Calculation of the Structure and Energy of Nematic Threads, *Phys. Rev. A*, **7**, 1737–1748 (1973).

[212] Orsay Group, Dynamics of Fluctuations in Nematic Liquid Crystals, *J. Chem. Phys.*, **51**, 816–822 (1969).

[213] Orsay Group, Simplified elastic theory for smectics C, *Solid State Commun.*, **9**, 653–655 (1971).

[214] C.W. Oseen, Beiträge zur Theorie der anisotropen Flüssigkeiten, *Arkiv För Matematik, Astronomi Och Fysik*, **19**A, part 9, 1–19 (1925).

[215] C.W. Oseen, The Theory of Liquid Crystals, *Trans. Faraday Soc.*, **29**, 883–899 (1933).

[216] M.A. Osipov, T.J. Sluckin and E.M. Terentjev, Viscosity coefficients of smectics C*, *Liq. Cryst.*, **19**, 197–205 (1995).

[217] O. Parodi, Stress Tensor for a Nematic Liquid Crystal, *J. de Physique*, **31**, 581–584 (1970).

[218] L.A. Pars, An Introduction to the Calculus of Variations, Heinemann, London, 1962.

[219] G. Pelzl, P. Schiller and D. Demus, Freedericksz transition of planar oriented smectic C phases, *Liq. Cryst.*, **2**, 131–148 (1987).

[220] P. Pieranski, F. Brochard and E. Guyon, Static and Dynamic Behavior of a Nematic Liquid Crystal in a Magnetic Field, *J. de Physique*, **34**, 35–48 (1973).

[221] P. Pieranski and E. Guyon, Shear-flow-induced transition in nematics, *Solid State Commun.*, **13**, 435–437 (1973).

[222] P. Pieranski and E. Guyon, Transverse Effects in Nematic Flows, *Physics Letters*, **49**A, 237–238 (1974).

[223] P. Pieranski and E. Guyon, Cylindrical Couette Flow Instabilities in Nematic Liquid Crystals, *Adv. Chem. Phys.*, **32**, 151–161 (1975).

[224] H. Pleiner and H.R. Brand, Nonlinear hydrodynamics of strongly deformed smectic C and C* liquid crystals, *Physica A*, **265**, 62–77 (1999).

[225] G. Porte and J.P. Jadot, A phase transition-like instability in static samples of twisted nematic liquid crystal when the surfaces induce tilted alignments, *J. de Physique*, **39**, 213–223 (1978).

[226] J. Prost and H. Gasparoux, Determination of twist viscosity coefficient in the nematic mesophases, *Phys. Lett. A*, **36**, 245–246 (1971).

[227] A. Rapini and M. Papoular, Distortion d'une lamelle nématique sous champ magnétique. Conditions d'ancrage aux parois, *J. de Physique Colloq.*, **30 C4**, 54–56 (1969).

[228] A. Rapini, Instabilités magnétique d'un smectique C, *J. de Physique*, **33**, 237–247 (1972).

[229] F. Reinitzer, Beiträge zur Kenntnis des Cholesterins, *Monatsh. Chem.*, **9**, 421–441 (1888).

[230] F. Reinitzer, Contributions to the knowledge of cholesterol. Translation of Reference [229]. *Liq. Cryst.*, **5**, 7–18 (1989).

[231] R. Ribotta, Experimental Study of the Elasticity of Smectic Liquid Crystals: Undulation Instability of Layers, in Molecular Fluids, R. Balian and G. Weill (Eds.), 353–371, Gordon and Breach, London, 1976.

[232] R. Ribotta and G. Durand, Mechanical Instabilities of Smectic-A Liquid Crystals Under Dilative or Compressive Stresses, *J. de Physique*, **38**, 179–204 (1977).

[233] T.P. Rieker, N.A. Clark, G.S. Smith, D.S. Parmar, E.B. Sirota and C.R. Safinya, "Chevron" Local Layer Structure in Surface-Stabilized Ferroelectric Smectic-C Cells, *Phys. Rev. Lett.*, **59**, 2658–2661 (1987).

[234] C.S. Rosenblatt, R. Pindak, N.A. Clark and R.B. Meyer, The parabolic focal conic: a new smectic A defect, *J. de Physique*, **38**, 1105–1115 (1977).

[235] D.E. Rutherford, Vector Methods, 9th Edition, Oliver and Boyd, Edinburgh, 1957.

[236] W. van Saarloos, M. van Hecke and R. Holyst, Front propagation into unstable and metastable states in smectic-C* liquid crystals: Linear and nonlinear marginal-stability analysis, *Phys. Rev. E*, **52**, 1773–1777 (1995).

[237] H. Sackmann, Smectic liquid crystals: A historical review, *Liq. Cryst.*, **5**, 43–55 (1989).

[238] H. Sagan, Introduction to the Calculus of Variations, Dover, New York, 1992.

[239] A. Saupe, On Molecular Structure and Physical Properties of Thermotropic Liquid Crystals, *Mol. Cryst. Liq. Cryst.*, **7**, 59–74 (1969).

[240] H. Schad and M.A. Osman, Elastic-Constants and Molecular Association of Cyano-Substituted Nematic Liquid-Crystals, *J. Chem. Phys.*, **75**, 880–885 (1981).

[241] M. Schadt and W. Helfrich, Voltage dependent optical activity of a twisted nematic liquid crystal, *Appl. Phys. Lett.*, **18**, 127–128 (1971).

[242] M. Schadt, The history of the liquid crystal display and liquid crystal material technology, *Liq. Cryst.*, **5**, 57–71 (1989).

[243] B.S. Scheuble, Liquid Crystal Displays with High Information Content, *Kontakte (Darmstadt)*, **1**, 34–48 (1989).

[244] P. Schiller and G. Pelzl, Discontinuous Freedericksz Transition in the Smectic C Phase, *Crystal Res. Technol.*, **18**, 923–931 (1983).

[245] P. Schiller, Zur Modellierung rheologischer Eigenschaften der smektischen C-Phase, *Wiss. Z. Univ. Halle*, **XXXIV'85 M, H.** 3, S. 61–78 (1985).

[246] P. Schiller, G. Pelzl and D. Demus, Bistability and domain wall motion in smectic C phases induced by strong electric fields, *Liq. Cryst.*, **2**, 21–30 (1987).

[247] F. Schneider and H. Kneppe, Flow Phenomena and Viscosity, in Handbook of Liquid Crystals, Volume 1, D. Demus, J. Goodby, G.W. Gray, H.-W. Spiess and V. Vill (Eds.), 454–476, Wiley-VCH, Weinheim, Germany, 1998.

[248] V. Sergan and G. Durand, Anchoring anisotropy of a nematic liquid crystal on a bistable SiO evaporated surface, *Liq. Cryst.*, **18**, 171–174 (1995).

[249] S. Shibahara, J. Yamamoto, Y. Takanishi, K. Ishikawa and H. Takezoe, Critical fluctuations at the untilted-tilted phase transition in chiral smectic liquid crystals, *Phys. Rev. E*, **62**, R7599–R7602 (2000).

[250] S. Shibahara, J. Yamamoto, Y. Takanishi, K. Ishikawa and H. Takezoe, Layer Compression Modulus in Smectic Liquid Crystals, *J. Phys. Soc. Japan*, **71**, 802–807 (2002).

[251] S.J. Singer, Buckling induced by dilative strain in two- and three-dimensional layered materials, *Phys. Rev. E*, **62**, 3736–3746 (2000).

[252] K. Skarp, S.T. Lagerwall and B. Stebler, Measurements of Hydrodynamic Parameters for Nematic 5CB, *Mol. Cryst. Liq. Cryst.*, **60**, 215–236 (1980).

[253] T.J. Sluckin, D.A. Dunmur and H. Stegemeyer, Crystals That Flow: Collected Papers from the History of Liquid Crystals, Taylor and Francis, London, *to appear.*

[254] V.I. Smirnov, A Course of Higher Mathematics, Volume 4, English Edition, Pergamon Press, Oxford and London, 1964.

[255] G.F. Smith and R.S. Rivlin, The Anisotropic Tensors, *Quart. Appl. Math.*, **15**, 308–314 (1957).

[256] A.J.M. Spencer, Continuum Mechanics, Longman, Harlow, 1980.

[257] S. Stallinga and G. Vertogen, Elasticity theory of smectic and canonic mesophases, *Phys. Rev. E*, **51**, 536–543 (1995).

[258] M.J. Stephen and J.P. Straley, Physics of Liquid Crystals, *Reviews of Modern Physics*, **46**, 617–704 (1974).

[259] I.W. Stewart, On the parabolic cyclide focal-conic defect in smectic liquid crystals, *Liq. Cryst.*, **15**, 859–869 (1993).

[260] I.W. Stewart, F.M. Leslie and M. Nakagawa, Smectic liquid crystals and the parabolic cyclides, *Q. Jl. Mech. Appl. Math.*, **47**, 511–525 (1994).

[261] I.W. Stewart, Painlevé analysis for a semi-linear parabolic equation arising in smectic liquid crystals, *IMA J. Appl. Math.*, **61**, 47–60 (1998).

[262] I.W. Stewart, Stability of traveling waves in smectic-C liquid crystals, *Phys. Rev. E*, **57**, 5626–5633 (1998).

[263] I.W. Stewart, Layer Undulations in SmC* Liquid Crystals Induced by an Electric Field, *Ferroelectrics*, **245**, 165–173 (2000).

[264] I.W. Stewart, Layer distortions induced by a magnetic field in planar samples of smectic C liquid crystals, *Liq. Cryst.*, **30**, 909–920 (2003).

[265] I.W. Stewart and R.J. Atkin, Derivation of a dynamic theory for smectic C liquid crystals, *to appear.*

[266] I.W. Stewart and F.M. Leslie, unpublished details.

[267] I.W. Stewart and G. McKay, *to appear.*

[268] K. Tarumi and M. Heckmeier, Viscous properties of nematics for applications, Section 11.4 in Reference [68].

[269] M.J. Towler, D.C. Ulrich, I.W. Stewart, H.G. Walton and P. Gass, Permeative flow and the compatability of smectic C zig-zag defects with compressive and dilative regions, *Liq. Cryst.*, **27**, 75–80 (2000).

[270] C. Truesdell and W. Noll in Handbuch der Physik, S. Flügge (Ed.), Volume III/3, Springer-Verlag, Berlin, 1965.

[271] H.C. Tseng, D.L. Silver and B.A. Finlayson, Application of the Continuum Theory to Nematic Liquid Crystals, *Phys. Fluids*, **15**, 1213–1222 (1972).

[272] P.D.S. Verma, Couette Flow of Certain Anisotropic Fluids, *Arch. Rat. Mech. Anal.*, **10**, 101–107 (1962).

[273] E.G. Virga, Variational Theories for Liquid Crystals, Chapman and Hall, London, 1994.

[274] D. Vorländer, Einfluß der molekularen Gestalt auf den krystallinisch-flüssigen Zustand, *Ber. Deutsch. Chem. Ges.*, **40**, 1970–1972 (1907). A translation of this article will be available in Referecnce [253].

[275] T. Wada and M. Koden, Liquid crystal flat panel displays, *Optoelectronics –Devices and Technologies*, **7**, 211–219 (1992).

[276] J. Wahl and F. Fischer, Elastic and Viscosity Constants of Nematic Liquid Crystals from a New Optical Method, *Mol. Cryst. Liq. Cryst.*, **22**, 359–373 (1973).

[277] H.G. Walton, I.W. Stewart and M.J. Towler, Flow past finite obstacles in smectic liquid crystals: permeative flow induced S_A to S_C phase transition, *Liq. Cryst.*, **20**, 665–668 (1996).

[278] G.N. Watson, A Treatise on the Theory of Bessel Functions, 2nd Edition, Cambridge University Press, Cambridge, 1996.

[279] C. Williams, P. Pieranski and P.E. Cladis, Nonsingular S=+1 Screw Disclination Lines in Nematics, *Phys. Rev. Lett.*, **29**, 90–92 (1972).

[280] P.D. Woods, N.J. Mottram and I.W. Stewart, Flow induced by director relaxation in smectic C* materials, *Ferroelectrics*, **277**, 251–258 (2002).

[281] Xue Jiu-Zhi, M.A. Handschy and N.A. Clark, Electrooptic response during switching of a ferroelectric liquid crystal cell with uniform director orientation, *Ferroelectrics*, **73**, 305–314 (1987).

[282] F. Yang, G.W. Bradberry and J.R. Sambles, Optical determination of the twist elastic constant of a smectic-C* liquid crystal, *Phys. Rev. E*, **53**, 674–680 (1996).

[283] H. Yokoyama and H.A. van Sprang, A novel method for determining the anchoring energy function at a nematic liquid crystal-wall interface from director distortions at high fields, *J. Appl. Phys.*, **57**, 4520–4526 (1985).

[284] H. Yokoyama, Surface Anchoring of Nematic Liquid Crystals, *Mol. Cryst. Liq. Cryst.*, **165**, 265–316 (1988).

[285] J.A. Zasadzinski, L.E. Scriven and H.T. Davis, Liposome structure and defects, *Philosophical Magazine A*, **51**, 287–302 (1985).

[286] H. Zocher, Über die Einwirkung magnetischer, elektrischer und mechanischer Kräfte auf Mesophasen, *Physik. Zietschr.*, **28**, 790–796 (1927).

[287] H. Zocher, The Effect of a Magnetic Field on the Nematic State, *Trans. Faraday Soc.*, **29**, 945–957 (1933).

[288] V. Zwetkoff (V.N. Tsvetkov), Bewegung anisotroper Flüssigkeiten im rotierenden Magnetfeld, *Acta Physicochimica URSS*, **10**, 555–578 (1939).

Index

Printed and bound by CPI Group (UK) Ltd, Croydon, CR0 4YY

01/11/2024

01782614-0011